Developmental Biology

The INSTANT NOTES series

Series editor
B.D. Hames
School of Biochemistry and Molecular Biology, University of Leeds, Leeds, UK

Animal Biology
Ecology
Genetics
Microbiology
Chemistry for Biologists
Immunology
Biochemistry 2nd edition
Molecular Biology 2nd edition
Neuroscience
Developmental Biology
Psychology

Forthcoming titles
Plant Biology

The INSTANT NOTES Chemistry series
Consulting editor: Howard Stanbury

Organic Chemistry
Inorganic Chemistry
Physical Chemistry

Forthcoming titles
Analytical Chemistry
Medicinal Chemistry

Developmental Biology

R.M. Twyman

Molecular Biotechnology Unit,
John Innes Institute, Norwich, UK

Springer

R.M. Tywman

Molecular Biotechnology Unit, John Innes Centre, Norwich, UK

Published in the United States of America, its dependent territories and Canada by arrangement with BIOS Scientific Publishers Ltd, 9 Newtec Place, Magdalen Road, Oxford OX4 1RE, UK

First published 2001

A CIP catalogue record for this book is available from the British Library.

ISBN 0-387-91610-5   Springer-Verlag New York Berlin Heidelberg   SPIN 10760238

Springer-Verlag New York Inc.
175 Fifth Avenue, New York
NY 10010-7858, USA

Production Editor: Andrea Bosher
Typeset by Phoenix Photosetting, Chatham, Kent, UK
Printed by Biddles Ltd, Guildford, UK www.biddles.co.uk

CONTENTS

ABBREVIATIONS

3'-UTR	3' untranslated region	IP$_3$	inositol-1,4,5-trisphosphate
ABA	abscisic acid	JAK	Janus kinase
ADMP	antidorsalizing morphogenetic protein	JNK	Jun N-terminal kinase
		LGN	lateral geniculate nucleus
AEC	apical ectodermal cap	LH	leutenizing hormone
AER	apical ectodermal ridge	LRD	left–right dynein
ALCs	anterior-like cells	MAP	microtubule associated protein
ANT-C	Antennapedia complex	MAR	matrix attachment region
AVE	anterior visceral endoderm	MBT	midblastula transition
BDNF	brain-derived neurotrophic factor	MTOC	microtubule organizing center
BMP	bone morphogenetic protein	N-CAM	neural cell adhesion molecule
BWS	Beckwith–Weidemann syndrome	NGF	nerve growth factor
BX-C	Bithorax complex	NIC	node-inducing center
CAM	cell adhesion molecule	PGC	primordial germ cell
CDK	cyclin dependent kinase	PKA	protein kinase A
CNS	central nervous system	PKC	protein kinase C
DAG	diacylglycerol	PMC	primary mesenchyme cell
DIF	differentiation inducing factor	PMZ	posterior marginal zone
ECM	extracellular matrix	PSF	prestarvation factor
EDTA	ethylene diamine tetra-acetic acid	RAGE	recombinant activation of gene expression
EGF	epidermal growth factor		
EMS	ethylmethane sulfonate	RAM	root apical meristem
ENU	N-ethyl-N-nitrosourea	REMI	restriction enzyme-mediated integration
FGF	fibroblast growth factor		
FLIP	floral initiation program	SAM	shoot apical meristem
FSH	follicle stimulating hormone	SMC	secondary mesenchyme cell
GA	gibberellic acid	STAT	signal transducer and activator of transcription
GAP	GTPase activating protein		
GDNF	glial-derived neurotrophic factor	TCR	T-cell receptor
GEF	guanine nucleotide exchange factor	Ti plasmid	tumor-inducing plasmid
		TIFs	transcriptional initiation factors
GPCR	G-protein-coupled receptor	UV	ultraviolet
GSK-3	glycogen synthase kinase-3	VPCs	vulval precursor cells
HOM-C	homeotic complex	Xa	active X
ICM	inner cell mass	Xi	inactive X
IFAP	intermediate filament-associated protein	*Xist*	X-inactive-specific transcript
		ZPA	zone of polarizing activity

PREFACE

Developmental Biology now unites all the biological sciences, and is assuming more and more importance in fields such as medicine and agriculture. For this reason, many more students are required to learn about development as a short module forming part of a larger course. Conscious of the extra pressure this places on students and their teachers, I have tried to write a book that provides a clear summary of the principles of development in a concise and easily manageable format. Development has always been perceived as a difficult subject, and this reflects its origins in the study of diverse organisms using different approaches: simple observation, experimental embryology, genetics and molecular biology. Over the last 10 years, these disparate elements have come together resulting in a new era for developmental biology, which is gene-based and in which the complex cell behaviors that guide development are being dissected at the molecular level. This has shown that fundamentally similar mechanisms guide the development of all organisms, a theme that is emphasized throughout the book.

This book focuses largely on the principles of embryonic development, while coverage of post-embryonic development is limited, except in plants. There is also no attempt to describe development in evolutionary terms. This slight reduction in coverage has been necessary to keep the book to an acceptable length and price, but it should still meet most of your course requirements.

This book would not have been possible without the valuable help and advice of many friends and colleagues. In particular I would like to thank those who took the time to read and comment on individual topics and sections: Gavin Craig, Dylan Sweetman, Phil Gardner, Jane Pritchard, Christine Holt, Victoria James, Manuela Costa, Mark Leech, Eva Stöger, Ajay Kohli, James Drummond, Roz Friday, Derek Gatherer, and especially Clare Hudson. I would also like to thank my many past and present students at Gonville and Caius, Newnham, New Hall, Girton, Peterhouse, and King's Colleges in Cambridge, who acted as guinea pigs for various parts of the book and helped to shape its content and presentation. I would like to thank my teachers, who cultivated my interest in development: Helen Baker, Trevor Jowett, Liz Jones, David Stott, and Bob Old. I would also like to acknowledge Bob Old (University of Warwick), Nigel Unwin (MRC Laboratory of Molecular Biology, Cambridge) and especially Paul Christou (Molecular Biotechnology Unit, John Innes Centre, Norwich) for their support and encouragement at different stages of the project. Finally of course, thanks to the dedicated team at BIOS (Jonathan Ray, Victoria Oddie, Lisa Mansell, Ailsa Henderson, Andrea Bosher, and Rachel Offord) and for the advice and encouragement of David Hames.

Cover images. Left *Drosophila* larval neuromuscular junction courtesy of James Drummond. Right: Ascidian larva showing localization of *Otx* mRNA courtesy of Clare Hudson. Upper inset panel: Section of mouse brain showing hippocalcin gene expression courtesy of Bill Wisden and Alison Jones. Lower panels: *Arabidopsis thaliana* wild type flower (left), *apetala2* mutant (middle) and *apetala2/agamous* double mutant (right) courtesy of Robert Soblowski.

To my parents, Peter and Irene,
and to my children, Emily
and Lucy

A1 BASIC CONCEPTS IN DEVELOPMENTAL BIOLOGY

Key Notes

Development

Development is the process by which a complex multicellular organism arises from a single cell. It involves an increase in cell number, differentiation, pattern formation and morphogenesis, as well as net growth. Development is a gradual process, so the complexity of the embryo increases progressively.

Overview of animal development

Although animals show great diversity, the development of most animals proceeds through a number of common stages. These include fertilization, cleavage to form a group of blastomeres, gastrulation to reorganize the structure of the embryo and generate the three germ layers, neurulation (formation of the nervous system) and organogenesis (the development of individual organs).

Overview of plant development

Flowering plants undergo double fertilization to produce the zygote and the endosperm tissue that nourishes the embryo. Embryonic development in plants comprises a series of stereotyped cell divisions to generate a mature embryo comprising three fundamental cell layers and containing regions corresponding to the shoot, root and cotyledons of the seedling. After germination, all parts of the mature plant are produced from two small populations of proliferative cells established in the embryo, the shoot and root meristems.

Growth and cell division

Although development involves both cell division and growth, these processes can occur independently. During the cleavage divisions that occur in early animal development, there is an increase in cell number without growth, so that the egg is divided into a series of progressively smaller cells. Later in development, cell division and growth occur together, although growth may occur without cell division through changes in cell size and the deposition of materials such as bone into the extracellular matrix.

Differentiation

Differentiation is the process by which different cell types are generated. Cells become structurally and functionally specialized by synthesizing different proteins. Differentiation thus reflects the activation and maintenance of different patterns of gene expression.

Pattern formation

All embryos of a given species have a similar structure or body plan because the cells become organized in the same way. The body plan is progressively filled with detail as development proceeds. To organize themselves in this way, cells must know where they are in relation to other cells in the embryo. This is achieved by giving each cell a positional value in relation to the principle embryonic axes, often in response to a morphogen gradient. Later in development, pattern formation generates the fine details of emerging organs. A variety of different mechanisms is involved, some intrinsic and others requiring cell–cell interactions.

Morphogenesis

Morphogenesis is the creation of shapes and structures. This is governed by different forms of cell behavior, including differential rates of cell proliferation, changes in cell shape and size, the movement of cells relative to each other, cell fusion and cell death.

Genes in development

All the major processes necessary for development are dependent, either directly or indirectly, on proteins. Genes therefore control development by determining where and when proteins are synthesized in the embryo, and thus how different cells behave. The analysis of developmental mutants has shown that many genes with important roles in development encode transcription factors and components of signaling pathways.

Related topics

Mechanisms of developmental commitment (A3)
Maintenance of differentiation (A5)
Pattern formation and compartments (A6)

Morphogenesis (A7)
Gene expression and regulation (C1)
Plant *vs* animal development (N1)

Development

Development is the process by which a **multicellular organism** arises, initially from a single cell. Development is **progressive**, i.e. a simple embryo, comprising few cell types organized in a crude pattern, is gradually refined to generate a complex organism with many cell types showing highly detailed organization. This gradual developmental strategy is known as **epigenesis**, in contrast to the earlier theory of **preformation**, which suggested that the early embryo comprised a miniature version of the adult, and development consisted entirely of growth. In fact, development involves five major overlapping processes: **growth, cell division, differentiation, pattern formation**, and **morphogenesis** (*Fig. 1*). Early development is usually studied in the context of the embryo as a whole. Later, developmental biologists concentrate on particular **model developmental systems**, which may be used to illustrate specific principles. For example, mammalian kidney development is used as a model for reciprocal cell–cell interactions (Topic K4), while vertebrate limb development is a useful model for pattern formation (Section M).

Overview of animal development

Much of this book is devoted to animal development, so this first Topic provides an opportunity to summarize some of the developmental stages common among animals and introduce some key terms. Animals show great diversity, but early development in most animals involves a common series of events (*Fig. 2*). Development usually begins with **fertilization**, when a **sperm** makes contact with the **egg** and the male and female **nuclei** fuse generating a **diploid zygote** (Topic E4). Fertilization is followed by **cleavage**, a series of rapid and synchronous cell divisions that occurs in the absence of growth and divides the large egg into a number of smaller cells, termed **blastomeres** (Topic F1). Uniquely in insects, the zygotic nucleus initially divides a number of times without cell division, so a **multinucleate syncytium** is generated. The blastomeres form when the nuclei migrate towards the periphery of the embryo and become

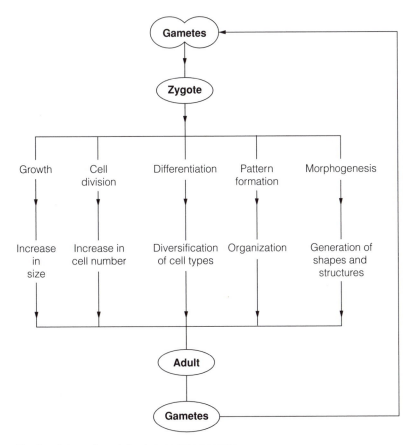

Fig. 1.   An overview of developmental processes.

enclosed by newly formed cell membranes. In all animal embryos, the result of cleavage is the formation of a ball or disc of blastomeres, often enclosing a fluid-filled cavity. Different terms are used to describe the embryo at this stage, depending on the species (e.g. blastula, blastodisc, blastocyst, cellular blasto-derm). The next stage of development is **gastrulation**, which involves a complex series of cell movements that reorganizes the embryo into three fundamental cell layers (Topics F2 and F3). These are called the **germ layers (ectoderm, meso-derm** and **endoderm)** and are present in all animal embryos.[1] Gastrulation is followed by **neurulation**, the development of the central nervous system (Section J). Once these fundamental processes have occurred, individual organs begin to develop according to their own programs (**organogenesis**).

Overview of plant development

Plant and animal development are discussed separately in this book because there are some important and fundamental differences between plants and animals in terms of their developmental biology (Topic N1). However, it is also

---

[1]Most animals are **triploblastic**, meaning that the embryo produces three germ layers. Some primitive animals (including coelenterates such as *Hydra*) are **diploblastic**, lacking the mesoderm layer. In these animals, a jelly-like secretion called the **mesogloea** separates the ectoderm and endoderm.

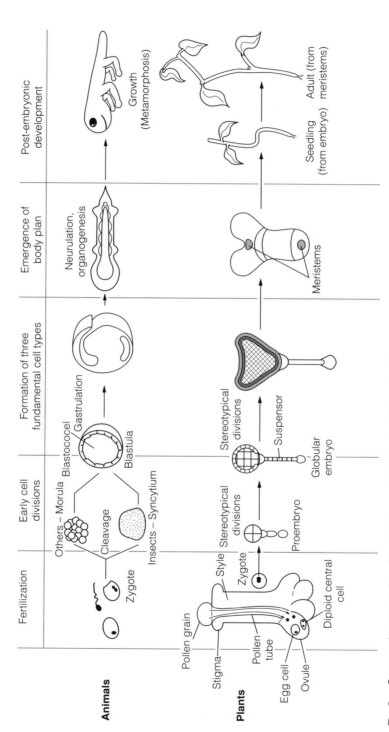

Fig. 2.   Comparative summaries of the development of animals and flowering plants.

useful to consider some similarities between animals and plants in early development (*Fig. 2*). The summary provided here refers to flowering plants. Development begins with **double fertilization**. The **male gametophyte (pollen grain)** produces two **sperm cells** and a vegetative cell. When a pollen grain adheres to the stigma of a flower, the vegetative cell elaborates into a **pollen tube**, which grows down through the style and penetrates one of the ovules. The two sperm cells migrate down the tube into the **female gametophyte** (the **embryo sac**) which is inside the ovule. One sperm cell fuses with the **egg cell** to generate a **diploid zygote**, while the other fuses to the diploid **central cell** to generate the triploid **endosperm cell**. The endosperm layer of the seed is derived from this cell, and nourishes the growing embryo. Fertilization is followed by a series of stereotyped cell divisions to generate a ball of cells called a **proembryo**, attached to the ovule by a **suspensor** (Topic N2). There is no gastrulation-like stage in plants because relative cell movement is prevented by the rigid cell walls. Further growth and cell division produces an embryo organized radially into three principal cell layers, **L1**, **L2** and **L3**, and organized axially into a series of organ-forming regions that correspond to the future **cotyledons** (storage organs) **shoot** and **root** of the seedling, and the proliferative **shoot and root meristem** populations that give rise to all adult structures. The function of the plant embryo is to produce the seedling, which emerges from the seed under environmental conditions suitable for **germination**, pushing a shoot towards the surface of the soil and the root downwards (Topic N3). All the organs of the adult plant (stem, leaves, flowers, roots) arise from the shoot and root meristems during post-embryonic development (Topic N4).

Growth and cell division

Development usually begins with a single cell, the **fertilized egg**, and generates an organism with many cells. Although the egg is a comparatively large cell, it is still smaller than the mature organism that develops from it. Development therefore involves both growth and an increase in cell number. In very simple organisms, such as the nematode *Caenorhabditis elegans*, the cell number in the adult is less than 2000. Conversely, large animals such as humans comprise many trillions of cells. An increase in cell number is necessary to allow cells to specialize, organize into patterns and carry out different functions.

In the earliest stages of animal development, the cell number increases in the absence of growth. During these **cleavage** divisions, the large egg is divided into a number of smaller cells as the cell cycle alternates rapidly between DNA replication and mitosis without any interstitial gap phases. The nature of these divisions is often instrumental in setting up the fundamental body plan of the embryo (see Topic F1). Later in development, growth can occur in the absence of cell division, as growth can be achieved not only by increasing cell number, but also by increasing cell volume, changing cell shape or by the growth of extra-cellular structures generated by materials secreted from cells (e.g. cartilage and bone). Cell growth and cell division do not therefore contribute solely to the size of the developing organism, but also to its shape and pattern.

Differentiation

Differentiation is the process by which cells become structurally and function-ally specialized, allowing the formation of distinct cell types (e.g. neurons, erythrocytes and keratinocytes in animals; xylem, phloem and parenchyma in plants). The structural and functional specificity of a cell depends on the proteins it synthesizes. For example, the ability of red blood cells to carry oxygen reflects the synthesis of hemoglobin specifically in these cells. With very few exceptions, all cells of a given organism contain the same genetic information; thus differentiated cells are genetically identical but express different genes. The process of differentiation therefore involves the control and maintenance of differential gene expression. Cells may begin to express different genes if they inherit different cytoplasmic components at cell division, or if they receive signals from other cells or the environment. Such interactions are considered in more detail in Topic A3. The maintenance of differential gene expression is discussed in Topic A5.

Pattern formation

Pattern formation is the process by which cells become organized in the embryo, initially to form the basic rudiments of the **body plan**, and later to generate the fine structure of individual organs. For development to succeed, each cell must behave in a manner appropriate for its position in the embryo, so that the correct differentiated cell types arise in the correct places and cells of similar type form regionally-appropriate structures (such as bones of different shapes and sizes in different parts of the body). The process by which cells become specialized according to their position is called **regional specification**.

The first patterning event that takes place in the embryo is **axis specification**, the establishment of the **principal body axes** (anteroposterior, dorsoventral etc; see Topic A6). These axes may be pre-determined by the distribution of maternal gene products in the egg (as occurs in *Drosophila* development; Section H), or may require a physical cue from the environment to influence gene expression or protein activity in the embryo itself (e.g. the point of sperm entry in *Xenopus*; Topic I1). Once the principal axes are established, the position of cells along each axis must be specified. This is often achieved by dividing an axis into individual **developmental compartments** and allocating each a **positional value**. Several experimental systems have shown that positional values may be assigned according to the position of a cell in a **morphogen gradient**. This is covered in more detail in Topic A6, and several well-characterized developmental systems based on this **morphogen model** are discussed in this book (e.g. early *Drosophila* development, Section H; vertebrate limb development, Section M). A diversity of patterning mechanisms occurs later in development, including asymmetric cell division, specific localized interactions between cells, and the generation of regular spacing patterns by lateral inhibition. In each case, the cell pattern that emerges reflects a *pre-existing pattern of gene expression* in the developing organ. However, a number of developmental systems also show spontaneous pattern-forming ability when known patterning genes are mutated, and this may reflect an underlying **reaction-diffusion mechanism** (Topic A6).

Morphogenesis

Morphogenesis means the creation of structure and form, and reflects the many different types of cell behavior that help to shape tissues and organs. Cells may change their size and shape, they may adhere or disperse, they may remain quiescent or divide, they may move relative to other cells, they may fuse together and they may even die in the process of sculpting a particular developing organ. In later development, morphogenesis can be regarded as a *response* to the developmental program, i.e. a differentiated cell aware of its position in the embryo will behave in such a manner as to generate a regionally-appropriate structure. Earlier in development, however, morphogenesis is instrumental in *driving* the developmental program, e.g. by bringing cells together to undergo inductive interactions. This is illustrated by the complex morphogenetic movements of gastrulation (Section F). The cellular basis of morphogenesis is discussed in more detail in Topic A7.

Genes in development

All the processes that are essential for development – cell division, growth, differentiation, pattern formation and morphogenesis – are mediated ultimately by **proteins** (acting either directly, or as enzymes to produce other molecules). Proteins are encoded by genes, so development is controlled to a large degree by gene expression. The pattern of gene expression in the embryo determines where, when and in what quantity particular proteins are made and therefore governs the properties of each cell (Topic C1). The analysis of developmental mutants (Topic B2) can provide clues as to the functions of particular genes. Many genes shown to have important roles in development encode two particular types of protein, **transcription factors** and components of **signaling pathways**. Transcription factors are important because they control gene expression, and therefore act as coordinators of developmental processes. Signaling pathways are necessary for cells to perceive external signals, either from other cells or from the environment. We return to the subject of transcription factors and signaling pathways many times in the course of this book.

A2 CELL FATE AND COMMITMENT

Key Notes

The developmental hierarchy

The number of different cell types in the embryo increases as development proceeds, but new cell types arise from particular pre-existing cell types through a hierarchical series of decisions. In animals, but not plants, such decisions tend to be irreversible, so there is a progressive restriction of cell fate and potency during development.

Cell fate and potency

Cell fate describes the range of cell types a particular cell can give rise to during normal development, while cell potency describes the range of cell types that a cell can give rise to in all possible environments. The fate of a cell is dependent on both its potency (which is an intrinsic quality) and its interactions with other cells in the embryo.

Fate maps

A fate map is a map of an embryo showing which parts of the embryo give rise to particular adult tissues. Fate maps are generated by marking or labeling cells in the early embryo and following their progress through development. In species that undergo stereotyped cell divisions, fate maps are accurate. In most species, there is a certain degree of random cell division and mixing, so fate maps are probabilistic.

Levels of developmental commitment

In animals, cells become committed to certain fates in stages, first reversibly and then irreversibly. A naïve cell may receive information in the form of cytoplasmic determinants or inductive signals that specifies its fate, but this fate can be altered by placing the cell in a different environment. A cell becomes determined when its fate is irreversible. This coincides with the loss of competence to follow alternative developmental pathways.

Related topics

Mechanisms of developmental commitment (A3)
Mosaic and regulative development (A4)
Maintenance of differentiation (A5)

Early animal development by single cell specification (G1)
Plant *vs* animal development (N1)

The developmental hierarchy

As development proceeds, the number of cell types in the embryo increases. However, new cell types do not arise randomly, but through a *hierarchical series of decisions* involving specific pre-existing cell types. One of the earliest events in animal development is the division of the embryo into the three **germ layers: ectoderm, mesoderm** and **endoderm.** Each of the germ layers then gives rise to a specific range of differentiated cell types as shown in *Fig. 1.* Similarly, the plant embryo is divided into three fundamental cell layers, **L1, L2** and **L3.** Each

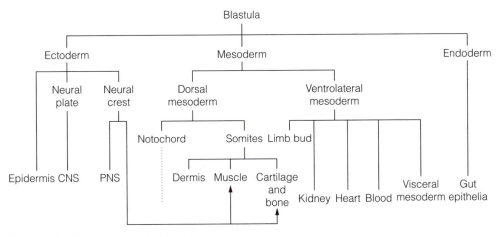

Fig. 1.   Simplified developmental hierarchy in vertebrates.

of these also gives rise to a particular range of differentiated cell types as shown in *Fig. 2*. In animals, these developmental decisions are usually irreversible. In vertebrates for example, once a cell has differentiated as ectoderm, its fate is fixed. It will eventually form either neural or epidermal tissue or a derivative of the neural crest, but can never give rise to liver or kidney tissue, which are specific to the endoderm and mesoderm lineages, respectively. Development in animals therefore comprises a series of decisions that *progressively and irreversibly restrict cell fate*. Conversely, plants show much greater plasticity in their developmental hierarchy, and cells originally in one of the layers can switch fates relatively easily.

Cell fate and potency

The **fate** of a cell is all the different cell types its descendants can become during normal development. In this context, **normal development** means development undisturbed by experimental manipulation. A cell may differentiate in an abnormal way if it is placed in an unusual environment e.g. by grafting, so it is exposed to other cells with which it normally does not come into contact. The term **potency** is used to describe the entire repertoire of cell types a particular cell can give rise to in all possible environments. The potency of a cell is an intrinsic property and is greater than or equal to its fate; the fate of a cell

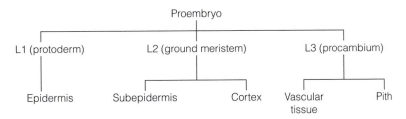

Fig. 2.   Simplified developmental hierarchy in flowering plants.

depends on its potency and its environment (e.g. its contact with other cells in the embryo).

In animal development, cell fate *and potency* are progressively restricted. For example, at the 16-cell stage, each blastomere of a mouse embryo is **totipotent**; i.e. it can potentially give rise to every cell type in the adult if transferred to another embryo. Later, the morula differentiates to form the inner cell mass and the trophectoderm (Topic F1). The inner cell mass gives rise to the embryo, and differentiates to form the three germ layers as well as other extra-embryonic structures. At this point, individual cells are still **pluripotent** since they can generate several different cell types, but they are no longer totipotent, since certain fates are now unavailable. Cell fate becomes increasingly restricted until a cell is **terminally differentiated** (can form only a single cell type). Some cells, e.g. hepatocytes, continue to divide but only produce identical daughters. Other cells, e.g. neurons, become quiescent (i.e. they exit from the cell cycle and do not divide further). **Stem cells** are exceptional because they are never terminally differentiated. Instead of dividing to produce two identical daughters, stem cells produce dissimilar daughter cells, only one of which undergoes terminal differentiation (*Fig. 3*). The other daughter is a replacement stem cell, and in this way stem cells can continually replenish the organism with a given cell type. This is useful for cells that are depleted in the course of their normal function (e.g. skin cells, cells lining the gut and genitourinary lumena, and blood cells).

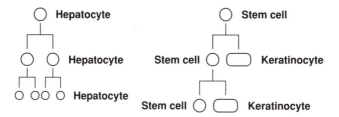

Fig. 3.   Terminally differentiated cells compared to stem cells.

In plants, cell fates are progressively restricted but differentiation does not correspond with terminal restriction in potency. Differentiated plant cells can change fates relatively easily if moved to a new position, and even fully differentiated plant cells can regenerate an entire new plant if placed in isolation. Therefore, while the fates of plant cells may be restricted during development, the cells remain totipotent even when they have differentiated. This is one of the fundamental differences between animal and plant development (Topic N1). Certain animal cells also show a limited developmental plasticity, which allows e.g. limb regeneration in insects and amphibians (Topic M3). In these particular cases, certain differentiated cells remain pluripotent even when they are differentiated.

Fate maps   A **fate map** is a map of an embryo or other developing system showing which adult tissues are formed from each region during normal development (*Fig. 4*). The term **'presumptive'** is used to describe the fate of each region. For example, the part of the *Xenopus* blastula-stage embryo that will become the neural plate

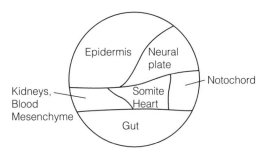

Fig. 4.   Fate map of the Xenopus *blastula – lateral view.*

is called the presumptive neural plate. Fate maps can be generated by labeling particular cells and following their progress through development. This can be achieved by staining the cell surface with a vital dye, injecting a fluorescent dye into the cytoplasm, or using cells grafted from a different species that can be distinguished from those of the host (e.g. chick–quail chimeras). Alternatively, genetic markers can be used, e.g. reporter genes or mutations that generate pigmentation defects. This is especially useful where the tissue is not accessible, e.g. plant meristems (Topic B4). Later, each labeled or marked cell will give rise to a clone of differentiated cells, confirming the fate of the original cell.

In some systems, such as the *Caenorhabditis elegans* embryo and the developing root of *Arabidopsis thaliana*, early fate maps are accurate because development involves a sequence of stereotyped cell divisions. In other organisms, there is a greater or lesser degree of random cell division and cell mixing, and fate maps are **probabilistic**, i.e. cells may have a choice of fates and likelihoods are attached to alternatives. Where cell mixing is totally random, as in the early zebrafish embryo, it is impossible to derive a fate map. In organisms such as *C. elegans*, where every cell division during development is stereotyped and each individual has the same number of cells, it is also possible to construct a diagram of **cell lineage** (this resembles a family tree and can trace every cell in the adult back to the egg). Such diagrams are useful, especially in the study of lineage mutants (Topic G1) but unlike fate maps, do not show the relative positions of the cells in the embryo. Cell labeling can also show whether cells have been allocated to particular **developmental compartments**, a subject considered in more detail in Topic A5.

Levels of developmental commitment

As cell fate becomes restricted following each decision in the developmental hierarchy, cells are said to become **committed** to a certain fate (or **developmental pathway**). In animals, commitment occurs in stages, initially reversible and then permanent (*Fig. 5*). In plants, commitment appears always to be reversible, so the following discussion relates only to animals.

An uncommitted cell can be described as **naïve**, meaning that it has received no instructions directing it along a particular developmental pathway. The fate of a cell is said to be **specified** if the cell is directed to follow a certain developmental pathway and does so when placed in isolation, which should provide a neutral environment. Specification may occur if a cell inherits a particular cytoplasmic determinant or receives inductive signals from another cell. However, that same cell placed in a different environment, such as in contact with other cells, may be **respecified** by its interaction with those cells. This shows that the

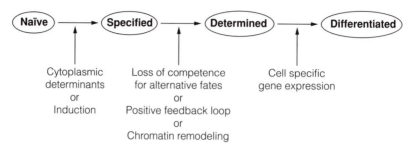

Fig. 5.   Stages of developmental commitment.

commitment at this stage is *reversible*. The fate of a cell is said to be **determined** if it cannot be changed, regardless of the cell's environment. This shows that the commitment has now become *irreversible*.

Determination to follow one developmental pathway coincides with loss of **competence** to follow alternative pathways. Cytoplasmic determinants, inductive signals and the concept of competence are discussed in more detail in Topic A3.

A3 Mechanisms of developmental commitment

Key Notes

How cells differentiate	To differentiate, cells must begin to synthesize new proteins. This requires the selective expression or activation of regulatory molecules such as transcription factors that control which proteins are synthesized in the cell. Such regulatory molecules could be inherited as cytoplasmic determinants or activated by external signals.
Cytoplasmic determinants	Cytoplasmic determinants are molecules in the cytoplasm of a cell that help to determine cell fate. The polarized distribution of such molecules in a mother cell can result in their inheritance by only one of the daughters following cell division, thus promoting differentiation of the daughter cells. This strategy is used frequently in early development, when maternal gene products localized to particular regions of the egg are asymmetrically distributed to different blastomeres during cleavage.
Induction	Induction is a process whereby one cell or group of cells can influence the developmental fate of another, and is a common strategy to control differentiation and pattern formation in development. Induction involves cell–cell signaling and can occur over various ranges. Responding cells may show a single stereotyped response to the inductive signal, or a graded response dependent on its concentration, in which case it is called a morphogen.
Competence	Induction relies both on the ability of the inducing cell to produce the signal and the ability of the responding cell to receive the signal and react in the appropriate manner. The ability of the responding cell to react to an inductive signal is termed competence. The loss of competence is one mechanism by which cells become irreversibly committed to a given developmental pathway.
Instructive and permissive induction	Two types of induction can be distinguished on the basis of the choices available to the responding cell. Instructive induction occurs when the responding cell has a choice of fates, and the inductive signal instructs the cell to adopt one of those fates in preference to the other. Permissive induction occurs when the responding cell is already committed to a certain developmental pathway, but needs the inductive signal to continue towards differentiation.
Lateral inhibition and the community effect	These are special types of induction in which single cells behave differently to cell populations. In lateral inhibition, differentiated cells arise in a regularly spaced pattern within a group of equivalent undifferentiated cells due to random fluctuations in levels of a ubiquitous

signal that inhibits differentiation. In the community effect, populations of cells can change their collective fate by secreting enough of an inductive signal to reach a critical concentration threshold, whereas isolated cells follow a different developmental pathway because they cannot produce enough of the signal on their own.

| **Related topics** | Cell fate and commitment (A2) | Maintenance of differentiation (A5) |
| | Mosaic and regulative development (A4) | |

How cells differentiate

The specialized properties of different cell types are conferred by the proteins they contain. The process of differentiation must therefore involve the synthesis of different sets of proteins in different cells. Exceptionally, this may be achieved by DNA rearrangement, as in the differentiation of antibody-producing blood cells (Topic C1). However, most cells contain the same DNA, and different sets of proteins are made by the selective expression or activation of particular gene products. The synthesis of a functional protein is dependent on a series of steps including transcription, RNA processing, protein synthesis and post-translational protein modification (Topic C1). Any or all of these stages can be regulated, so differentiation usually begins with the activation of a particular regulator molecule, such as a transcription factor. What is of interest to developmental biologists is how such regulator molecules are controlled and how this fits in with the concept of developmental commitment as discussed in Topic A2. There appear to be two major strategies for establishing commitment and hence initiating the series of events that results in cell differentiation: the inheritance of **cytoplasmic determinants** and the perception of external **inductive signals**. These strategies are discussed in more detail below.

Cytoplasmic determinants

A cell can divide to produce two daughters committed to different fates in the absence of any external influences. Stem cells provide an excellent example of this process. Each division of a stem cell produces one daughter cell committed to differentiate, and a replacement stem cell. This stereotyped division program can occur in isolation, indicating that the mechanism of differentiation is entirely intrinsic. One way in which this could be achieved is through the asymmetric distribution of **cytoplasmic determinants** (molecules in the cytoplasm that can help to determine cell fate). If a mother cell contains a cytoplasmic determinant that is localized to one pole as the cell undergoes division, that determinant will be inherited by only one of the daughters. In the simplest scenario, the determinant could be a transcription factor that would activate certain genes in only one daughter cell (*Fig. 1*).

Cytoplasmic determinants are encountered in many developmental systems. One example is the *Drosophila* nervous system, where stem cells divide to produce further stem cells and progenitor cells that are committed to differentiate into neurons. This reflects the asymmetric distribution of at least two proteins (a transcription factor called Prospero and a signaling protein called Numb) which are inherited by the neuronal progenitor. Further discussion of this system can be found in Topic J5.

Many examples of segregating cytoplasmic determinants are found in early animal development, where maternal gene products are used as determinants to

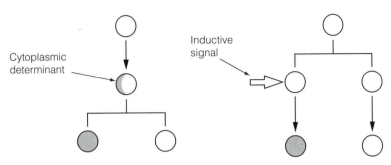

Fig. 1.   Cytoplasmic determinants and induction.

polarize and pattern the early embryo. Many animal eggs are polarized with respect to the animal–vegetal axis, reflecting the distribution of the yolk and the consequent displacement of the nucleus towards the animal pole (Topic F1), but there is often additional hidden polarity at the molecular level. For example, the maternal mRNA for the transcription factor VegT is localized to the vegetal pole of the *Xenopus* egg. Active VegT is therefore synthesized only in the vegetal cells in the blastula-stage embryo, and this plays an important role in early patterning, specifically in mesoderm/endoderm specification and the formation of the embryonic organizer (Topic K1). The distribution of maternal mRNAs in the *Drosophila* egg is critical in the establishment of the anteroposterior axis. Maternal *bicoid* mRNA is located at the anterior pole and maternal *nanos* mRNA is located at the posterior pole. The proteins encoded by these messages control the establishment of gradients of transcription factors in the early syncytial embryo, which in turn control downstream genes that divide the embryo into segments and confer upon each segment its positional identity (Section H). The role of maternal gene products in early development is discussed in more detail in Topic A4.

Induction

Two identical cells can follow alternative fates if one is exposed to an external signal (often secreted by a different cell) while the other is not. The process where one cell or group of cells changes the developmental fate of another is termed **induction** and animal development is rich in examples of such **inductive interactions**. The inductive signal may be a protein or another molecule secreted from the inducing cell. It usually interacts with a receptor on the surface of the responding cell (although some inductive signals diffuse through the membrane and interact with cytosolic receptors). The signal initiates a signal transduction cascade inside the responding cell that alters the activity of transcription factors and/or other proteins, and eventually alters the pattern of gene expression (Topic C3). Unlike the segregation of cytoplasmic determinants, induction is an extrinsic process that depends on the position of a cell in the embryo (*Fig. 1*).

Induction can occur over different ranges (*Fig. 2*). In the most extreme case, **endocrine signaling** involves the release of inductive signals by one type of cell located a considerable distance from the responding cell population, and the molecule must travel to its target through the vascular system. This is the mechanism utilized by many **hormones**. Most inductive interactions in early development occur locally. This can involve the release of a diffusible substance (**paracrine signaling**) or a substance that is deposited in the extracellular matrix.

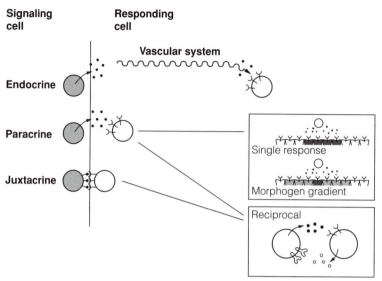

Fig. 2.   Ranges and types of induction.

The range of such signals varies, but rarely exceeds a few cell diameters. The range of a signal can be extended by a **relay mechanism**, in which one signal causes the responding cell to release another signal, which similarly induces a third and so on. It is therefore possible for the responding tissue to take on the inductive properties of the inducer, a phenomenon termed **homeogenetic induction** (e.g. see Topic J1 for a discussion of homeogenetic induction in the vertebrate nervous system). In other cases, the signal may only reach the cell adjacent to the inducer, particularly in cases where direct contact between cells is required because the inductive signal is fixed to the cell membrane, or it travels through gap junctions (**juxtacrine signaling**). Cells may also reciprocally induce each other, i.e. each cell acts as the inducer of the other, and responds to the other's inducing signal. Such **reciprocal induction** occurs in the kidney, where the ureteric bud induces the metanephric blastema to differentiate into the adult kidney structures, while the blastema induces the ureteric bud to proliferate and branch. This system is discussed in more detail in Topic K4.

   An important distinction can be drawn between inductive signals that elicit a single, stereotyped response in responding cells, and those that elicit different responses at different concentrations. The former type of induction is prominent in situations where two tissues come together for the interaction to take place, e.g. in neural induction (Topic J1). The latter type of inductive signal is known as a **morphogen**[1]. Morphogens are of great significance in development because a **morphogen gradient** can be set up within a given tissue, and different responses can be induced depending on the distance from the source of the morphogen. This is one way in which cells can receive positional information,

---

[1]Do not confuse a morphogen with morphogenesis. **Morphogenesis** is the process by which structures form during development, reflecting different types of cell behavior. A **morphogen** is an inducer that elicits different responses at different concentrations; the term literally means 'form-generating substance'.

and differentiate according to their position along an axis. The different, concentration-dependent responses in terms of alternative patterns of gene expression give cells positional values that enable them to form the correct structures in the body plan. This subject is considered in more detail in Topic A6.

Competence

Competence is a property of the cell responding to induction. A cell is described as competent if it can respond to the inductive signal by undergoing all the appropriate molecular changes that allow it to follow the 'induced' developmental pathway. In the absence of induction, the cell eventually becomes determined to an alternative pathway, and this coincides with its loss of competence to respond to induction (Topic A5). In the case of endocrine or paracrine signaling, competence depends on the synthesis of all the components of the signal transduction pathway that link the inductive signal to its target, such as a transcription factor, in the responding cell. If any of these components is lost, e.g. the cell surface receptor, the signal transduction apparatus, or the downstream target transcription factor itself, the cell loses competence. In the case of juxtracrine signaling, a cell may also lose competence simply by breaking contact with the inducing cell. This may reflect the movement of cells away from each other, or the disassembly of gap junctions.

Instructive and permissive induction

These two categories of induction reflect the choices available to the responding cell. **Instructive induction** occurs where the responding cell has a choice of fates and will follow one pathway in response to induction but an alternative pathway in the absence of induction. For example, in the early *Xenopus* blastula, ectoderm will form epidermis in the absence of induction, but neural plate in the presence of inductive signals from the notochord. **Permissive induction** occurs where the responding cell is already committed to a certain developmental fate, and simply requires the inducing signal to continue down that developmental pathway. An example is muscle development, where myoblasts continue to proliferate until growth factors are withdrawn, when they differentiate into myotubes and eventually muscle fibers. Growth factor withdrawal prevents the synthesis of an important class of inhibitor protein, and allows expression of the transcription factor MRF4, which drives muscle differentiation. The molecular basis of muscle differentiation is discussed in more detail in Topic K3. This example also demonstrates that induction may occur just as well through the *withdrawal of an existing signal* as it does through the secretion of a new signal.

Instructive and permissive induction can be distinguished by grafting experiments. For example, in mammals, the cardiogenic mesenchyme (future heart) is required to induce hepatocyte development in the presumptive liver-forming region of the foregut. The signals could be instructing the foregut to form hepatocytes instead of other cell types or could be permitting the differentiation of hepatocyte cells that have already been specified by other mechanisms. If the cardiogenic mesenchyme is grafted under the hindgut, no hepatocytes are induced. The inductive signals from the cardiogenic mesenchyme are therefore permissive rather than instructive (Topic K6).

Lateral inhibition and the community effect

Lateral inhibition is the inhibition of a particular developmental process in one cell by signals from an adjacent cell. Lateral inhibition can be used as a special form of induction, which involves an initially equivalent field of cells, yet results in the differentiation of individual cells in a regularly spaced pattern (*Fig. 3*).

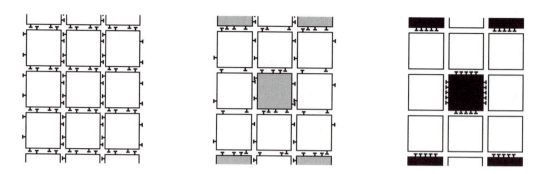

Fig. 3.   How spacing patterns are generated by lateral inhibition.

Since all the cells are initially equivalent, they all have the potential to differentiate in the same way. However, in the undifferentiated state, they all signal to each other to repress differentiation. As individual cells begin to differentiate, their ability to repress the differentiation of neighboring cells increases while their tendency to be repressed diminishes. This could be achieved, for example, by increasing the production of the inhibitor signal and decreasing the synthesis of its receptor as part of the process of differentiation. The cells that happen to produce more of the inhibitor signal through random fluctuation would therefore begin to differentiate and suppress the differentiation of the cells around them by producing more of the inhibitor signal. The spacing of the differentiated cells would be governed by the range of the signal and the strength of its effect (*Fig. 3*). Lateral inhibition is thought to control the choice of neuronal progenitors in *Drosophila* and vertebrates (Topic J5), the choice between alternative cell fates in the nematode vulva (Topic L1) and perhaps the distribution of trichomes on the leaves of flowering plants (Topic N5).

The **community effect** is a mechanism whereby a *population* of cells must be present to change the collective cell fate (*Fig. 4*). This special form of induction

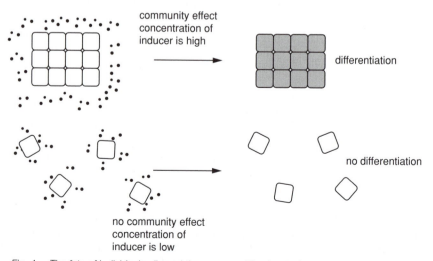

Fig. 4.   The fate of individual cells and tissues may differ due to the community effect.

involves **autocrine signaling**, where the cell secretes a signal that acts on one of its own receptors. A single cell in isolation does not produce enough of the signal for an inductive effect to occur, because the signaling molecule does not reach sufficient concentration to trigger the effect. However, a population of cells produces enough of the signal, with the result that the fate of the entire population is changed *en masse*. The community effect occurs during both mesoderm induction (Topic K1) and neural induction (Topic J1) in *Xenopus*. For example, isolated animal cap (ectoderm) tissue from the *Xenopus* blastula-stage embryo will differentiate into epidermis, whereas individual cells can form neurons. This is because the ectoderm secretes certain **bone morphogenetic proteins** that inhibit neural differentiation, but isolated cells do not produce enough of the signal to prevent this outcome.

A4 MOSAIC AND REGULATIVE DEVELOPMENT

Key Notes

Definitions of mosaic and regulative development	Cytoplasmic determinants and inductive signals can both be used to control cell fates during development. The exclusive use of either mechanism defines two extreme strategies, mosaic development and regulative development. Most organisms use a combination of these strategies.
Mosaic development	Mosaic development is controlled entirely by cytoplasmic determinants. Each cell undergoes autonomous specification, so that its fate is determined by its lineage irrespective of its position. Isolated blastomeres develop into the complete corresponding parts of the embryo as shown by the fate map, and embryos with missing blastomeres develop with the corresponding missing parts.
Regulative development	Regulative development is controlled entirely by inductive interactions. Each cell is specified conditionally, according to its interactions with other cells. Cell fate is therefore determined by position, irrespective of lineage. Isolated blastomeres do not develop into the corresponding parts of the embryo (as defined by the fate map) because they lack the appropriate interactions. However, embryos with missing blastomeres can regulate to replace their missing parts.
Prevalence of mosaic and regulative development	Most species use a combination of mosaic and regulative mechanisms in their developmental program. Certain animals, such as ascidians, rely on cytoplasmic determinants for many aspects of early development and produce mosaic-like embryos. Conversely mammals appear to possess no determinants, and early development is entirely regulative. Other animals fall between these extremes, with both cytoplasmic determinants and inductive interactions playing important roles in early development. The development of plant embryos and meristems is also highly regulative.
Maternal and zygotic genes	Maternal gene products are often used as cytoplasmic determinants, and these are placed in the egg during oogenesis. In such cases, early development is controlled by the maternal genome, and developmental mutations show a maternal effect (i.e. the phenotype of a mutant individual is seen in her offspring). Zygotic genes become active during development, and this may be a more or less gradual process depending on the species.
Related topics	Mechanisms of developmental commitment (A3) Early animal development by single cell specification (G1) Early patterning in vertebrates (I1)	Molecular aspects of embryonic pattern formation in *Drosophila* (H1) Development of the plant embryo (N2) Shoot and root meristems (N4)

Definitions of mosaic and regulative development

Cytoplasmic determinants and inductive signals can both be used to control cell fates (Topic A3). If development was controlled entirely by cytoplasmic determinants, the fate of every cell would depend on its lineage, while its position in the embryo would be irrelevant. This is the definition of **mosaic development**. Conversely, if development was controlled entirely by inductive interactions, the fate of every cell would depend on its position in the embryo and its lineage would be irrelevant. This is the definition of **regulative development**. The development of most organisms involves a combination of these mechanisms. The two mechanisms are discussed in more detail below, and their key features are compared in *Fig. 1*.

Mosaic development

Properties of mosaic embryos
In **mosaic development**, the fate of every cell is governed entirely by its intrinsic characteristics, i.e. the cytoplasmic determinants it inherits at cell division, and fate is therefore equivalent to potency. Cell fate depends on **lineage** (line of descent) rather than position in relation to other cells. During development, each cell is said to undergo **autonomous specification**, i.e. if removed from the embryo each cell should, in principle, develop according to its intrinsic instructions and differentiate into the appropriate part of the embryo whether or not the rest of the embryo is there. Furthermore, the remainder of the embryo should also develop normally, but would lack the parts specified by the missing cell (*Fig. 1*). The fate map of an embryo undergoing mosaic development can therefore be constructed by looking at the fate of individual cells developing in isolation. Formally, the fate map is equivalent to a specification map (Topic A2).

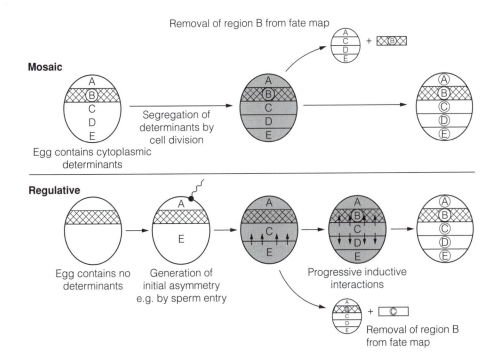

Fig. 1. Behavior of embryos and explants during mosaic and regulative development.

Problems with mosaic development as a sole strategy

Many animals use mosaic mechanisms in early development but it is not practical to base the entire developmental program on this strategy unless the organism is very simple. For even moderately complex organisms, the egg would need to be divided into thousands of zones that would eventually segregate into different cells. Furthermore, continuation between generations would require the gamete forming region of the embryo to be predetermined in the egg, and prepatterned with the cytoplasmic determinants of the subsequent generation. Each generation would have to be packaged and patterned in the gamete-forming region of the last, like a series of Russian dolls, in a manner reminiscent of preformation (Topic A1).

Regulative development

Properties of regulative embryos

In **regulative development**, the fate of every cell is governed entirely by its interactions with other cells. Cell fate depends on position in the embryo and is independent of lineage. The potency of each cell is therefore much greater than its fate. During regulative development, each cell is said to undergo **conditional specification**, i.e. conditional on the presence of other cells. Therefore, if removed from the embryo, a given cell will not fulfil its normal fate because it lacks the necessary interactions. Furthermore, the remainder of the embryo can **regulate** to replace missing parts, because the appropriate inductive interactions have yet to take place, and other cells can be respecified to fill in the missing pattern (*Fig. 1*). The fate map of a regulative embryo is not the same as a specification map, since cells in isolation will not develop in the same way as those in the embryo.

Problems with regulative development as a sole strategy

Regulative development is suited to a developmental program based on epigenesis, because sequential inductive interactions could progressively increase the complexity of an embryo by increasing the number of cell types. One problem with regulative development, however, is starting the process off, since development begins with a single cell. Regulative embryos need an alternative process to introduce asymmetry into the system and set up the initial source of inductive signals that sets off the cascade of conditional specification. This asymmetry could be generated by cytoplasmic determinants, or it could come from an external source. Vertebrate embryos provide examples of both. In *Xenopus*, for example, there are asymmetrically-distributed maternal gene products in the egg, which act as cytoplasmic determinants, while the site of sperm entry provides an external symmetry-breaking cue. These signals help to establish the initial source of inductive signals, which is called the **organizer**. The establishment of the organizer is discussed in detail in Section I, and in the context of mesoderm induction, in Topic K1.

Prevalence of mosaic and regulative development

All organisms use both cytoplasmic determinants and inductive interactions during development, with some cells specified autonomously and some conditionally. The predominance of each strategy varies in different species. In ascidians, the epidermis, endoderm and most muscle cells are specified by the distribution of cytoplasmic determinants in the egg, so that embryos show striking mosaic behavior. This is most obvious in species such as *Halocynthia roretzi*, where the determinants and the tissues they give rise to are different colors (Topic G3). However, the specification of other tissues, such as the notochord and central nervous system, is conditional upon cell–cell interactions. In contrast to ascidians, the early development of mammals is purely regulative. There appear to be no determinants in the egg, so each of the blastomeres in the morula is equivalent and totipotent. The first differentiation separates the inner cell mass from the outer trophectoderm, and this appears to be dependent only on position within the morula, with those outer cells exposed to the environment differentiating into trophectoderm (Topic F1). Cytoplasmic determinants appear much later in mammalian development, e.g. during neuronal differentiation (Topic J5). The development of plant meristems is also highly regulative (Topic N4). Insects employ a specialized strategy, **syncytial specification**, where the egg nucleus divides a number of times to generate a syncytium containing thousands of nuclei. Cell fates are determined prior to cellularization by interactions between diffusible regulators and individual nuclei. *Drosophila* development is illustrative of this strategy and is described in Section H.

Maternal and zygotic genes

Cytoplasmic determinants often feature in early animal development, where **maternal gene products** are localized in the egg to help pattern the early embryo. In such cases, early development becomes dependent on the **maternal genome**, while the paternal genome plays no part. This leads to an interesting class of mutants called **maternal effect mutants**, whose phenotypes are manifest in the embryos of mutant females, but not in the mutant females themselves. Furthermore, if the mutation is recessive and the female is homozygous for the mutation, embryonic development cannot be rescued by mating to a wild type male.

Maternal gene products may be critical in the earliest stages of development, but they may be active for only a short time, or they may become depleted. Development then becomes dependent on **zygotic gene products**, and **zygotic mutations** that affect development become apparent. The beginning of zygotic gene expression can be a landmark in development, accompanied by changes in the cell cycle (Topic C4) and in cell behavior. In *Xenopus*, this is called the **midblastula transition** and occurs just prior to gastrulation (Topic F1). A similar landmark occurs in *Drosophila* development, coinciding with the cellularization of the blastoderm (see Topic C4 for a discussion of cell cycle regulation in this system). However, unlike the *Xenopus* midblastula transition, the sudden change in *Drosophila* development does not mark the onset of zygotic gene expression, but rather the depletion of one or more key maternal gene products that regulate the cell cycle. In fact, zygotic gene expression in *Drosophila* commences gradually, as it does in most species. In mammals, zygotic gene expression begins during early cleavage, as early as the two-cell stage in some species.

A5 MAINTENANCE OF DIFFERENTIATION

Key Notes

Genomic equivalence	With few exceptions, all differentiated cells in a given organism contain the same DNA as the fertilized egg. This means that most differentiated cells possess the information required to produce an entire organism.
Potency of differentiated cells	Although they carry all the necessary DNA, differentiated animal cells cannot recapitulate the developmental program and produce a new animal. Conversely, differentiated plant cells can regenerate whole plants, either by the direct production of shoots and roots, or by somatic embryogenesis. This shows that the differentiation of animal cells is generally irreversible, although there are some examples of limited plasticity, such as the regeneration of limbs in urodele amphibians.
Totipotency of nuclei	Nuclei from differentiated animal cells can support development when injected into an enucleated egg, although the process is more efficient using embryonic nuclei than nuclei from adult cells. This confirms that nuclei retain all the genetic information required for development, but suggests the DNA is modified in some way as cells differentiate.
Maintenance of differentiation	Cell fate in animal development is progressively restricted and usually irreversible. This means that animal cells retain a memory of previous developmental decisions, which is perpetuated through successive cell divisions. Both cytoplasmic and nuclear mechanisms are involved in developmental memory and the maintenance of differentiation. Transcription factors can regulate their own production to maintain the expression of a particular set of genes. Furthermore, by controlling chromatin structure and the extent of DNA methylation, active and repressed chromatin domains can be maintained through successive cell divisions. Plants may use similar mechanisms to regulate gene expression, but their developmental memory can be erased by placing cells in a new environment.
Related topics	Cell fate and commitment (A2) Somitogenesis and patterning (K2) Mechanisms of developmental commitment (A3) Plant *vs* animal development (N1) Chromatin and DNA methylation (C2)

Genomic equivalence

With very few exceptions, all differentiated cells in a given organism contain the same DNA, which is the same as the DNA in the fertilized egg. This means that almost all differentiated cells in principle contain all the information required to generate a complete new individual. Evidence for genomic equivalence comes from molecular analysis including **Southern blot hybridization** and **PCR amplification** of genomic DNA isolated from embryos and various adult tissues. When probed for different developmental marker genes and cell type-specific genes, all banding patterns and PCR products are the same. There are a few notable exceptions to this rule, which are listed in *Table 1*.

Table 1.   *Examples of genomic non-equivalence during development*

Mechanism	Examples
DNA amplification	Selective amplification of rRNA genes in *Xenopus* oocytes. Amplification of the chorion genes during oogenesis in *Drosophila* (Topic E3).
DNA loss	In mammals, loss of nuclei in differentiated erythrocytes and keratinocytes.
DNA rearrangement	In vertebrates, VDJ recombination and class switching during B- and T-cell development (Topic C1)

Potency of differentiated cells

If all differentiated cells carry the same genetic information, it should be possible in principle to use differentiated cells to generate new individuals. In plants this is generally true. Differentiated plant cells can be isolated and will, under the appropriate culture conditions, form undifferentiated tissue called **callus**. This can be used to regenerate whole plants, either by **organogenesis** (direct formation of shoots and roots) or **somatic embryogenesis** (full recapitulation of development, including the embryonic stage). This makes it possible to produce whole plants from cuttings, callus or even single cells in culture, and has greatly facilitated the production and exploitation of transgenic plants (Topic B3).

Conversely, differentiated animal cells are almost always irreversibly committed to a particular fate, and as development proceeds, this fate becomes more and more restricted until the cell is terminally differentiated (Topic A2). There are certain cases in which cell fate in animals shows a limited plasticity. *Drosophila* **imaginal discs** are undifferentiated larval structures that produce specific appendages in the adult fly (Topic L2). Their fates are usually determined during embryogenesis, but if larval discs are serially cultured in the abdomen of adult flies, they can undergo **transdetermination** and later differentiate to form alternative structures. As the imaginal discs represent an early stage of commitment, it is not really surprising that their fates can be altered when the cells are continuously exposed to such an artificial environment. A more striking example of developmental plasticity is the ability of urodele amphibians to regenerate limbs and tails (Topic M3). In this case, fully differentiated cells at the wound site can dedifferentiate to give pluripotent progenitor cells, which then redifferentiate to form the various different types of replacement cells. This process can be termed **transdifferentiation**.

Although the transdetermination of insect imaginal discs and the transdifferentiation of cells in the urodele limb represent examples of how differentiated cells can reverse their fate and recapitulate at least part of the earlier developmental program, these cells are still unable to give rise to an entire new animal. This indicates that animal cells remember developmental decisions taken by their ancestors and, except in the limited cases discussed above, their fates are fixed by mechanisms that maintain differentiation.

Totipotency of nuclei

Although differentiated animal cells cannot recapitulate the entire developmental program, it is possible for the nuclei from those differentiated cells to do so. This type of experiment, where the nucleus of a differentiated cell is used to replace the nucleus of a fertilized egg, shows that all the information required to generate the animal is retained in the nuclei of differentiated cells.

The earliest experiments of this kind were performed on *Xenopus*. Nuclei taken from tadpole cells or from different kinds of adult cell were injected into UV-irradiated eggs (UV treatment destroys the endogenous DNA). In many cases, the eggs containing tadpole nuclei developed into normal swimming tadpoles, and in a few cases these produced viable adults. Adult cell nuclei from many different cell types were also able to support full development but at a much lower efficiency. However, nuclei from some adult cell types consistently failed to allow development. The results from such experiments firstly confirm the totipotency of adult cell nuclei, showing there is no irreversible change to genetic information in the nucleus during development. However, it becomes more difficult to recapitulate the entire developmental program as development proceeds, and indeed becomes impossible in certain cell types, such as neurons. Thus, although there is no change to the genetic information in most differentiated nuclei, the DNA does undergo some change that reduces the ability of the nucleus to be **reprogrammed** by the intracellular environment of the egg. These results have been confirmed in other species, including insects and mammals. It has been possible for a long time to produce mammals by nuclear transfer from other embryos, but only recently has it been possible to produce e.g. sheep and cows by nuclear transfer from adult cells. There appear to be certain poorly-understood species-specific properties that affect the success of nuclear transfer procedures. For example, the technique as currently used is very inefficient in mice and pigs.

Maintenance of differentiation

The results of experiments using differentiated cells and their nuclei suggest that the differentiated state is maintained by a combination of cytoplasmic and nuclear factors. The role of the cytoplasm in differentiation is demonstrated by nuclear transfer experiments, where the egg cytoplasm is sometimes able to reprogram the nucleus of a differentiated cell to behave in the manner of an egg nucleus. Similarly, cell fusion experiments demonstrate that the cytoplasm from one cell can influence the nucleus of the other. For example, fusion between virtually any cell type and a muscle cell will induce the expression of muscle-specific genes in the non-muscle cell nucleus; this is a limited type of transdifferentiation. The cytoplasm is the site of protein synthesis, so it is likely to be full of newly synthesized transcription factors awaiting import into the nucleus. The arrival of a new nucleus therefore attracts these transcription factors to the sites they would occupy in their own nucleus, inducing similar patterns of gene expression. This would be particularly true in the egg, which has a larger volume of cytoplasm than any other cell, and a stockpile of proteins.

The maintenance of differentiation through successive rounds of cell division could therefore be achieved by transcription factors. In one mechanism, the production of large amounts of a given transcription factor would provide enough of the regulator to persist in the cytoplasm through numerous cell divisions. A more refined mechanism would be for the transcription factor to exercise positive feedback regulation on its own gene. This appears to be the case in muscle differentiation, where the transcription factor MyoD activates not only a range of muscle-specific genes encoding structural proteins, but also its own gene and the genes for other muscle-specific transcription factors (Topic K3). In principle, however, this type of cytoplasmic loop could be broken by transferring the nucleus to a different cytoplasm.

The role of the nucleus in differentiation is demonstrated by the inability of nuclei from certain cell types to recapitulate the developmental program in the egg, and the decrease in the efficiency of recapitulation as cells become progressively more differentiated. As discussed in Topic C2, both **DNA methylation** and **chromatin structure** have profound effects on gene expression, and there are mechanisms for perpetuating both types of modification through successive rounds of DNA replication. Hence, in female mammals, the X-chromosome inactivated in one cell remains inactive in all descendants in part because the DNA methylation pattern is perpetuated by DNA methyltransferases that preferentially recognize hemimethylated DNA (DNA methylated on one strand only). In *Drosophila* it has been shown that homeotic selector gene expression patterns are initiated by transcription factors encoded by the upstream gap genes, but the patterns are maintained through successive rounds of cell division by reorganization of chromatin structure, mediated by the *Polycomb* and *trithorax* gene products (Topic H5). Related genes have been identified in vertebrates, and their products maintain *Hox* gene expression.

Differentiated plant cells use transcription factors to regulate gene expression, and in the case of *Arabidopsis*, it has been shown that homeotic selector gene expression is also maintained by chromatin reorganization. As plants and animals use similar mechanisms of gene regulation, how do differentiated plant cells remain totipotent? Developmental responses in plants, such as the growth of shoots and roots and the production of callus tissue, can all be elicited by treatment with different hormones. This suggests that the differentiated state of the plant cell is maintained predominantly by hormonal signaling, and that despite their similarities, maintenance mechanisms involving transcription factors or DNA/chromatin modification in plants are inherently reversible.

A6 PATTERN FORMATION AND COMPARTMENTS

Key Notes

Regional specification

Regional specification describes any mechanism that tells a cell where it is in relation to other cells in the embryo. Regional specification is necessary for pattern formation, where cells become organized correctly and generate appropriate structures in different parts of the embryo.

First coordinates

Pattern formation begins with a basic body plan. The organization of cells to form a body plan requires a polarized framework of coordinates, represented by the principal body axes. In animals these are the anteroposterior, dorsoventral and left–right axes. The axes are specified and polarized either by the localization of cytoplasmic determinants, or the perception of external physical cues.

Positional information

The location of any cell in the embryo can be defined by its position along each of the principal axes. A key question in developmental biology is how cells become aware of their position, i.e. what is the nature of the positional information they receive? Several model systems suggest that cells may be assigned positional values on the basis of their distance from the source of a morphogen.

Morphogen gradients

A morphogen is an inductive signal that elicits different responses at different concentration thresholds. A morphogen synthesized at one end of an axis would establish a concentration gradient along the axis, and cells could be assigned positional values according to the concentration of morphogen. The gradient of morphogen would be translated into a gradient of intracellular signal transduction and ultimately differential gene expression. Morphogens can establish gradients by free diffusion but alternative mechanisms are required for extracellular secreted proteins such as Hedgehog, Wingless/Wnt and Decapentaplegic/BMP.

Compartments and segmentation

Once an axis has been established, it is often divided into repetitive units that allow the same basic developmental structure to be used in different ways. These units may exist only in terms of boundaries of gene expression, or they may become manifest as physical structures such as segments. They are often developmental compartments, i.e. structures within which the clonal expansion of a particular cell line is constrained.

Homeotic genes and mutations

Homeotic genes give cells their positional identity. Different combinations of homeotic genes are expressed in response to different morphogen concentrations, and in this way cells can be assigned positional values along an axis. The homeotic genes encode transcription factors that regulate downstream effector genes controlling differentiation and morphogenesis. Homeotic mutations cause cells to be assigned incorrect positional identities, resulting in the development of one body part with the likeness of another.

Pattern, pre-pattern and spontaneous pattern formation

There is evidence for spontaneous pattern formation in systems such as the developing limb and plant meristems, since these can form normal organs when disaggregated, and extra organs when oversized, whereas in both cases they would be expected to produce disorganized tissues. Patterns in developing organs are likely to be built up in layers, starting with a weak pre-pattern that is refined and strengthened by later acting genes.

Related topics

Axis specification and patterning in *Drosophila* (Section H)
Axis specification in vertebrates (Section I)
Patterning the anteroposterior neuraxis (J2)
Patterning the dorsoventral neuraxis (J3)

Mesoderm induction and patterning (K1)
Somitogenesis and patterning (K2)
Patterning and morphogenesis in limb development (M2)
Flower development (N6)

Regional specification

Regional specification is any mechanism for telling a cell *where* it is in relation to the rest of the embryo, so that it can behave in a manner appropriate for its position. Regional specification allows cells to adopt the correct spatial organization, and is therefore an essential requirement for **pattern formation**, the process by which cells act on their instructions and become organized into appropriate structures. In some cases, pattern formation involves telling a cell to undergo the correct form of differentiation. For example, hepatocytes should differentiate only in the liver-forming region of the foregut and nowhere else. In other cases, the same cell type is found throughout the body, but the cells must behave differently at different positions in order to generate regionally appropriate structures. For example, bone tissue is found throughout the vertebrate spinal column, but the architecture differs according to position so that ribs form only in the thorax and the atlas and axis form only in the neck.

First coordinates

Development is progressive (Topic A1), so pattern formation occurs step by step, starting with an initially simple **body plan** that is gradually filled with detail. The establishment of a body plan requires each cell in the embryo to be aware of its position in relation to other cells, and this in turn requires the definition of a basic framework for positional coordinates. The framework comprises the principal **axes** of the embryo. All vertebrates and many other animals share a similar basic body plan that is, at least superficially, bilaterally symmetrical. The central nervous system runs along this plane of symmetry and defines the **anteroposterior axis**. In vertebrates, the same axis may be termed the **craniocaudal** or **rostrocaudal** axis, literally meaning 'head to tail', so that it can be applied equally to fish, birds and terrestrial vertebrates whether they walk on four legs or upright. The other principal axis, the **dorsoventral axis**, runs from back to belly; the dorsal side faces upwards and the ventral side downwards. Once the anteroposterior and dorsoventral axes are in place, the remaining axis, the **left–right axis**, is specified by default.

The names of all three principal axes infer **polarity**, i.e. the anteroposterior axis runs from the anterior end of the embryo to the posterior end. The brain is found at the anterior end, and this is usually the end of the animal that moves forwards. Polarization is an important part of axis specification and requires a symmetry-breaking process, such as the asymmetric distribution of cytoplasmic determinants in the egg, or an external physical cue. For example, the antero-posterior axis of the *Drosophila* embryo is polarized by the distribution of maternal gene products (Section H). The anteroposterior axis of the vertebrate embryo, however, is polarized by a physical cue, such as gravity in the case of chickens (Topic I1). The left–right axis is established by default, but must still be polarized. This appears to involve an intrinsic mechanism based on molecular asymmetry (Topic I3).

In particular developmental systems, the principal axes may have different names. When considering the development of individual structures such as limbs or eyes, the **proximodistal axis** refers to the axis running from the body towards the extremity of the structure (see Section M). Animal eggs have an **animal–vegetal axis** if the nucleus is displaced to one end; this often occurs if the egg is yolky, as in amphibians, reptiles and birds (Topic I1). Plant embryos have an **apical–basal axis**, which corresponds to the shoot to root axis of the seedling (Topic N2). Leaves have both apical–basal and dorsoventral axes, the dorsal side facing upwards towards the sun (Topic N5).

Positional information

Once the principal axes of the embryo have been established, the location of any cell in the embryo can be absolutely defined according to its position along each of the axes, rather in the same way that a town can be found on a map by finding its coordinates. In this manner, each cell has an 'address' telling it where it is, and thus how to behave to generate regionally appropriate structures. This address is known as a **positional value** or a **positional identity**.

An important question addressed by developmental biologists is how cells become aware of their positional identities. What is the nature of the **positional information** that tells a cell where it is in the embryo? Several experimental systems, including early *Drosophila* development and limb development, suggest that positional information may be generated by a morphogen gradient.

Morphogen gradients

A **morphogen** is a substance that can influence cell fate, but has different effects at different concentrations. In its simplest form, the **morphogen model** suggests that positional information along an axis can be generated by the synthesis of a morphogen at a **source** at one end of the axis, and diffusion away from the source would set up a **morphogen gradient**. Cells along the axis would receive different concentrations of the morphogen and this would induce different patterns of gene expression at different **concentration thresholds** (*Fig. 1*). Such

Lines of cells, non polarized

External signal specifies one cell as "posterior" by inducing the synthesis of a morphogen

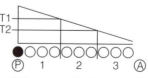

Cells respond to the concentration of the morphogen (**T1, T2**) and become patterned along the anteroposterior axis (**regions 1, 2 and 3**)

Fig. 1.   Principle of a morphogen gradient.

concentration-dependent patterns of gene expression would represent the 'address' or positional identity of the cell.

A number of systems have been studied that appear to involve morphogen gradients. In early *Drosophila* development, the Hunchback protein is distributed in a gradient along the anteroposterior axis of the syncytial embryo. Hunchback is a transcription factor and activates downstream genes such as *Krüppel*, *knirps* and *giant* in a concentration dependent manner. The latter genes are expressed in stripes defining broad sectors of the anteroposterior axis (Topic H2). Similarly in vertebrate limb development, the signaling protein Sonic hedgehog is expressed in the posterior zone of polarizing activity (ZPA) of the limb bud, and establishes a concentration gradient running from posterior to anterior. Although the mechanism is complex (Topic M2) the result is that different *HoxD* genes are activated in concentric rings surrounding the ZPA, resulting in the formation of different digits from essentially the same cell types. In both these systems, interfering with the level and/or position of the morphogen has profound effects on development. The data suggest that such experimental manipulations cause cells to receive the *wrong* positional information, consequently they adopt the *wrong* positional identities, resulting in the appearance of regionally inappropriate structures.

In the early *Drosophila* embryo, transcription factors can act as morphogens by diffusing freely through the cytoplasm to establish concentration gradients. Similarly, the lipid signaling molecule retinoic acid may be able to establish a concentration gradient across a field of cells by free diffusion. It is thought unlikely, however, that secreted proteins of the Hedgehog, Wingless/Wnt and Decapentaplegic/BMP families, all of which function as morphogens in a variety of developmental systems, can establish concentration gradients in this manner. Recent evidence has emerged to suggest that such proteins could act as morphogens in a variety of ways (*Fig. 2*):

- Initiating a signal relay, where one signaling protein diffuses over a short range, but induces adjacent cells to produce another signaling protein, which acts in the same way, with diminishing effect across a field of cells. For example, Sonic hedgehog has been shown to induce BMP2 expression in adjacent cells in the vertebrate limb.
- Responding cells could extend cytoplasmic processes (**cytonemes**) to contact the inducing cell, and the longest cytonemes would be extended by more distant cells. Ligands binding to receptors on the end of the cytonemes would induce the same signal transduction pathway in all cytonemes, but the response to signaling might be diminished as a function of increased distance from the cell nucleus. Such cytoplasmic processes have been observed in *Drosophila*.
- The signaling protein could be propagated by a carrier. Proteoglycans in the extracellular matrix have been implicated in this role, and mutations affecting proteoglycan synthesis such as *sugarless* and *tout velu* have been shown to inhibit Hedgehog and Wingless signaling in *Drosophila*.
- The signaling protein could be transported from cell to cell by cyclical endocytosis and exocytosis. Evidence for the intracellular transport of Wingless has been reported in *Drosophila*.

Finally, it should be noted that concentration-dependent induction can also be established by an 'anti-morphogen' gradient. For example, a morphogen can be secreted uniformly, but the level of its activity can be graded by the diffusion of

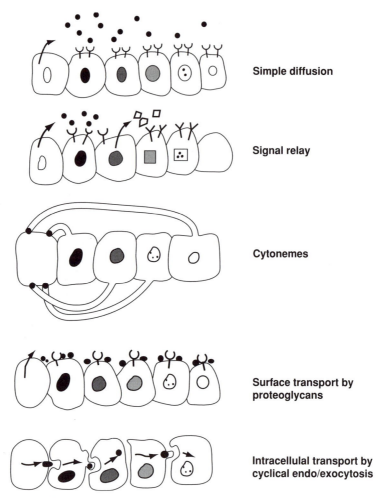

Simple diffusion

Signal relay

Cytonemes

Surface transport by proteoglycans

Intracellulal transport by cyclical endo/exocytosis

Fig. 2.   Mechanisms by which secreted proteins could generate morphogen gradients.

an inhibitor. This is apparent in systems such as the vertebrate organizer, which secretes molecules that inhibit BMPs and Wnts (Topic I2). Similarly, differential responses to a uniform concentration of inducer can be achieved by varying the levels of the receptor in the responding cells. However, this requires pre-existing polarity in the system.

Compartments and segmentation

The establishment of an axis is often followed by **segmentation**, so that the axis is divided into a repetitive series of similar but independent developmental units. This is known as **metamerism**. It allows the same basic structure, generated by the same set of genes, to be used over and over again for different purposes, and allows evolution of new functions by adding or deleting segments. Segments are exploited by many species, from the obvious segmentation of insect and annelid bodies, to the rhombomeres and somites of vertebrate embryos, and the repetitive organ primordia of flowers. However, the overt segments seen in the fully developed embryo or adult may not correspond to the underlying metameric units defined by the developmental program. For example, the *Drosophila* anteroposterior axis is patterned on the basis of **paraseg-ments** that correspond to the anterior two thirds of one segment and the posterior third of the next. Similarly, the vertebrae of the eponymous vertebrates are formed from the anterior compartment of one somite and the posterior compartment of the next.

A *Drosophila* parasegment or a vertebrate rhombomere can be viewed as a developmental **compartment**, i.e. a structure within which a clone of cells is constrained (this is termed **lineage restriction**[1]). Clonal analysis (Topic B4) can be used to determine when compartments form (*Fig. 3*). For example, a cell labeled in the early chicken hindbrain, before the rhombomeres become apparent, may give rise to a clone that spans two rhombomeres. Conversely, a cell labeled after the boundaries have formed gives rise to a clone that is restricted to a single rhombomere. In some cases this may occur simply because the labeled cell was centrally placed and the progeny did not spread far enough to reach the putative boundary. However, if the labeled cell was near the edge of the compartment, a sharp boundary forms between labeled and unlabeled cells. The term **allocation** means that a cell and its clonal descendants are confined to a compartment. Allocation may occur simply because there is a physical barrier to cell mixing. Alternatively, compartments may be defined by patterns of gene expression in the absence of any obvious boundary.

Homeotic genes and mutations

A **homeotic transformation** occurs when one body part develops with the like-ness of another, indicating that the cells involved have the wrong positional identities and have therefore developed into a regionally inappropriate struc-ture. In *Drosophila*, large scale screens of mutagenized flies revealed occasional **homeotic mutations**, i.e. individuals carrying mutations that caused a homeotic transformation between segments. The existence of such mutant flies indicated a genetic basis for positional identity, and allowed the mapping and cloning of the **homeotic selector genes** involved.

The *Drosophila* homeotic selector genes map to two clusters, and encode transcription factors with a helix-turn-helix DNA-binding motif termed a **homeodomain** (the part of the gene encoding the homeodomain is called a

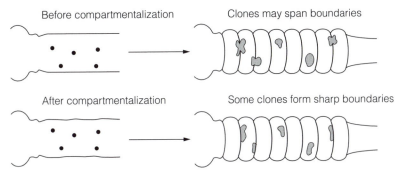

Fig. 3. *Clonal analysis of developmental compartments.*

[1]Lineage restriction means the (physical) restriction of a cell lineage to a developmental compartment. It does not necessarily mean that the *fate* of the cell is restricted in any way.

homeobox). Similar clusters, known as *Hox* **gene clusters**, are found in all animals, suggesting a common genetic basis for pattern formation in the animal kingdom. Although homeobox genes are also found in plants, there are no clusters involved specifically with pattern formation. Rather, they appear to encode transcription factors involved in other functions, just as homeobox genes unrelated to patterning are also found in animals. Homeotic genes involved in flower patterning encode another family of transcription factors characterized by a motif termed a **MADS box**. In animals and plants, non-expression or misexpression of homeotic genes can lead to homeotic transformations.

The *Hox* genes and MADS-box genes are discussed in more detail later (see Topics H5, K2 and N6). The purpose of this introductory topic is to illustrate the principle of homeotic gene function using a simple model. This model is sometimes termed the **epigenetic address model**, and shows how positional values could be established by a morphogen gradient and how those positional values could be altered by homeotic mutations. The model is shown in *Fig. 4*. It assumes that a morphogen is synthesized at the source, S, and establishes a gradient across a field of four equivalent cells (C1, C2, C3, C4) that eventually form segments with four different structures. The cells can respond to the morphogen gradient at four concentration thresholds (T1, T2, T3, T4) by expressing one or more homeotic genes (H1, H2, H3, H4). At a low concentration threshold (T1) only H1 is expressed. At a higher concentration threshold, H2 is expressed. Note however, that H1 is also expressed because its concentration threshold has been exceeded. Similarly, H3 and H4 are expressed at the higher thresholds T3 and T4. This establishes a pattern of gene expression that is nested at the source of the morphogen, i.e. nearest the morphogen all four homeotic genes are expressed while furthest away, only one is expressed. Such nested patterns are characteristic of anteroposterior axis patterning in animal embryos (see Topics H5 and K2, and for a similar pattern in limb development, see Topic M2). In this way, each cell can be given an 'address' or positional identity in the manner of a simple binary code, depending on whether each gene is on or off. Cell C4 has the code 1000, while cell C1 has the code 1111. Now consider what happens in a loss-of-function homeotic mutant, when H4 is not expressed. In this case, the code for cell C1 becomes 1110, the same as C2. The result is that both C1 and C2 would have the same positional identity and would develop into the same structure, i.e. there would be a transformation of segment C1 into C2. Such homeotic transformations have been identified in vertebrates, *Drosophila* and *Arabidopsis*. Also consider what happens when one of the genes is overexpressed. If H2 is expressed in all four cells, C3 and C4 become equivalent in terms of their positional identities, and segment C4 is transformed into C3. Again, such gain of function homeotic mutations have been generated in *Drosophila* and other systems. Finally, consider the effect of overexpressing H4 in all cells. In this case, some entirely novel codes are produced, such as 1001. The effects of this type of mutation are unpredictable, and can lead to lethality, disorganization or the development of novel structures.

The epigenetic address model is simple and it fits data provided by gene expression patterns and mutant phenotypes in some developmental systems. However, in nature, the situation is much more complicated. For example, homeotic selector genes also regulate each other so that the misexpression of one gene also alters the expression patterns of several others. In vertebrates, it appears that the levels of *Hox* gene expression as well as the patterns are impor-

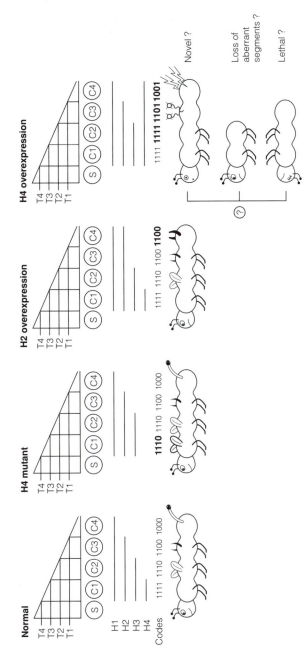

Fig. 4. Model for the generation of positional values using homeotic genes, and the basis of homeotic transformations.

tant in conferring positional values. Such complex interactions are beginning to be unraveled and are discussed in more detail in Topics H5, K2 and M2.

What does a cell do with its positional information? The fact that the homeotic genes all appear to encode transcription factors suggests that the effect of positional information is to regulate the expression of downstream genes, probably those whose function is to control cell differentiation and morphogenesis to ensure that the correct structures are formed in the appropriate positions. This does in fact appear to be the case where downstream effector genes have been found, but surprisingly few downstream targets of homeodomain transcription factors have been identified (Topic H5).

Pattern, pre-pattern and spontaneous pattern formation

From the discussion above it is obvious that genes play an important role in pattern formation, and in many of the developmental systems used as models of pattern formation (e.g. vertebrate limb development and neurogenesis), genes with patterning roles have been identified. However, in cases where such genes are mutated, it is often found that certain elements of the pattern remain. This indicates that pattern formation occurs in stages, starting with a **pre-pattern** that is refined and strengthened by the major patterning genes. In the developing vertebrate limb, for example, there are three signaling centers that polarize the three limb axes (Topic M2). In mutants where two of these signaling centers are abolished and the third is no longer polarized, a certain degree of normal patterning still occurs, suggesting that an underlying pre-patterning mechanism exists to establish the organizing centers in the first place.

There is also evidence that certain developmental systems, such as the vertebrate limb bud and the meristematic tissues of plants, have an intrinsic pattern-forming capacity that is independent of gene expression. This is revealed by the unexpected formation of normal structures when these tissues are disrupted and reorganized. For example, the mesenchyme cells of the limb bud can be disaggregated and randomly mixed, and then replaced within the limb ectoderm. After a period of disorganized growth, these buds will give rise to normal looking digits in the absence of the gene expression patterns that accompany normal limb patterning. This may reflect an underlying **reaction–diffusion**

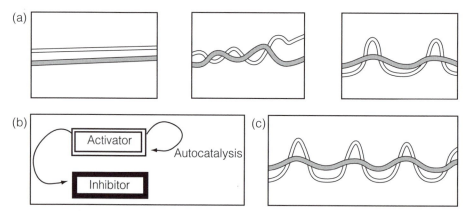

Fig. 5. (a) A reaction diffusion model in which the activator (white) and inhibitor (gray) are initially uniformly distributed across a field of cells, but spontaneously generate regular peaks of the activator. (b) This is achieved if the activator stimulates its own synthesis and that of the inhibitor, and the inhibitor diffuses away faster than the activator. (c) If the size of the field increases, so does the number of activator peaks.

mechanism, a chemical system comprising two or more interacting diffusible components that can generate spontaneous patterns. Given that many patterning systems depend on diffusion (e.g. morphogen gradients, lateral inhibition), it would not be surprising if other developmental mechanisms were also based on diffusion. A number of models based on this type of mechanism have been proposed. In one model, there is an activator and an inhibitor. The activator stimulates its own synthesis (i.e. it is autocatalytic) as well as that of the inhibitor, and the inhibitor suppresses the activator. If the inhibitor diffuses at a greater rate than the activator, a repetitive wave-like pattern forms spontaneously with peaks of activator activity (*Fig. 5*). The system is initially unstable, with some peaks disappearing, but it eventually stabilizes into a regular pattern. Importantly, if the size of the field is increased, the pattern adjusts accordingly and extra peaks are generated. This may explain the formation of extra digits in oversized limb buds, and extra floral organs in oversized floral meristems, rather than the disorganized tissue that would be expected if each organ was established independently.

A7 MORPHOGENESIS

Key Notes

Mechanisms of morphogenesis

In the dynamic embryo, cells assemble into tissues and organs with many different shapes and structures, and this process is termed morphogenesis. Morphogenesis reflects different aspects of cell structure and behavior, including cell division, cell shape and size, cell–cell adhesion, interaction between cells and the extracellular matrix, cell fusion, and cell death.

Cell division

Cell division influences morphogenesis in two ways. The rate of cell division contributes to differential growth in different parts of the embryo, while the position and plane of cell division influences the size and orientation of the daughter cells.

Cell size and shape

Changes in cell shape and size drive many folding and buckling movements, such as those occurring during gastrulation and the formation of the neural tube. Cell size and shape can change due to reorganization of the cytoskeleton, or the uptake of water or lipids.

Cell fusion

Cell fusion generates a multinucleate syncytium. This is important in the morphogenesis of developing muscle tissue and that of the trophectoderm during implantation in mammals.

Cell death

Programmed cell death by apoptosis in development is responsible for creating cavities (such as the proamniotic cavity in mouse development) and gaps (e.g. between the digits in vertebrate limb development). Extensive cell death also occurs in the developing nervous system as neurons compete for innervation targets.

Cell adhesion

Tissue reorganization (transitions between epithelial and mesenchyme cells, delamination, intercalation) occur through changes in the pattern of cell adhesion molecules. The expression of different classes of cell adhesion molecules helps to keep like cells together in tissues, and maintain boundaries between different tissues.

The extracellular matrix and cell migration

The extracellular matrix is a network of macromolecules secreted by cells into their local environment. Interactions between cells and the matrix can maintain epithelial sheets, provide a substrate for migration, and induce differentiation. The matrix is also the predominant component of some tissues, such as bone and cartilage.

Related topics

The cell division cycle (C4)

The cytoskeleton, cell-adhesion and the extracellular matrix (C5)

Mechanisms of morphogenesis

The developing embryo is dynamic, and forms all manner of shapes and structures. This is achieved through various different types of cell behavior. The term **morphogenesis** (creation of form) is used to describe how embryonic structures arise. Without morphogenesis, the embryo would never progress beyond a simple ball of cells, and dynamic processes such as gastrulation, embryonic folding and organogenesis would not be possible. A number of morphogenetic mechanisms are listed in *Table 1*, together with examples from different developmental systems. The importance of each of these processes is discussed below.

Table 1.   *Morphogenetic processes in development and examples from model developmental systems*

Process	Examples
Differential rates of cell proliferation	Selective outgrowth of vertebrate limb buds by proliferation of cells in the progress zone. Growth of shoots and roots in mature plants by proliferation of meristems.
Alternative positioning and/or orientation of mitotic spindle	Different embryonic cleavage patterns in animals. Stereotyped cell divisions in nematodes and *Arabidopsis* roots. Switch from anticlinal to periclinal divisions in leaf development
Change of cell size	Cell expansion during leaf development in *Arabidopsis*
Change of cell shape	Change from columnar to wedge-shaped cells during neural tube closure in birds and mammals.
Cell fusion	Formation of trophoblast and myotubes in mammals.
Cell death	Separation of digits in vertebrate limb bud. Selection of functional synapses in the mammalian nervous system.
Gain of cell–cell adhesion	Condensation of cartilage mesenchyme in vertebrate limb bud.
Loss of cell–cell adhesion	Delamination of cells from epiblast during gastrulation in mammals.
Cell–matrix interaction	Migration of neural crest cells and germ cells. Axon migration.
Loss of cell–matrix adhesion	Delamination of cells from basal layer of the epidermis.

Cell division

Cell division is important for morphogenesis in two ways. Firstly, the *rate* of cell division contributes to the differential growth of parts of the embryo. This can be seen in the vertebrate embryo for example, in the selective growth of the limb bud rather than adjacent regions of the lateral plate mesoderm. Such growth results from the secretion of growth factors firstly by the mesoderm itself, and then by the apical ectodermal ridge overlying the limb bud. The growth factors diffuse only a short distance, resulting in a small but rapidly dividing growth zone at the tip of the limb bud, from which all the skeletal and connective tissue elements of the limb derive (Topic M1).

Secondly, the position and orientation of the mitotic spindle within the cell at cell division determines the direction of cleavage and the relative sizes of the two daughter cells. There are many examples of how this can influence the structure of a particular tissue or even the embryo as a whole. In a single layer of epithelial cells, for example, cell division in the plane of the sheet will cause it to expand laterally, while division perpendicular to the plane of the sheet will increase the number of layers. In plants, divisions within the plane of an epithelial sheet are known as **anticlinal**, while perpendicular divisions are **periclinal**. A switch from anticlinal to periclinal division is important for outgrowth of lateral organs, such as leaves, from the stem (Topic N5). During the cleavage stage of animal development, the orientation of the cleavage plane determines the relative sizes and organization of the blastomeres. There are many variations in cleavage patterns, including radial, discoidal and spiral cleavage (Topic F1). Furthermore, asymmetric cell division during cleavage in nematodes is responsible for setting up the principal body axes (Topic G2). The cell cycle, its regulation, and mechanisms controlling the formation of the mitotic spindle apparatus are discussed in Topic C4.

Cell size and shape

The shape of a cell can change by reorganization of the cytoskeleton (Topic C5) and this can have profound effects on tissue structure. In the developing vertebrate nervous system, the neural tube forms from the flat neural plate in part because the columnar cells of the plate constrict at the apical surface and become wedge-shaped, therefore helping to pull the two neural folds together (Topic J4). A similar process occurring in any epithelial sheet will make the sheet buckle (**invaginate**). The invagination of epithelial sheets occurs in many different developmental systems, and is particularly important during gastrulation in e.g. sea urchins and amphibians (Topics F2 and F3).

Cell size and shape can also change by varying the cell volume. In animals, for example, adipocytes grow in size as they accumulate storage lipids. Changes in cell volume are important in plant development because the cell walls make relative cell movement impossible. In the developing leaf, cells can increase in size by up to 50% by importing water into vacuoles. This can be used to dilate cells in a specific direction (**directional dilation**) (Topic N5).

Cell fusion

Cell fusion generates **syncytia** (multinucleate cells) and is important in several developmental processes. During muscle development, myoblasts (undifferentiated proliferating cells) become aligned, and then fuse to generate multinucleated myotubes. These cells begin to synthesize muscle-specific proteins such as actin and myosin, and then assemble into muscle fibers (Topic K3). Cell fusion is also important in the development of the human placenta. Trophectoderm cells begin to fuse soon after implantation to generate the multinucleate syncytiotrophoblast. This invasive tissue sends active processes between uterine cells and secretes an enzyme that breaks down the contacts between them, facilitating embryonic implantation (Topic F1). Note that probably the most critical event in development, fertilization, is also dependent on cell fusion (Topic E4).

Cell death

Cell death often accompanies trauma or infection. This type of cell death is termed **necrosis**, and is characterized by swelling and membrane lysis, and tissue inflammation resulting from the release of cytoplasmic proteins. A different type of cell death, termed **apoptosis**, occurs as a physiological process.

Table 2. *Some terms used to describe the morphogenetic behavior of cells in animal embryos, reflecting changes in cell shape, cell–cell adhesion or cell–matrix adhesion.*

Movement of individual cells (migration)

Ingression, egression	The movement of individual cells into the embryo, or out towards its surface.
Delamination	The movement of individual cells out of an epithelial sheet, particularly when one epithelial sheet splits into two or more layers.
Intercalation	The movement of individual cells into an epithelial sheet, particularly when two or more sheets merge into one.
Condensation	A **mesenchymal to epithelial transition**, where dispersed cells congregate to form an epithelium.
Dispersal	An **epithelial to mesenchymal transition**, where an epithelial sheet breaks up to form dispersed mesenchyme cells.

Movement of cell sheets en masse

Invagination, evagination	The folding of a sheet of cells to form an indentation or protrusion. Invagination and evagination are essentially the same process, but in one case the sheet folds inwards and in the other it folds outwards.
Involution	The inward turning of an expanding sheet of cells, so that the cells continue to spread over the internal surface of the sheet and create a second layer.
Epiboly	The spreading of a sheet of cells, may be driven by intercalation or cell proliferation.
Convergent extension	A combination of processes that occurs during gastrulation resulting in cells converging at the organizer and then extending along the future anteroposterior axis of the embryo. Convergence involves epiboly and involution, extension involves extensive intercalation.

Apoptosis is particularly important during development, where the programmed death of particular cells helps to generate the structure and organization of various organs. Unlike necrosis, apoptosis involves cell shrinkage and nuclear condensation, often without the loss of membrane integrity. The cell breaks up into membrane-bound **apoptotic bodies** that are, in animals, engulfed by macrophages, and this avoids inflammation.

Cell death plays an important role in *C. elegans* development, where the cell lineage is invariant and particular cells must undergo apoptosis to generate the correct cell number and cell organization in the adult (Topic G2). The importance of programmed cell death is also illustrated by vertebrate limb development. In the limb bud, apoptosis occurs in several discrete regions, including the inappropriately named **interdigital necrotic zones**, resulting in the separation of the digits (Topic M2). Programmed cell death is also important in the development of the vertebrate nervous system, where up to 50% of the neurons may die while competing for innervation targets (Topic J7). The regulation of apoptosis is discussed in Topic C4.

Cell adhesion

Animal cells bind together using membrane-bound **cell adhesion molecules** (Topic C5). The adhesion molecules synthesized by a particular cell type determine the cells it can and cannot adhere to. This helps organize cells into tissues and to maintain the boundaries between different tissues.

Many developmental processes in animals involve the reorganization of cells. For example, the condensation of mesenchyme cells to form cartilage in the limb (Topic M2), or nephrotomes in the kidney (Topic K4), and the dispersal of epithelial cells that occurs during the degeneration of the mesonephric duct (Topic E5). Major reorganization of the entire embryo occurs during gastrulation (Section F). All these processes occur through changes in cell affinity caused by changing patterns of adhesion molecules (for definitions of some terms, see *Table* 2). In this way, cells can spontaneously aggregate or disperse, delaminate from a cell layer, or associate with different neighbors. The power of cell adhesion molecules in tissue integrity and stability can be demonstrated by **dissociating** the cells from different tissues and mixing them. The cells spontaneously reassociate and form homogenous tissues by randomly testing bonds with neighboring cells, and trading weaker bonds for stronger ones. Similarly, introducing genes encoding adhesion molecules into cultured cells causes them to aggregate, with those cells expressing the highest and lowest levels of adhesion molecules tending to adhere to cells like themselves. The importance of cell adhesion molecules in morphogenesis can be demonstrated by expressing them in ectopic tissues. For example, during normal vertebrate development, the neural tube separates from the epidermal ectoderm and sinks beneath it. This happens in part because the epidermal ectoderm and neural ectoderm begin to express different cell adhesion molecules. The injection of mRNA for N-cadherin (which is usually restricted to the neural tube) into epidermal cells (expressing E-cadherin) results in the failure of the two tissues to separate. The common expression of N-cadherin maintains them as a single tissue.

The extracellular matrix and cell migration

The extracellular matrix is a network of molecules secreted by cells, which may form a barrier between different tissues or provide a substrate over which cells can migrate. Cells interact with molecules in the extracellular matrix through membrane-bound proteins termed **integrins** (Topic C5). Many different integrins are synthesized, and these have differential affinity for matrix proteins such as laminin, collagen and fibronectin. Cells may gain or lose contact with the extracellular matrix by altering the expression of integrin proteins. This happens, for example, when epidermal stem cells produce differentiating keratinocytes. The stem cells are tightly bound to the basal lamina, but differentiation involves the downregulation of integrin synthesis so the young keratinocytes detach and begin to migrate into the upper layers.

Interactions between cells and the extracellular matrix can allow cells to move over relatively large distances during development. This occurs in animals but not plants, as the movement of plant cells is restricted by the rigid cell wall. Vertebrate development provides many examples of extensive relative cell movement, which can occur either as the migration of individual cells, or the orchestrated movement of entire cell sheets. Long distance migration by individual cells occurs in the neural crest lineage (Topic J6) and the germ line (Topic E2). Cell migration may be guided by diffusible chemoattractants and chemorepellants, but adhesion to the extracellular matrix (and other cells) provides important local guidance cues. It is thought that cells can migrate along a path with an increasing adhesive gradient (**haptotaxis**), i.e. cells may migrate by randomly testing the matrix in all directions, and move in the direction where weak bonds can be exchanged for stronger ones. Such factors also influence the growth of axons in the developing nervous system (Topic J7). The movement of entire cell sheets occurs in gastrulation, either by involution or epiboly (*Table 2*). In each case, this may involve a combination of migration-like behavior at the expanding front of the cell sheet, and other forces such as cell proliferation or intercalation within the body of the cell sheet to drive its expansion.

Finally, the matrix itself becomes the primary determinant of structure in vertebrate skeletal tissues. Bone and cartilage contain cells, but most of the tissue mass is made up of substances secreted by these cells into the extracellular matrix. Cells that secrete cartilage and bone and those that break it down continually remodel the tissue.

B1 MODEL ORGANISMS

Key Notes

Model organisms
There is a great diversity of life on Earth, but research on development has concentrated on a small number of representative species. These model organisms have been chosen because they are easy to obtain and breed in the laboratory, and each has particular advantages for experimental study, e.g. genetic amenability or accessibility for surgical manipulation.

Unicellular model organisms
Unicellular organisms undergo predictable changes in phenotype that resemble developmental processes in metazoans, and thus provide simple models of development. Among the best-studied systems are sporulation in the bacterium *Bacillus subtilis*, mating type switching in yeast and aggregation in the cellular slime mold *Dictyostelium discoideum*. All these processes occur in response to an unfavorable environment, e.g. nutrient depletion.

Model invertebrates
The principle invertebrate models in use today are the fruit fly *Drosophila melanogaster* and the nematode worm *Caenorhabditis elegans*. Both are genetically amenable, and easy to maintain and breed in the laboratory. They both have well characterized developmental pathways, and robust embryos that stand up to surgical manipulation.

Model vertebrates
The major vertebrate developmental models are the frog *Xenopus laevis*, the chicken, the mouse and the zebrafish (*Danio rerio*). *Xenopus* and chick embryos are easy to study and manipulate, but genetic analysis and genetic modification are more difficult. The mouse is genetically amenable, but embryos are more difficult to study because they develop *in utero*. The zebrafish is the most recently-adopted vertebrate model organism and combines ease of study with genetic amenability. The mouse is the model most closely related to human development.

Model plants
Developmental processes appear highly conserved among the flowering plants, and a single organism, the wall cress *Arabidopsis thaliana*, has risen as the principle developmental model. *Arabidopsis* is small, diploid, has a short life-cycle and is amenable to genetic analysis and manipulation.

Related topics

Cleavage (F1)
Gastrulation in invertebrate
   embryos (F2)
Gastrulation in vertebrate embryos
   (F3)
Cell specification and patterning
   in *Caenorhabditis elegans* (G2)
Molecular aspects of embryonic
   pattern formation in *Drosophila*
   (H1)

Early patterning in vertebrates (I1)
Organogenesis in invertebrate
   model systems (Section L)
Vertebrate limb development
   (Section M)
Plant *vs.* animal development (N1)

Model organisms There are too many different species for scientists to individually study the development of every single life form, and related species are likely to use similar developmental mechanisms. For this reason, researchers have focused on a small number of **model organisms**, which represent the major divisions of multicellular life, and from which the *general* principles of development can be learned. The choice of model organisms reflects the ease with which each can be obtained, bred and studied, as well as more specific experimental advantages each may possess. Many of the model organisms are extremely amenable to genetic analysis – they have short generation intervals, can be maintained and bred in large numbers, and have dense genetic maps. The technology for genetic manipulation in these species is usually advanced. Other model organisms, lacking genetic amenability, have robust, easy to access embryos that yield to microsurgical manipulation techniques. In each model organism, the process of normal development has been studied in great detail, allowing even minor deviations to be identified.

Unicellular model organisms Certain bacterial cells and unicellular eukaryotes undergo predictable changes in phenotype that resemble developmental processes in multicellular life forms (**metazoans**). Such changes often occur when the environment becomes hostile or unfavorable for growth. The bacterium *Bacillus subtilis* forms spores under conditions of starvation and other stresses. This is a simple form of differentiation, during which the normal vegetative cell divides asymmetrically to form a prespore cell and a mother cell that encloses it. Sporulation thus results in the production of two very different cells with different functions. Similarly, some strains of the yeasts *Saccharomyces cerevisiae* and *Schizosaccharomyces pombe* can switch mating types in response to starvation, to allow sexual reproduction and the production of spores. One of the most intriguing unicellular life forms is the cellular slime mold *Dictyostelium discoideum*. Under vegetative conditions, it exists as free living cells called **myxamobae**, but under conditions of nutrient depletion up to 10^5 cells can aggregate to form a large multicellular mound, which tips on its side and migrates as a single organism, termed a **slug**. The slug demonstrates many of the developmental principles of certain metazoan embryos, i.e. it contains differentiated cells organized in a specific pattern, and regulates for missing parts. If the slug moves into a suitable environment, further differentiation and reorganization of cells occurs to generate a **fruiting body** from which spores are released. Unicellular developmental models are discussed in more detail in Section D.

Model invertebrates The two major invertebrate developmental models are the fruit fly *Drosophila melanogaster* and the nematode worm *Caenorhabditis elegans*. *Drosophila* was chosen because of its early pivotal role in the study of genetics. Saturation mutagenesis screens are relatively straightforward, allowing mutants for any system of biological interest to be isolated. It has become established as a developmental model over the last 20 years through the identification and cataloging of many hundreds of developmentally important genes. Genetic manipulation in *Drosophila* is also a simple procedure, involving the injection of recombinant P-elements (*Drosophila* transposons; see Topic B3) into the egg. Early *Drosophila* development is described in Topics F1 and F2. The molecular basis of axis specification and patterning in *Drosophila* is the subject of Section H (also see Topics J5, L2 and L3).

C. elegans is a remarkably simple organism, containing approximately 1000 somatic cells and a similar number of germ cells. Like *Drosophila*, *C. elegans* is amenable to genetic analysis and modification, and the embryos are transparent. Further advantages of *C. elegans* include the fact that adults can be stored as frozen stocks and recovered later by thawing, and that the species is hermaphrodite, so that one individual can seed an entire population. One remarkable feature of *C. elegans* is that the somatic cell lineage is invariant, i.e. every cell division and inductive interaction between cells is programmed and predictable. This has allowed the entire somatic cell lineage from egg to adult to be mapped out. There is also a complete wiring diagram of the *C. elegans* nervous system. *C. elegans* development is discussed in Topics G1 and G2 (also see Topic L1).

Although most research on invertebrate development has focused on *Drosophila* and *C. elegans*, they represent only two branches of the extremely diverse phylogeny within the invertebrates. Other species, such as the sea urchin, the leech, ascidians, and snails of the genera *Lymnaea* and *Nassarius* have also been used as models. The sea urchin in particular has been used as a model to study fertilization, radial cleavage and the cell movements of gastrulation (see Topics E4, F1 and F2). The others have embryos with a small cell number, and use single cell specification in early development. They are discussed in Section G.

Model vertebrates

Compared to invertebrates, the vertebrates represent a relatively narrow phylogeny. All vertebrates pass through similar stages of development, including cleavage of a large egg, gastrulation to generate the three germ layers, formation of a notochord, neural induction and consequent formation of a neural tube, segmentation of mesodermal blocks along the anteroposterior axis to form somites, and the formation of limb buds. This leads to a **phylotypic stage**, where the body plan of all vertebrates is the same (Topic I1). The major differences between the vertebrate classes include the structure of the embryo before gastrulation (representing different nutritional strategies) and the final detailed anatomy of the adult (reflecting adaptations to fill particular evolutionary niches). The reason that vertebrates occupy such a predominant position in the study of developmental biology is that they provide models for human development. The first vertebrate model organisms were amphibians, because their development occurs in water (allowing observation at all stages) and the embryos are robust, allowing surgical manipulation. A number of different organisms have been exploited, including newts and axolotls. The amphibian model of choice today is the South African clawed frog *Xenopus laevis*, whose eggs are large, easily obtainable in large quantities, and can be fertilized *in vitro*. The large eggs can be microinjected with DNA, RNA or proteins using inexpensive equipment, in order to determine gene function. Embryos and tadpoles can be surgically manipulated or exposed to specific reagents, and embryonic tissue can be cultured easily to investigate inductive interactions. The major disadvantage of *Xenopus* is its lack of genetic amenability, mainly due to its long generation interval. Only recently has it become possible to generate transgenic frogs. The other established vertebrate models, the chicken and the mouse, provide close parallels to human development. Chicken embryos are easily obtainable and can be observed throughout development by creating a window in the eggshell. Early chicken embryos can also be cultured outside the egg. It is easy to expose developing chickens to controlled amounts of specific reagents, and they are amenable to surgical manipulation, making them particularly suitable for the study of limb development (Topics M1 and M2). It is possible to create **chick/quail chimeras** for cell tracing and lineage studies. The major disadvantage of the chicken is its lack of genetic amenability, and techniques for generating transgenic birds are not widely available. Conversely, the mouse is amenable to genetic analysis and there is highly developed technology for DNA transfer to mouse embryos, including the **microinjection of DNA** into the egg and the use of **embryonic stem cells** for techniques such as gene knockout and allele replacement (see Topic B3). The disadvantage of the mouse is that, due to its intrauterine development, the embryos are difficult to observe and manipulate. It is possible to culture mouse embryos only briefly. The most recent addition to the family of vertebrate model organisms is the zebrafish *Danio rerio*. This species has many advantages – it has a short life cycle, fish can be kept in large numbers and females produce up to 1000 eggs per day. The embryos are transparent and, particularly advantageous, haploid embryos can be produced and develop normally up until the pre-hatching stage. This allows the isolation of informative recessive mutants without breeding to homozygosity. A saturation mutagenesis screen was recently performed on zebrafish to identify developmentally important genes.

Early vertebrate development is described in Topics F1 and F3. Section I considers the early patterning of vertebrate embryos, while Sections J and K are largely devoted to the further diversification of the three germ layers in vertebrate embryos.

Model plants   Plant development differs fundamentally from that of animals (see Topic N1) but the organization of the plant embryo, at least within flowering plants, is highly conserved across species. Early models of plant development were used mainly for descriptive studies, and included Shepherd's purse (*Capsella bursapastoris*). The wall cress *Arabidopsis thaliana* has risen as the principle model plant for a range of biological systems (including development) due to its amenability for genetic analysis and manipulation, allowing combined cellular, genetic and molecular approaches. It is small, has a life cycle of about 6 weeks, produces hermaphrodite, self-fertilizing flowers and is diploid (many plants are polyploid, making mutational analysis difficult). Transgenic *Arabidopsis* plants can be generated using *Agrobacterium*-derived Ti vectors or particle bombardment (see Topic B3) and the same strategy can be used for insertional mutagenesis and cloning developmentally important genes by tagging. In addition to *Arabidopsis*, *Antirrhinum majus* provides a useful model for studying genes involved in flower development since its flowers show dorsoventral polarity while those of *Arabidopsis* are more or less radially symmetrical (Topic N6). *Arabidopsis* and *Antirrhinum* are both dicot plants. More recently, rice and maize have emerged as model species for monocot development.

B2 DEVELOPMENTAL MUTANTS

Key Notes

Developmental mutants

The genetic basis of development can be investigated by obtaining mutants that are deficient in particular developmental systems. Since developmental genes control fundamental processes such as pattern formation, the mutant phenotypes can be striking and even bizarre. Mutant phenotypes may be informative not only in terms of their morphology, but may also provide clues as to the molecular function of the genes involved, e.g. if the mutation alters cell fate or disrupts the expression patterns of other genes.

Classes of mutation

Mutations may be classed as dominant or recessive, depending on whether or not the phenotype is manifest in the heterozygous state. Mutations are also classed as loss- or gain-of-function, depending on their effect on gene activity. Owing to their fundamental roles, the disruption of many developmental genes results in a lethal phenotype. However, many developmental genes encode signaling components and transcription factors with multiple functions at different stages. These later functions can be investigated using conditional mutants.

Obtaining developmental mutants

Developmental mutants may occur spontaneously, but a more systematic approach is to carry out saturation mutagenesis in an attempt to mutate all genes in the genome and isolate all mutants affecting a given developmental system. Mutagenesis may by carried out by irradiation, chemical mutagenesis or insertional mutagenesis using transposable elements. The effects of such mutations on development have to be studied in the subsequent generation.

Mosaics and chimeras

A number of important questions can be addressed by combining mutant and wild type tissues in the same organism, which is termed a mosaic or a chimera depending on the origins of each tissue. This strategy is particularly useful for elucidating the roles of different genes in signaling pathways. Chimeras can also be used for cell tracing experiments.

Genetic pathways and epistasis

Genes often act sequentially in pathways to achieve a particular outcome, such as the transduction of an inductive signal. The order of gene activity in such pathways can be established by combining different types of mutants in the same individual, because loss of function mutations in later acting genes quash the effects of gain of function mutations in early acting genes. Early mutant phenotypes can also be rescued by later acting gene products.

Developmental phenocopies

A phenocopy resembles a mutant phenotype but is caused by an epigenetic (outside the DNA sequence) effect rather than by mutation. Phenocopies can be generated by interfering with gene function at the protein level, e.g. by injecting antibodies that bind to a protein and disable it, or a dominant negative version of a receptor that blocks signal transduction. Phenocopies are particularly useful in systems that are not amenable to gene transfer.

Related topics   Transgenic organisms in   Gene expression and regulation
                       development (B3)   (C1)

Developmental mutants   The genetic basis of any biological system can be investigated by obtaining **mutants** in which that system is impaired, and using such mutants to map and clone the genes involved. Developmental mutants can generate highly informative phenotypes. As they often affect fundamental processes like pattern formation, the phenotypes can be dramatic and bizarre. Hence, developmental genes are often given fanciful and descriptive names (*Table 1*). As well as their

Table 1.   A selection of colorfully-named developmental genes, with explanations.

Species: gene name	What the name means
Drosophila: nanos, pumilio, oskar	These genes control abdominal specification in the *Drosophila* embryo, and each mutant is much smaller than usual because of its missing abdominal segments. Nanos and pumilio are Greek and Latin, respectively, for 'dwarf'. Oskar was the boy who refused to grow any more, in a book called *The Tin Drum* by Günter Grass. The *Drosophila* gap genes *Krüppel* and *knirps* are named in a similar vein, because the mutants lack several segments: Krüppel is German for cripple, and Knirps means 'little fellow'.
Drosophila: tudor, vasa, valois	These genes contribute to the specification of pole cells (the *Drosophila* germline), and the mutant embryos lacking germ cells are therefore sterile. Tudor, Vasa and Valois are European royal families that died out because they failed to produce children.
Vertebrates: *sonic hedgehog*	The original *Drosophila hedgehog* gene was so-called because the mutant phenotype disrupted segment polarity resulting in extra denticle belts; similar phenotypes are seen e.g. in *goooseberry* and *engrailed* mutants. Mutants in other genes lose their denticle belts, hence names such as *smooth* and *naked*. Vertebrates have more than one *hedgehog* gene homolog, so they were given names representing different types of hedgehog, e.g. *sonic hedgehog* from the Sega computer game, and *tiggywinkle hedgehog*, from the Beatrix Potter character.
Zebrafish: *techno trousers*	A gene involved in the development of the nervous system, the mutants twitch continuously, reminiscent of the remote-controlled trousers in the Wallace and Grommit animation.
Arabidopsis: GURKE, KEULE, KNOLLE, WUSCHEL	These genes are involved in plant embryonic development. All the names are German words describing the shape of the embryo. Gurke is a cucumber, Keule is a club, Knolle is a root swelling or nodule, Wuschel means disorganized and scruffy, like a mop of unruly hair.
Arabidopsis: SUPERMAN	This is a flower patterning gene. In the mutant plants, there are extra whorls and the carpels are replaced by stamens. The flowers have multiple sets of male sex organs and no female sex organs.
Drosophila: dorsal, tube, pipe, nudel, windbeutel, spätzle	These genes are involved in dorsoventral axis specification in *Drosophila*. The normal embryo has ventral denticle belts and smooth dorsal epidermis. Loss of function mutations cause dorsalization, so the embryo becomes a tube of epidermis (hence *tube, pipe*). Windbeutel, Nudel and Spätzle are German words for cream puff, noodle and pasta, respectively. Mutations in another gene, *cactus*, have a ventralizing effect, producing all-round denticle belts.

morphological phenotypes, developmental mutants often show interesting cellular phenotypes (e.g. aberrant differentiation) and molecular phenotypes (e.g. the disruption of other gene expression patterns) that provide important clues to their function and link them into genetic pathways and hierarchies. Although some model organisms are not amenable to genetic analysis, developmental genes are often conserved between species, so the identification of a particular gene in one species can be used to isolate a homologous gene from another.

Classes of mutation

Mutations can be categorized according to the effect of a particular **mutant allele** in a heterozygous individual also carrying a normal **wild type allele** (*Fig. 1*). Mutations are described as **dominant** or **recessive** depending on whether the mutant phenotype is manifest in the heterozygote. An individual heterozygous for a recessive allele will normally show the wild type phenotype because the wild type allele is able to compensate fully for the mutation, whereas an individual homozygous for the same recessive allele will show the mutant phenotype. Conversely, an individual heterozygous for a dominant allele will show the mutant phenotype despite the presence of the wild type allele. Some dominant mutations are described as **semi-dominant** because the severity of the mutant phenotype is greater in the homozygote than in the heterozygote (e.g. see Topic K1 for a discussion of the *Brachyury* mutation). Consequently, dominant mutations can be identified in heterozygous population, whereas recessive mutations must be bred to homozygosity.

Mutations can also be classed as loss-of-function or gain-of-function, according to the effect of the mutation on gene activity. A **loss-of-function mutation** causes reduction or abolition of gene activity, whereas a **gain-of-function mutation** increases gene activity, or confers an entirely novel function on the gene. The dominant/recessive and loss/gain-of-function classification systems are not mutually exclusive. Most loss of function mutations are recessive because the underlying effect of the mutation is quantitative, and the wild

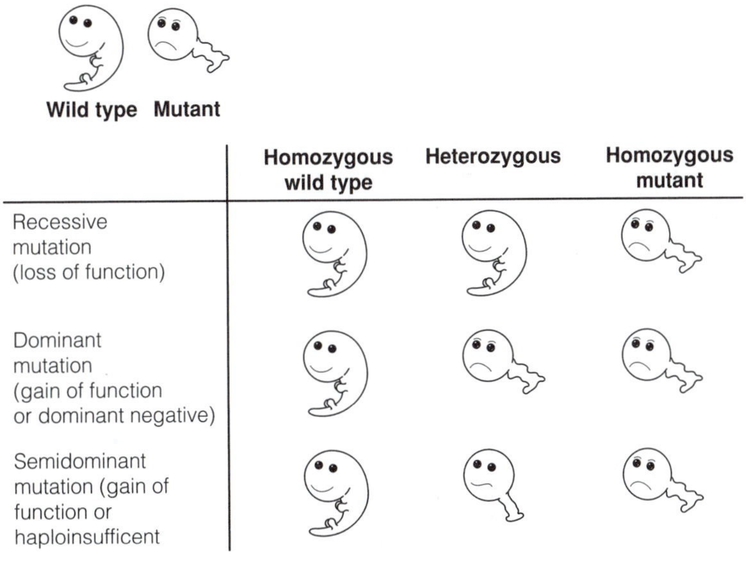

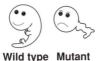

Wild type   Mutant

	Homozygous wild type	Heterozygous	Homozygous mutant
Recessive mutation (loss of function)			
Dominant mutation (gain of function or dominant negative)			
Semidominant mutation (gain of function or haploinsufficent			

Fig. 1. Classes of mutation.

type allele produces enough product for the system to function normally (in the most severe case of a **null allele**, gene expression is abolished all together, but the wild type allele alone produces 50% of the normal amount of gene product, and this is often enough). Where this is not sufficient, a loss-of-function mutation generates a semi-dominant phenotype and the condition is termed **haplo-insufficiency**. The phenotype is more severe in the homozygous state because the dosage of the gene product is reduced from 50% to zero. A loss-of-function allele can show full dominance if the mutation causes a qualitative rather than a quantitative effect, and the mutant gene product interferes with the function of the wild type gene product. This is also called a **dominant negative mutation**, and these are exploited in experimental systems to deliberately disrupt gene activity (Topic C3). Gain-of-function mutations are almost always dominant or semi-dominant. Homeotic mutations (Topic A6) often fall into this class.

Many mutations affecting development are described as **lethal**, because development is disrupted to such a degree that the embryo dies. Some genes have a range of different alleles, which generate phenotypes of varying severity. Such alleles may be described as **strong lethals** or **weak lethals**, depending on the likelihood of survival. Many developmental genes are **pleiotropic**, i.e. they have different functions at different developmental stages. This reflects the fact that many developmental genes encode transcription factors or signaling components, and the same signaling pathways and transcription factors tend to be used over and over again in different systems. Unfortunately, where mutation results in a strong embryonic lethal phenotype, it is impossible to see the effect of the same mutation at a later developmental stage. In such cases, **conditional mutants** are very useful. These mutations affect protein structure so that under normal **permissive conditions** the protein functions normally, but under abnormal **restrictive conditions** such as elevated temperature (**temperature-sensitive mutants**, represented by the initials *ts*), the protein denatures and the mutant phenotype is generated. Conditional mutants allow embryos to be raised under permissive conditions and then shifted to restrictive conditions at a specific stage of development. It is now possible to generate artificial conditional mutants using the Cre-*loxP* recombination system in transgenic animals and plants (Topic B3). See also the discussion of maternal effect and zygotic effect mutations in Topic A4.

Obtaining developmental mutants

Initially, developmental genes were identified through the occurrence of rare **spontaneous mutations**. However, it is much more efficient to **mutagenize** a population deliberately, to generate **induced mutations**, and then **screen** that population to identify individuals with informative mutant phenotypes. Mutagenesis can be carried out by bombardment with X-rays, or by feeding with chemical mutagens such as ethylmethane sulfonate (EMS) or *N*-ethyl-*N*-nitrosourea (ENU). In both cases, the DNA is damaged and mutations occur when it is incorrectly repaired. Alternatively, it is possible to generate **insertional mutations** in populations carrying transposons by stimulating the transposons to mobilize and interrupt endogenous genes. P-elements are used for this purpose in *Drosophila*, and *Ac-Ds* elements are used in plants (Topic B3). Conditions are usually chosen so that there is a good chance that every gene in the genome will be mutated at least once in the mutagenized population, in the hope of isolating all the genes that affect a particular developmental process. This is known as **saturation mutagenesis**.

In animals, the mutagenized population must be outbred because the effect of an induced developmental mutation is seen in progeny, and only if the mutation occurs in the germline. Dominant mutations can be seen in the F1 progeny, but to see recessive mutations, F1 siblings must be interbred to derive homozygous F2 individuals. There are various genetic tricks to simplify the isolation of informative mutants. In zebrafish, for example, haploid embryos develop normally until the pre-hatching stage, so recessive mutations that affect development can be identified without breeding to homozygosity. The identification of plant developmental genes is simplified if the plant is self-fertilizing. This is the case with *Arabidopsis*: irradiated seed gives rise to plants that may contain mutated gametes, and self-fertilization generates both homozygous and heterozygous plants in the next generation.

Mutations can also be deliberately introduced into cloned genes and reintroduced into the organism to investigate the effect of the mutation in development. This approach is discussed in Topic B3.

Mosaics and chimeras

Some mutations are **cell autonomous** in their effect, i.e. the phenotype is restricted to the cell carrying the mutation. For example, in mammals, if a mutation affecting hair pigment occurs spontaneously in one follicle cell, only the hair growing from that cell is affected, and hairs from surrounding cells will be normal. Other mutations are **cell non-autonomous**. For example, if the expression of a gene encoding a signaling molecule is abolished by spontaneous mutation in one cell, the surrounding cells are affected by the loss of signal even if they are wild type. Where such cell non-autonomous mutations occur, it is often impossible to determine the roles of individual cells if every cell carries the mutation. The failure of an inductive response could result from the loss of a signal, or the loss of a responding cell's ability to perceive or process that signal.

Such problems can be addressed by using mosaics and chimeras. A **mosaic** is an individual comprising cells of different genotypes derived from a common ancestor (i.e. a mutation occurs during the individual's lifetime resulting in clones of genotypically distinct cells). These can be generated in a number of ways. In *Drosophila* for example, mosaics can be produced by irradiating flies with X-rays, leading to **mitotic recombination**. A **chimera** is an individual comprising cells of different genotypes from different sources, and these are usually generated deliberately by grafting or embryo fusion, or by introducing

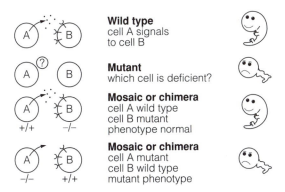

Fig. 2.   Mosaics and chimeras in mutation analysis. In this example, cell A is shown to be deficient in producing the signal.

DNA into a selected population of cells in an embryo. Mosaics and chimeras allow tissues of different genotypes to be juxtaposed. In the example discussed above, the lost inductive response could be investigated using individuals where the inducing cell was mutant but the responding cells wild type, or vice versa (*Fig. 2*). This type of experiment has been used to determine the role of different cells during the development of the *Drosophila* compound eye (Topic L3) and the nematode vulva (Topic L1).

Genetic pathways and epistasis

A **genetic pathway** is a group of genes acting sequentially to achieve a particular goal, e.g. the production of a specific product in a metabolic pathway, or the activation of a transcription factor in a signal transduction pathway. Genetic screens for developmental mutants often identify several to many mutations with identical or related phenotypes, suggesting that a genetic pathway is involved. The number of genes involved can be determined by **complementation analysis** or **crossing**, to see if it is possible to derive wild type individuals by complementation or recombination between two mutant genomes (*Fig. 3*).

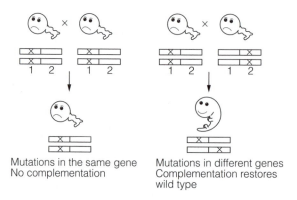

Mutations in the same gene
No complementation

Mutations in different genes
Complementation restores
wild type

Fig. 3.   Complementation analysis to determine the number of genes in a pathway.

In the case of a signal transduction pathway, a mutation at any point could result in failure to deliver the signal (loss-of-function) or constitutive signaling (gain-of-function). If it is possible to combine gain- and loss-of-function mutants from different genes in the pathway in the same individual, it may be possible to work out an order of gene activity because loss-of-function alleles of later acting genes will quash the effect of dominant gain of function alleles in earlier acting genes, an effect termed **epistasis** (*Fig. 4*). For example, the transmembrane receptor Toll and the signal transduction protein Pelle function in *Drosophila* dorsoventral axis specification (Topic H6). Toll usually responds to a ligand, Spätzle, which is activated only on the ventral side of the embryo. Constitutive *Toll* mutants are therefore **ventralized** (meaning ventralized all over, i.e. cells that would normally take on dorsal fates now also take on ventral fates), but when combined with homozygous loss of function *pelle* mutants, the embryos are **dorsalized**. This shows that the Pelle protein acts downstream of Toll. Another way to establish gene order is to attempt to **rescue** mutant embryos (restore their phenotype to normal) with the product of another gene.

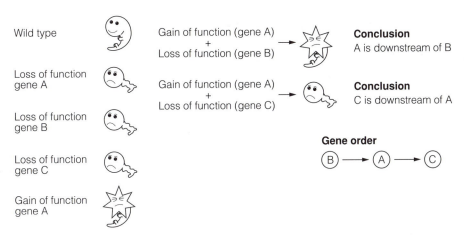

Fig. 4.   Determining gene order by epistasis.

For example, the loss of abdominal segments that occurs in the *Drosophila* posterior class mutants *nanos*, *oskar*, *staufen*, *tudor*, *vasa* and *valois* (Topic H2) can be rescued in each case by injecting *nanos* mRNA into the posterior of the egg. This shows that *nanos* is the executive control gene in posterior specification. The other genes are in fact required for the localization of *nanos* mRNA.

Developmental phenocopies

A **phenocopy** has the appearance of a mutant phenotype but is not caused by mutation. For example, pregnant rats starved of cholesterol give birth to pups that appear very similar to mouse embryos carrying a homozygous mutation in the *sonic hedgehog* gene. The similarity reflects the fact the Sonic hedgehog protein requires cholesterol to function correctly. Phenocopies are generated by interfering with gene function rather than gene expression: such effects are termed **epigenetic**[1] (meaning outside the genes, i.e. not caused by changes to the actual nucleotide sequence) and can reflect many processes including DNA methylation and chromatin structure, interactions at the RNA and protein levels, and interaction with the environment. Phenocopies can be just as informative as developmental mutant phenotypes, and are easier to generate in systems that are not amenable to transgenesis. For example, where it is impossible to investigate loss of gene function effects by gene knockout (targeted gene disruption), it may be possible to block the function of the gene by injecting **antisense RNA**, perhaps associated with a **ribozyme**, or **antibodies** against the protein. Conversely, injections of large amounts of wild type mRNA or protein can mimic the effects of overexpression or ectopic expression mutants. The use of 5-azacytidine to change DNA methylation patterns can also generate developmental phenocopies, known as **epimutations** (see Topic C2).

---

[1]Do not confuse the term *epigenetic* with *epigenesis*. As discussed above, **epigenetic** phenomena cover a wide range of influences outside the nucleotide sequence that affect the phenotype associated with a given genotype. **Epigenesis** describes development characterized by gradually-increasing complexity, as opposed to preformation (Topic A1).

B3 TRANSGENIC ORGANISMS IN DEVELOPMENT

Key Notes

Transgenic technology	Transgenic organisms contain foreign DNA in every cell. They are generated by microinjecting DNA into eggs, or transforming somatic cells that can produce new individuals. Individuals developing from transformed cells are often chimeric for transgene integration, but produce fully transgenic offspring.
Transgenic strategies in developmental biology	Developmental genes can be manipulated *in vitro* and then expressed in transgenic animals and plants where their effects can be studied in the context of the whole organism. In most organisms, transgenes integrate randomly and only dominant effects can be studied. In mice, gene targeting by homologous recombination also allows loss-of-function recessive effects to be analyzed. Through the use of site-specific recombination systems, it is possible to generate conditional mutants. Transgenic technology can also be exploited for cell ablation and insertional mutagenesis.
Nuclear transfer technology	Nuclear transfer can be used to generate clones of identical animals, including transgenic animals, which is useful if transgenesis from first principles is inefficient. Nuclear transfer is essential in the production of transgenic frogs.
Transgenic mice	Transgenic mice can be generated by microinjecting DNA into the pronuclei of fertilized eggs, or by infection with a retrovirus. Both strategies result in random integration. Another strategy is to transfect ES cells and inject these into mouse embryos where they may contribute to the germline. ES cells allow both random integration and homologous recombination (gene targeting).
Transgenic fruit flies	Foreign genes are introduced into *Drosophila* using transposons called P-elements. Defective recombinant P-elements (lacking the gene required for integration, but carrying the transgene) are injected into the pole plasm along with a helper P-element that supplies missing functions but cannot integrate itself. This allows the transgene to integrate into the germline and become stably incorporated.
Transgenic plants	Transgenic plants are generated either by exploiting the natural gene transfer ability of *Agrobacterium tumefaciens*, or by direct DNA transfer techniques such as transformation of isolated protoplasts or particle bombardment of callus. Callus tissue and even individual plant cells can regenerate whole plants.
Recent advances – chick and *Xenopus*	Only recently has it been possible to generate transgenic *Xenopus*, by restriction enzyme mediated integration of foreign DNA into male nuclei,

followed by nuclear transfer to the egg. Current strategies for generating transgenic chickens are inefficient, although progress is being made with chicken ES cells.

Related topics   Model organisms (B1)   Developmental mutants (B2)

Transgenic technology

A **transgenic** organism carries foreign DNA, usually the same piece of foreign DNA (the **transgene**), in every cell. In this context, **foreign DNA** can mean DNA from a different species, or DNA from the same species that has been manipulated *in vitro*. Transgenic organisms are generated by injecting recombinant DNA into gametes or eggs, or transforming somatic cells and using such cells to produce new individuals. Such individuals are termed **founders** or **primary transformants**, and they are usually **chimeric**, i.e. the transgene is not present in every cell. Fully transgenic individuals are produced in the next generation, if the primary transformant carries the transgene in gamete-producing cells (in animals this is termed **germline transmission**). Microinjection has been used to introduce DNA into most animals with some success, and is the method of choice in those systems not discussed specifically below, e.g. zebrafish and *C. elegans*. The transformation of *C. elegans*, involving the microinjection of DNA into oocytes or the syncytial ovary, usually results in extrachromosomal maintenance, although stable transformants arise at a lower frequency.

Transgenic strategies in developmental biology

Transgenic animals and plants are important tools in developmental biology because they allow developmental genes to be manipulated and the effects to be studied in the context of the whole organism. The simplest strategies involve the study of gain-of-function effects in developmental genes, such as dominant mutations, transgene **overexpression** (expressing the gene at a higher level than usual) and **ectopic expression** (expressing the gene in an inappropriate place, or at an inappropriate stage of development). The timing of transgene expression can be controlled externally using an **inducible promoter**. As the effects of all these manipulations are dominant to the wild type copy of the gene already present in the genome, it is sufficient for the transgene to integrate randomly into the genome by illegitimate recombination. However, transgene expression may be influenced by rearrangement upon integration, interaction between multiple transgene copies, and so-called **position effects** reflecting the local molecular environment. Position effects may cause in variable expression or silencing, mosaic expression and undesirable tissue-specific expression resulting from the influence of local *cis*-acting elements, chromatin structure and DNA methylation.

In mice, an alternative strategy involves homologous recombination between exogenous DNA and the endogenous genome. In this procedure, the transgene does not integrate randomly, but replaces a homologous gene in the mouse genome, as discussed below. Another recent development is the use of **site-specific recombination systems**, comprising an enzyme that catalyzes recombination between short specific DNA sequences. One of the most widely used systems is **Cre-*loxP*** (*Fig. 1*). The enzyme, **Cre recombinase** from bacteriophage P1, catalyzes recombination between short palindromic sequences termed *loxP*. If the *loxP* sites are used to flank a transgene, Cre recombinase can catalyze recombination between them and delete the transgene. Similarly, if a segment of

(a) **Function of protein**

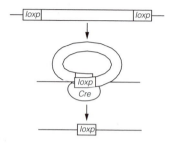

(b) **Transgene inactivation**

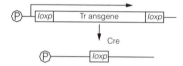

(c) **Transgene activation (RAGE)**

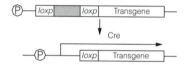

Fig. 1.   Cre-loxP site-specific recombination in transgenic organisms. P = promoter.

DNA between two *loxP* sites is placed between the transgene and its promoter to block expression, Cre recombinase can remove the segment and activate gene expression, a strategy termed **RAGE (recombinant activation of gene expression)**. The Cre-*loxP* system is extremely versatile, and by regulating the supply of recombinase it allows transgenes to be switched on and off in an inducible, cell type-specific or developmental stage-specific manner. In combination with gene targeting, the Cre-*loxP* system has facilitated the production of conditional developmental mutants in mice, allowing the effects of pleiotropic mutations to be studied at different developmental stages. Previously, unless a natural conditional mutant was available, only the earliest effects of embryonic lethal mutants could be studied (see Topic B2).

Transgenesis can also be used in place of microsurgical techniques. The ablation of particular cell types may be achieved by pricking, burning, laser ablation or surgical removal. An alternative is to express a **toxic transgene** under the control of a cell-specific promoter. A number of toxins are available that are both potent and restricted to the cells in which they are synthesized, e.g. **ricin**. For example, the expression of ricin under the control of a lens-specific promoter in mice significantly reduces the number of lens cells, and prevents them taking part in normal inductive interactions, resulting in serious eye defects in the transgenic mice. Finally, transgenic technology can be used to increase the versatility of mutagenesis screens using transposable elements for the identification of developmental genes.

Nuclear transfer technology

Nuclear transfer (removing the nucleus from one cell and introducing it into another where the nucleus has been removed or destroyed) is an important technique used to address the stability of the genome during development and mechanisms by which differentiation is maintained (see Topic A5). Nuclear transfer is also a method to generate **clones** of genetically identical animals, for nuclei can be removed from numerous cells in the same animal and each introduced into an **enucleated egg**. Initially, such **nuclear cloning** was demonstrated in *Xenopus*, but the technique is equally applicable to insects and mammals. Where the production of transgenic animals is inefficient, nuclear transfer can be used to produce many identical transgenics. Nuclear transfer is also a method to generate first generation transgenic animals, if the nuclei are subject to genetic manipulation before introduction into the egg. This is the only current strategy for the production of transgenic *Xenopus* (see below).

Transgenic mice   Transgenic mice can be produced by microinjecting DNA into the male pronucleus of the fertilized egg prior to nuclear fusion. The egg is then placed in the uterus of a pseudopregnant surrogate female. The foreign DNA does not integrate until one or two rounds of cleavage have taken place, so the primary transformant is usually chimeric for the presence of the transgene, or may comprise cells containing the transgene integrated at different loci. Similarly, chimeric mice can be generated by infecting early embryos with a recombinant retrovirus vector carrying the transgene. In each case, if cells containing the transgene contribute to the germline of the primary transformant, some of its offspring will be fully transgenic.

An alternative and favored strategy is to use **embryonic stem cells (ES cells)**[1] (*Fig. 2*). ES cells are derived from the inner cell mass of the mouse embryo and

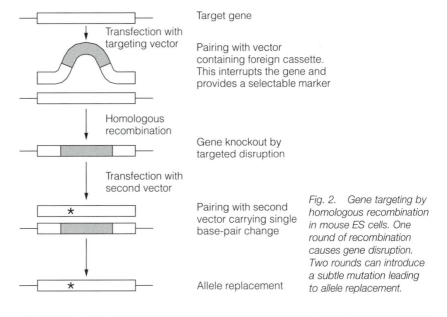

Target gene

Transfection with targeting vector

Pairing with vector containing foreign cassette. This interrupts the gene and provides a selectable marker

Homologous recombination

Gene knockout by targeted disruption

Transfection with second vector

Pairing with second vector carrying single base-pair change

Allele replacement

Fig. 2.   Gene targeting by homologous recombination in mouse ES cells. One round of recombination causes gene disruption. Two rounds can introduce a subtle mutation leading to allele replacement.

[1]**Embryonic stem cells** are derived from the inner cell mass of the mouse embryo. They are not stem cells in the strict sense, i.e. they do not divide to produce two dissimilar daughters, one a replacement stem cell and the other committed to differentiate. The name reflects their pluripotent nature, indicating that they derive from the 'stem' of the developmental hierarchy, before the three germ layers are specified.

they have two remarkable properties. Firstly, they are pluripotent, which means that if they are introduced into the inner cell mass of another embryo, they will mix with the host cells and can contribute to any tissue in the resulting mouse (including the germline). Secondly, these cells show unusually efficient homologous recombination, which means they can be used for gene targeting (Topic B2). This unique combination allows the introduction of specific mutations into any gene, and the production of mice carrying that mutation.

Transgenic mice are generated as follows (*Fig. 3*). The ES cells are transfected with DNA, which will either integrate randomly (allowing the production of a traditional transgenic mouse), or if the DNA sequence is so designed, it may undergo homologous recombination with the targeted endogenous gene and replace it. A drug-resistance marker is included in the DNA so transformed ES cells can be selected. This selection process eliminates non-transformed cells and thus provides a great advantage over the alternative strategy of microinjecting eggs, where there is no way to verify transgene integration other than by testing the primary transformants by PCR or Southern blot hybridization of genomic

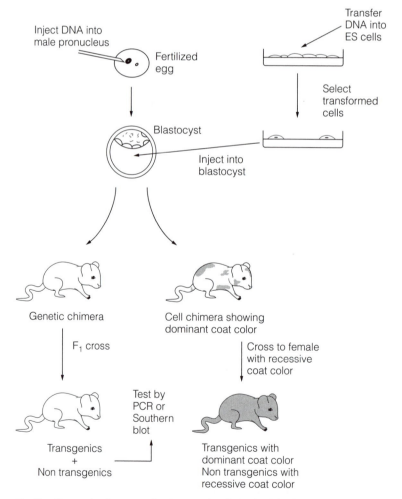

Fig. 3.   Two routes for generating transgenic mice: microinjection of fertilized eggs and transfection of ES cells. See text for details.

DNA. Transformed ES cells are injected into the blastocyst of another mouse embryo, where they contribute to the inner cell mass. The ES cells are usually derived from a mouse strain with a dominant coat color marker, and injected into the blastocyst of an embryo with a recessive coat color. The chimeric embryos are transferred to the uterus of a pseudopregnant female and the resulting mice have patchwork coats of both colors. Chimeric males are then crossed to females with the recessive coat color. If the ES cells have contributed to the germline, the transgenic mice in the next generation will have the dominant coat color. These will be heterozygous for the mutation and can be interbred to generate homozygotes.

Transgenic fruit flies

Foreign genes can be introduced into *Drosophila* using P-elements as vectors (*Fig. 4*). P-elements are transposons that exist naturally in some *Drosophila* populations. They integrate essentially at random in the genome and can mobilize in germ cells so that new integration events occur in each generation. To avoid unpredictable transposition events, transgenic fruit flies are generated by microinjecting a plasmid containing a defective P-element into the egg of a *Drosophila* strain lacking endogenous P-elements. The DNA is injected into the posterior pole of the egg, which is where the germ cells arise. The defective P-element lacks the gene required for transposition, but it is co-injected with a second plasmid containing the missing gene. The second plasmid cannot integrate, but transiently supplies the gene product required for the defective P-element to mobilize and integrate the transgene stably and permanently into the genome.

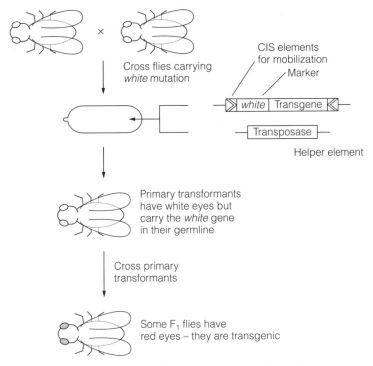

Fig. 4. Generation of transgenic Drosophila *using recombinant P-elements carrying a marker that restores normal eye color. See text for details.*

The P-element vector usually carries a marker allowing the selection or visual screening of flies carrying the insertion. For example, host flies often carry the *white* mutation, which affects eye development and results in the development of white rather than normal red eyes. The P-element vector carries a functional copy of the *white* gene along with the experimental transgene, and transformed flies can therefore be identified because of their red eyes. Flies developing from injected eggs have white eyes because the P-element integrates only in their germ cells. However, the transgenic offspring of these flies have red eyes, and are therefore heterozygous for the *white* marker gene and the transgene.

Transgenic plants

Plant tissues show remarkable regenerative properties. Plant tissues cultured on appropriate media will dedifferentiate to produce a mass of proliferating cells called a **callus.** In the presence of the appropriate hormones, the callus can redifferentiate into shoots and roots, to generate a new plant. In some species, it is possible to regenerate entire plants from single somatic cells. The production of transgenic plants therefore involves transforming individual cells or tissues with foreign DNA and using these cells or tissues for regeneration.

There are several methods for plant cell transformation, and the method chosen depends largely on species (*Fig. 5*). One popular method, which is particularly suitable for dicots, is to exploit the natural DNA transfer mechanism of

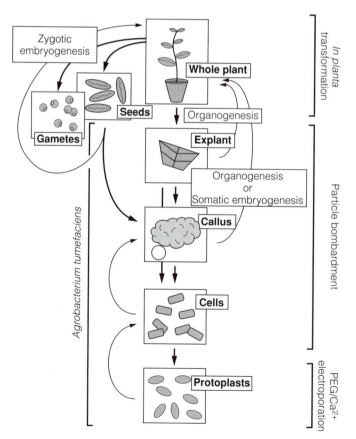

Fig. 5.   Summary of procedures used to generate transgenic plants.

Agrobacterium tumefaciens. This bacterium contains a **tumor-inducing plasmid (Ti plasmid)** that carries all the necessary functions to transfer part of itself, the **T-DNA**, into the plant genome. The T-DNA genes normally cause plant cells to proliferate and form callus, and to synthesize novel chemicals that the bacteria use as a carbon source. These genes can be removed from the T-DNA and replaced with a transgene. Wounded plant tissues, e.g. discs cut from leaves, can then be infected with bacteria carrying the modified Ti plasmid, and the transgene will be inserted into the genome. If the plasmid also carries a selectable marker gene, callus can be cultured and regenerated under selection, to produce transgenic plants with the transgene in every cell.

Other transformation strategies involve **direct DNA transfer**. In certain species, whole plants can be regenerated from transformed protoplasts. One of the most versatile techniques is **particle bombardment**, in which DNA-coated metal particles are fired into the nucleus using gunpowder, a blast of high-pressure gas or an electric discharge. This technique is applicable to any species, multiple genes can be introduced at the same time, and plasmid backbone sequences are not required. As above, plants are regenerated under selection and usually contain the transgene in every cell.

Recent advances – chick and *Xenopus*

Until recently, convenient transgenic technology was not available for two of the principal vertebrate model organisms, *Xenopus* and the chicken. It is now possible to produce transgenic frogs in large numbers by mixing linearized transgene constructs with decondensed sperm nuclei and treating the mixture with restriction enzymes to introduce breaks into the DNA (**restriction enzyme mediated integration, REMI**). The sperm nuclei are then transplanted into eggs and transgenic tadpoles develop, which are capable of developing into adult frogs and passing the transgene through the germline.

Current strategies for generating transgenic chickens are based on those used in mice and other mammals (microinjection into the egg, retroviral infection, transformation of primordial germ cells) but these are very inefficient. However, recent progress has been made with the transformation of chick ES cells, which could allow the rapid production of transgenic chicken lines, and birds carrying targeted mutations.

B4 CELLULAR AND MICROSURGICAL TECHNIQUES

Key Notes

Drugs and radiation	Drugs, and other disruptive treatments such as irradiation, can be used to interfere with development through the inhibition of fundamental processes including cytoskeletal reorganization, cell division, cell adhesion and cell signaling. Embryos can also be treated with putative developmental regulators to study their effects.
Grafting	Grafting involves moving cells and tissues from one part of the embryo to another. This can be used to study cell–cell interactions, the acquisition of cell fates and pattern formation. Grafting is also used to produce chimeric embryos.
Separating and recombining tissues *in vitro*	Embryonic tissues are robust and can be cultured *in vitro*. Isolated tissues can be used to study cell specification, while the co-culture of different tissues allows the investigation of inductive interactions. Embryo dissection and recombination experiments can be used to investigate regulation and cell sorting, as well as facilitating the production of chimeric embryos.
Microinjection	Microinjection is used to introduce a variety of substances into cells, including DNA, RNA, protein, and cell labels. The technique is also used for nuclear and cytoplasmic transfer, and to introduce cells into embryos.
Reorganizing cells and cytoplasm	Pattern formation and differentiation are often controlled by the distribution of determinants in the cytoplasm, and the spatial organization of cells in the early embryo. Both these factors can be experimentally manipulated by mechanically rotating cells in the embryo, or redistributing determinants within the cytoplasm e.g. by centrifugation.
Cell ablation	The influence of individual cells on development can be investigated by removing them from the embryo. In early embryos, this can be achieved by surgical ablation. Later, more sophisticated techniques are required, such as laser ablation or genetic ablation.
Barriers and beads	Inductive interactions involve signaling between cells in the embryo. Signaling can be perturbed or manipulated using permeable and impermeable barriers between tissues, and by inserting beads coated with putative signaling molecules to mimic endogenous sources of inductive signals.
Cell labeling and clonal analysis	Labeled cells can be traced during development, allowing fate mapping, cell lineage analysis and the characterization of developmental compartments. Cells can be labeled using radioactivity, vital stains, or

high molecular weight tracers. Alternatively, genetic or cytological labels can be used, or cells from a related species can be incorporated into the embryo.

Related topics	Cell fate and commitment (A2)	Pattern formation and
	Mechanisms of developmental	compartments (A6)
	commitment (A3)	The cytoskeleton, cell adhesion and
	Mosaic and regulative	the extracellular matrix (C5)
	development (A4)	

Drugs and radiation

Before tools such as antibodies and antisense RNA were available to inhibit the function of individual genes, drugs were widely used to interfere with particular cellular processes in order to determine their importance in development. Such drugs are still valuable tools in experimental developmental biology. Certain questions can be addressed by disrupting key cellular processes such as DNA replication, transcription and protein synthesis. For example, blocking the activity of RNA polymerase II with low doses of the fungal toxin **α-amanitin** can determine whether new gene expression is required for a particular developmental process, or whether it relies on stored mRNA. Another fungal toxin, **aphidicolin**, strongly inhibits eukaryotic DNA polymerase α, and therefore blocks DNA replication. Drugs can also be used to interfere with the cytoskeleton and thus inhibit developmental processes involving dynamic cytoskeletal reorganization. The fungal alkaloids **cytochalasin B** and **cytochalasin D** cause actin depolymerization, and thus break down the microfilament network responsible for, e.g. cell migration. Similarly, **colchicine** can break down microtubules at high concentrations, while at lower concentrations it blocks microtubule assembly, therefore causing cells to arrest at mitosis. Microtubules can also be disrupted by **UV-irradiation**, which is a technique used to prevent cortical rotation in *Xenopus* eggs, resulting in ventralization of the embryo (Topic I2). Conversely, **lithium ions** dorsalize *Xenopus* embryos, probably by inhibiting the action of a ventralizing signal, GSK-3 (Topic I2). Manipulating the availability of other ions can have dramatic effects. **Chelating agents** such as EDTA remove divalent cations (Ca^{2+}, Mg^{2+}) from solution. This can affect the morphogenetic behavior of developing organs because it influences cell adhesion, by disrupting the activity of calcium-dependent cell adhesion molecules (cadherins; see Topic C5). Many other substances have more specific effects on embryos and the tissues derived from them because they act as developmental regulators (e.g. retinoic acid, FGF, and cAMP).

Grafting

Grafting or **transplantation** means moving cells or tissues from one part of the embryo to another, or to another embryo. This type of experiment can answer many questions about cell–cell interactions and the acquisition of cell fate. Grafting is useful for establishing the timing of developmental commitment. For example, the ventral ectoderm of an early blastula-stage *Xenopus* embryo will normally form epidermis, but if transplanted to the dorsal surface of a gastrula-stage embryo it will differentiate into neural plate because it receives inductive signals from the notochord. However, if a similar experiment is carried out using ectoderm from a later blastula, the graft will differentiate as epidermis (i.e. according to its old position). Such experiments show that the ventral ectoderm becomes determined to form epidermis within the time window defined by the two grafts. Grafting experiments are discussed extensively in this book, e.g. in the definition of the organizer and its role in neural induction (Topics I2, and J1), in the analysis of neural crest lineages (Topic J6), in the regional specification of somites (Topic K2) and in the analysis of limb development and regeneration (Section M).

Grafting can also be used to generate chimeric embryos. Where a developmental mechanism involves signaling between two tissues such as ectoderm and mesoderm, grafting can be used to place mutant ectoderm over normal mesoderm and vice versa, to investigate the signaling mechanism. This has been used to study signaling between the apical ectodermal ridge and underlying mesoderm during limb development (Topic M1), and the role of mesoderm signaling and ectodermal competence in the development of cutaneous structures such as hair and feathers. Grafting can introduce cells carrying a genetic marker into an embryo, to facilitate lineage studies or fate mapping, or tracing the pathway of migrating cells. Similar experiments can be carried out using chimeras of different species, e.g. **chick–quail chimeras**, or chimeras of *Xenopus laevis* and *Xenopus borealis*. **Heterochronic grafts** involve the juxtaposition of tissues from different developmental stages. Such experiments have been carried out using chick limb buds to address the timing of proximodistal axis specification (Topic M2).

Single-cell transplants can show the influence of individual cells on the development of their neighbors. This is important in developmental systems with a small cell number, where single cells often have critical roles. For example, *C. elegans* vulval development depends on interactions between the anchor cell and the uterine precursor (Topic L1). Single-cell grafts have also been used in young vertebrate embryos, e.g. to study the timing of commitment of individual blastomeres.

Separating and recombining tissues *in vitro*

Many embryonic tissues are robust and can be maintained in relatively simple media. Therefore, as an alternative to grafting, tissues isolated from embryos can be combined in vitro and **co-cultured** to study inductive interactions. One example of this type of technique is the animal cap assay in *Xenopus* (Topic K1), which can be used to study mesoderm induction and neural induction (Topics K1 and J1). The reciprocal induction that occurs during kidney development can also be studied in this manner (Topic K4). Embryonic tissues can also be cultured in isolation, to reveal their innate specification (Topic A2).

Some important questions can be answered by cutting up whole embryos and observing how the various parts develop. For example, one can show whether an embryo is undergoing mosaic or regulative development by dissection and seeing whether the pieces develop into embryo parts, or whether some pieces regulate to form complete embryos (Topic A4). Experiments with **embryo fusion** can also confirm regulation. For example, chimeric **hexaparental mice** have been generated by fusing three morulae with different coat color markers. Chimeric embryos can also be produced by **disaggregating** embryos into single cells and **reaggregating** cells from different sources. Experiments where embryonic tissues have been disaggregated into single cells and allowed to reassociate spontaneously have shown the role of cell adhesion in tissue organization and the maintenance of tissue boundaries (see Topic C5) and provide evidence for spontaneous pattern-forming capacity in developing limbs (Topic M2).

Microinjection

Microinjection allows the introduction of foreign material into eggs and embryos through a fine needle. The procedure allows the introduction of cells into embryos (e.g. the injection of ES cells into the blastocoel of a mouse embryo), and the introduction of DNA, RNA, protein, dyes, toxins, viruses, isolated nuclei etc. into individual cells. Microinjection is therefore used for a variety of experimental procedures, including the production of transgenic or chimeric embryos (Topic B3), nuclear transfer, transient gene expression, cell labeling, cell ablation and the analysis of developmental regulators. The technique also allows material to be removed from cells or moved from one part of a cell to another, which can be used to demonstrate the importance of cytoplasmic determinants (see below).

Reorganizing cells and cytoplasm

The distribution of cytoplasmic determinants is critical for early development in many animals. This has been established by experimental techniques that reorganize the cytoplasm, and thus redistribute the determinants or block their transport. For example, the transfer of cytoplasm from the anterior pole to the posterior of the *Drosophila* egg by microinjection demonstrates the role of polarized maternal cytoplasmic determinants in axis specification, since the embryo develops with two heads and no abdomen (Topic H1). Similarly the role of asymmetric cell division in development can be investigated by reorganizing the cytoplasm and displacing the mitotic spindle. This can be achieved by a number of different techniques including centrifugation, squashing, displacement with a foreign object such as a glass bead, and constriction with a ligature. Centrifugation results in the reorganization of determinants that enjoy free movement in the cytoplasm, but not those attached to the cytoskeleton. The role of asymmetric divisions can also be investigated by mechanically changing the orientations of different cells in the embryo. For example, rotating the AB cell in the cleaving *C. elegans* embryo results in the inversion of the dorsoventral axis (see Topic G2).

Cell ablation

The role of individual cells in development can be investigated by selective ablation, and observing how this affects the fate of neighboring cells. Such experiments are useful for the investigation of cell–cell signaling, particularly in organisms with a small cell number and/or stereotyped cell divisions (e.g. *C. elegans*, Topic G2; *Arabidopsis* root development, Topic N4). There are a number of crude ways in which cells can be selectively killed, including pricking and burning, but only large and accessible cells such as those in the cleaving *Xenopus* embryo can be killed in this way. More refined techniques include **laser ablation**, where a finely focused laser beam is used to destroy individual cells in an embryo. This technique is applicable to small cells, but is only suited to clear embryos, i.e. those of nematodes, *Drosophila* and other invertebrates, and those of fish. For mammalian embryos, cell ablation can be achieved through the expression of a toxin under a cell-specific promoter, as discussed in Topic B3.

Barriers and beads

Many developmental processes rely on signaling between cells or groups of cells, and such processes can be studied by the use of barriers and beads. **Permeable barriers** (such as a Micropore filter) and **impermeable barriers** (such as a piece of foil) can be placed between tissues to determine whether development proceeds normally when signaling is physically interrupted. Impermeable barriers block signaling all together, while permeable barriers can be used to permit the passage of diffusible molecules while preventing cell–cell contacts, therefore distinguishing between paracrine and juxtracrine signaling mechanisms (Topic A3). Permeable barriers can also be used to filter molecules on the basis of size, e.g. allowing the passage of ions while preventing the passage of large glycoproteins. Signaling processes can be mimicked by implanting **inert plastic beads** coated with the appropriate signaling molecule. As an example, barriers and beads have been used extensively in the study of limb development (Topics M1 and M2). Beads soaked in retinoic acid, or the secreted protein Sonic hedgehog can both mimic the activity of the limb's zone of polarizing activity, while beads soaked in FGF8 can mimic the role of the apical ectodermal ridge. The combination of barriers and beads allows the components of signaling pathways to be defined. For example, an impermeable barrier placed beneath the limb field can prevent limb outgrowth, showing that signals from the underlying mesoderm are required. A bead soaked in FGF10 placed on top of the barrier can induce limb growth but not dorsoventral polarity. This approach therefore allowed the components required for limb development to be dissected (Topic M2).

Cell labeling and clonal analysis

Cells can be labeled in a number of ways, e.g. by external staining with a dye, incorporating a radioactive label, or injecting fluorescent labels or high molecular weight tracers such as horseradish peroxidase. Alternatively, genetic markers may be used such as reporter genes or cell-autonomous mutations that generate visible phenotypes such as pigmentation defects. Cells may also be recognized due to unusual cytological features (e.g. chromosome aberrations induced by irradiation, or polyploidy induced by treatment with colchicine). Finally, cells may also be recognized because they are from a different, although related, species (as in chick–quail chimeras). Cell labeling allows the descendants of a cell to be identified later in development and this has several important applications, including fate mapping (Topic A2), tracing cell lineage (especially where cell divisions are stereotyped such as in *C. elegans* development) and characterizing developmental compartments (Topic A6). Labeling also allows the route of migrating cells to be mapped, such as neural crest cells in vertebrates (Topic J6), germ cells in *Drosophila* (Topic E2) and individual myxamebae in the aggregating mound of *Dictyostelium* (Topic D3). Recently, the use of the reporter gene for green fluorescent protein has allowed cell labeling experiments to be followed in real time.

C1 GENE EXPRESSION AND REGULATION

Key Notes

Gene expression and regulation

Gene expression involves the transcription of DNA into RNA and the translation of RNA into protein. The flanking regions of the gene control gene expression at the level of transcription, but regulation may also be exercised at the levels of RNA processing and translation, so that the distribution of the final gene product differs from that of the mRNA. Regulation also occurs at the protein level, through covalent post-translational modifications and noncovalent interactions.

Immunoglobulin and T-cell receptor diversity

The synthesis of immunoglobulins and T-cell receptors is an exception to the rule of genomic equivalence. In the B- and T-cell lineages, DNA rearrangement brings together V, D and J segments encoding portions of the variable region of each antigen-binding molecule. Diversity is increased by variable junctions, loss of nucleotides, addition of random nucleotides by the enzyme terminal transferase and somatic hypermutation.

Transcription

Transcription is the synthesis of RNA on a DNA template. Transcription requires an enzyme, RNA polymerase, and a host of ubiquitous proteins that help to load the enzyme onto the promoter and establish the basal apparatus. Transcription factors binding at the promoter and distant enhancers help to stabilize the basal apparatus.

Transcription factors

Transcription factors influence gene expression either by interacting with the basal apparatus, modifying DNA structure, or controlling the activity of other transcription factors. Some transcription factors are constitutively active while others are synthesized in a cell- or stage-specific manner. Many transcription factors are dormant, and can be activated by signal transduction pathways.

mRNA processing

RNA is extensively processed before it leaves the nucleus. Modifications include capping, polyadenylation, intron splicing and editing. Once in the cytoplasm, the mRNA may be targeted to a particular location. All these processes, and mRNA degradation, can be regulated. There are many examples of regulation at the level of RNA processing in developmental systems.

Protein synthesis

In protein synthesis, ribosomes assemble on the mRNA and move along, assembling polypeptides according to the sequence of codons in the open reading frame. There are several examples of regulation at the level of protein synthesis in developmental systems, reflecting the inhibition of ribosome assembly by factors that bind to the 3′ noncoding region.

Post-translational modification

Once translated, proteins are extensively processed by chemical modification of side chains, glycosylation, conjugation to large chemical groups and proteolytic cleavage. These may be permanent modifications required for correct function or intracellular sorting, or may be reversible modifications used to regulate protein activity. Proteins may also be regulated by noncovalent interactions with other proteins and co-factors.

Related topics	Mechanisms of developmental commitment (A3)	Chromatin and DNA methylation (C2)
	Maintenance of differentiation (A5)	Signal transduction in development (C3)

Gene expression and regulation

With very few exceptions, each cell in a developing organism contains the same DNA. The diversification of the body during development therefore depends on the expression of different genes in different cells. **Gene expression** involves two major processes: the **transcription** of DNA into RNA and the **translation** of RNA into protein (*Fig. 1*). The information carried in a gene specifies the amino acids to be incorporated into the protein during protein synthesis, but the gene is flanked by further information in the form of *cis*-acting **regulatory elements** that control where and when the gene is transcribed, and the rate of transcription, so governing the amount of RNA produced and its spatial and temporal expression pattern. These flanking regions may also carry response elements that allow gene expression to be regulated by external signals, thus linking signal transduction to gene expression (Topic C3). **Transcriptional regulation,**

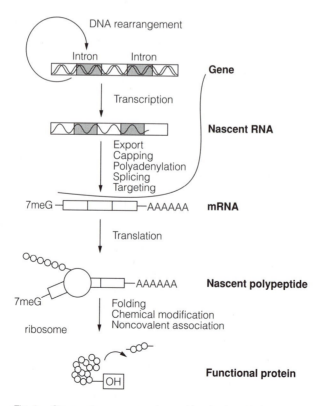

Fig. 1. Stages of gene expression and levels of regulation.

as mediated by these elements, is often the most important level of control in gene expression.

Once the gene is transcribed into RNA, further **post-transcriptional regulation** can take place to control the amount of functional protein produced in the cell. The RNA must be processed and exported from the nucleus, sometimes targeted to a specific region of the cell, translated, and then broken down. All these processes can be regulated. Furthermore, all proteins are processed in various ways before they can function properly: they must fold, undergo chemical modifications, some amino acids may be cleaved off, they may be conjugated to other molecules, and they may associate in multimers or with different proteins to form multisubunit complexes. Proteins may also be targeted to different parts of the cytoplasm, or imported into organelles, or they may be secreted. Finally, proteins must also be broken down. Once again, all these levels of gene expression can be regulated.

In this Topic, we briefly discuss mechanisms of gene regulation at the transcriptional and post-transcriptional levels, with examples from various developmental systems. First, however, we look at an exceptional system, where the rearrangement of DNA itself plays an important role in the regulation of gene expression.

Immunoglobulin and T-cell receptor diversity

During the development of the vertebrate immune system, two types of antigen-recognizing proteins are synthesized: **immunoglobulins** (when secreted, these are called **antibodies**) and **T-cell receptors (TCRs)**. These proteins are synthesized by B- and T-lymphocytes respectively, and have the potential to recognize 10^6–10^8 different antigens. This diversity is generated by **somatic recombination** in the corresponding genes. The recombination occurs only in these somatic cells, not in the germ line, therefore the differences generated are not passed to the next generation. Different somatic cells can rearrange their DNA in different ways so that many different antibodies can be produced in the same individual. The following discussion relates to immunoglobulin genes.

Immunoglobulins are Y-shaped molecules comprising two identical heavy protein chains and two identical light chains (*Fig. 2*). Most diversity occurs in the

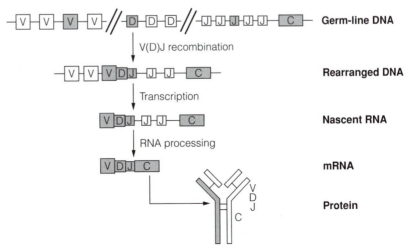

Fig. 2. Simplified representation of DNA rearrangement in the immunoglobulin heavy chain locus

C-terminal **variable region** of each chain, which contacts the antigen. In the corresponding genes, the coding region is not contiguous but built up of segments. In the heavy chain gene there are multiple **V (variable), D (diversity)** and **J (joining)** segments, while the light chain gene contains just the V and J segments. During B-cell maturation, recombination occurs between random V, D and J segments to bring them together as a contiguous coding unit. This is termed **V(D)J recombination.** Diversity is generated not only by the random combination of different V, D and J segments (and the inclusion of multiple D segments in some heavy chains), but also because the joining is imprecise. Recombination involves the cleavage of DNA strands between the segments, and these can be subject to a limited amount of degradation, so up to 10 nucleotides can be removed. Furthermore, an enzyme called **terminal transferase,** which is expressed only in these cells, can add random nucleotides to the free ends before they are joined together. Even when the coding unit is complete, further recombination can occur between the complete unit and upstream (unused) V segments. Finally, the entire coding region corresponding to the variable region of the immunoglobulin gene is also subject to **somatic hypermutation**, where random base changes occur at a high frequency. Once these changes to the DNA sequence have been made, the recombined gene is transcribed and functional immunoglobulins can be synthesized.

Transcription

Transcription is the synthesis of RNA using DNA as a template. It is the first level of gene expression and the predominant level of gene regulation. In eukaryotes there are three **RNA polymerases** (enzymes that carry out transcription), but only **RNA polymerase II** transcribes protein-encoding genes, and these genes are of the most interest in developmental biology.

RNA polymerase II binds upstream of the gene at a control region called the **promoter**. RNA polymerase II promoters show great diversity, reflecting the many different expression profiles of protein-encoding genes. However, RNA polymerase II is unable to bind on its own and requires a collection of **basal transcription factors** or **transcriptional initiation factors (TIFs)**, present in every cell, which must bind first to assist recognition. The complex containing RNA polymerase II, the TIFs and the promoter is called the **basal apparatus**. On its own, the basal apparatus assembles slowly and is unstable, and therefore supports only minimal transcription. Other **transcription factors** bind near the promoter, or at more distant *cis*-acting sites termed enhancers, and interact with the basal apparatus to increase the speed and stability of its assembly. Some transcription factors are present in every cell and provide general transcriptional activation. Other transcription factors are only synthesized or activated in certain cell types, or at certain stages of development, or in response to certain external stimuli, and this is how transcription is regulated. Most transcription factors act positively, but others act as inhibitors to decrease the stability of the basal apparatus, either directly or by blocking the access of another transcription factor. In many cases, differentiation involves the activation or synthesis of new transcription factors, therefore stimulating the assembly of the basal apparatus on new genes, and changing patterns of gene expression in the cell.

An important component of transcriptional gene expression is chromatin structure. The DNA in a eukaryotic cell is packaged with an approximately equal amount of protein to form a highly ordered nucleoprotein complex called chromatin. The proteins in chromatin control the way in which DNA is packaged and this has a major effect on transcription. Many transcription factors therefore function by interacting with chromatin proteins and making chromatin structure either more or less favorable for transcription. Such interactions are discussed in more detail in Topic C2.

Transcription factors

Transcription factors can influence gene expression in three ways: (a) directly, by contacting the basal apparatus and affecting its stability; (b) directly, by binding to DNA or chromatin and altering its structure; (c) indirectly, by affecting the activity of a second transcription factor. Transcription factors are modular proteins. The principal modules are the **DNA-binding domain** (which recognizes *cis*-acting elements) and the **transactivation domain** (which interacts with components of the basal apparatus), although they may also possess further domains allowing interaction with other cellular components, such as signal transduction proteins. Transcription factors can be grouped into families according to the nature of their DNA-binding and transactivation domains (*Table 1*) and this often suggests how they work, and which components of the basal apparatus they interact with. Transcription factors that inhibit gene expression may carry a specific silencing domain in place of the transactivation domain, or they may function in an indirect manner, by blocking interactions between other transcription factors and the basal apparatus. Changes in gene expression patterns reflect changes in the availability of active transcription factors. Such changes can be brought about by synthesizing new transcription factors, or by activating/inactivating those already present in the cell. The latter

Table 1.   *Major families of transcription factors with roles in development*

Family (DNA-binding domain)	Examples
Basic (helix-loop-helix dimerization domain)	*Drosophila* Twist, and products of the *achaete-scute* complex; vertebrate MyoD and MASH families; *Arabidopsis* GLABROUS1.
Basic (leucine zipper dimerization domain)	*Drosophila* Giant; *Arabidopsis* HY5.
High mobility group	Mammalian Sry and Sox proteins. Sox17α and Sox17β in endoderm specification.
Homeodomain	The products of all animal HOM-C/*Hox* genes. Knotted-1 in maize and SHOOT MERISTEMLESS in *Arabidopsis*.
MADS box	*Arabidopsis* APETALA1-3, PISTILLATA and AGAMOUS. Animal MEF2 proteins.
Paired domain	*Drosophila* Paired and Gooseberry; vertebrate Pax proteins.
POU domain	Vertebrate Pit-1, Brn-1 and Brn-2, controlling neural and endocrine cell fates.
Rel homology	*Drosophila* Dorsal.
T-box	Vertebrates: Brachyury, VegT, Tbx family
Winged-helix	*Drosophila* Forkhead and Sloppy paired; vertebrate HNF3β
Zinc finger (Cys)	Steroid receptors, retinoic acid receptor.
Zinc finger (Cys-His)	*Drosophila* Krüppel, Knirps and Hunchback; vertebrate Krox-20.

Family (transactivation domain)	Examples
Acidic	Vertebrate glucocorticoid receptor
Proline-rich	Vertebrate Jun
Glutamine-rich	*Drosophila* Antennapedia, Engrailed
Inhibition domains	*Drosophila* Even-skipped

strategy can provide a rapid response to induction, and there are many available mechanisms. For example, transcription factors can be activated by phosphorylation or dephosphorylation (e.g. MAP kinase phosphorylates Elk-1 in the signal transduction pathway responding to growth factor stimulation; see Topic C3), the disassociation of an inhibitory protein (e.g. Cactus dissociates from Dorsal to allow nuclear uptake of Dorsal and the expression of ventralizing genes in the *Drosophila* embryo; see Topic H6) or ligand binding (e.g. retinoic acid binds to its receptor, and this facilitates nuclear uptake and binding to retinoic acid response elements in target genes).

Transcription factors that do not interact with the basal apparatus may activate transcription by modulating the structure of DNA. Several transcription factors are known that control or maintain chromatin structure, e.g. by displacing nucleosomes or recruiting enzymes that modify histones (such factors are discussed in Topic C2). Other transcription factors function by introducing bends into DNA and facilitating the interaction of other proteins. Examples include the Sry and Sox transcription factors that have prominent roles in sex determination in mammals (Topic E5).

mRNA processing

In eukaryotes, mRNAs are extensively processed before export and translation. All pre-mRNAs are **capped** (a 7-methylguanosine residue is added to the 5′ end; this improves stability and is required for translation) and most are **polyadenylated** (a string of about 200 adenosine residues is added to the 3′ end). Most eukaryotic genes also contain introns that must be spliced out of the pre-mRNA before translation. In many cases this is a constitutive process (i.e. all mature mRNAs have the same structure) but very often the pre-mRNA may be **alternatively spliced** so that different mature transcripts are made. This may generate different proteins containing common modules, or some splice variants may prevent protein synthesis by incorporating a stop codon. Capping, polyadenylation and splicing have all been shown to play important roles in certain developmental systems.

Capping is sometimes used to regulate gene expression during development. For example, mRNAs in the egg of the tobacco hornworm moth have unmethylated caps that cannot be recognized by ribosomes. Fertilization activates the methyltransferase enzyme that completes the modification and allows the ribosome to bind to the mRNA. In this way, translation is initiated only after fertilization. Polyadenylation is also used to regulate the translation of developmentally-important genes. For example, translation of *bicoid* mRNA in the *Drosophila* embryo does not commence until after fertilization, when the short polyadenylate tail is extended (Topic H2).

Alternative splicing is central to the regulation of a number of developmental processes. One of the most interesting is sex-determination and X-chromosome dosage compensation in *Drosophila*, because it involves a hierarchy of sequentially acting RNA-splicing proteins (this is discussed in detail in Topic E5).

Another form of processing, termed **RNA editing**, involves the chemical modification of residues in the mRNA to alter the sense of a codon, or the addition of extra residues (usually uridine residues) to radically change the coding region. There are isolated examples of replacement editing in mammalian mRNAs, but most cases are restricted to organelle genomes and do not have overt effects on development.

RNA targeting is the process by which mRNAs are localized to specific regions in the cell. This has particular significance in development, because mRNAs can be used as cytoplasmic determinants in axis specification and differentiation. Examples include maternal *bicoid*, *oskar* and *nanos* mRNA localization in the *Drosophila* egg (Topic H2) and the localization of maternal *Vg-1* and *VegT* mRNA in the *Xenopus* egg (Topic I1).

Finally, **mRNA turnover** plays an important role in gene regulation because it contributes to the steady state mRNA level, i.e. reflecting the balance between synthesis and degradation. The stability of most mRNAs is sequence dependent, probably reflecting secondary and tertiary structures formed by the RNA in the cell. Particularly unstable mRNAs, such as those transcribed from rapid response genes like c-*fos* and c-*jun*, have AU-rich elements in their 3′ noncoding region, which can form stem loop structures that may be recognized by RNases. The stability of some RNAs is regulated. For example, transferrin receptor mRNA contains an iron-response element. A protein binds to this element when intracellular iron levels fall, increasing transcript stability and thus the steady state mRNA level in the cell.

Protein synthesis

Protein synthesis begins when a **ribosome** assembles at the 5′ cap of the mRNA, and scans along the 5′ noncoding region until an **AUG codon** is detected (however, some mRNAs have an **internal ribosome entry site**, and translation is not cap dependent; an example is the *Drosophila* homeotic selector gene *Antennapedia*). The ribosome then moves along the transcript, tRNAs carrying amino acids are matched to each codon in the open reading frame, and amino acids are added to the growing polypeptide chain. Protein synthesis ends when a nonsense codon is detected, and **release factors** bind to the ribosome causing disassembly and release of the polypeptide chain.

Gene expression may be regulated at the level of protein synthesis by factors that bind to the mRNA and prevent ribosome assembly or scanning. For example the function of the *Drosophila* protein Nanos is to repress the translation of *hunchback* mRNA by associating with another protein, Pumilio, which binds to sequences in the *hunchback* 3′ noncoding region. Similarly the Bicoid protein acts as a translational inhibitor of *caudal* mRNA (Topic H2). Similar translational repression occurs in *C. elegans*, involving mRNAs for the *tra-2* and *fem-3* genes that control germ cell differentiation (Topic E3). In the chick embryo, translation of the *fgf2* mRNA in some mesodermal cells is regulated by a small **antisense RNA** molecule that binds to the transcript and targets it for degradation.

Post-translational modification

Proteins begin to fold as they are being synthesized, to adopt the most energetically favorable conformation. However, spontaneous self-assembly may not produce the **native conformation** (the shape required for correct function) or it might be necessary to delay folding so that the protein can be transported through an internal membrane. Newly synthesized proteins therefore associate with **molecular chaperones**, whose function is to control folding.

Most proteins also undergo some manner of covalent modification during or after synthesis. Almost all cytoplasmic proteins, for example, are cleaved to remove the first N-terminal amino acid, the **initiator methionine**. Many proteins also require specific modifications before they can function correctly. In some cases, individual amino acid residues undergo minor chemical modifications. For example, some hormones require sulfation of tyrosine residues, and the hydroxylation of proline residues in collagen stabilizes the triple-helical tertiary structure. Such chemical modifications can be used to regulate protein activity, e.g. phosphorylation of tyrosine, threonine and serine residues in signal transduction proteins and transcription factors (Topic C3), and the acetylation of lysine residues in histones (Topic C2). Another common modification is the formation of disulfide bonds, which stabilize protein tertiary structure. Other proteins undergo more extensive side chain modification so that the amino acid residues are augmented with large chemical groups, e.g. cholesterol is added to Hedgehog family proteins (Topic C3). Such modifications may be required for protein trafficking in the cell, e.g. glycosylation of proteins in the secretory pathway, and the conjugation of GPI anchor groups to lodge proteins in the cell membrane. Many proteins are also synthesized as inactive precursors that must be activated by proteolytic cleavage. For example, the signaling protein Spätzle, which functions in *Drosophila* dorsoventral specification, is activated by a **proteolytic cascade** involving the serine proteases Snake, Easter and Gastrulation defective (Topic H6).

Protein function can also be modified by noncovalent interactions. For example, many proteins co-operate by assembling into multisubunit complexes. Conversely, interaction between the transcription factor Dorsal and the inhibitory protein Cactus blocks the function of Dorsal by masking its nuclear localization signal and preventing nuclear uptake (Topic H6). There are also many examples where noncovalent interactions can regulate the activity of a protein by changing its conformation, e.g. when an extracellular ligand binds to a cell surface receptor and causes an internal conformational change that results in signal transduction (Topic C3). The path from nascent polypeptide to functional protein may therefore be a long and complex one.

C2 CHROMATIN AND DNA METHYLATION

Key Notes

Chromatin

DNA exists in association with proteins to form chromatin. The DNA is wound into nucleosomes containing histone octamers, and strings of nucleosomes are packed into a 30-nm fiber. Higher orders of chromatin structure are poorly understood, but chromatin can exist in an open conformation (euchromatin) that is accessible to transcription factors and a repressed conformation (heterochromatin) that is transcriptionally inactive.

Chromatin and gene regulation

The chromatin structure of active and repressed genes is distinct. Open chromatin is sensitive to DNase and may contain hypersensitive sites where nucleosomes have been displaced. It is rich in acetylated histones. Repressed chromatin is insensitive to DNase, the histones are deacetylated and the DNA itself may be methylated. In order to activate gene expression, chromatin structure must be disrupted. Some transcription factors appear to modulate chromatin structure, e.g. by displacing nucleosomes or recruiting histone acetylases.

Maintenance of chromatin structure

Chromatin structure is used to restrict the potency of cells and maintain the differentiated state. It must therefore be perpetuated through successive rounds of DNA replication. This may be achieved by reassembly of chromatin proteins on daughter strands immediately after replication, aided by proteins such as Polycomb and Trithorax that maintain open and repressed chromatin domains.

DNA methylation and gene regulation

DNA cytosine methylation is associated with transcriptional repression, although methylation is thought to maintain rather than initiate gene silencing. DNA methylation may block the access of transcription factors directly, or may recruit proteins that displace transcription factors and sequester the DNA into repressed chromatin.

Parental imprinting

Most genes in mammals are expressed on both chromosomes, but in a few cases the parental origin in important, and only the allele from one parent is expressed. Genes showing parental imprinting often have sex-specific methylation patterns established during gametogenesis. These methylation imprints are not affected by global changes in methylation that occur in early development.

X-chromosome inactivation

In female mammals, one of the X-chromosomes is inactivated in each somatic cell. The choice is made early in development. It is random but perpetuated through successive cell divisions, so females are mosaics of cell clones in which different X-chromosomes are inactivated. Inactivation is controlled at a single locus, *Xist*, which is expressed on the inactive X-chromosome. During gametogenesis, the maternal *Xist* allele is imprinted, so the paternal allele is expressed and the paternal X inactivated. However, this imprint is erased by global demethylation and a new random imprint is established by an unknown mechanism.

Maintenance of DNA methylation   DNA methylation occurs mostly at symmetrical sites (CpG in mammals, and both CpG and CpNpG in plants). This allows the methylation pattern to be perpetuated at each round of DNA replication by enzymes that methylate DNA preferentially at hemimethylated sites (methylated on one strand).

Related topics   Maintenance of differentiation   Gene expression and regulation
                        (A5)   (C1)

Chromatin

The DNA in a eukaryotic cell is associated with an equal mass of protein, forming a highly ordered nucleoprotein substance termed **chromatin** (*Fig. 1*). The protein component of chromatin consists mainly of **histones**, basic proteins that bind strongly to DNA and help to organize it firstly into fundamental structural units termed **nucleosomes**, and then into higher order structures. Each nucleosome comprises 146 bp of DNA wrapped around a histone octamer (two molecules each of histones H2A-H2B and H3-H4). Individual nucleosomes are separated by a species-specific amount of linker DNA. Linked nucleosomes are wound into a more compact structure, termed the **30-nm fiber**, which requires a further **linker histone** (histone H1). This fiber is organized into **chromatin loops** containing 30–100 kbp of DNA, with their bases attached to the nuclear matrix. These loops are thought to represent functional *chromatin domains*, as discussed below. Chromatin contains other proteins in addition to histones, including transcription factors, DNA repair enzymes, proteins involved in replication and recombination, and **high-mobility group proteins** with various functions in structural organization and gene regulation.

At interphase, two distinct forms of chromatin can be seen in the cell nucleus: diffuse **euchromatin**, which is thought to adopt the 30-nm fiber organization; and dense **heterochromatin**, which is more compact than euchromatin and tends to congregate at the nuclear periphery. Euchromatin is thought to be accessible and potentially available for transcription, whereas heterochromatin is sometimes termed **repressed chromatin** because it is inaccessible, and transcription is inhibited. There are many biochemical differences between euchromatin and heterochromatin including differences in histone structure, DNase-sensitivity, and the amount of DNA methylation.

Chromatin and gene regulation

The highly ordered structure of chromatin excludes other proteins. There are over 70 000 genes in the vertebrate genome, so chromatin may provide an important mechanism to shut down transcription on a global basis, and prevent potentially damaging 'noise' from the background transcription of many genes at a low level. Thus, the transcription of individual genes requires specific activation, and this must involve the disruption of chromatin structure.

The importance of chromatin structure can be demonstrated by the differential **DNaseI sensitivity** of genes in different tissues. For example, globin DNA isolated from chicken erythrocytes (which retain their nuclei, unlike their mammalian counterparts) is sensitive to DNAseI digestion, but globin DNA

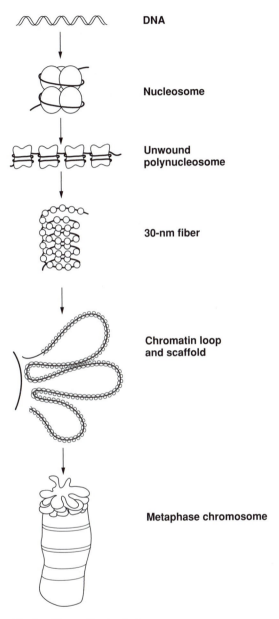

DNA

Nucleosome

Unwound
polynucleosome

30-nm fiber

Chromatin loop
and scaffold

Metaphase chromosome

Fig. 1. Chromatin structure.

isolated from brain tissue is not. Furthermore, globin genes isolated from erythrocytes also contain **hypersensitive sites** (sites exquisitely sensitive to DNAseI digestion), which appear to be devoid of nucleosomes all together. These results indicate that transcriptionally active genes have a chromatin structure that is accessible, and may contain **nucleosome-free sites** where transcription factors have displaced nucleosomes, while transcriptionally inactive genes may have inaccessible chromatin structure. However, transcription itself is not necessary for the formation of accessible chromatin, since the DNase-sensitivity extends for several kilobase pairs beyond the transcription unit. The extent of DNase sensitivity defines a conceptual **open chromatin domain**, whose structure is independent from other chromatin domains. It is thought that chromatin domains may correspond to the physical division of chromatin into loops attached to the nuclear matrix. Evidence supporting this includes the existence of **matrix attachment regions (MARs)** at the boundaries of known chromatin domains (e.g. the β-globin locus), plus the fact that MARs can isolate genes from the effect of enhancers in transfection experiments, suggesting they act as **boundary elements** or **insulators.** Open chromatin is rich in **acetylated histones**, a form of post-translational modification that may perturb histone interactions and disrupt nucleosome structure. There is a strong link between enzymes that remove acetyl groups from histones and those that add methyl groups to cytosine residues in DNA, so repressed chromatin structure and DNA methylation (see below) may act together to prevent transcription. Open chromatin is also rich in certain classes of high mobility group proteins and is relatively depleted in linker histones. Some genes are associated with distant *cis*-acting elements called **locus control regions**. Transcription factors binding to these sites are thought to help maintain an open chromatin domain.

Maintenance of chromatin structure

During DNA replication, nucleosomes are displaced from DNA and reassemble immediately on the two daughter duplexes. Displaced histones show no preference for either daughter strand, so patterns of open and repressed chromatin domains would be maintained as long as the newly synthesized histones were modified after mixing with the 'old' histones and assembling into nucleosomes. However, the transient displacement of nucleosomes during replication could allow competition between nucleosomes and transcription factors for strategic regulatory sites, resulting in an opportunity for changes in patterns of gene expression following DNA replication. At the molecular level, this might represent the moment at which cells become committed to certain developmental fates.

A number of DNA-binding proteins have been shown to help establish patterns of open and repressed chromatin and maintain them through successive rounds of DNA replication. The function of some eukaryotic transcription factors is to displace nucleosomes and either clear the way for other transcription factors to bind at sites previously occupied by nucleosomes, or relieve repression caused by higher order chromatin structure. A number of such transcription factors have been identified, including the *Drosophila* **GAGA factor**, which binds at many purine-rich sites, and the **SW1/SNF complex**, which is associated with RNA polymerase II. Other transcription factors are associated with enzymes controlling histone acetylation and DNA methylation. Therefore, there are a number of mechanisms by which transcription factors could regulate gene expression by controlling chromatin structure. Established chromatin domains are maintained through successive cell divisions by members of the *Polycomb* and *trithorax* gene families. In *Drosophila*, *Polycomb* and *trithorax* were first identified as mutants that failed to maintain normal homeotic selector gene expression patterns. Both Polycomb and Trithorax function in multiprotein complexes, which have counterparts in many species ranging from vertebrates to plants. Polycomb proteins maintain patterns of repressed chromatin while Trithorax proteins maintain open chromatin domains.

DNA methylation and gene regulation

Cytosine residues in DNA can be **methylated** by enzymes termed **DNA methyltransferases**, generating a modified base called **5-methylcytosine**. In mammals, about 5% of cytosine residues are methylated, and methylation usually occurs in the dinucleotide sequence CpG. In plants, up to 30% of cytosine residues can be methylated, and this can occur at either CpG or CpNpG sequences (where N is any nucleotide). Conversely, invertebrates have much lower levels of DNA methylation, and methylated DNA has not been detected in *C. elegans* or *Drosophila*.

There is evidence that DNA methylation contributes to differential gene expression during development. Firstly, gene expression is in some cases inversely correlated with the amount of DNA methylation in the promoter. For example, during the differentiation of erythrocytes, the ε- and γ-globin genes are sequentially expressed. When ε-globin is expressed, the γ-globin gene promoter is methylated but the ε-globin gene promoter is not. When the γ-globin gene is expressed the reverse applies. Furthermore, both promoters are methylated in cells that do not synthesize globin at all. Secondly, genes can be repressed or activated by adding or removing methyl groups. For example, fully methylating a transgene before introducing it into a cell usually abolishes transcription. Conversely, incorporating the cytosine analog **azacytosine**, which cannot be methylated, can reactivate previously repressed genes. Thirdly, in both mammals and plants, the genome undergoes global changes in methylation during early development, followed by cell-specific changes in methylation patterns associated with individual genes. In mammals, DNA methylation patterns are erased by a global wave of demethylation activity in the zygote, and methylation patterns are generated *de novo* in the epiblast just prior to gastrulation. In plants, changes in methylation are revealed by the tendency for transposons (which are repressed by methylation) to jump to new sites and generate mutations in somatic cells.

Methylated DNA could repress transcription in a number of ways. Methylation could directly prevent the binding of a transcription factor with cytosines in its recognition site. Alternatively, methylated DNA could recruit proteins that cover the DNA and block access to transcription factor binding sites or sequester the DNA into higher order chromatin structure. Several **methyl-CpG-binding proteins** of this nature have been isolated. There is also some evidence that such proteins associate with enzymes that control histone acetylation, providing a link between DNA methylation and repressed chromatin.

Although DNA methylation correlates with transcriptional repression, it is thought to be a consequence of (rather than an initiator of) silencing. DNA methylation helps to maintain DNA in a repressed state through successive rounds of replication (*Fig. 2*). There are different theories concerning the primary role of DNA methylation: it could represent a mechanism to globally control gene expression and prevent low level background transcription, or it could represent a defense against invasive DNA. In both animals and plants, transgenes are often subject to *de novo* methylation, resulting in loss of transgene activity. Whatever its general role, DNA methylation is predominant in the control of two specific processes, parental imprinting and (in mammals) X-chromosome inactivation, discussed below.

Parental imprinting

Parental imprinting has been documented in mammals and in certain flowering plants. In plants, the developing embryo appears not to be affected, and the

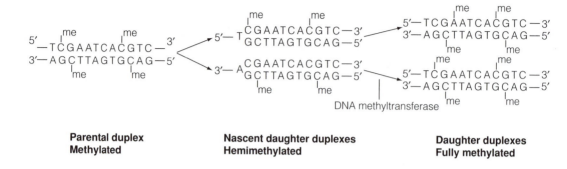

Parental duplex   **Nascent daughter duplexes**   **Daughter duplexes**
Methylated   **Hemimethylated**   **Fully methylated**

Fig. 2.   Perpetuation of DNA methylation.

greatest impact of parental imprinting occurs in the endosperm tissue that nourishes the embryo prior to germination. Conversely, in mammals, parental imprinting has a profound influence on embryonic development. Mammalian somatic cells are predominantly diploid and both copies of each gene are usually expressed. However, some genes carry an **imprint** that prevents transcription from the chromosome inherited from one particular parent. About 20 genes showing this phenomenon of **parental imprinting** have been discovered so far and many of them appear to control embryonic growth. The importance of such genes in mammalian development is demonstrated by the failure of parthenogenesis (development from unfertilized eggs) and the aberrant development of **uniparental diploids** (**androgenotes** and **gynogenotes**, zygotes containing two male or two female nuclei, respectively). Androgenous embryos show poor growth, although the extraembryonic membranes are normal. Conversely, gynogenous embryos show normal early development but the extraembryonic membranes do not form properly.

Imprinted genes are grouped in several gene clusters, and specific *cis*-acting sites called **imprinting boxes** are required to regulate parent-specific expression. These regions show **parent-specific DNA methylation patterns** that are established during gametogenesis, and they are not affected by the global wave of demethylation and *de novo* methylation that occurs in the embryo prior to gastrulation. The best-characterized imprinted cluster is associated with the fetal overgrowth disorder **Beckwith–Weidemann syndrome (BWS)** and is discussed below. Importantly, the analysis of genes in this cluster shows that methylation of a single imprinting box can differentially regulate two genes, and that multiple imprinting boxes with different methylation patterns can exist within the same gene (*Fig. 3*).

In humans, the BWS cluster is found on chromosome 11 and contains about 10 genes showing parental imprints. We concentrate on two of these genes, *IGF2* and *H19* to demonstrate the importance of DNA methylation. *IGF2* is maternally imprinted (only the paternal copy is expressed) although this does not correspond to a particular methylation pattern. The *IGF2* gene encodes a growth factor, and most cases of BWS involve aberrant biallelic expression of this gene. *H19* is paternally imprinted (only the maternal copy is expressed) and this correlates with paternal-specific methylation in the *H19* locus. The *H19* gene

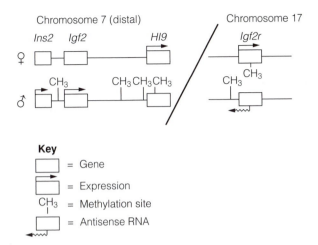

Fig. 3.   Control of gene expression by parental imprinting in the mouse.

produces an RNA that is not translated, and its sole function appears to be the inhibition of *IGF2* transcription. Therefore, the paternal-specific methylation of the *H19* gene prevents *H19* expression, which in turn allows the expression of *IGF2*. The same methylation site can therefore repress one gene but activate another. The *IGF2R* gene is found on a different chromosome and encodes the IGF2 receptor. In humans, this gene shows biallelic expression, whereas in mice, the homologous gene is paternally imprinted (only the maternal copy is expressed). In this case, there are two methylation sites in the locus, a paternal-specific site in the promoter and a maternal-specific site in an intron. The intron methylation site marks the position of an inverted promoter, which produces antisense *igf2r* RNA. On the maternal chromosome, the intron site is methylated and the *Igf2r* gene is expressed. On the paternal chromosome, antisense RNA is produced that either prevents transcription of the sense transcript or inhibits the sense RNA at the post-transcriptional level.

X-chromosome inactivation

In female mammals, one of the two X-chromosomes is inactivated in every somatic cell and is condensed into repressed chromatin. This is the mechanism of **dosage compensation** for X-linked genes chosen by mammals (see Topic E5 for other mechanisms). The **inactive X (Xi)** can be seen as a densely-staining **Barr body** throughout the cell cycle, while the **active X (Xa)** is seen only at mitosis. DNA methylation is associated with X-chromosome inactivation in two ways. Firstly, most of the Xi is methylated and transcriptionally inactive. This is thought to be no more significant than the association between DNA methylation and transcriptional repression on other chromosomes, i.e. it is used as a mechanism to perpetuate inactivity rather than initiate it. However, DNA methylation is also thought to play a major role in initiating X-chromosome inactivation. This involves a locus on the X-chromosome called *Xist* (for **X-inactive-specific transcript**). As the name suggests, this gene is expressed only on the Xi. Its product, an untranslated RNA molecule, coats the inactive chromosome and appears to play the predominant role in the chromatin remodeling process that accompanies inactivation. During gametogenesis, the *Xist* locus undergoes sex-specific methylation like the other imprinted genes discussed above. In most placental mammals (but not humans), the maternal allele is methylated, so this results in preferential expression of the paternal *Xist* allele and hence preferential inactivation of the paternal X-chromosome. Unlike the other imprinted genes, however, the methylation pattern on *Xist* is erased by the global demethylation that occurs in the zygote, and both X-chromosomes are active in pre-gastrulation embryos. Just before gastrulation, *de novo* methylation establishes a new imprint *randomly* in the epiblast so that one X-chromosome is inactivated in each cell of the embryo. It is still unclear how this choice is made, but different cells inactivate different X-chromosomes and maintain the same inactive X through successive rounds of replication. Therefore, the female embryo develops as a mosaic of clones of cells with alternative active and inactive X-chromosomes.

Maintenance of DNA methylation

DNA methylation occurs predominantly at symmetrical sites (CpG and CpNpG) so that the same sequence is present on both DNA strands although running in opposite directions. This provides a mechanism for maintaining patterns of DNA methylation through successive cell divisions. The global changes in methylation patterns that accompany early development in mammals indicate the existence of a *de novo* **methylase activity**. However, once methylation is established, the same pattern can be perpetuated through successive rounds of DNA replication by another DNA methyltransferase that recognizes **hemimethylated DNA**, i.e. DNA methylated on one strand only, and specifically methylates cytosines on the unmethylated strand. The mechanism is shown in *Fig. 2*. Several DNA methyltransferases of this nature have been isolated in mammals and plants.

C3 SIGNAL TRANSDUCTION IN DEVELOPMENT

Key Notes

Signal transduction

Cells respond to inductive signals by changing patterns of gene expression. Many signaling molecules are ligands that bind to transmembrane receptors. Ligand binding to the extracellular domain induces a conformational change in the cytoplasmic domain that stimulates a latent enzyme activity, such as a kinase. This allows the signal to be propagated inside the cell as a cascade of kinase activity, eventually resulting in the activation of dormant transcription factors.

Signals and receptors

Several families of secreted proteins are commonly used as signals in development, including growth factors, cytokines, polypeptides of the TGF-β superfamily, Hedgehog and Wnt proteins, and ephrins. Steroid-like molecules represent another important class of diffusible signals. Other signaling molecules are immobilized in the cell membrane or the extracellular matrix. These different classes of signaling molecules act as ligands for a number of different families of transmembrane and intracellular receptors. Receptors often act as dimers or multimers, and their roles in development can be probed using constitutive and dominant negative mutants.

G-protein-coupled receptors

G-protein-coupled receptors bind to a diverse range of ligands. Ligand binding stimulates cytoplasmic G-proteins to exchange GDP for GTP, allowing them to dissociate into α- and βγ-subunits that can modify the levels of downstream second messengers.

Growth factor signaling – Ras, Raf and MAP kinase

Growth factors act as ligands for receptor tyrosine kinases. Ligand binding stimulates dimerization and activates the cytoplasmic tyrosine kinase activity intrinsic to the receptor. Phosphotyrosine residues can recruit enzymes that modulate the activity of Ras. Activated Ras recruits Raf to the membrane where its kinase activity is stimulated. Raf then activates MEK, which in turn activates MAP kinase. Both MEK and MAP kinase activate latent transcription factors therefore changing patterns of gene expression in the nucleus.

Cytokine signaling – the JAK-STAT pathway

Cytokines bind to receptors that associate with cytoplasmic Janus kinases. Ligand binding stimulates dimerization and activates the Janus kinases, which phosphorylate the receptor. Inactive transcription factors called STATs are recruited to the phosphotyrosine residues and are phosphorylated by the Janus kinases. This allows them to translocate to the nucleus and regulate gene expression.

TGF-β signaling – SMADs

The TGF-β superfamily of secreted proteins includes several families of signaling molecules used in development, including the TGF-β family, activins and bone morphogenetic proteins (BMPs). Although the precise mechanisms vary, ligand binding stimulates receptor oligomerization and the activation of an intrinsic serine/threonine kinase activity. An important class of downstream targets is the SMAD proteins, which can translocate to the nucleus and act as transcriptional regulators.

Hedgehog, Wingless (Wnt) and Decapentaplegic (BMP)

Proteins of the Hedgehog, Wingless (Wnt) and Decapentaplegic (BMP) families act as morphogens, and are often found in reciprocal signaling pathways. Hedgehog and Wingless maintain each other's expression at the parasegment boundaries in *Drosophila*, and are also involved in the patterning of imaginal discs. Homologous molecules in vertebrates are involved in many developmental systems.

Steroid receptors

Steroid-related molecules such as retinoic acid can diffuse through cell membranes, so their receptors are found inside the cell. Ligand binding induces a conformational change that allows the receptors to dimerize, translocate to the nucleus and act as transcription factors.

Second messengers

Signals arriving at the cell surface may be converted into simple biochemical changes in the cytoplasm, a process mediated by molecules called second messengers. Second messengers can have profound effects on cell physiology, e.g. by releasing calcium ions from the endoplasmic reticulum, which is important in fertilization, and by activating various second messenger-dependent kinases with roles in many developmental processes.

Networks and cross-talk

Signal transduction pathways are regulated by feedback and cross-talk from other pathways, so that cells respond to the sum of signals arriving at their surfaces yet do not become saturated with information. In development, the dramatic changes that accompany inductive interactions reflect the strong and unambiguous nature of the inductive signals, allowing changing patterns of gene expression to establish a new equilibrium in the cell.

Related topics

Mechanisms of developmental commitment (A3)
Mosaic and regulative development (A4)

Gene expression and regulation (C1)
The cell division cycle (C4)

Signal transduction

In animal development, cell fates are often specified by **induction**, which requires communication between cells (Topic A3). Such inductive interactions involve one cell secreting or displaying a molecular signal, which is perceived and acted upon by the responding cell. In most cases, the signal acts as a **ligand** for a **membrane-spanning** (or **transmembrane**) **receptor** on the surface of the responding cell. When a ligand binds to the extracellular **ligand-binding domain** of its receptor, this causes a conformational change in the **cytoplasmic domain** that stimulates a dormant enzymatic activity, often a protein kinase. The receptor is then able to phosphorylate cytosolic proteins. Thus, the extracellular signal is turned into a different type of intracellular signal and the ligand never has to enter the cytoplasm. This process is called **signal transduction**.

Inductive interactions result in changes in patterns of gene expression and protein activity in the responding cell. The ultimate targets of signal transduction are often transcription factors and other regulatory proteins that have a direct influence on gene expression. Following receptor activation, there is further *intracellular* signal transduction, often in the form of a **cascade** of sequential enzyme activation: one kinase phosphorylates and activates a second, which performs the same process on a third and so on until the target transcription factor is activated. There are many such signal transduction pathways in the animal cell, which vary in the number of steps between receptor and target, providing many opportunities for amplification, branching and integration. Some of these pathways are discussed below.

Signals and receptors

Cells can respond to a diverse range of external signals, including both physical stimuli (e.g. light, heat) and molecules in the environment. Plant cells appear to have a limited signaling repertoire, restricted to a few hormones and polypeptides with specific roles. This may reflect the preferred use of cytoplasmic communication channels called plasmodesmata, which allow communication by molecules that are transferred through the common cytoplasm. Conversely, animal cells respond to a vast array of molecules including secreted polypeptides, large glycoproteins, steroid-like molecules, ions, amino acids, small peptides, nucleotides and bioactive amines. In early animal development, secreted polypeptides and steroid-like molecules play a predominant role in local inductive interactions, while large glycoproteins in the extracellular matrix or on the surface of other cells are important for contact-dependent signaling. In some cases, contact-dependent signaling between cells involves two receptors that act as ligands for each other. This occurs in the Delta/Notch signaling pathway (see Topic J5) and perhaps the Boss/Sevenless signaling pathway in the *Drosophila* compound eye (see Topic L3). Cells perceive diverse signals with an equally diverse range of receptors, although the receptors can be grouped into a small number of structurally-related families (*Table 1*). Since the cell has a smaller repertoire of responses than there are stimuli, the receptors channel incoming signals into a number of common signal transduction pathways. The response to a given signal therefore depends on which signaling components, transcription factors and other proteins are available in the cell. This enables the same types of signal and receptor to be used over and over again in different developmental systems. Consequently many signal transduction pathways are highly conserved in different systems and in different organisms (*Table 1*).

Many receptors are dimeric or multimeric and this provides an opportunity to determine their function. Mutant receptors that signal constitutively (in the absence of a ligand) generate dominant gain of function effects, while mutant nonfunctional receptors that sequester wild type receptors into inactive complexes generate dominant loss of function effects (dominant negative effects). The consequences of gain or loss of function can be investigated by transgenesis or injecting mRNA for the appropriate mutant receptor into the embryo (*Fig. 1*). This is discussed in Topic B3.

G-protein-coupled receptors

G-protein-coupled receptors (GPCRs) are single polypeptides whose central hydrophobic region spans the membrane seven times (*Fig. 2*). The external, N-terminal domain is the ligand binding domain and the internal, C-terminal domain is associated with a trimeric **G-protein** (guanine nucleotide-binding protein). In its inactive state, the G-protein is associated with GDP. Ligand binding to the receptor increases the rate at which GTP replaces GDP, and this

Table 1.   Major classes of signaling molecules and receptors with important roles in development

Signal	Receptor	Example
Secreted signals		
Various	G-protein coupled receptor	Most classic GPCRs are involved in physiological functions rather than specific roles in development. *Drosophila* Bride-of-sevenless and Frizzled are atypical receptors with homology to GPCRs.
Growth factor family	Receptor tyrosine kinase	Vertebrates: FGFs in limb development and neurulation. EGF homologs in *Drosophila* axis specification. *C. elegans*: LIN-3 in vulval specification
Cytokine family	Receptors with associated Janus kinases	Mammals: erythropoietin, colony stimulating factors and interleukins in the differentiation of blood cells
TGF-β superfamily (includes TGF-β, activin, BMP and nodal-related proteins)	Receptors with associated serine/threonine kinase activity	*Xenopus*: Xnr1, Xnr2, Xnr4, Vg-1 and activin in mesoderm induction. Decapentaplegic (*Drosophila*) and BMPs (vertebrates) in dorsoventral axis specification and limb development. Mammals: Nodal in left-right axis specification
Hedgehog family	Patched	*Drosophila*: Hedgehog in segment polarity, and patterning of legs and wings. Vertebrates: Sonic hedgehog in patterning the DV neuraxis and limbs
Wnt family	Frizzled	*Drosophila*: Wingless in compartment boundaries, segment polarity, and patterning of imaginal discs. Vertebrates: Wnt family in dorsoventral axis specification and patterning of neural tube and limbs
Ephrin family	Receptor tyrosine kinase	Vertebrates: compartment formation along the anteroposterior neuraxis
Retinoids	Nuclear receptor	Vertebrates: anteroposterior axis patterning
Immobilized signals		
Delta/Serrate	Notch	*Drosophila*: neural cell specification
Extracellular matrix glycoproteins	Integrins	Perception of extracellular environment

results in the dissociation of the α-subunit from the β- and γ-subunits. Depending on the type of G-protein associated with the receptor, the α-subunit can stimulate or inhibit adenylate cyclase therefore modulating cAMP levels in the cell, or it can induce phospholipase C, increasing the levels of lipid second messengers (see below). Other α-subunits can modulate ion levels by interacting with ion channels. The βγ-dimer may also mediate downstream responses. Classical GPCRs are generally involved in physiological responses rather than specific developmental functions, but there are a number of atypical seven-pass transmembrane receptors with important roles in development, including the Frizzled family, which act as receptors for Wingless/Wnt secreted proteins.

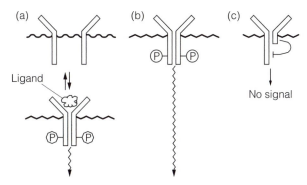

Fig. 1.   Activity of multimeric receptors. (a) The activity of wild type receptors is ligand-dependent. (b) Constitutive receptors are always active and generate gain of function phenotypes. (c) Dominant negative monomers sequester wild type monomers into inactive multimers and generate loss of function phenotypes.

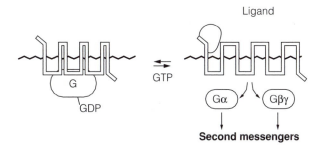

Fig. 2.   G-protein coupled receptors are associated with a trimeric G protein. Ligand binding causes GTP to replace GDP and the G protein to dissociate into α and βγ subunits.

Growth factor signaling – Ras, Raf and MAP kinase

Growth factor receptors are **receptor tyrosine kinases**, i.e. the receptors have intrinsic, ligand-activated tyrosine kinase activity. Ligand binding stimulates receptor dimerization and **autotransphosphorylation** (the phosphorylation of each receptor monomer by the kinase activity of the other). The signal can then be propagated by the phosphorylation of downstream target proteins, but it is first necessary to recruit these target proteins to the membrane. This is possible because many signaling proteins can bind specifically to phosphorylated tyrosine residues. Such proteins usually contain either an **SH2** or **PTB phosphotyrosine-binding domain**, and may themselves function as signaling proteins, or they may act as adaptors to recruit other proteins to the membrane-associated signaling complex.

Receptor tyrosine kinases can propagate the intracellular signal in a number of ways, one of which is through the small nucleotide-binding protein **Ras**. This is a membrane associated protein that cycles between active (GTP-associated) and inactive (GDP-associated) states, a process regulated by **GTPase activating proteins (GAPs)** that stimulate the intrinsic GTPase activity of Ras (thus inactivating Ras) and **guanine nucleotide exchange factors (GEFs)** which replace GDP with GTP (thus activating Ras). Both GAPs and GEFs can interact directly or indirectly with phosphotyrosine residues, thus signal transduction can either stimulate or inhibit Ras activity depending on the abundance of the active form of each enzyme.

Activated Ras interacts with a number of downstream proteins, but the best-characterized is **Raf**, a cytoplasmic protein that is activated by recruitment to the membrane. Activated Raf is a kinase, which phosphorylates another kinase called **MEK**. MEK phosphorylates and activates another kinase, **MAP kinase**, which can also activate a number of downstream transcription factors and other target proteins (*Fig. 3*).

The Ras-Raf-MAP kinase pathway can be activated by different receptor tyrosine kinases in different cells and can therefore be used in different contexts within the same organism. For example, a *Drosophila* MAP kinase, **Rolled**, can be activated by the receptor Torso as part of the terminal specification system in early development (Topic H2), by the receptor Torpedo in the regionalization of follicle cells during oogenesis (Topic H2) and by the receptor Sevenless in the specification of R7 photoreceptor cells in the eye (Topic L3). It is not entirely clear how the specificity of responses to different ligands and receptors is achieved, although this may involve the presence of different cell-specific coactivators and inhibitors of the pathway, the presence of different cell-specific transcription factors, or differences in the intensity or duration of the signal. There are also different genes encoding many of the signaling components. In *Drosophila*, a second MAP kinase called **Jun N-terminal kinase (JNK)** is required for dorsal closure and the immune response. JNK is activated by Rho, a protein involved in changes in the cytoskeleton and cell motility (e.g. see Topics F2 and J6). In vertebrates, there are several different genes encoding each component of the pathway, and the proteins have different substrate affinities and expression patterns. For this reason, the response to growth factor signaling can be diverse.

Cytokine signaling – the JAK-STAT pathway

Cytokine receptors are dimeric or oligomeric, with each polypeptide spanning the membrane once. The receptors possess no intrinsic tyrosine kinase activity, but they are constitutively associated with cytoplasmic tyrosine kinases of the Janus family (**Janus kinases, JAKs**). Ligand binding causes the receptors to dimerize. This results in reciprocal autotransphosphorylation of the associated JAKs, which become active and phosphorylate the receptor itself. The receptor is then able to recruit inactive transcription factors called **STATs (signal transducers and activators of transcription)** that bind to phosphotyrosine residues through their SH2 domains. The STATs are then phosphorylated by the JAKs, allowing them to dimerize and translocate to the nucleus, where they activate downstream genes (*Fig. 4*). In mammals, the **JAK-STAT pathway** is used extensively during the differentiation of blood cells, and is also activated by prolactin during milk production to stimulate transcription of the caesin gene.

TGF-β signaling – SMADs

The TGF-β superfamily of signaling molecules comprises several different families, e.g. the prototype TGF-β family, the activins, the bone morphogenetic proteins, nodal-related proteins and hormones such as Müllerian inhibiting substance. Receptors for the TGF-β family of signaling proteins are multimeric and have intrinsic serine/threonine kinase activity, but the mechanism of ligand binding and activation may vary. For example, TGF-β itself initially binds to a class II receptor, which has constitutive kinase activity. Binding recruits a class I receptor, which is phosphorylated by the class II receptor and activated. Conversely, most BMPs bind directly to class I receptors, which are activated by oligomerization and autotransphosphorylation. Downstream targets include inhibitors of the cell cycle (see Topic C4) and proteins of the **SMAD family**.

SMAD proteins are phosphorylated by activated type I receptors, form heterodimers and translocate to the nucleus. In the nucleus, SMAD proteins may act directly as transcription factors, or they may influence the activities of other transcription factors (*Fig. 5*).

Hedgehog, Wingless (Wnt) and Decapentaplegic (BMP)

Hedgehog and Wnt (Wingless) signaling proteins are often secreted by adjacent groups of cells at compartmental boundaries. This is because the two proteins establish a reciprocal signaling system in which they become mutually dependent, and this helps to stabilize developmental compartments. For example, the dependence of *hedgehog* expression on Wingless signaling and of *wingless* expression on Hedgehog signaling helps to stabilize and reinforce the segmental boundaries in the *Drosophila* embryo (Topic H4). Both proteins also act as morphogens, as does an important downstream target of Hedgehog signaling, Decapentaplegic. Hedgehog, Wingless and Decapentaplegic are therefore used in many different contexts in *Drosophila* development such as in dorsoventral patterning (Topic H6), segmentation and segment polarity (Topic H5), limb development (Topic L2) and eye development (Topic L3). In vertebrates, each of the three morphogens has several counterparts, comprising the Hedgehog, Wnt and BMP families of secreted proteins. These are employed in complex interdependent pathways in many of the processes discussed in this book, including establishment of the organizer (Topic I2) dorsalization of the mesoderm and ectoderm (Topic J1, Topic K1), patterning the anteroposterior and dorsoventral neuraxes (Topics J2 and J3), development of the neural crest (Topic J6), kidney (Topic K4), heart (Topic K5) and gut (Topic K6), and limb development (Section M). The Hedgehog-Wingless signaling pathway in *Drosophila* is shown in *Fig. 4*, Topic H4.

Steroid receptors

Steroid-related ligands consist of the steroid and thyroid hormones, and vitamins A and D and their derivatives (including retinoic acid). These molecules can diffuse through the cell membrane, so they have intracellular receptors, some located in the cytoplasm, others in the nucleus. The receptors are dormant transcription factors. Ligand binding causes a conformational change that allows the receptors to dimerize, translocate to the nucleus (if appropriate) and bind to DNA. The steroid receptors usually act as homodimers and bind palindromic recognition sites, while the thyroid hormone, vitamin D and retinoic acid receptors act as heterodimers with a common partner (in mammals, the **retinoid X receptor**) and bind to direct repeats in the DNA. Retinoic acid is a potent teratogen and has wide ranging effects on pattern formation in early embryos (e.g. see Topics K2, M2, M3). In *Drosophila*, the steroid hormone **ecdysone** regulates molting. The ecdysone receptor is a heterodimer of the products or the *ecdysone receptor* (*ecr*) and *ultraspiracle* (*usp*) gene products.

Second messengers

Signal transduction may elicit a response from **second messengers** in the cell, which allows the complex signals arriving at the cell surface to be converted into simple biochemical events inside the cell. There are three major types of second messenger: **cyclic nucleotides** (e.g. cAMP), the lipids **inositol-1,4,5-trisphosphate (IP$_3$)** and **diacylglycerol (DAG),** and **calcium ions**. The intracellular levels of second messengers can have a profound effect on cellular physiology, which can be very important in certain aspects of development as discussed below.

The levels of cAMP in the cell are controlled by **adenylate cyclase**, which can be stimulated or inhibited by signaling through G-protein coupled receptors. G-protein coupled receptors may also activate certain isoforms of **phospholipase C**, the enzyme that splits phosphatidyl inositiol 4,5-bisphosphate (PIP$_2$) into IP$_3$ and DAG. A different isoform of phospholipase C is stimulated by receptor tyrosine kinases because it contains an SH2 domain (*Fig. 6*). One of the functions of IP$_3$ is to release Ca^{2+} from the endoplasmic reticulum. Calcium release is required for a number of processes, e.g. the acrosome reaction in sperm and the corticle granule reaction in the egg (Topic E4). The cell also contains many proteins that are dependent on second messengers for their activity. **Protein kinase A (PKA)** is stimulated by cAMP; this enzyme has important roles in *Drosophila* oogenesis, axonal outgrowth in mammals and differentiation in *Dictyostelium* (Topic D3). **Protein kinase C (PKC)** is stimulated by calcium ions and DAG; this enzyme is responsible for the pH changes that occur after fertilization, and also has roles in compaction and neurulation in vertebrates. **Calmodulin** is a calcium-binding protein that activates a number of downstream proteins, including **CaM kinase II**, which regulates the meiotic cell cycle.

Networks and cross-talk

It is important that signaling pathways can be switched off as well as on, otherwise the pathways would become saturated. The duration of a signal depends only in part upon how long a ligand binds to its receptor on the outside of the cell. For every active component of a signaling pathway there is also an inhibitor, and the balance of these activating and inhibiting forces determines the outcome of signaling. Many signaling pathways involve kinase cascades, but for every kinase there is a corresponding phosphatase. When ligand binding

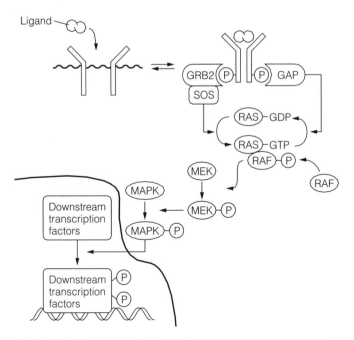

Fig. 3.   Growth factor signaling and the Ras-Raf-MAP kinase pathway.

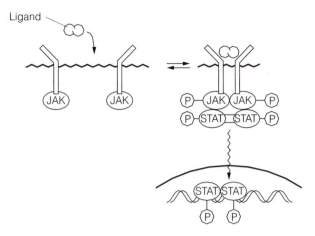

Fig. 4.   The cytokine receptor and JAK-STAT signaling pathway.

activates a cascade of kinase activity, the corresponding phosphatases are also activated, with a slight delay, so that there is a pulse of information transfer followed by shut-down.

Furthermore, the signaling pathways discussed above do not operate in isolation, but are interconnected at many levels. This is a useful strategy as cells are bombarded with many different signals, some of which will reinforce each other, and some of which will produce opposite effects. Cross-talk between pathways allows cross-regulation, so the response of a cell to its environment may be the net response to all incoming signals.

Signaling pathways are therefore controlled by a complex system of feedback and cross regulation that reflects the balance of opposing forces in the cell. The arrival of a signal at the cell surface causes a disruption to the equilibrium, allowing a brief pulse of information transfer, before the equilibrium is restored.

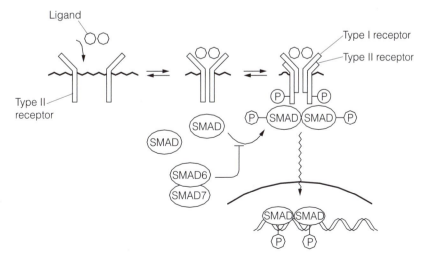

Fig. 5.   Signaling by TGF-β involves the recruitment of an oligomeric receptor complex and the activation of SMAD proteins that can regulate gene expression.

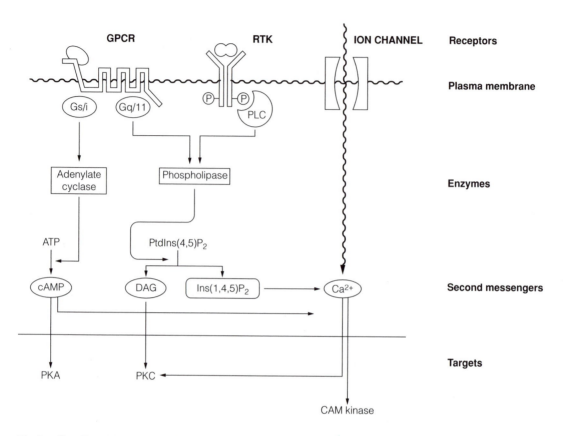

Fig. 6.   Simplified representation of pathways leading to the activation of second messengers in the cell, and some of their important targets.

In this way, cells can compensate for changes in their environment and maintain homeostasis. In development, signaling is used to achieve change. Inductive interactions produce a definitive response in the responding cell, such as differentiation. Such changes are possible because inductive signals are strong and unambiguous: the changing patterns of gene expression elicited by an inductive signal allow a new equilibrium to be established.

C4 THE CELL DIVISION CYCLE

Key Notes

The cell division cycle

The typical cell division cycle comprises S-phase (DNA replication) and M-phase (mitosis), separated by two gap phases (G_1 and G_2). However, development involves a number of variant cell cycles that fulfil specific purposes, such as the rapid divisions without growth that occur during cleavage. Many important developmental genes encode cell cycle regulators.

Cell-cycle control

The cell cycle is organized as a dependent series of events, so that one stage must be complete for the next to start. The cell cycle is controlled by dimeric complexes of cyclins and cyclin-dependent kinases (CDKs) that facilitate transitions between successive cell cycle stages. Upstream regulators of the cyclins and CDKs receive feedback from the cell cycle itself (so that DNA replication does not commence until mitosis is complete and vice versa) and from signaling pathways (e.g. responding to extrinsic signals such as growth factors) to co-ordinate the cell cycle with the developmental program. These regulators are called cell cycle checkpoints.

Maternal control of the zygotic cell cycle

In *Drosophila*, 13 rounds of rapid synchronous nuclear divisions occur prior to cellularization. Then, cell division becomes asynchronous, and some cells cease to proliferate and begin to differentiate. These events reflect the depletion of maternal cell cycle components stored in the egg, and the synthesis of zygotic cell cycle regulators.

Cell cycle regulation by growth factors

Animal cells proliferate in response to growth factors. In their absence, the cells withdraw from the cell cycle and enter a quiescent state known as G_0. Growth factors work by stimulating the production of an unstable cyclin, cyclin D, which is required in early G_1 to facilitate the G_1 to S-phase transition. Once S-phase is initiated, the cell will undergo a complete round of division and will arrest in the next G_1 phase until cyclin D is synthesized again.

Control of cytokinesis

Cell division involves the formation of a contractile ring of actin filaments at the mid-point of the mitotic spindle; constriction of the ring generates a cleavage furrow that bisects the cell. The various symmetrical and asymmetrical cleavage patterns that occur during animal development thus reflect the controlled localization of the centrioles that give rise to the mitotic spindle. Experiments have shown that the microtubule asters, but not the spindle apparatus itself, are necessary for the formation of a cleavage furrow.

Apoptosis

Apoptosis is programmed cell death. Unlike cell death resulting from injury or infection (necrosis), apoptosis is under genetic control; it occurs in isolated cells and does not cause inflammation. In many systems, apoptosis is an important part of the developmental program. A family of proteins that regulates cell death has been identified in *C. elegans* and vertebrates. These proteins activate downstream proteases that control nuclear condensation and other apoptotic processes.

Related topics

Signal transduction in   Cleavage (F1)
   development (C3)

The cell division cycle

The **cell division cycle** is the series of events between successive cell divisions. In eukaryotes, the 'typical' cell cycle comprises four non-overlapping phases: G_1, S, M and G_2 (*Fig. 1*). **S-phase** represents **DNA synthesis**, where the DNA is replicated. **M-phase** represents **mitosis**, where the replicated DNA is equally divided between two daughter nuclei; this is closely followed by **cytokinesis** (cell division). These discrete events or landmarks are separated by two **gap phases**, G_1 and G_2, during which there is no overt change to the DNA content of the cell, although mRNAs and proteins continue to be synthesized, and the cell continues to grow. To a casual observer, only M-phase involves any striking visible changes to the cell's appearance. The remainder of the cell cycle (G_1, S and G_2) is therefore grouped together as **interphase**.

The eukaryotic cell cycle serves two functions. Firstly, it duplicates the genome precisely and divides it between two daughter nuclei. Secondly, it coordinates genome duplication and mitosis with cell growth and division. In a plate of cultured cells, this ensures there is no progressive loss or gain in cytoplasm through successive divisions. In a living organism, the cell cycle is regulated by the developmental program, so that a variety of different outcomes is possible. During cleavage, for example, there is rapid cycling between S- and M-phases so that division occurs in the absence of growth and cells get progressively smaller. Later in development, morphogenesis occurs through differential rates of cell proliferation in different parts of the embryo, and this may be regulated intrinsically (by maternal gene products) or through dependence on growth factors and other extrinsic signals. Later still, certain cells withdraw from the cell cycle as part of their program of differentiation, e.g. muscles and neurons. Many genes with fundamental roles in development are therefore regulators of the cell cycle.

Cell-cycle control

To achieve the elementary function of the cell cycle (precise genome duplication and division), cell-cycle events must proceed in the correct order and each stage must be completed before the next commences. This is achieved by organizing

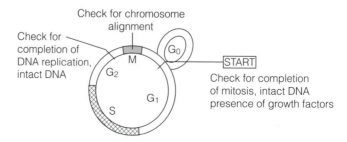

Fig. 1.   The eukaryotic cell cycle showing stages and principle checkpoints.

the cell cycle as a *dependent series of events* that relies entirely on information intrinsic to the nucleus: DNA replication is dependent on the successful completion of mitosis because signals produced during mitosis inhibit DNA replication, and mitosis is dependent on the successful completion of DNA replication because signals produced during DNA replication inhibit mitosis. Such signals are channeled into regulatory systems termed **cell cycle checkpoints** that determine whether the cell cycle can proceed to the next stage (a **cell-cycle transition**). If the conditions for a transition are not met, the checkpoint blocks progress to the next stage, resulting in **cell-cycle arrest**. External signals that regulate the cell cycle feed into the same system of checkpoints (*Fig. 1*).

The basis of cell-cycle regulation is a family of protein kinases, termed **cyclin-dependent kinases (CDKs)**. These facilitate cell cycle transitions by phosphorylating downstream target proteins, resulting in the transcription of genes and the activation of proteins that mediate cell-cycle events. The kinases work in concert with another class of proteins, termed **cyclins** because their availability oscillates with the cell cycle. The cyclins and cyclin-dependent kinases are themselves regulated by other proteins that represent the cell-cycle checkpoints and respond to intrinsic and external signals, ensuring that cell-cycle transitions are not triggered prematurely or in inappropriate circumstances. Two examples of cell-cycle regulation during development are discussed below.

Maternal control of the zygotic cell cycle

In the *Drosophila* embryo, the first 13 nuclear divisions occur rapidly and synchronously in a syncytium, with no obvious gap phases. This is because all components of the basic cell-cycle machinery are provided in the egg as maternal gene products. The 14th cell cycle, however, incorporates a G_2 phase and coincides with cellularization of the blastoderm. Thereafter, cells in different parts of the embryo begin to divide at different rates. These events occur because maternal stocks of a key regulator of the cell cycle, **String**, become depleted. The maternal String protein is distributed evenly throughout the egg, so that the initial nuclear divisions are synchronous. However, by cycle 14, the maternal String protein becomes limiting, and the cycle lengthens by incorporating a G_2 phase to allow expression of the zygotic *string* gene. When the zygotic *string* gene is activated, it is expressed at different levels in different parts of the embryo, under the control of the anteroposterior and dorsoventral patterning genes described in Section H. This is responsible for the sudden loss of synchronization. A few cycles later, another maternal gene product (cyclin E) becomes limiting in epithelial cells. This protein is required for the G_1–S phase transition, so the cells arrest in G_1 and begin to differentiate.

Cell-cycle regulation by growth factors

In mammals, entry into S-phase requires the presence of growth factors. The regulated secretion of growth factors in the embryo therefore provides a mechanism for selective cell proliferation in specific regions of the embryo, such as the limb bud. Growth factors reaching the cell in early G_1 stimulate the synthesis of **cyclin D,** which associates with different CDKs, including **CDK4** and **CDK6**. These phosphorylate a protein called **pRb**, the product of the **retinoblastoma susceptibility gene**. The function of pRb is to block the cell cycle by forming repressive heterodimers with transcription factors of the **E2F family**, which normally activate genes encoding S-phase proteins. The phosphorylation of pRb by the CDK/cyclin D complexes causes it to be released from the heterodimer, allowing the E2F transcription factor to act as a transcriptional activator of S-phase genes. Once S-phase commences, the cell does not need any further external signals and will proceed through DNA replication and mitosis. The cell arrests at the next G_1 phase if growth factors are not present.

If growth factors are withdrawn from the cell during the critical G_1 period, the cell exits from the cell cycle, entering a quiescent state termed G_0. This is because the D cyclins are very unstable, and break down soon after synthesis. A constant supply of growth factor is therefore required to maintain cell division. Growth factors stimulate the synthesis of cyclin D e.g. through activation of the MAP kinase signaling pathway (see Topic C3). The targets of MAP kinase include transcriptional regulators that process mitogenic signals by activating so-called immediate early genes, some of which also encode transcriptional regulators that may activate transcription of the *cyclinD* gene. Some cells, such as neurons, exit from the cell cycle all together because cyclin D is never synthesized.

Control of cytokinesis

In animal cells, **cytokinesis (cell division)** involves the formation of a **contractile ring** of actomyosin microfilaments, which slowly draws in to generate a **cleavage furrow**, and eventually pinches the cell in two. The position and plane of cell division is stereotyped in the early development of many animals and contributes to axis specification and early patterning of the embryo (Topic F1). Later in development, the plane of cell division contributes to tissue structure. For example, cell division in the plane of an epithelial sheet will cause the sheet to expand, while division perpendicular to the sheet will increase the number of layers.

The position and plane of cleavage is determined by the positions of the asters of the mitotic spindle. Different planes of cleavage occur in different animal embryos because the position of the centrioles is controlled by cytoplasmic factors probably acting in concert with the cytoskeleton. This can be demonstrated by the disruption to normal cleavage patterns that occurs through reorganization of the egg cytoplasm e.g. by compression or centrifugation, and the abnormal cleavage furrows that form in eggs carrying multiple sperm nuclei (and therefore multiple centrioles). An elegant series experiments to demonstrate the role of the microtubule asters was carried out in 1960. A glass bead was introduced into sea urchin eggs, displacing the mitotic spindle to the periphery of the cell. The cleavage furrow of the first division began to form normally, but only reached as far as the bead, producing a horseshoe-shaped cell with two nuclei. At the second division, mitotic spindles formed at right angles to the first plane of cleavage, but cleavage furrows appeared not only around the mitotic spindle apparatus, but also in the region above the bead separating the two adjacent asters (*Fig. 2*). The extra cleavage furrow showed that asters are required for cell division, but not necessarily the mitotic spindle apparatus itself.

Apoptosis

Apoptosis is programmed cell death, a physiological process under genetic control that occurs in response to a number of intrinsic and external stimuli, and is also an important part of the developmental program. Unlike **necrosis** (cell death caused by trauma or infection), apoptosis does not involve cell lysis and does not usually result in inflammation. Apoptosis is characterized by cell shrinkage and nuclear condensation without the loss of membrane integrity. The cell breaks up into membrane-bound **apoptotic bodies** that are, in animals, engulfed by macrophages.

Apoptosis plays a major role in a number of developmental systems,

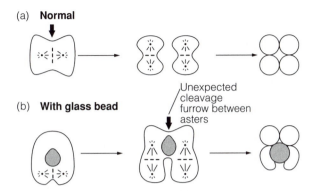

Fig. 2.   Role of asters in cell division. Cleavage furrows normally form between the asters of the mitotic spindle. If the spindle apparatus is displaced with a glass bead, cleavage furrows still form between the asters even if there is no spindle between them.

including the vertebrate nervous system (where about 50% of emerging neurons are programmed to die; Topic J7) and the limb bud, where apoptosis in the interdigital regions separates the digits (Topic M2). In *C. elegans*, apoptosis is written into the invariant cell lineage, so that precisely 131 somatic cells are eliminated from the embryo. Several genes that control cell death have been identified in *C. elegans*. The *ced-3* and *ced-4* genes are expressed in all cells that undergo apoptosis. In *ced-3* and *ced-4* mutants, cells that normally undergo apoptosis survive and differentiate, generating an embryo with an abnormally large cell number. In wild type embryos, surviving cells express *ced-9*, which inhibits the expression of *ced-3* and *ced-4*. Loss-of-function *ced-9* mutants show extensive cell death, while in gain-of-function mutants show the opposite effect.

In mammals, apoptosis is regulated by the **Bcl-2 protein family**. Bcl-2 itself is a **survival factor** (a protein that inhibits cell death) and is homologous to the *C. elegans* CED-9 protein. Other members of the family can also act as survival factors (e.g. Bcl-X$_1$) or can promote cell death (e.g. Bax, Bad and Bak). These proteins can form combinatorial heterodimers. Bcl-2 inhibits Apaf-1, the mammalian homolog of CED-4, which in turn activates a central component of the apoptotic pathway called **caspase**. Caspase is initially synthesized as an inactive precursor, but activated caspase is a protease that modifies a number of downstream proteins to directly mediate apoptotic cell behavior. Caspase activity is stimulated by Apaf-1/CED-4 but inhibited by the **IAP family** of proteins (Inhibitors of APoptosis). In *Drosophila*, a second developmental pathway to apoptosis has been identified, which involves the inhibition of the IAP proteins. A number of cell death activators have been identified in this pathway, such as the proteins Grim and Reaper.

C5 THE CYTOSKELETON, CELL ADHESION AND THE EXTRACELLULAR MATRIX

Key Notes

Basis of morphogenesis

Many aspects of morphogenesis are dependent on the cytoskeleton and cell adhesion. There is a close interaction between the cytoskeleton and transmembrane proteins that mediate adhesive interactions with other cells and the extracellular matrix.

The cytoskeleton

The cytoskeleton is a proteinaceous array of filaments that provides stability, an intracellular transport network, and generates contractile forces. There are three major polymers: microfilaments, microtubules and intermediate filaments.

Microfilaments

Microfilaments are polarized actin polymers. They may be arranged in bundles and lattices to provide strength and support, and may associate with myosin to generate contractile fibers that facilitate changes in cell shape and cell movement, as well as providing the contractile ring of the cleavage furrow in cell division.

Microtubules

Microtubules are polarized structures comprising a circular arrangement of polymerized tubulin protofilaments. Microtubules have many important structural roles in the cell, e.g. in the maintenance of axons, microvilli, cilia and flagella. The mitotic spindle apparatus is made of microtubules, and microtubules also facilitate intracellular transport using the motor proteins kinesin and dynein.

Intermediate filaments

Intermediate filaments fulfil a mainly structural role in the cell. There are a number of different types, which are expressed in a cell-specific manner, e.g. the neurofilaments, and the epidermal-specific protein keratin.

Cell adhesion molecules

Cell adhesion molecules can stick cells together and to the extracellular matrix. In this way, they determine the structure and composition of tissues and maintain tissue boundaries. There are three major types of cell adhesion molecule: cadherins and immunoglobulin-type CAMs, which are involved in cell–cell adhesion, and integrins, which bind cells to the extracellular matrix.

Cadherins

Cadherins are calcium-dependent cell adhesion molecules that bind to other cadherins of the same type. The major role of the cadherins is therefore to stick cells of the same type together and maintain boundaries between dissimilar cells.

Immunoglobulin-type CAMs	CAMs of the immunoglobulin superfamily are calcium-independent adhesion molecules. They usually bind to similar adhesion molecules on other cells, but some may interact with integrins.
The extracellular matrix	The extracellular matrix is a network of macromolecules secreted by cells into their immediate surroundings. It comprises collagen, proteoglycans and proteins such as fibronectin and laminin that interact with integrins displayed on the cell surface. The relative amounts of the different matrix components determine its properties, allowing it to form tight basal laminae, loose reticular networks and tissues such as cartilage and bone.
Integrins and glycosyltransferases	Integrins are dimeric transmembrane receptors that bind extracellular matrix proteins such as laminin and fibronectin. There are many types of integrins with differing affinities for matrix proteins. The cytoplasmic domain of integrins is associated with the microfilament cytoskeleton, allowing information about the matrix environment to be interpreted by the cell. Glycosyltransferases bind to carbohydrate groups on matrix proteins and may play an important role in cell migration.
Related topics	Morphogenesis (A7) Gastrulation in vertebrate embryos (F3) Cleavage (F1) Gastrulation in invertebrate embryos (F2) Shaping the neural tube (J4)

Basis of morphogenesis

As discussed in Topic A7, **morphogenesis** (the creation of structures in the developing embryo) is dependent on many different forms of cell behavior. Some of these behavioral processes, such as controlling the rate of cell division and the orientation of the mitotic spindle apparatus, are discussed in more detail in Topic C4. This Topic describes the cytoskeleton (which governs cell shape and helps to facilitate cell migration) and cell adhesion molecules (which bind cells to each other, and to the extracellular matrix, and hence determine the structure of cell layers and tissues).

The cytoskeleton

The **cytoskeleton** is a framework of polymeric protein filaments inside the cell that provides stability, generates forces that allow changes in cell shape and cell movement, facilitates the intracellular transport of organelles, and allows communication with the extracellular matrix. The cytoskeleton comprises three major types of polymer: microfilaments, microtubules and intermediate filaments (*Fig. 1*). Drugs such as **cytochalasin D** (which prevent actin polymerization) and **colchicine** (which block the assembly of microtubules) can demonstrate the importance of the cytoskeleton in development.

Microfilaments

Microfilaments are polymers of **actin**. They are polar filaments with a diameter of 7 nm, which can be organized into parallel **bundles** to provide great strength, and into lattice-like **networks** to provide support and stability. **Actin-binding proteins** hold individual filaments together, and the organization of the filaments into bundles and lattices reflects the flexibility of these binding proteins. Other proteins interact with actin filaments to control polymerization and depolymerization, enabling the cell to assemble and disassemble filaments rapidly in different locations, a process essential for changes in cell shape and

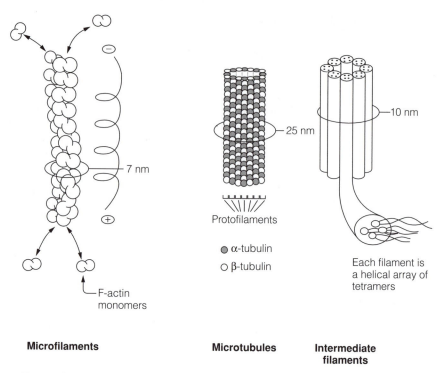

Fig. 1.   Components of the cytoskeleton.

motility. Actin filaments associate with another protein, **myosin**, which enables
bundles of actin fibers to generate contractile forces. Such **contractile fibers** are
used to generate changes in cell shape. For example, contraction of the microfilament network at one end of a cell causes **apical constriction**, which drives the
invagination of epithelial sheets and contributes to the forces that close the
neural tube (see Topics F1 and J4). Contractile fibers are also required for cell
movement. The assembly of actomyosin bundles pointing out towards the cell
membrane allows the extension of fine cytoplasmic processes termed **filopodia**.
These structures become anchored to the substratum at **focal points**, where
actin bundles associate with integrins that are bound to the extracellular matrix
(see below). Once filopodia are anchored at the leading edge of cell movement,
the release of contacts at the trailing edge together with contraction of the microfilament network can draw the cell forwards. Highly organized parallel arrays
of actomyosin are present in mature muscle cells, and confer their extraordinary
contractile ability. Actomyosin fibers are also assembled at the plane of cell
division to form the contractile ring that constricts the cell during cytokinesis.

Microtubules   **Microtubules** are polymers of α- and β-tubulin. They are polar structures
consisting of 13 **protofilaments** arranged in a ring to generate a hollow tubule
about 25 nm in diameter. Microtubules have many important roles in the cell,
including the structural support of cytoplasmic processes such as microvilli and
axons, intracellular transport, the formation of structures such as cilia and
flagella, and construction of the mitotic spindle apparatus. Most microtubules
have one end attached to a **microtubule organizing center (MTOC)** which in

mitosis is called a **centriole**. Microfilaments are dynamically unstable, i.e. they are constantly growing and shrinking, and their net length depends on the balance of polymerization and depolymerization activities. They are associated with a number of proteins termed **MAPs (microtubule associated proteins)** that control their assembly and organization.

The importance of microtubules in cell division, cell shape and intracellular transport, reflects their intrinsic polarity. Two classes of motor proteins, **kinesin** and **dynein**, are able to move along microtubules in opposite directions. This facilitates chromosome separation during mitosis, unidirectional vesicular transport and directional beating of cilia. The last process has recently been shown to play an important role in left–right axis specification in vertebrates, as discussed in Topic I3.

Intermediate filaments

Intermediate filaments have a predominantly structural role in the cell, as they have no polarity and there are no known motor proteins like myosin, kinesin and dynein associated with them. All intermediate filaments are polymers of about 10 nm in diameter, but there are numerous different kinds, which are expressed in a tissue-specific manner. Examples include vimentin, three different classes of neurofilaments, and keratin. Intermediate filament proteins are diverse in structure, but all possess a conserved α-helical region that dimerizes to form a coiled coil, as well as unique globular N-terminal and C-terminal domains. Dimers assemble to form antiparallel tetramers that join end to end by interactions between the globular domains. Intermediate filaments are associated with a number of proteins in the cell, termed **intermediate filament-associated proteins (IFAPs)** which organize their structure and in some cases attach them to microtubule arrays. The developmental roles of intermediate filament proteins are limited to their tissue specific effects in differentiation.

Cell adhesion molecules

Cell adhesion molecules are responsible for sticking cells to each other and to the extracellular matrix. In this manner, adhesion molecules dictate which cells can associate and which cannot, and therefore help to organize cells into tissues and maintain tissue boundaries. Changing the patterns of cell adhesion molecules allows the reorganization of cells to form new tissues, and the migration of

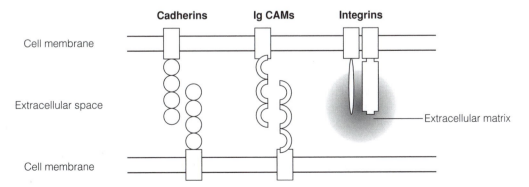

Fig. 2. Cell adhesion molecules.

cells from one part of the embryo to another (see Topic A7). There are two major classes of adhesion molecule involved in cell–cell interactions, the calcium-dependent cell adhesion molecules (**cadherins**) and the calcium-independent cell adhesion molecules (**CAMs**). Another class of adhesion molecules, the **integrins**, is responsible for binding cells to the extracellular matrix (*Fig. 2*). All three classes of adhesion molecules can interact, directly or indirectly, with the cytoskeleton. This means that the cell can relay information about the composition of the extracellular matrix, allowing it to respond by moving, or changing its shape, or changing its pattern of gene expression.

Cadherins

Cadherins are calcium-dependent transmembrane adhesion molecules that predominantly recognize and bind to other cadherin molecules of the same type (**homophilic binding**). A major role of the cadherins is therefore to stick cells of the same type together in tissues, and the expression of different cadherins is primarily responsible for the spontaneous cell sorting that occurs in mixtures of disaggregated cells, and in normal developmental processes that involve tissue separation (Topic A7). Cadherin–cadherin binding can be disrupted with **chelating agents**, such as EDTA, that remove calcium from solution. The intracellular domain of the cadherins associates with β-catenin, providing a link to the actin microfilaments of the cytoskeleton. These cadherin–catenin complexes form the **adherens** junctions that join epithelial cells together.

Approximately 30 different cadherins have been identified in vertebrate cells. Well characterized examples include **E-cadherin** (expressed early in development and later restricted to epithelial tissues, also known as **uvomorulin** or **L-CAM**), **N-cadherin** (expressed in differentiated mesodermal cells but predominantly in the central nervous system) and **P-cadherin** (expressed in the placenta, and possibly involved in adhesion between the trophoblast of the embryo and the uterine wall).

Immunoglobulin-type CAMs

Calcium-independent cell adhesion is mediated by cell adhesion molecules of the immunoglobulin superfamily. Such molecules often recognize similar molecules on other cells in the manner of cadherins, although some CAMs bind to integrins (see below). The first such molecule to be discovered, **N-CAM (neural cell adhesion molecule)**, plays an important role in the development of the nervous system, as it helps axons attach to muscle cells. However, N-CAM is also expressed on other cell types, including mesoderm cells, and this is thought to help keep them separate from epidermal cells, which express E-cadherin. For example, the role of N-CAM in somitogenesis is discussed in Topic K2.

The extracellular matrix

The **extracellular matrix** is a network of molecules secreted by cells into their local environment. Depending on the nature of these molecules, the matrix may take several forms, e.g. a tight layer of macromolecules termed a **basal lamina** or a loose network termed an **extracellular reticulum**. The matrix may form an insoluble barrier between tissues, or a loose network in which cells are scattered. It may form a substratum to which cells can bind and form a sheet, or over which cells can migrate. In the case of cartilage and bone, the extracellular matrix may be the major determinant of tissue structure.

The matrix comprises three types of molecules: **collagens**, **proteoglycans** and **integrin-binding proteins** such as laminin and fibronectin, which facilitate cell-matrix adhesion. The properties of the matrix depend on the nature and relative amounts of these molecules. For example, the loose reticular matrix of mesenchymal cells comprises a mixture of collagens that form a network of fibrils, while the basal laminae of epithelial cells contain predominantly collagen type IV arranged in a fine mesh. The softness of cartilage reflects predominance of proteoglycans, whereas the toughness of tendons reflects the predominance of collagen fibers. The organization of collagen and proteoglycan molecules is mediated by other proteins, including **fibronectin, laminin** and **tenascin**. By interacting with different integrins on the cell surface, these molecules promote adhesion in some cells, and detachment in others. For example, tenascin can stimulate the secretion of proteases by some cells, resulting in the cleavage of adhesion proteins and release of the cell from the matrix. It is possible that gradients of adhesiveness may be generated by the relative proportions of these molecules in the matrix. The extracellular matrix may also play an important role in pattern formation, since proteoglycans are implicated in the carrier-assisted transport of secreted morphogens, such as Hedgehog, Wingless and Decapentaplegic (Topic A6).

Integrins and glycosyltrans-ferases

Integrins are membrane-spanning cell adhesion molecules that mediate contacts with the extracellular matrix by binding to proteins such as fibronectin, laminin and tenascin. Integrins are dimeric proteins and there are many different receptor monomers in vertebrates, allowing the formation of many combinatorial heterodimers that show different affinities for different components of the extracellular matrix. Like cadherins, integrins are also associated with the microfilament cytoskeleton, in this case because their intracellular domains bind to actin-binding proteins such as **talin** and **α-actinin**. This is one way in which integrins pass information about the extracellular environment into the cell. However, integrins have also been shown to bind to growth factor receptors on other cells. This allows them to initiate the same signaling cascades elicited by the normal ligands (e.g. fibroblast growth factor), and hence to play a significant role in the control of cell growth and differentiation.

Glycosyltransferases are membrane-bound enzymes that add sugar groups to proteins to generate glycoproteins. There are many different glycosyltransferases with different substrate affinities. These enzymes can function as substrate adhesion proteins by binding to carbohydrate groups on extracellular matrix proteins. When activated sugar molecules become available, the catalytic reaction is completed and the adhesion is broken. Cycles of adhesion and catalysis have been observed in migrating cells, suggesting that glycosyltransferases may play an important role in this process.

D1 SPORULATION IN *BACILLUS SUBTILIS*

Key Notes

Overview of sporulation

In response to starvation and other stresses, vegetative cells of the bacterium *Bacillus subtilis* undergo sporulation, a process characterized by asymmetrical division and differentiation to produce a large mother cell and a smaller forespore. The forespore is engulfed by the mother cell and encased in a tough outer coat, allowing it to survive in a harsh environment. The mother cell is lysed when the spore is released.

Morphological stages

Sporulation is divided into seven stages, each dependent on a specific set of genes. The stages are represented by Roman numerals, with 0 indicating the vegetative state. Stage I is characterized by axial chromosome alignment, septation occurs at stage II, engulfment at stage III, and stages IV, V and VI represent consecutive steps of spore maturation. Release of the spore occurs at stage VII.

Initiation of sporulation

Sporulation begins when internal or external stress signals activate a kinase signaling cascade. This culminates in the phosphorylation and activation of a transcription factor called Spo0A, which alters the specificity of the vegetative RNA polymerases and initiates transcription of the early sporulation genes.

The regulatory cascade

The forespore and mother cell differentiate in parallel, with transcription factors activated in one cell initiating signals that cause further transcription factors to be activated in the other cell. Two of the genes activated by Spo0A encode novel sigma factors, σF and σE, that control later sporulation genes. σF and σE are synthesized in both cells but are specifically activated in the forespore and mother cell, respectively. σF activated in the forespore permits the synthesis of a signal that leads to the activation of σE in the mother cell. σE activated in the mother cell then permits the synthesis of a signal that activates σG in the forespore. Finally, a signal dependent on σG in the forespore activates σK in the mother cell. This sequential cross regulation allows batteries of cell-specific genes to be activated in a temporal sequence.

Relevance to development

Sporulation is a simple form of differentiation: it is a response to signals in the environment and involves the production of two distinct cell types expressing different genes and with different functions. However, other characteristics of the system, including the asymmetrical cell division and the hierarchy of regulatory factors, are reminiscent of development in higher organisms. The fate of the mother cell is to be lysed when the spore is released. Cell death is also a feature of the developmental program of higher organisms.

Related topics	Mechanisms of developmental commitment (A3)	The cell division cycle (C4)
	Model organisms (B1)	

Overview of sporulation

In an optimal environment, the bacterium *Bacillus subtilis* undergoes vegetative growth by binary fission, a process involving rapid symmetrical cell divisions. However, harsh environmental conditions (such as starvation, overcrowding or exposure to mutagens) induce **sporulation**, where the cell divides asymmetrically and the smaller cell (the **forespore**) is engulfed by the larger **mother cell**, encased in a tough protective coat and released as a mature **spore** (*Fig. 1*). Sporulation is a simple form of differentiation as two cells are formed that are morphologically distinct, express different proteins and have different functions, but they carry the same genetic material. The mature spore is released from the mother cell by lysis, and the mother cell is destroyed. The process of sporulation is controlled primarily at the level of transcription and involves the synthesis of alternative components of the bacterial RNA polymerase (sigma factors), and other transcriptional regulators.

Morphological stages

The sporulation process can be divided into seven morphologically distinct stages, each controlled by a particular set of genes (*Fig. 1*). The stages are designated by Roman numerals. Stage 0 is the vegetative state. In stage I, the chromo-

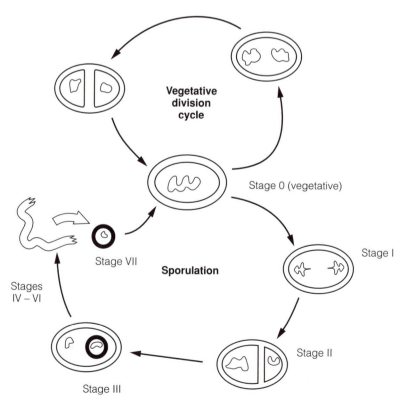

Fig. 1   Life cycle of Bacillus subtilis, *showing the vegetative division cycle and sporulation.*

somes align on the long axis of the cell to form **axial filaments**; this is the first overt morphological change associated with sporulation. In stage II, the cell divides asymmetrically (**polar septation**) to form the forespore and the mother cell. In stage III the forespore is engulfed by the mother cell. Stages IV–VI represent steps in spore maturation as the various layers of its coat are synthesized. Spore release occurs at stage VII. Genes controlling sporulation are named according to the convention *spoXYZ*, where *spo* indicates a sporulation function, *X* refers to the morphological stage at which the gene is first active, and *Y* and *Z* identify the operon (if appropriate) and individual gene. For example, *spoIIIGA* is the first gene of the sixth operon active at stage III, and *spo0A* is active in vegetative cells, and the gene is not part of an operon.

Initiation of sporulation

Sporulation begins when environmental or internal signals initiate a protein kinase signaling cascade, culminating in the phosphorylation of a transcriptional regulator called Spo0A. This interacts with the sigma factors of vegetative RNA polymerase and alters their specificity so that a new set of **sporulation-specific genes** is transcribed (sigma factors confer specificity to RNA polymerase by recognizing the promoter: different signal factors recognize the promoters of different subsets of genes[1]). The new genes include those controlling axial chromosome alignment, polar septation and translocation of one chromosome into the small compartment, which becomes the forespore. Two further genes induced at this stage encode novel sigma factors, σF and σE, which initiate the regulatory cascade controlling cellular differentiation.

The regulatory cascade

Differentiation of the forespore and mother cell occur as an interdependent series of stages, reflecting the sequential activation of different regulatory proteins. The cascade is summarized in *Fig. 2*.

● *Activation of σF specifically in the forespore.* σF controls the expression of forespore-specific genes, but it is initially evenly distributed throughout the forespore and mother cell, although as an inactive complex with an inhibitor protein called SpoIIAB. This inhibitor is displaced in the forespore by another protein, SpoIIAA, which can exist in either a phosphorylated or dephosphorylated form. Only dephosphorylated SpoIIAA can free the σF protein, but the SpoIIAB inhibitor is the kinase that phosphorylates SpoIIAA. SpoIIAB thus acts in two ways to inactivate σF, firstly by sequestering it into an inactive complex, and secondly by phosphorylating and inactivating the SpoIIAA protein that can release σF from that complex. The kinase activity of SpoIIAB is blocked by a membrane-bound inhibitor called SpoIIE. Although SpoIIE is displayed on both sides of the septum, the much smaller size of the forespore means that SpoIIE is relatively much more highly concentrated in the forespore, leading to preferential inactivation of SpoIIAB and the release of σF.

● *Activation of σE in the mother cell.* σE controls mother cell gene expression but it too is initially distributed uniformly, although as an inactive precursor. The protease that cleaves and activates σE is a membrane-bound protein called SpoIIGA. This requires a ligand, SpoIIR, whose synthesis is σF-dependent.

[1]Bacterial RNA polymerases work in a fundamentally distinct way to those in eukaryotes. The bacterial RNA polymerase recognizes the promoter directly, through its σ-factor, while eukaryotic RNA polymerases require a host of transcription initiation factors (TIFs) to bind to the promoter first (Topic C1).

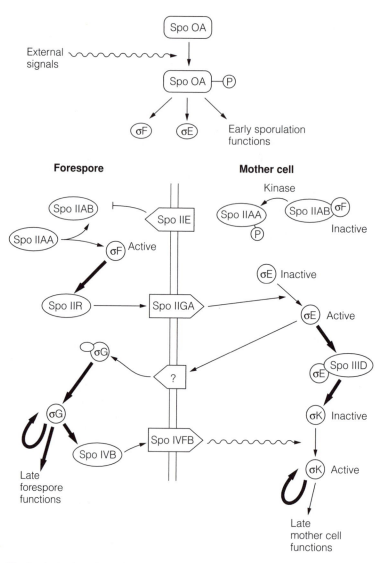

Fig. 2   Activation of the sporulation pathway in B. subtilis, and a summary of the sequence of interdependent events in the forespore and mother cell. See text for details.

σE is therefore activated in the mother cell only after σF is activated in the forespore. σE in the forespore is degraded.

- *Activation of σG in the forespore.* Once σE is activated in the mother cell, a signal is transmitted to the forespore resulting in the production of factors that allow σF to transcribe a new set of genes. These include *spoIIIG*, which encodes a further sigma factor, σG. σE-dependent signaling is required both for transcription of the σG-encoding gene and activation of the protein. σG is therefore only activated in the forespore after σE is activated in the mother cell. σG maintains its own synthesis and activates genes for spore maturation.
- *Activation of σK in the mother cell.* One of the first σE-dependent genes to be expressed in the mother cell is *spoIIID*, which encodes a transcriptional regu-

lator. SpoIIID alters the specificity of σE and allows it to transcribe a new set of genes, including a gene encoding a further sigma factor, σK. σK is also expressed as an inactive precursor, and requires the activity of a membrane bound protease SpoIVFB to cleave and activate it. The activity of SpoIVFB is dependent on a signal from the forespore, the SpoIVB protein, whose synthesis is under the control of σG. Therefore σK in the mother cell is activated only after σG is activated in the forespore. σK maintains its own synthesis and directs the expression of genes encoding late mother cell functions.

Relevance to development

- Sporulation is a response to signals from the environment and involves a signal transduction cascade culminating in the expression of new sets of genes. This is analogous to induction in metazoans.
- Sporulation involves the sequential activation of different classes of transcriptional regulators. Hierarchies of genetic regulators are also seen in animal development, e.g. in the early patterning of the *Drosophila* embryo (Section H).
- Sporulation results in the formation of two distinct cell types by asymmetrical division. Asymmetrical divisions are important in the early stages of both animal and plant development (Topics F1 and N2; Section G). The mother cell and prespore have the same genome but express different genes and have different phenotypes. This is a simple form of differentiation (Topic A5).
- Sporulation involves the repositioning of a cell wall. In plant development, the position of the cell wall also plays an important role in morphogenesis (Topic N1).
- The mother cell always undergoes lysis in the process of spore release, so cell death is the end result of the mother cell's developmental program. Programmed cell death is also an important aspect of animal development, although the apoptotic cell does not undergo lysis (Topic C4).

D2 MATING TYPE SWITCHING IN YEAST

Key Notes

Yeast mating types

Yeast cells propagate by budding, either as haploid or diploid cells. Haploid cells have one of two mating types, termed a and α, conferred by the particular allele, *MATa* or *MATα*, present at the *MAT* locus. Only cells of opposite mating type can fuse to form diploid cells. The alternative *MAT* alleles encode different transcriptional regulators that activate the expression of mating type-specific genes.

Mating type switching

Haploid cells can switch mating types in every generation. Mating type switching involves gene conversion between the active *MAT* locus and one of two loci on the same chromosome *HMLα* and *HMRa*. These contain transcriptionally repressed copies of the *MATα* and *MATa* alleles. HO endonuclease cuts the DNA at the *MAT* locus and the free ends recombine with one of the two silent loci, usually the one containing the alternative *MAT* allele. The specificity of mating type switching is regulated by transcriptional control of the *HO* gene. The *HO* gene promoter binds factors that confer haploid cell-, mother cell- and G_1 phase-specificity.

Relevance to multicellular development

Mating type switching involves a change in phenotype caused by the activation of new genes, and is therefore a simple form of differentiation. Although the mechanism of expressing new transcription factors by gene conversion is not used in higher organisms, DNA rearrangement is involved in the differentiation of B- and T-cells in animals. The restriction of mating type switching to the larger of the two products of asymmetric cell division produces a cell lineage that is similar to that of stem cells. The repression of gene expression at the *HMLα* and *HMRa* loci is reminiscent of chromatin silencing by proteins such as Polycomb in *Drosophila*.

Related topics

Mechanisms of developmental
  commitment (A3)
Model organisms (B1)

Gene expression and regulation
  (C1)
Chromatin and DNA methylation
  (C2)

Yeast mating types

The yeast *Saccharomyces cerevisiae* can grow vegetatively as either haploid or diploid cells, each reproducing asexually by budding (after DNA replication, a small daughter cell containing one copy of the genome buds off from the larger mother cell, which contains the other copy of the genome). Under certain conditions such as starvation, haploid yeast cells fuse (mate) to form diploid cells. Haploid yeast cells have one of two **mating types**, termed **a** and α, and only cells of opposite mating types can fuse. The recognition of opposite mating types involves the perception of **pheromones**: a-cells secrete one type of pheromone recognized by a receptor specific to α-cells, while α-cells secrete another type of pheromone recognized by a receptor specific to a-cells. Both pheromones are small peptides that activate a G-protein-coupled receptor. Although the pheromones and their receptors are mating type-specific, they activate the same signal transduction pathway and this stimulates mating behavior common to all haploid cells. The components of this signal transduction pathway are absent from diploid cells, thus preventing mating between diploids.

The mating type is controlled at a single locus, called the **MAT locus**. Cells with the allele *MATa* are a-cells and those with the allele *MATα* are α-cells. The two alleles have a similar structure comprising three regions: X, Y and Z. The X and Z regions are the same in both alleles so mating type specificity is conferred by the sequence of the Y region. The *MATa* allele encodes a single transcription factor called **a1**, while the *MATα* allele encodes two transcription factors, called α**1** and α**2**. These work in concert with a dimeric constitutive transcriptional regulator **MCM1** to activate particular sets of genes, as shown in *Fig. 1*.

Mating type switching

Haploid yeast cells do not maintain the same mating type indefinitely, as this would generate populations of cells with the same mating type within which mating would not be possible. In fact, haploid yeast cells undergo **mating type switching** in every generation, to produce an unbiased population. Mating type switching occurs by the replacement of one *MAT* allele with the other. It occurs only in haploid cells and only in the G₁ phase of the cell cycle. Furthermore, mating type switching occurs only in the mother cell, never in the daughter cell.

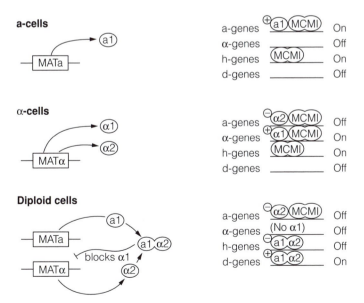

Fig. 1 Regulation of gene expression by transcription factors encoded by the MAT *locus. Different* MAT *genotypes produce different combinations of transcription factors, which act in concert with the constitutive regulator MCM1 to differentially regulate a-genes (a-cell specific), α-genes (α-cell specific), h-genes (haploid cell specific, both mating types) and d-genes (diploid cell specific).*

The result is that every cell division produces one cell of each mating type so that, regardless of the make-up of the starting population, after one round of division there are approximately equal numbers of cells of each mating type (*Fig. 2*).

Mating type switching in yeast is achieved by directed **gene conversion**, a process involving recombination between homologous DNA sequences. Switching involves recombination between the *MAT* locus and one of two sites on the same chromosome termed **HMLα** and **HMRa** (*Fig. 3*). These contain copies of the *MATα* and *MATa* alleles respectively, but they are transcriptionally repressed by silencer elements that sequester the loci into repressed chromatin (Topic C2). During the G_1 phase of the cell cycle in each mother cell, the information at the transcriptionally active *MAT* locus is replaced by that stored at one of the silent loci. This occurs through the activity of a site-specific endonuclease, **HO endonuclease**, which cuts DNA within the *MAT* locus and initiates recombination (*Fig. 3*). HO endonuclease cleaves the *MAT* locus between regions Y and Z. This is followed by degradation of the DNA in the Y region, leaving a gap in the chromosome. The free DNA ends then associate with either of the silent loci and recombination-mediated repair of the gap copies the sequence from the silent locus into the *MAT* locus. In over 90% of repair events, the cleaved *MAT* locus associates with the *opposite* silenced locus, so that mating type switching occurs (in the remaining ~10% of cases, the *MAT* locus is replaced by the same original sequence and no switching occurs). It is not known how this switch specificity is achieved.

Mating type switching is restricted by the transcriptional activity of the gene encoding HO endonuclease. The *HO* gene is activated specifically in haploid mother cells at the G_1 phase of the cell cycle. Haploid-specific, mother cell-specific and G_1 stage-specific expression are regulated by independent sets of transcription factors (*Fig. 3*). Haploid cell-specificity is achieved through repression by the **a1α2 regulator** produced when there are both *MAT* alleles active in

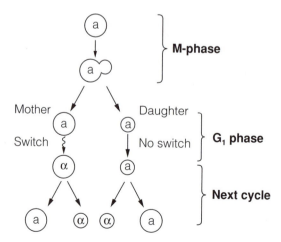

Fig. 2   Mating type switching in one product of each cell division ensures equal numbers of a-cells and α-cells in the population after one round of division.

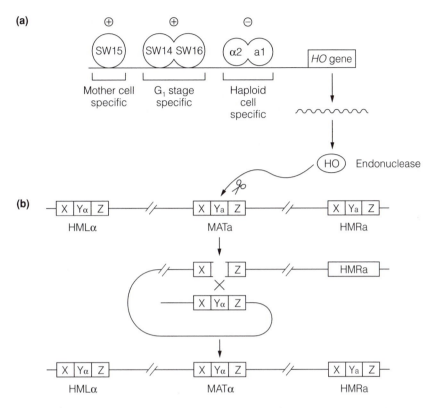

Fig. 3. Mating type switching in yeast. (a) Regulation of HO endonuclease expression so that switching occurs only in the haploid mother cell during G$_1$ phase. (b) The switching mechanism, involving repair of the MAT locus using information from one of the silent flanking loci, HMLα and HMRa.

the cell. Cell cycle-specificity is achieved through the binding of a dimeric transcription factor comprising the SWI4 and SWI6 proteins, whose activities are restricted to G$_1$ by cell cycle regulators. Mother cell-specificity is achieved through the transcription factor SWI5, which is active only in the mother cell. However, SWI5 is present in both the mother and the daughter cells: the mechanism by which its cell-specific activity is regulated is not understood.

Relevance to multicellular development

- Mating behavior is brought about by mutual signaling between two different cells. This is reminiscent of reciprocal induction in animal development, as illustrated by the metanephric mesenchyme and ureteric bud in the kidney (Topic K4).
- Cell division in haploid yeast produces one cell with the same phenotype as the mother cell and another with a different phenotype (ironically, the budding daughter cell is the one that remains the same mating type as the mother, while the mother cell itself switches). This is reminiscent of stem cell behavior in metazoans (Topic A5).
- Mating type switching involves the production of new transcription factors and results in a change of phenotype. This is a simple form of differentiation (Topic A3).

- Mating involves signaling between cells, but mating type switching is an autonomous process. Yeast therefore demonstrate autonomous specification of mating types (Topic A4).
- The *HMLα* and *HMRa* loci remain silent because they are sequestered into repressed chromatin. Chromatin structure is also used to regulate gene expression in animals and plants (Topic C2).
- Mating type switching involves a change to the yeast genome. This is an uncommon mechanism in the development of higher eukaryotes, although DNA rearrangement does occur in the maturation of B- and T-cells (Topic C1).
- The *MAT*-encoded transcription factors contain a homeodomain. Homeodomain transcription factors are central to many developmental processes in animals, and also in plants (Topic C1).

D3 AGGREGATION AND CULMINATION IN *DICTYOSTELIUM DISCOIDEUM*

Key Notes

Overview of life cycle

In its vegetative state, *Dictyostelium discoideum* exists as single cells. However, under conditions of nutrient depletion, up to 10^5 cells may aggregate to form a multicellular structure sharing many properties with a developing animal embryo. The aggregate can migrate as an independent multicellular organism (slug) and undergoes differentiation, pattern formation and morphogenesis to produce a fruiting body for spore dispersal.

Aggregation

The decision to aggregate reflects nutrient availability and population density. Starved cells release cAMP, which causes nearby cells to migrate towards the source and release their own cAMP to relay the signal to more distant cells. As incoming cells migrate, they begin to synthesize cell adhesion molecules, initially allowing them to form streams and then a large mound.

Differentiation and pattern formation in the mound

Aggregating cells stop proliferating and differentiate into prestalk cells (future structure of the fruiting body) and prespore cells (future spores), according on the stage of the cell cycle at which growth arrest occurred. The prestalk cells migrate to the tip of the mound while the prespore cells sort passively to the body. Sorting depends on differential migration in response to cAMP and requires prestalk differentiation in response to cAMP and differentiation inducing factor (DIF). Prestalk cells form a number of regionally-distributed subpopulations - PstA, PstAB, Pst0 and PstB cells - which express different extracellular matrix proteins and are destined to form different parts of the fruiting body.

Cell movement during grex migration

During slug migration, PstAB cells move back through the slug and are lost. These are replaced by PstA cells, which are themselves replaced by Pst0 cells. To maintain a constant ratio of cell types, Pst0 cells are replaced by anterior like cells among the prespore cell population in the body of the slug. These events mimic the dramatic cell movements that occur during culmination.

Culmination

In humid darkness, the slug produces lots of ammonia, which prevents further differentiation. In light or reduced humidity (conditions suitable for spore dispersal) ammonia levels fall and the genes responsible for terminal differentiation are derepressed. The prestalk cells undergo extensive morphogenetic movements to generate a fruiting body with a spore case, stalk and base. The PstB cells form a basal disc. The PstAB cells push down through the fruiting body into the basal disc to elevate the other cells on a long stalk. The PstA and Pst0 cells form the spore case. The prespore cells form mature spores and are dispersed. The spores become new myxamebae.

Relevance to metazoan development

The *Dictyostelium* life cycle involves the formation of a multicellular organism. Unlike animal and plant development, this occurs by aggregation rather than growth and division, but many of the principles of development in higher organisms are preserved. Such principles include differentiation, cell sorting, pattern formation using morphogen gradients, regulation for missing parts, and cell death. Mutants impaired for many of these mechanisms have been isolated.

Related topics

Mechanisms of developmental commitment (A3)
Pattern formation and compartments (A6)

Morphogenesis (A7)
Model organisms (B1)
Developmental mutants (B2)

Overview of life cycle

The cellular slime mold ***Dictyostelium discoideum*** is unique among the unicellular model organisms considered in this section, because its life cycle includes a multicellular stage (*Fig. 1*). In its vegetative state, *Dictyostelium* exists as unicellular haploid cells called **myxamebae** that undergo asexual reproduction. However, when the local nutrient supply is exhausted, up to 10^5 cells **aggregate** to form a multicellular structure called a **mound**. Within the mound, **prespore cells** and several classes of **prestalk cells** differentiate and undergo spontaneous self-organization to specify the anteroposterior axis. The mound then tips onto

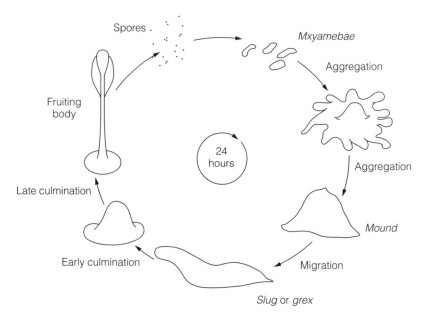

Fig. 1.   The life cycle of Dictyostelium discoideum.

its side and migrates *en masse* as a single organism, variously called a **pseudo-plasmodium**, **grex** or **slug**. Under suitable conditions, the slug stops and the cells undergo terminal differentiation. This final **culmination** stage results in the formation of a **fruiting body** comprising a long thin stalk and an elevated spore case containing mature spore cells. These are dispersed to form new myxamebae.

Metazoans and *Dictyostelium* use opposite strategies to become multicellular, but the *Dictyostelium* aggregate shares many properties and developmental mechanisms with metazoan embryos. It is clear that cells of the aggregate undergo differentiation, regional specification and morphogenesis. Furthermore, the *Dictyostelium* aggregate demonstrates remarkable powers of regulation, suggesting that cell–cell interactions play a critical role in the multicellular phase of the life cycle. *Dictyostelium* provides a particularly useful model for cell movement during morphogenesis, as the aggregation of single cells and their relative movement within the mound, slug and fruiting body are representative of similar processes occurring in animal embryos. However, *Dictyostelium* provides a much more convenient analytical tool, as cells can be labeled as myxamebae and then followed through multicellular development. Furthermore, combining populations of mutant and wild type cells prior to aggregation is a convenient way of generating chimeras to investigate whether gene functions are cell autonomous or otherwise.

Aggregation

Aggregation is triggered by nutrient depletion but also depends on cell density. All myxamebae secrete a protein called **prestarvation factor (PSF)**, and cells can determine the density of the local population by monitoring the levels of PSF in the environment. Starved cells release **cAMP** into their surroundings, and if the levels of PSF are also high enough, neighboring cells begin to move towards them to form an aggregate. Initially, cAMP is released by scattered individual cells that have depleted their food supply. The cAMP binds to G-protein-coupled receptors on the surface of neighboring cells and activates the enzyme adenyl cyclase, stimulating the synthesis of more cAMP. Cells can be activated by cAMP for about a minute, and then become insensitive to the molecule for the next 5 minutes until local cAMP is degraded by membrane-bound phosphodiesterases. In this way, the signal is relayed to more peripheral cells as a series of discontinuous pulses. The period of resolution is essential for the signal to stimulate cell movement *towards* the source of the signal. cAMP induces the polymerization of F-actin at the leading edge of the cell, resulting in the extension of lamellipodia in the direction of the signal. As the myxamebae migrate towards the source of the signal, they begin to synthesize new cell adhesion molecules, enabling the cells to first to form streams and then the mound.

Mutants that fail to aggregate fall into three classes: (1) those that cannot perceive cAMP; (2) those than cannot produce cAMP; (3) those lacking an essential component of cell motility. Chimeras show whether these effects are cell automous. In mixtures of wild type cells and mutants with either defective cell motility or an inability to perceive cAMP, the mutants get left behind while the wild type cells aggregate. However, mutants that fail to produce cAMP undergo normal chemotaxis towards a cAMP source in a mixed population. The failure of this process in homogeneous mutant populations is due to the fact that the signal is not propagated.

Differentiation and pattern formation in the mound

During aggregation, migrating cells undergo growth arrest and differentiate into **prestalk cells** (which later form the structure of the fruiting body) and **prespore cells** (which later produce spores). The prestalk cells sort the tip of the mound, which becomes the anterior of the migrating slug, while the prespore cells form the body. The architecture of the very early mound reflects the order in which cells arrive, the central cells acting as the source of the cAMP pulses. However, as more cells arrive, the tip of the mound becomes populated with prestalk cells, which migrate fastest in response to the cAMP. Newly arriving prestalk cells first migrate into the center of the mound and then spiral upwards into the tip. The initial decision to differentiate into either prestalk or prespore cells may depend on the stage of the cell cycle at which growth arrest occurs. Cells arrested at G_1 preferentially form prestalk cells, while those arrested at G_2 are more likely to form prespore cells. Sorting is thought to be mostly dependent on differential rates of migration in response to cAMP, although it has not been ruled out that differential cell adhesion also plays a role (see below). The importance of cAMP in patterning the mound can be shown by adding a source of cAMP to the substratum on which the mound forms. In such experiments, the prestalk cells sort to the base of the mound instead of the tip. The tip of the mound can therefore be thought of as an **organizer** (Topic A6).

The analysis of temperature sensitive mutants shows that many of the genes required for aggregation are also required for cell sorting in the mound. This is not surprising as cells that cannot respond to cAMP during aggregation will be unable to respond to the organizer, while cells with deficiencies in motility will not be able to move at all. In chimeras, both types of conditionally mutant cell sort passively as others move around them, and the mutant cells therefore tend to end up in the prespore population. There are other mutants that aggregate normally but fail to sort properly in the mound, resulting in developmental arrest at the mound stage. The first class comprises mutants with motility defects that are not apparent until the mound stage. It is likely that aggregation towards the mound is easier than pushing between packed cells in the mound itself, so mutations with silent phenotypes during aggregation could show stronger phenotypes during morphogenesis. The second class comprises mutants lacking the ability to differentiate into prestalk cells, suggesting that prestalk cells are the driving force behind cell sorting, while the prespore cells are sorted passively. This can be demonstrated using chimeras. The LIM2 protein is a marker of prestalk cell differentiation and is required for reorganization of the cytoskelton. In mutants for the corresponding *limB* gene, the cells aggregate normally but fail to organize themselves in the mound. Chimeras of *limB* and wild type cells also aggregate normally, and undergo morphogenesis, but the *limB* mutant cells fail to penetrate the mound.

cAMP is important not only for cell guidance in the aggregate, but also for differentiation. cAMP is required for the differentiation of both prespore and prestalk cells, but the differentiation of prestalk cells also requires a number of related organic molecules collectively called **differentiation inducing factor (DIF).** The initial sorting of prestalk and prespore cells on the basis of differential migration is reinforced when the prestalk cells begin to synthesize different classes of extracellular matrix adhesion molecules, and this may also help to organize the different classes of prestalk cells (*Fig. 2*). Most prestalk cells synthesize **EcmA**. Cells that express the *ecmA* gene strongly sort to the tip and are termed **PstA cells**, while those with weaker expression levels fill the anterior

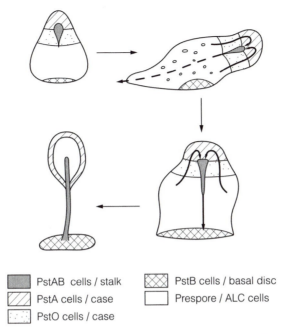

▓	PstAB cells / stalk	▨	PstB cells / basal disc
▨	PstA cells / case	☐	Prespore / ALC cells
░	PstO cells / case		

Fig. 2.   Distribution and relative movement of different cell types during the multicellular phase of the Dictyostelium *life cycle.*

third of the slug and are termed **Pst0 cells**. Within the PstA cells, a subpopulation begins to synthesize another matrix protein, **EcmB**. These **PstAB cells** become arranged in a column, which eventually initiates the formation of the mature stalk of the fruiting body. Some prestalk cells synthesize EcmB alone. These **PstB cells** are positioned just posterior to the PstA and Pst0 cells, and eventually form the basal disc of the fruiting body.

Cell movement during slug migration

In darkness, the mound tips onto its side so that the tip, containing the PstA and Pst0 cells, becomes the anterior end of the slug. As the slug moves, some PstAB cells move from the posterior end of the column back through the prespore cells and are lost. They are replaced by PstA cells in the anterior tip, which begin to synthesize EcmB and move into the column at the anterior end. The PstA cells are in turn replaced by adjacent Pst0 cells. Scattered among the prespore cells in the posterior of the slug is another class of cells called **anterior-like cells (ALCs).** These are very similar to Pst0 cells and also express *ecmA*. Some of these cells migrate forwards to replace the true Pst0 cells as the slug moves. Prespore cells differentiate into ALCs to maintain the correct ratio of cell types (*Fig. 2*). The fate of cells in the slug therefore appears to change continuously. The plasticity of cell fate is confirmed by experiments in which mounds and slugs are divided into smaller populations or several aggregates are combined. In such experiments, the aggregate regulates to form a single aggregate.

Culmination

The migrating slug seeks out a more suitable environment by moving up a heat and light gradient, and moving away from sources of ammonia. Ammonia is produced by the slug's own metabolism and limits the level of available cAMP, probably by activating a phopshodiesterase called RegA. Terminal differentiation requires the inhibition of RegA, allowing cAMP levels to rise. RegA is inhibited when the slug reaches a suitable environment, through an unknown signal transduction pathway responding to external cues. The cells begin to reorganize and undergo terminal differentiation. This process, termed **culmination**, results in the construction of the fruiting body.

Light and reduced humidity both increase the rate at which ammonia diffuses away from the slug. Ammonia depletion restores the ability of the cells to produce cAMP, and as cAMP levels rise, a cAMP-dependent protein kinase (**protein kinase A**) is activated. This is thought to phosphorylate and inactivate one or more transcriptional repressors, resulting in the activation of a battery of genes required for culmination. One of these genes encodes a transcription factor called CudA. If the genes for either CudA or protein kinase A are mutated, the slug is unable to culminate (mutants with this phenotype are known as **'sluggers'**). Chimeras of slugger and wild type cells have shown that the PstA cells at the anterior of the slug control its migration and the decision to culminate.

Culmination involves the migration of PstA cells towards the column of PstAB cells at the tip of the aggregate. Here, the PstA cells begin to synthesize EcmB and become converted into PstAB cells. At the same time, PstB cells migrate to the point at which the aggregate contacts the substratum and form the base of the fruiting body. The terminal differentiation of the PstB cells involves a different pathway to the stalk and spore cells, which requires the inhibition of **glycogen synthase kinase-3 (GSK-3)** since *gsk3* mutants have few spore cells, and the fruiting bodies have very large basal discs. Other genes have been identified with specific roles in terminal differentiation. For example, the terminal differentiation of prespore cells into mature spores begins at the base of the spore case, adjacent to the prestalk cells. These cells are a source of spore differentiation factors (SDF1 and SDF2). The transcription factor **StalkyA** is required for spore cell differentiation. In normal culminates, 80% of the cells form spores. In *stalkyA* mutants, none of the cells form spores and the fruiting body comprises an unusually long stalk. As the population of PstAB cells grows, they push through the prespore cells into the base and elevate the spores on a stalk so they can be effectively distributed (*Fig. 2*). The PstAB cells synthesize cellulose, become vacuolated and die to form a rigid stalk. Pst0 cells, and any remaining PstA cells migrate to the periphery of the aggregate to form the spore case and the associated upper and lower cusps.

Relevance to metazoan development

- *Dictyostelium* aggregation is a model of directed cell movement in development, with relevance to cell sorting and individual cell migration in animals.
- Cells respond to external cAMP and change their behavior. This is a simple form of induction.
- The cells at the tip of the mound control the morphogenesis of the surrounding cells to establish the anteroposterior axis of the slug. The tip of the mound is therefore analogous to an organizer in metazoan development.
- The aggregated cells undergo differentiation, generating at least five distinct cell types with different gene expression patterns and functions.
- The mound and slug can regulate for missing parts and form the same structure with the same relative proportions of different cell types over a range of

aggregate sizes. This is similar to the ability of certain metazoan embryos to regulate, where position in the embryo is more important than lineage. Indeed lineage is irrelevant in *Dictyostelium* because the multicellular organism is formed by aggregation.

- Terminal differentiation of the stalk involves lignification (which occurs in plant development) and cell death (which is an important process in animal development).

E1 GERM LINE SPECIFICATION

Key Notes

Germ cells and somatic cells	Multicellular organisms contain germ cells that give rise to gametes, and somatic cells that give rise to the rest of the cells of the body. In many animals, the somatic and germ cell lineages segregate very early in development, and committed somatic cells cannot give rise to further germ cells. In contrast, plants do not set aside a germ line. Rather, germ cells can arise from somatic cells in the adult floral meristem.
Overview of germ-line specification	In animals that set aside a germ line in early development, germ-cell specification occurs during cleavage or gastrulation. In many cases, this involves a special area of cytoplasm, the germ plasm, which is distinguishable in the egg. In such animals, the formation of germ plasm is dependent on asymmetrically distributed maternal gene products. In birds and mammals, the germ line is not specified by cytoplasmic determinants, but by inductive interactions controlled by zygotic genes.
Germ-line specification in *Drosophila*	In *Drosophila*, the germ plasm is located at the posterior pole of the egg, and undergoes cellularization a few divisions before the somatic cells are formed. This produces about 15 primordial germ cells called pole cells. The pole plasm is distinguishable because it contains distinct polar granules, comprising a number of proteins that control localization of the pole plasm and germ cell specification. The *oskar* gene plays a predominant role in pole plasm localization, since polar granules form wherever *oskar* mRNA is localized.
Germ-line specification in *C. elegans*	The first four cleavage divisions of the *C. elegans* embryo are reminiscent of stem cell divisions, each producing the founder of a somatic cell lineage and another stem cell (a P-cell). The fourth division produces a somatic founder cell and the P_4 cell, which is the source of the germ line. The unfertilized egg contains germ cell determinants called P-granules. These are initially distributed uniformly in the egg, but become restricted to the P-cells during cleavage and are eventually restricted to the P_4 cell. Mutations that disrupt the embryonic cleavage pattern also affect germ line specification, while other mutations block germ cell specification without influencing embryonic polarity.
Specification of the vertebrate germ line	In *Xenopus*, the germ line is specified by a vegetal germ plasm comprising distinct yolk-free islands. This migrates to the vegetal pole and is incorporated into cells that form the endodermal lining of the blastocoel, from which the germ cells are derived. Inactivation of the germ plasm by UV-irradiation produces tadpoles that lack germ cells but are otherwise normal. In birds and mammals, there are no cytoplasmic determinants in the egg, and germ cells arise during gastrulation by largely uncharacterized cell–cell interactions.

Related topics	Mechanisms of developmental commitment (A3)	Anteroposterior axis specification in *Drosophila* (H2)
	Germ-cell migration (E2)	Plant *vs* animal development (N1)
	Cell specification and patterning in *C. elegans* (G2)	

Germ cells and somatic cells

In multicellular eukaryotes, two fundamentally different categories of cell can be distinguished. **Germ cells** are specialized cells that can undergo **meiosis** and give rise to **gametes**. All other cells are called **somatic cells**, and they can divide only by mitosis.

In most animals, the lineage that gives rise to germ cells, the **germ line**, separates from the lineage of somatic cells very early in development. As discussed in Topic A2, the acquisition of cell fate in animals involves a series of irreversible decisions, so once this segregation has taken place, the germ line is the sole source of gametes – somatic cells cannot produce more germ cells later in development. Thus, the continuity of life in most animals is dependent on germ cells. Conversely, in plants (and some invertebrates), there is no early segregation of a germ line. Germ cells arise from somatic cells in the adult organism, so gamete-producing cells can be replenished in post-embryonic development.

Overview of germ-line specification

In many animals (but not birds or mammals) germ-line specification occurs during cleavage and involves the inheritance of particular cytoplasmic determinants asymmetrically distributed in the egg. These are often located in a visually distinct region of cytoplasm, called **germ plasm**. In *Drosophila* and *C. elegans*, mutants have been identified in which germ line specification is disrupted, either because germ cell determinants are missing or because the germ plasm is not positioned correctly (in both species germ cell specification is linked to the polarization of the anteroposterior axis). The germ plasm is established by maternal gene products, and mutants lacking a functional germ plasm therefore show a maternal effect. A maternal effect mutation that disrupts germ cell specification in the embryo results in sterility of the embryo and hence the **absence of grandchildren**. The removal of germ plasm from the egg also results in sterility. Only germ plasm, and not ordinary cytoplasm, injected into embryos developing from such eggs can rescue the germ line. In contrast, germ line specification in birds and mammals does not involve cytoplasmic determinants. Rather, germ cells arise in response to cell–cell interactions just prior to gastrulation, and such interactions are blocked by zygotic mutations. Germ cell specification in several model organisms is discussed below.

Germ-line specification in *Drosophila*

As discussed in Topic F1, the early development of insects is characterized by rapid division of the zygotic nucleus to form a multinucleate syncytium. The *Drosophila* germ plasm is located at the posterior pole of the egg, and is therefore termed **pole plasm**. It can be distinguished by its relatively large organelles and the presence of **polar granules**, which consist of RNA and protein. The pole plasm undergoes cellularization a few hours before the rest of the embryo, so that approximately 15 **pole cells** (*Drosophila* germ cells) form before the emergence of somatic cells. The appearance of pole cells is dependent on the pole plasm, since if the pole plasm is transplanted to another region of the embryo, germ cells develop at the new site instead. Flies produced from such manipulated embryos are not fertile because the ectopic germ cells cannot migrate to the gonads. However, if these germ cells are transferred to the gonads artificially they are functional and can produce gametes. The appearance of pole cells is also dependent on migration of nuclei into the pole plasm, a process that is blocked in mothers carrying the *grandchildless* mutation.

The polar granules contain proteins required for germ plasm localization and germ cell specification (*Table 1*). The product of the *oskar* gene appears to play the primary role in germ plasm localization, as the polar granules form wherever *oskar* mRNA is localized. In normal embryos, *oskar* mRNA is targeted to the posterior of the egg, a process dependent on the recognition of specific motifs in its 3′ untranslated region and mediated (indirectly) by the protein Staufen. If the *oskar* message is targeted to the anterior of the egg by substituting the 3′ UTR from *bicoid* mRNA, the pole plasm forms there instead. Furthermore, the dosage of Oskar protein appears to control the amount of pole plasm and hence the number of germ cells formed, as introducing extra copies of the *oskar* gene into transgenic flies produces embryos with increased numbers of germ cells. Oskar probably functions by helping to assemble other components of the pole plasm that facilitate germ cell specification, and this requires the *vasa*, *valois* and *tudor* gene products. A number of proteins recruited to the pole plasm have been shown to specify germ cell fate or facilitate germ cell migration. One important component of the germ plasm is Nanos, a translational regulator. Nanos plays an important role in abdominal specification by inhibiting the translation of mRNA for the anterior morphogen Hunchback (Topic H2). However, Nanos also translationally regulates germ cell-specific genes, and the lack of Nanos protein renders primordial germ cells unable to migrate.

Table 1. *A selection of genes required for germ cell specification in* Drosophila

Gene	Function
cappuccino, spire, par-1	Localization of Staufen protein to posterior of egg
staufen	Localization of *oskar* mRNA to posterior of egg
oskar	Assembly of polar granules, localization of Vasa protein
vasa	Assembly of polar granules
tudor, valois	Maintenance of polar granules
grandchildless	Migration of nuclei into pole plasm
germ cell-less	Nuclear protein involved in germ cell specification
nanos	Necessary for germ cell migration; regulates germ cell-specific gene expression at the level of translation.

Germ-line specification in *C. elegans*

Germ cell specification in *C. elegans* occurs during the fourth cleavage division about 1 hour after fertilization. As discussed in Topic G2, the first cleavage division produces the somatic founder cell AB and the P_1 cell. The next three divisions in the **P-cell lineage** are like those of stem cells, each producing a somatic founder cell and another stem cell (P_2-P_4). The fourth division produces the somatic founder cell D (that gives rise to muscle cells) and the P_4 **cell**, from which all germ cells are derived (See *Fig. 1* in Topic G2).

As is the case for *Drosophila*, *C. elegans* germ cells arise in the posterior of the embryo and their specification involves the inheritance of a particular cytoplasmic component. The egg initially contains a number of **P-granules (germ cell granules)** that are uniformly distributed. When the egg is fertilized, however, the P-granules sort to the posterior of the egg (defined by the site of sperm entry) and segregate into the P_1 cell at the first cleavage division. The P-granules are uniformly distributed in the P_1 cell until the onset of mitosis, when they again sort to the posterior so they segregate into the P_2 cell. This process continues for another two rounds of cell division, so all the P-granules are eventually inherited by the P_4 cell.

Mutations that affect germ cell specification in *C. elegans* either disrupt anteroposterior polarization or the process of cell specification itself. A number of proteins, including PIE-1, and two homologs of Nanos, Nos1 and Nos2, appear to directly affect germ cell specification. **PIE-1** is a nuclear protein that is restricted to P-cells, and appears to function as a general repressor of RNA polymerase II. In mutants lacking this protein, the P cells become transcriptionally active and differentiate into somatic cells. Transcriptional repression may therefore be a general feature of germ cell development, perhaps in order to prevent the cells becoming committed to somatic fates and losing their totipotency. In support of this, two other genes essential for germ-line specification in *C. elegans* (*mes-1* and *mes-2*) encode inhibitory chromatin proteins related to the Polycomb protein of *Drosophila*, and are probably required to maintain gene silencing through successive rounds of DNA replication.

Mutants with the most severe anteroposterior polarity defects disrupt the early cleavage pattern of the embryo and are therefore called *par* **(partition-defective) mutants**. The role of these genes in polarizing the embryo is discussed in Topic G2. However, with specific reference to the germ line, two of the proteins encoded by these genes, PAR-1 and PAR-2, are restricted to P-cell lineage and may help to establish germ-cell fate. In *par-1* mutants, for example, the P cell divisions are symmetrical and the P granules do not segregate to the posterior cell. It is thought that PAR-1 could help to establish the restricted expression of germ-cell determinants such as PIE-1, through the inhibition of intermediate regulators such as MEX-5 (see Topic G2 for details).

Specification of the vertebrate germ line

In *Xenopus*, germ plasm forms at the vegetal pole of the egg after fertilization, by the aggregation of small cortical cytoplasmic regions to form larger yolk-free islands. These islands migrate to the vegetal pole along microtubule arrays, so that the germ plasm is distributed to the most vegetal cells during the cleavage divisions. The importance of the germ plasm in germ-cell specification can be demonstrated by UV-irradiating the ventral region of the egg. This disrupts the microtubule array, the germ plasm cannot migrate to the vegetal pole and the resulting tadpoles lack germ cells. During gastrulation (Topic F3), these cells form the endodermal floor of the blastocoel, but they are not determined as germ cells until after gastrulation, i.e. if transplanted to other regions of the embryo before gastrulation, they can still become somatic cells. It therefore appears that both cytoplasmic determinants and cell–cell interactions play a role in *Xenopus* germ line specification. Several gene products in the *Xenopus* germ plasm are homologs of those found in *Drosophila* polar granules, including Xcat2 (homolog of Nanos) and Xcat3 (homolog of Vasa). Homologous proteins have also been found in the zebrafish, which until recently was not thought to possess germ plasm because no visually distinct region of cytoplasm exists in the egg.

In mammals and birds, there is no germ plasm in the cleaving embryo. Instead, mammalian and avian primordial germ cells arise in the epiblast of the gastrulating embryo (Topic F3) and their specification involves cell–cell interactions. In fact, other amphibians such as salamanders also do not have a defined germ plasm, and germ cells are thought to arise through inductive interactions as they do in mammals. In the mouse, germ cells are first detected in the epiblast just posterior to the primitive streak at 7 days post fertilization. These 8–10 cells are larger than the surrounding cells and can be detected easily because they stain positive for the enzyme **alkaline phosphatase**. If they are removed, the resulting embryo lacks germ cells. The nature of the inductive interactions that give rise to germ cells is unknown, but there is evidence that BMP4 secreted by the extra-embryonic ectoderm is an important component. In *bmp4*-knockout mice, there are no germ cells, whereas in mice heterozygous for the mutation, the number of germ cells is reduced. The BMP4 signaling pathway may maintain the expression of the transcription factor **Oct-4**. This factor is initially expressed throughout the inner cell mass but becomes restricted to the epiblast cells that give rise to primordial germ cells during gastrulation, and is thereafter restricted to the germ cells themselves.

E2 GERM-CELL MIGRATION

Key Notes

Germ-cell migration

In animals, the germ line is usually specified before the position of the gonads has been established. The gonads arise later in development, often at some distance from the origin of the primordial germ cells (PGCs). The PGCs must be transported to the developing gonads before forming gametes, and this involves a combination of passive movements and active migration. During migration, PGCs may undergo several rounds of division, so that the population of cells colonizing the gonads is larger than the population that started the journey.

Germ-cell migration in _Drosophila_

During gastrulation, the _Drosophila_ pole cells are passively moved from their peripheral origin to the posterior midgut inside the embryo. The cells then actively migrate through the gut wall and into the adjacent mesoderm, where they split into two populations and continue their journey to the gonads. Genes such as _nanos_ are required for germ cell migration, while _wunen_ and _columbus_ encode guidance signals that direct the migrating cells to the correct destination.

Germ-cell migration in vertebrates

In _Xenopus_ and mammals, PGCs move from their site of origin to the endodermal cells of the hindgut during gastrulation. They then migrate along the hindgut, across the dorsal mesentery to the lateral plate mesoderm of the abdominal body wall, where they split into two populations and move across to the paired gonadal ridges. Migration may involve a polarized network of fibronectin molecules in the extracellular matrix, and factors secreted by cells along the migratory route stimulate the PGCs to proliferate. In reptiles and birds, germ cells are transported to the gonadal ridges by a completely different route involving the vascular system, but migration from the capillaries into the gonadal ridge may also involve chemoattraction.

Related topics

Morphogenesis (A7)	Neuronal connections (J7)
Germ line specification (E1)	The neural crest (J6)

Germ-cell migration

In animals that set aside a germ line early in development (Topic E1), the **primordial germ cells** (**PGCs**; germ cells that have yet to reach the gonad) are specified before or during gastrulation, before the position of the **gonads** is established. The PGCs therefore often arise some distance away from the future gonads and must **migrate** to the gonads as they develop, in order to differentiate and form gametes. This involves a combination of passive movements brought about by the reorganization of cells during gastrulation, and active migration on the part of the PGCs. Only a few PGCs begin the journey, although they proliferate _en route_ so that the emerging gonads are colonized by a larger number of germ cells.

In plants, germ cells arise from somatic cells in the floral meristem, at the site of gonad development. This is necessary because plant cells are unable to migrate due to their rigid cell walls. Cell migration is also limited in some animals. In C. *elegans*, for example, most cells are born where they are needed.

Germ-cell migration in *Drosophila*

The first phase of germ-cell migration in *Drosophila* is passive, brought about by the infolding of the cellular blastoderm during gastrulation, at the **posterior furrow**. This transports the population of pole cells from their site of origin at the posterior pole to the posterior midgut inside the embryo (*Drosophila* gastrulation is discussed in Topic F2). From there, the germ cells actively migrate through the gut wall and into the adjacent mesoderm (active migration requires the activity of Nanos protein in the pole plasm). Migration through the gut wall is induced by the differentiation of the endodermal cells and facilitated by the product of the *wunen* gene, which is expressed specifically in the posterior midgut. The Wunen protein is a chemorepellant that drives the germ cells away from the endoderm and into the mesoderm. The migrating germ cells then split into two populations that move laterally and migrate in parallel towards the gonads. This is facilitated by the product of the gene *columbus*, a chemo-attractant that is expressed specifically in the gonadal mesoderm (*Fig. 1*). The opposing effects of Wunen and Columbus are confirmed by misexpression experiments. Germ cells migrate away from any cells misexpressing *wunen*, including the gonad, while they migrate towards any cells misexpressing *columbus*. Although the significance is not clear, both *wunen* and *columbus* encode proteins involved in lipid metabolism. Columbus is the *Drosophila*

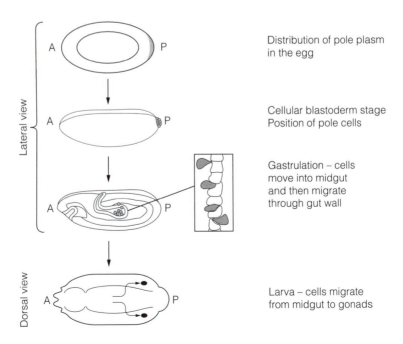

Fig. 1   Route of migrating germ cells in the Drosophila *embryo.*

homolog of HMG CoA reductase, while Wunen is a phosphatidic acid phosphatase.

Germ-cell migration in vertebrates

Germ-cell migration has been studied in a number of amphibians. *Xenopus* is discussed as an example below, although it is apparent that the route and mechanism of germ-cell migration is different in other amphibians and even in other frogs and toads. Germ-cell determinants in the *Xenopus* egg segregate to the

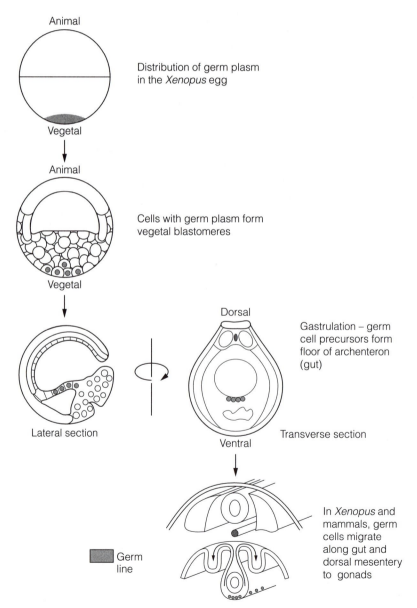

Fig. 2   *Route of migrating germ cells in* Xenopus. *The latter stage of the journey is similar in mammals (see text for details).*

most vegetal blastomeres during cleavage (Topic E1). Gastrulation brings these presumptive PGCs from the vegetal pole to the endodermal floor of the blastocoel, although they are not determined as germ cells until gastrulation is complete. After gastrulation, the PGCs are concentrated in the endodermal lining of the hindgut. They migrate along the gut and across the dorsal mesentery. They then split into two populations and migrate along the lateral plate mesoderm of the abdominal wall to reach the paired gonadal ridges. Only three to four germ cells begin the journey. During migration they undergo three rounds of cell division so that approximately 30 PGCs colonize the gonads. During this process, the PGCs are in close contact with the cells over which they migrate. The migratory route of the cells may be specified by the orientation of the underlying cells and the molecules they secrete into the extracellular matrix. In particular, fibronectin is thought to play an important role in migration because antibodies against fibronectin cause loss of cell adhesion between PGCs and the dorsal mesentery, and inhibit migration. The gene *Xdaz1*, which is homologous to the *Drosophila* gene *boule* (Topic E3), is also required for germ-cell migration in *Xenopus*.

In mammals, the route of the migrating germ cells is very similar to that in *Xenopus* (*Fig. 2*). PGCs arise in the epiblast posterior to the primitive streak. During gastrulation, they move into the mesoderm and then the endoderm, collecting at the posterior of the yolk sac. From here they migrate along the hindgut to the dorsal mesentery and into the gonadal ridges. Again, the migratory route may be mapped out by a polarized network of fibronectin and other extracellular matrix components. However, the number of cells making the journey to the gonadal ridge in mammals is substantially greater than in *Xenopus*. About 50 PGCs begin the journey, and 2–5000 cells colonize the gonads. Cell proliferation is dependent on signaling between the PGCs and the cells over which they travel. The migrating cells express a receptor called Kit, which is the product of the *White spotting* gene. The cells lining the migratory route secrete the ligand for this receptor, which is called Steel. In mice with mutations in either the *White spotting* or *Steel* genes, a reduced number of germ cells arrive at the gonad. The same genes also affect the proliferation of

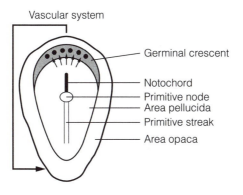

Fig. 3 Route of migrating germ cells in the chicken.

melanocytes and hemopoieteic stem cells (Topic J6). It has recently been possible to purify migrating germ cells during mouse development, investigate their repertoire of cell and matrix adhesion molecules, and design experiments to disrupt their function. Such experiments have shown specific requirements for a number of integrins (Topic C5) including a role for $\beta1$-integrin in the colonization of the gonad.

Germ-cell migration in birds and reptiles involves a completely different mechanism to that used in mammals. The PGCs arise in the epiblast at the anterior end of the embryo and migrate from the area pellucida into the hypoblast to form an anterior extraembryonic structure termed the **germinal crescent** (*Fig. 3*). From here, the PGCs squeeze through the endothelial lining of the blood vessels and are carried to the gonadal ridges through the vascular system. At the site of the gonadal ridge, the PGCs collect on the internal endothelial lining and once again squeeze through to colonize the ridge. It is thought that cell-adhesion molecules on the endothelial cells in the region of the gonadal ridge cause the PGCs to stick to the vascular wall, and that chemoattractants secreted by the ridge itself may cause the PGCs to migrate through the wall and colonize the gonads. This is supported by the ability of quail genital ridges to attract migrating germ cells even when grafted to ectopic sites in the chicken embryo.

E3 GAMETOGENESIS

Key Notes

Comparison of oogenesis and spermatogenesis

Sperm and eggs are haploid gametes that unite at fertilization to generate a diploid zygote. Their development from diploid primordial germ cells has certain features in common indicating homology (e.g. the requirement for meiosis and the derivation of specialized secretory organelles from the Golgi apparatus). There are also major differences, including the asymmetry and interrupted nature of oogenic meiosis, and differences in the timing of maturation.

Spermatogenesis in mammals

Diploid germ cells arriving at the embryonic testis are arrested in the mitotic cell cycle until the animal is born. Now known as spermatogonia, the cells proliferate to form a population of stem cells that continuously generate primary spermatocytes committed to enter meiosis. Meiosis produces four cells termed spermatids. These undergo extensive differentiation, involving nuclear condensation, ejection of the cytoplasm, and formation of the acrosomal vesicle and flagellum, to form functional sperm. Mammalian spermatogenesis is representative of most of the animal kingdom.

Oogenesis in mammals

Diploid germ cells arriving at the embryonic ovary (oogonia) divide several times by mitosis and then many die, the remainder entering meiosis as primary oocytes. Meiosis arrests at prophase I until sexual maturity, when periodic ovulation begins. Groups of oocytes are then stimulated to continue meiosis and mature, and eventually one is ovulated as a secondary oocyte, arrested at prophase II. This block is removed at fertilization. Oogenic meiosis is asymmetrical, producing only one egg (the other products being discarded as polar bodies). Oogenesis in mammals is relatively specialized, and different strategies are used in other species.

Germ cell sex determination

Primordial germ cells of both sexes are indistinguishable, but begin either spermatogenesis or oogenesis upon arrival at the gonad. In species with individual sexes, it has been shown that germ cell sex determination is controlled by both the intrinsic genetic sex of the germ cell and the influence of the surrounding somatic tissue of the gonad. Germ cells of one sex placed in gonads of the opposite sex do not differentiate properly. In hermaphrodites such as *C. elegans*, germ cell sex determination is under genetic control and mutations can be identified that switch germ cells from one sex to the other.

Gene expression during gametogenesis

Gametogenesis is an unusual form of differentiation because the egg must store a large amount of RNA and protein, and the nuclei of both sperm and egg undergo modifications that render them inactive. *Xenopus* oocytes show intense transcriptional activity prior to meiosis, involving highly extended lampbrush chromosomes and extrachromosomal amplification of rRNA genes. Conversely, insect oocytes are transcriptionally inactive. Surrounding nurse cells synthesize and export most of the gene products required in the oocyte, while follicle cells secrete the vitelline layer and chorion. In both *Xenopus* and *Drosophila*, yolk is made by maternal cells and imported into the oocyte. Most transcription is also shut down during sperm development, although some genes are transcribed after meiosis. In both eggs and sperm, gene regulation often involves the control of translation using messages transcribed before meiosis.

Related topics   Mechanisms of developmental   Gene expression and regulation
                      commitment (A3)   (C1)
                                                           Sex determination (E5)

Comparison of oogenesis and spermatogenesis

Gametogenesis is the differentiation of primordial germ cells to produce gametes. The development of **eggs (ova)** is termed **oogenesis**, and the development of **sperm (spermatozoa)** is termed **spermatogenesis** (*Table 1*). Eggs and sperm are homologous cells and their development shares certain common features:

- Both oogenesis and spermatogenesis involve **meiosis**, a specialized form of cell division in which **recombination** occurs between **homologous chromosomes** and in which the chromosome number is halved (*Fig. 1*). The mature egg and sperm are therefore **haploid** cells, and when their nuclei fuse at fertilization, the normal **diploid chromosome number** of the species is restored.
- Both oogenesis and spermatogenesis involve extensive morphological differentiation to equip each gamete for its particular role in fertilization. In sperm, such adaptations include the growth of a flagellum, nuclear condensation, ejection of the cytoplasm and the derivation of the acrosomal vesicle from the Golgi apparatus. In eggs, such adaptations may include a massive increase in size, the accumulation of nutrients to nourish the growing embryo, the secretion of proteins to form a protective outer layer, the assembly of cortical granules and the redistribution of cytoplasmic determinants that control embryonic development.
- Both eggs and sperm are unable to survive for very long if fertilization does not occur.

Table 1. Nomenclature of germ cells at various stages of gametogenesis

	Female	Male
Process	Oogenesis	Spermatogenesis
Diploid cell *en route* to gonads	Primordial germ cell	Primordial germ cell
Diploid cell colonizing gonad	Oogonium	Spermatogonium
Diploid cell committed to meiosis	Primary oocyte	Primary spermatocyte
After meiosis I	Secondary oocyte and first polar body	Two secondary spermatocytes
Haploid cell after meiosis II	Egg (ovum) and second polar body	Four spermatids
Mature gamete	Egg (ovum)	Sperm (spermatozoa)

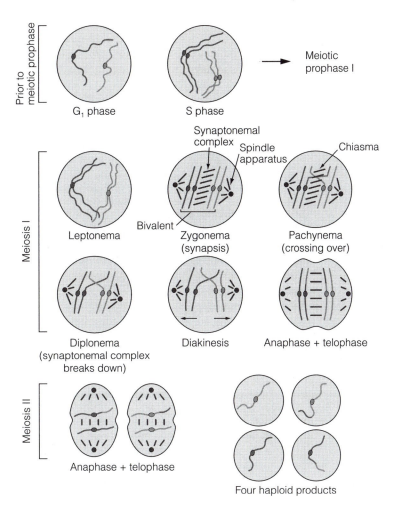

Fig. 1   Chromosome behavior during meiosis.

Although oogenesis and spermatogenesis are homologous processes, there are also major differences between them:

- The meiotic divisions in spermatogenesis are equivalent, resulting in the production of four equal products, **spermatids**, which differentiate to form four functional sperm. Conversely, the meiotic divisions that occur during oogenesis are highly asymmetrical. Only one egg is produced (containing all the cytoplasm) together with a number of **polar bodies**, which act as vessels for discarded chromosomes.
- Spermatogenic meiosis is uninterrupted, and spermatids are rapidly produced from spermatogonia (in mice this process takes about 2 weeks). Conversely, oogenic meiosis is arrested at one or more stages, and the oocyte may remain frozen in mid-meiosis for a considerable length of time. In many vertebrates, oogenic meiosis begins in the embryo but is arrested in the prophase of meiosis I until the animal reaches sexual maturity. In most

animals, meiosis is arrested once again awaiting fertilization. The stage at which this occurs is highly variable. In dogs, the diploid primary oocyte is fertilized, and fertilization heralds the onset of meiosis. In many insects, fertilization occurs during the first meiotic division, while in most amphibians and mammals, fertilization occurs during the second meiotic division. In sea urchins, the mature egg is fertilized (i.e. meiosis is complete before fertilization).

- The egg matures gradually, often taking advantage of the long period of meiotic arrest. Conversely, the maturation of round spermatids into spermatozoa occurs rapidly, after the completion of meiosis, in a concerted phase of differentiation termed **spermiogenesis**.

In most animals, both the spermatogonia and oogonia are self-renewing stem cells, so that the population of spermatocytes and oocytes is continuously replenished and both gametes are available continuously. However, oogenesis in mammals involves the production of a finite number of eggs that are rationed throughout the animal's life. Oogenesis in mammals is unique for several reasons, all in some way related to the placental nourishment of the embryo: the relatively small size of the egg, the small number of eggs produced and the lack of yolk. Mammalian eggs also lack cytoplasmic determinants and show a rapid onset of zygotic gene expression during development (Topic F1). Spermatogenesis and oogenesis in mammals are discussed below and compared in *Fig. 2*. The diversity of oogenesis in animals is discussed using *Xenopus* and *Drosophila* as examples.

Spermatogenesis in mammals

Diploid primordial germ cells arriving at the male gonad (testis) are called **spermatogonia**. They undergo several rounds of mitotic division and then arrest at the G_1 phase of the mitotic cell cycle, where they remain until after birth. During this time, the gonad matures: the epithelial cells of the immature gonad differentiate into **Sertoli cells**, which provide nourishment to the spermatogonia and remain closely attached to them through mutual expression of the cell adhesion molecule **N-cadherin**.

The cell cycle block is lifted after birth, and the spermatogonia undergo further rounds of mitotic division. The spermatogonia are stem cells, each division producing one further stem cell as well as a cell that will go on to differentiate. Differentiating cells are not completely separated during subsequent mitotic divisions, producing a syncytial network joined by thin cytoplasmic bridges. This allows free diffusion of molecules between the cells, enabling them to mature at the same rate. The fourth division produces another stem cell as well as the first cell that is committed to enter meiosis, the **intermediate spermatogonium**. Intermediate spermatogonia undergo two more rounds of mitotic division to produce **primary spermatocytes** that undergo meiosis. After the first meiotic division, the cells are called **secondary spermatocytes**. After the second mitotic division, the round haploid cells are called **spermatids**. During their maturation, the spermatocytes, and later the spermatids, migrate towards the luminal surface of the seminiferous tubule.

Spermatids have a large volume of cytoplasm, no flagellum and they are joined in a syncytial network. They must therefore undergo extensive differentiation during spermiogenesis to form functional mature sperm. Spermiogenesis involves five major processes, and differentiation continues *after* the sperm are released into the lumen of the seminiferous tubules.

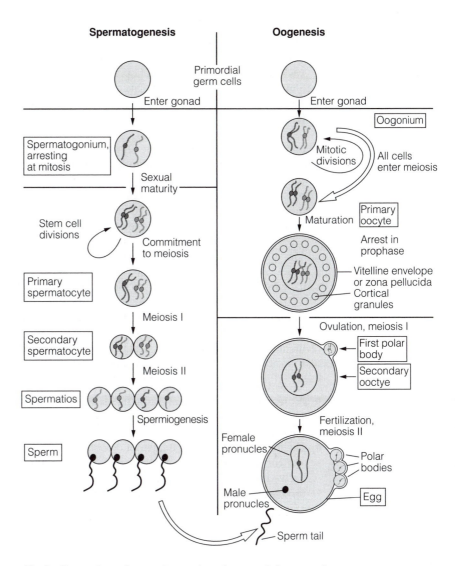

Fig. 2   Comparison of spermatogenesis and oogenesis in mammals.

- The formation of the **acrosomal vesicle** from the Golgi apparatus. The acrosomal vesicle forms a cap over the sperm nucleus. It contains enzymes that digest the outer layers of the egg. In sea urchins and some other species, **globular actin** is stored behind the acrosomal vesicle to form the **acrosomal process** at fertilization (Topic E4).
- The growth of the **flagellum**. This involves polymerization of **tubulin** monomers from the centriole, which becomes localized to the base of the sperm nucleus. The proximal region of the flagellum, which forms the neck and midpiece of the sperm, is called the **axoneme** and is associated with many mitochondria, which provide the energy for sperm propulsion.
- The extrusion of the cytoplasm. The cytoplasm is ejected from the cell as a

cytoplasmic droplet, causing the remaining mitochondria to form a ring around the axoneme.

- The condensation and remodeling of the nucleus. This involves the layered packaging of DNA using sperm-specific DNA binding proteins termed **protamines**, that largely replace the histones found in normal chromatin.
- The presentation of egg-binding proteins on the sperm plasma membrane (see Topic E4).

Oogenesis in mammals

Primordial germ cells entering the female gonad (ovary) are termed **ooogonia**. They undergo several rounds of mitotic division to form a relatively large population (5–7 million in humans). However, after this brief period of proliferation, many of the cells die, and the remaining 50 000 or so enter meiosis and become **primary oocytes**. In mammals, oogonia are not self-renewing stem cells, so the population of primary oocytes in the embryo represents the maximum number of eggs that any individual can produce. The primary oocytes arrest at diplonema in the prophase of the first meiotic division and remain in that state until sexual maturation. Each oocyte is encapsulated in an epithelial layer of somatic **follicular granulosa cells**, to form a **primordial follicle**.

Sexual maturation in most mammals involves the onset of **periodic ovulation**, in which meiosis resumes in small groups of oocytes and one oocyte is released from the ovary into the reproductive tract at regular intervals. Ovulation is preceded by a short period of **follicular growth**, in which the follicular granulosa cells surrounding the oocyte proliferate rapidly, and the oocyte itself matures. Oocyte maturation involves the following processes:

- A great increase in size, reflecting increased RNA and protein synthesis.
- The secretion of a number of proteins that form the outer covering of the egg, the **zona pellucida**.
- The formation of a cortical layer of cytoplasm, which contains **globular actin** and **cortical granules**. The latter, like the acrosomal vesicle of the sperm, are derived from the Golgi apparatus, and contain digestive enzymes as well as proteins with other functions in fertilization (see Topic E4).

Follicular growth and oocyte maturation are induced by **follicle stimulating hormone (FSH)**, secreted from the pituitary gland. Although a group of follicles is stimulated to grow, most break down, and only one oocyte is ovulated. This reflects a positive feedback loop in which **estrogen** is produced by **thecal cells** surrounding each follicle. Estrogen stimulates the production of **FSH receptors** on the follicular granulosa cells but at the same time causes the pituitary gland to reduce FSH secretion. Therefore, there is competition for a dwindling amount of FSH and only the follicle with the most FSH receptors can survive. Under the control of **lutenizing hormone** (LH) secreted from the pituitary gland, meiosis resumes in this remaining oocyte and the first meiotic division occurs. Meiosis is inhibited by signals from the surrounding follicular granulosa cells, arriving through gap junctions that extend through the zona pellucida. The resumption of meiosis occurs because FSH and LH can inhibit communication between follicle cells and oocyte, by blocking these junctions. At the first meiotic division, both the secondary oocyte and the first polar body are retained within the zona pellucida, and ovulation occurs at this stage. Many of the follicular granulosa cells remain on the outside of the oocyte as it is ovulated, forming the **cumulus layer**. Meiosis arrests once again at the prophase of meiosis II, awaiting fertilization (see Topic E4).

Germ-cell sex determination

Sexual reproduction depends on the fusion of male and female gametes. In some species, these are produced by separate **male and female individuals**, while in other species both types of gamete can be produced by the same individual, which is termed a **hermaphrodite**. Upon arrival at the gonad, male and female primordial germ cells are phenotypically indistinguishable, but they soon begin either oogenesis or spermatogenesis. In species with individual sexes where sex is determined on the basis of the differential distribution of sex chromosomes, it is important to ask how much the development of germ cells is governed by the sexual phenotype of the surrounding somatic tissue in the gonad, and how much by the genotype of the germ cell itself. Gonad-dependent differentiation would involve signaling from gonadal cells, whereas genotype-dependent differentiation would reflect a cell autonomous property of the germ cells. In hermaphrodites, it is important to ask how the same gonad can induce identical germ cells to differentiate along two such very different pathways.

The mechanism of germ cell sex determination in species with individual sexes can be addressed by introducing germ cells of one sex into gonads of the opposite sex. This can be achieved either mechanically by transplantation, or by the generation of mosaic or chimeric embryos that contain germ cells of one sex and gonadal somatic cells of the other (see Topic E5). In mammals, early gametogenesis appears to be controlled by signals from the surrounding gonadal tissue. Primordial germ cells of either sex become oocytes in the ovary and spermatocytes in the testis. This sexual differentiation is controlled in part by regulation of the cell cycle: the ovary provides an environment that encourages germ cells to enter meiosis immediately, while the testis causes cell-cycle arrest before meiosis. Germ cells of either sex that colonize somatic tissue outside the gonad become oocytes, because there is no signal to arrest the cell cycle. However, ectopic germ cells and germ cells of the 'wrong' sex for the gonad they colonize do not produce mature gametes. It appears that later in differentiation, the genotype of the germ cell itself also plays a role in gametogenesis. In *Drosophila*, male germ cells incorporated into the ovary undergo spermatogenesis, so it appears that the genotype drives early gametogenesis in male germ cells. Conversely, female germ cells incorporated into the testis also undergo spermatogenesis, so it appears that signals from the testis can overrule the influence of the female genotype. Again, however, neither process results in mature gametes. This suggests that, at least in those species where sex is genetically determined, both genotype and environment act in concert to control gametogenesis.

In hermaphrodite *C. elegans*, where sperm and eggs are produced in the same gonad, germ-cell differentiation is regulated by signals from somatic cells. All germ cells develop from the P_4 blastomere that contains P-granules originally distributed uniformly in the egg (see Topic E1). The proliferation of these germ cells is controlled by a signal from the **distal tip cell** of the gonad, the LAG-2 protein that also controls cell specification during vulval development (Topic L1). In germ cells, LAG-2 is thought to interact with the GLP-1 receptor, which is homologous to the *Drosophila* protein Notch. Germ cells near the distal tip cell undergo mitosis and proliferate, but as the cell population grows some cells move outside the range of the signal and these enter meiosis. The first cells to enter meiosis become sperm, while the more distal cells that undergo meiosis later become eggs. This switch between male and female germ cell differentiation is controlled by two principal sets of genes: *fem* (feminizer) and *fog* (feminization of germline). Loss of function *fem* and *fog* mutations cause all germ cells to become eggs, so hermaphrodites are converted into females. The FEM proteins are thought to activate the *fog* genes, whose products directly control spermatogenesis. The control of sex determination is mediated by the *C. elegans* homologs of Nanos and Pumilio, which repress *fem* translation.

Gene expression during gametogenesis

As discussed in Topic A5, cell differentiation requires the expression of new sets of genes to produce the proteins that confer new characteristics on the differentiating cells. Gamete differentiation is no exception, so new genes must be expressed during gametogenesis. However, gametogenesis is unusual for two reasons. First, the egg is very large compared to somatic cells, and must store large amounts of RNA and protein to be used during development. Second, the nuclei of both eggs and sperm undergo extensive changes that might interfere with transcription: meiosis occurs during the differentiation of both types of gamete, and in the case of sperm, the nucleus is condensed for packing into the sperm head. A number of mechanisms have therefore evolved to facilitate gene expression during gametogenesis.

Rapid accumulation of RNA and protein – oogenesis in **Xenopus** *and* **Drosophila**

The egg is much bigger than somatic cells in all animals, because eggs store large amounts of RNA and protein, some of which is needed for oogenesis, some for fertilization, and some for development. The human egg is relatively small, although still about 1000 times larger than a somatic cell. Conversely, many amphibian, reptile and bird eggs are immense compared to somatic cells. Such differences exist due to diverse strategies for nourishment and development. Some species (e.g. sea urchins) rapidly produce a feeding larval stage, while mammalian embryos are nourished through a placenta. In other animals the egg must nourish the embryo until it can feed independently, and this is accomplished by the storage of large amounts of protein as **yolk**. Similarly, in some species, the mRNA and protein required for development is present as maternal gene products in the egg, while in others (such as mammals) much of this mRNA and protein is expressed from the zygotic genome during development. As discussed below, transcription in the egg usually shuts down during meiosis, so where there is a huge demand for RNA and protein, this must be met during oogenesis prior to the onset of meiosis.

In *Xenopus*, eggs are produced from stem cells and new batches develop each year. Oogenic meiosis is arrested in diplotene of the first meiotic prophase perhaps in order to allow for the accumulation of RNA and protein in the oocyte. During this phase, the oocyte shows intense transcriptional activity. The chromosomes are completely decondensed and transcription occurs continuously, generating extensively unwound structures called **lampbrush chromosomes** (because they resemble the little brushes on a wire that are used to clean test tubes). Even with this elevated transcriptional activity there are certain genes that are still not transcribed fast enough to provide sufficient material for the egg. Gene expression occurs in two stages, transcription and translation, so that the amount of protein can be increased by maximizing the efficiency of both processes. However, the embryo also needs large amounts of rRNA to form ribosomes, and as rRNA is produced directly by transcription, its synthesis cannot be increased to the same magnitude. Elevated rRNA synthesis is achieved by **gene amplification**, involving the extrachromosomal replication of the **rRNA genes** to form many DNA circles. These are actively transcribed as

rolling circles, to produce large amounts of rRNA in a very short time. *Xenopus* oocytes also accumulate large amounts of yolk, which contributes greatly to their increase in size. Unlike other proteins required for development, yolk proteins are made by somatic cells in the liver, and imported into the oocyte by micropinocytosis.

In *Drosophila*, both eggs and sperm are produced continuously from stem cell populations. A distinct strategy for RNA and protein accumulation is used in *Drosophila* oogenesis because, in insects, the oocyte is transcriptionally inactive. Oogenesis in *Drosophila* is described as **meroistic** because when the oogonium divides, the daughter cells remain connected to each other by narrow cytoplasmic bridges called **ring canals**. The *Drosophila* ovariole contains a population of stem cells undergoing mitotic divisions, each division producing another stem cell and an oogonium. Each oogonium undergoes four mitotic divisions to produce 16 interconnected cells termed **cystocytes**, which are surrounded by hundreds of somatic **follicle cells** to form a **germinal cyst**. The ovariole thus comprises a series of germinal cysts at various stages of maturity. Of the 16 cystocytes, only one becomes the oocyte, while the others become **nurse cells**. The posterior-most cell in the germinal cyst always becomes the oocyte, because this retains a proteinaceous organelle, the **fusosome**, which was present in the original oogonium. This asymmetric structure helps to organize the cytoskeleton such that both microtubules and microfilaments are orientated to facilitate the transport of proteins, mRNAs and organelles from the nurse cells into the oocyte.

Oogenesis in *Drosophila* takes less than 2 weeks and the nurse cells are transcriptionally very active during this time. This is facilitated by their **polytenization** (the cells undergo several rounds of DNA replication without intervening mitosis, so that they contain 512 copies of each chromosome), allowing the transport of large amounts of RNA and protein to the oocyte. The vitelline membrane and the hard shell of the egg, the chorion, are secreted by the follicle cells. Uniquely in *Drosophila melanogaster*, the gene encoding the chorion protein is amplified within the chromosome to meet the especially high demand for this protein. In other *Drosophila* species, the chorion gene is already present as tandem multiple copies, and specific amplification within the follicle cells is not necessary. The yolk proteins for the *Drosophila* egg are made in other somatic cells in the ovary and the fat body, and are taken into the oocyte by receptor-mediated endocytosis.

Gene expression during and after meiosis

Nuclear reorganization interferes with transcription, requiring special mechanisms to allow the production of proteins required at each stage of gametogenesis. One strategy, used in both sperm and egg development, is to transcribe the genes required for gametogenesis before meiosis and store the products as inactive mRNAs. At the appropriate time, the mRNAs can be reactivated by regulatory factors that act at the RNA rather than the DNA level (see Topic C1). In the maturing oocyte, genes are often transcribed whose products are not needed until after ovulation, or until after fertilization. The mRNAs for these genes may be inactivated by shortening or removing the polyadenylate tail, or by associating the mRNA with a protein that inhibits translation. Similar mechanisms are found in sperm development. The mRNA for protamine, the major DNA-binding protein in the sperm head, is transcribed before meiosis, but not translated until late spermiogenesis. A number of genes encoding **RNA-binding**

proteins have been identified as important regulators of sperm development. For example, mutations in mammalian *DAZ* genes, and the homolog *boule* in *Drosophila*, block spermatogenesis before meiosis.

In most animals, transcription in the oocyte ceases during the first meiotic division and does not commence again until after fertilization. In sperm development, however, a number of genes are known to be transcribed specifically in the haploid spermatids before nuclear condensation. For example, one of the zona-binding proteins found on the sperm plasma membrane is synthesized from mRNA transcribed in the spermatids. This is known to be the case because individuals heterozygous for a mutation in the corresponding gene produce 50% sperm that can bind the zona pellucida and 50% sperm that cannot. If the gene was expressed in diploid cells, the mRNA from the functional allele would be distributed to all haploid sperm cells, and all would bind to the zona.

E4 GAMETE RECOGNITION, CONTACT AND FERTILIZATION

Key Notes

Requirements for fertilization

Fertilization, the fusion between egg and sperm, involves a number of different steps. The egg and sperm must first be brought together, which involves a combination of passive movement and active motility on the part of the sperm. Once the gametes are together they must recognize each other and make contact. This is followed by the sperm penetrating the protective layers of the egg and fusing with the egg plasma membrane enabling the nucleus to enter the egg. One a sperm nucleus has entered the egg, the participation of further sperm must be blocked.

Attraction between sperm and egg

The egg attracts sperm by releasing potent chemoattractant signals that both direct the sperm to the egg and increase sperm activity. Such signals are often species-specific and are usually timed to coincide with egg maturity, to prevent unproductive interactions between gametes. Chemoattraction is crucial in aquatic environments but may also be used in species that utilize internal fertilization.

Gamete contact in sea urchins

The jelly coat of the sea urchin egg induces the sperm's acrosome reaction, allowing the sperm to penetrate through the jelly coat and the vitelline envelope. The release of the acrosomal vesicle allows the acrosomal process to form, containing a protein called bindin that interacts with receptors on the vitelline envelope. Both induction of the acrosome reaction and binding to the vitelline envelope may involve species-specific interactions.

Gamete contact in mammals

In mammals, surface properties of the sperm membrane must be modified in order for the sperm to interact with the zona pellucida, a process termed capacitation. Interaction with the zona pellucida induces the acrosome reaction and the sperm can penetrate this layer and contact the membrane.

Gamete fusion and sperm entry

After penetrating the protective layers of the egg, the sperm and egg plasma membranes fuse together, allowing the sperm nucleus to enter the egg cytoplasm. Membrane fusion may require specific fusogenic proteins carried by the sperm and exposed by the release of the acrosomal vesicle.

Prevention of polyspermy

Most animals prevent more than one sperm nucleus fusing with the egg nucleus by blocking the entry of additional sperm once the first sperm has entered the egg. Sea urchins use a fast blocking mechanism in which the first sperm induces depolarization of the egg membrane and prevents later arriving sperms from fusing, as well as a slow mechanisms in which cortical granules in the egg fuse with the plasma membrane and release enzymes and other molecules that change the properties of the vitelline envelope. A similar cortical granule reaction occurs in many mammals, resulting in a zona reaction that releases sperm from the zona pellucida.

The zygote

After entry, most of the sperm breaks down, leaving the nucleus and centriole. The nucleus decondenses and the centriole initiates an aster of microtubules that contact the egg nucleus. The two nuclei are drawn together as the microtubules contract. Depending on the species, fusion may occur immediately or may require the completion of meiosis by the egg nucleus.

Related topics   Signal transduction in   Cleavage (F1)
                            development (C3)

Requirements for fertilization

Fertilization is the fusion of haploid gametes, a sperm and an egg, to form a diploid zygote. This marks the beginning of development. Although simple in concept, fertilization presents a number of challenges, which can be summarized in the following list of questions:

- How are the egg and sperm brought together?
- How do the egg and sperm recognize each other and make contact?
- How does the sperm get inside the egg?
- How is it possible to prevent the same egg being fertilized more than once?
- How is the specificity of fertilization achieved?

These challenges must be addressed in the context of two fertilization strategies. **External fertilization** occurs when both sperm and eggs are deposited into water. In this case the gametes of each sex may move about on random currents and there is the added complication that the gametes of many species mix together. **Internal fertilization** occurs when the egg is retained in the body of the female, and the sperm are deposited there by the male. In this case there is a definitive pathway to the egg, and gametes from other species are usually not a consideration. In external fertilization, large numbers of both gametes are available, although often more sperm than eggs. In internal fertilization, there are often very few eggs, but large numbers of sperm.

Attraction between sperm and egg

Regardless of whether a species utilizes internal or external fertilization, sperm are rarely if ever deposited immediately on top of the egg. The gametes must therefore possess a mechanism to attract each other at a distance. In aquatic environments, where external fertilization is common, there is the added complexity that the sperm and eggs of many species mix together. Where sperm and eggs are shed into water, it has been shown that sperm are attracted to eggs because the latter release potent chemoattractant signals. Sea urchin eggs, for example, release substances that have profound effects on sperm behavior even at very low concentrations. In response to such signals, sperm not only change direction but also increase their activity, so they swim faster. It appears that sperm can swim up a concentration gradient towards the egg. This mechanism is beneficial because it can control both the timing and specificity of fertilization. Many eggs are released before they are ready to be fertilized, and interactions with sperm would be unproductive. By timing the release of chemoattractant signals to coincide with the onset of maturity, the egg attracts sperm only when ready to be fertilized, and sperm swimming near an immature egg will safely

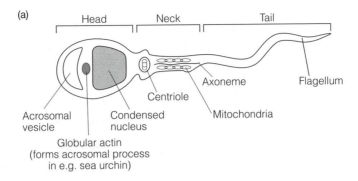

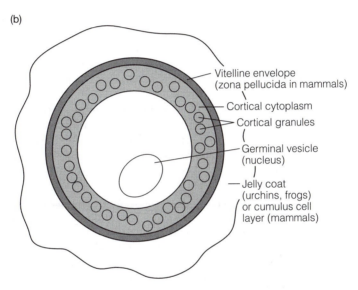

Fig. 1.   Simplified structure of the gametes. (a) The sperm. (b) The egg.

ignore it. The chemoattractants are also species-specific, preventing unproductive interactions with sperm from other species living in the same environment.

With internal fertilization, the female reproductive tract often plays an important role in the early part of the sperms' journey. In mammals, for example, muscular contractions of the uterus are primarily responsible for transporting sperm from the neck of the cervix to the oviduct. Only within the oviduct does the sperms' own motility become important, and once again the sperm may respond to signals, either from the egg itself or from the ovarian follicle, which increase motility and provide a directional cue.

Gamete contact in sea urchins   Sea urchin eggs are surrounded by a fibrous **vitelline envelope** and enclosed in a coat of **egg jelly**. Once the sperm reaches the egg, it must penetrate the egg jelly and the vitelline envelope to make contact with the egg plasma membrane. This is achieved by the **acrosome reaction**. In all animals, sperm maturation involves the formation of an **acrosomal vesicle** from the Golgi apparatus of the spermatocyte. The acrosomal vesicle contains enzymes that digest the protective layers surrounding the egg, allowing the sperm access to the egg membrane. In sea urchins, contact with oligosaccharide groups in the egg jelly opens calcium channels in the outer acrosomal membrane of the sperm, causing an influx of calcium into the sperm head. This induces the acrosomal membrane and the sperm plasma membrane to fuse, releasing the contents of the acrosomal vesicle into the egg jelly. Globular actin molecules stored behind the acrosomal vesicle then polymerize to form the **acrosomal process**. A protein called **bindin** is displayed on the acrosomal process and this is recognized by a specific receptor on the surface of the vitelline envelope. After contact is established, the contents of the acrosome continue to work on the vitelline envelope, until there is room for the sperm to penetrate the envelope and make contact with the egg membrane. Recognition between bindin and its receptor is species-specific, so contact between the sperm and eggs of different species is prevented. In some sea urchins, induction of the acrosome reaction is also species-specific. Unproductive interactions with foreign sperm may therefore be avoided at three levels: chemoattraction between egg and sperm; induction of the acrosome reaction; and bindin-mediated attachment to the vitelline envelope.

Gamete contact in mammals

Mammalian eggs are enclosed within a proteinaceous envelope called the **zona pellucida**, which is surrounded by a **cumulus layer** comprising granulosa cells of the ovarian follicle. Sperm can push through the cumulus layer but the acrosome reaction is required to penetrate the zona pellucida. In mammals, sperm are not competent to undergo the acrosome reaction immediately after release. They must spend some time in the female reproductive tract undergoing a process called **capacitation**. The duration of capacitation varies from species to species but probably involves modification of the lipid composition in the sperm cell membrane and the removal of proteins or oligosaccharide groups that block interaction between sperm and the zona pellucida. After capacitation, the sperm can bind to the zona pellucida and this induces the acrosome reaction. The zona pellucida comprises two major proteins, ZP2 and ZP3, which are cross-linked by another protein, ZP1. Initial binding between the sperm and the zona pellucida is mediated by ZP3. This interaction is only moderately species-specific, probably because foreign sperm are unlikely to be present in species that use internal fertilization as a mating strategy. ZP3 can bind a number of different proteins on the sperm, one or more of which are involved in initiating the acrosome reaction. Unlike the situation in sea urchins, the mammalian acrosome reaction occurs *after* sperm bind to the egg, so that the enzymes are released precisely at the site required. However, the acrosome reaction is similarly initiated by calcium influx into the sperm, but the mechanism by which ZP3 causes this is not precisely understood. The outer membrane of the sperm head is shed during the acrosome reaction, together with the proteins that bind ZP3. Contact between the sperm and egg is maintained by **secondary binding** to the other major constituent of the zona pellucida, ZP2. Sperm proteins that bind ZP2 have been identified in some mammals. Experiments in which binding between ZP3 or ZP2 and their ligands on the sperm have been disrupted have shown that fertilization is prevented. Both these strategies are being explored as possible means of contraception.

Gamete fusion and sperm entry

Once contact has been established between the gametes, enzymes from the acrosome digest away part of the vitelline envelope (sea urchins) or zona pellucida (mammals), allowing direct contact between the sperm and egg plasma membranes. In many species, fusion is induced by specific **fusogenic proteins** on the inner acrosomal membrane of the sperm, such as the **fertilins** in mammals. Such proteins recognize receptors on the egg surface and contain long stretches of hydrophobic amino acid residues that facilitate membrane fusion. Once the membranes have fused, the sperm nucleus can enter the egg, along with the associated centriole, mitochondria and flagellum.

Prevention of polyspermy

Only one sperm is required to fertilize the egg, but many sperm are usually found on the outside of the egg at the time of fertilization. In some animals, it is normal for several sperm to enter the egg, and in some unknown way all but one are broken down and inactivated. In most animals, however, the entry of multiple sperm into the egg (**polyspermy**) results in abnormal ploidy because more than two haploid nuclei fuse together to generate the zygotic nucleus. Furthermore, due to the presence of additional centrioles, the zygote undergoes abnormal cleavage (Topic C4). The egg must react quickly to avoid the entry of supernumerary sperm.

In the sea urchin embryo, there are at least two mechanisms to prevent polyspermy. The **fast block to polyspermy** is activated immediately after the first sperm fuses with the egg, and involves a change in electrical potential. Before fertilization, the egg membrane has a resting potential of about –70 mV, which is maintained in part by the active export of sodium ions. This negative potential is required for sperm fusion, since artificial depolarization of the membrane has been shown to prevent fertilization. Sea urchin sperm carry a protein that activates sodium channels in the egg membrane, therefore allowing an influx of sodium ions and increasing the resting potential to +20 mV. The first sperm to fuse with the membrane therefore reverses membrane polarity and prevents further fusion. Membrane depolarization is transient, lasting about one minute. However, about 20 seconds after sperm entry, a second **slow block to polyspermy** begins, termed the **cortical granule reaction**. Fusion between egg and sperm causes specialized granules in the actin-rich cortical region just below the egg membrane at the site of sperm entry to fuse with the membrane and release their contents into the space between the membrane and the vitelline envelope. The cortical granules contain a cocktail of enzymes and other proteins with multiple effects that prevent further sperm entry:

- Peroxidases cross-link tyrosine residues in the vitelline envelope to block the progress of sperms making their way through.
- Proteases cleave the bindin receptor, so that sperm on the outside of the vitelline envelope are released from their attachment points.
- Proteases cleave the proteins that attach the vitelline envelope to the cell membrane and mucopolysaccharides increase the osmotic potential under the vitelline envelope, causing an influx of water. This increases the distance between the vitelline envelope and the cell membrane, forming the **fertilization envelope**.
- **Hyalin** forms a layer around the egg, which plays several roles in cleavage (Topic F1).

A cortical granule reaction also occurs in mammalian eggs, although a fertilization envelope and hyalin layer are not created. The enzymes released modify both ZP3 and ZP2 so that primary and secondary-binding sperm are shed from the zona pellucida (**zona reaction**).

The cortical granule reaction is induced by the release of calcium ions from the endoplasmic reticulum of the egg. This begins at the point of sperm entry and propagates across the surface of the egg. It is thought that sperm binding initiates a signal transduction cascade culminating in the production of **inositol trisphosphate (IP₃)** (there are several pathways leading to this second messenger, discussed in Topic C3). There are **IP₃-gated calcium channels** in the endoplasmic reticulum, so an increase in IP_3 could activate these channels flooding the cytoplasm with calcium ions. Calcium itself can activate further calcium channels, so the release of calcium would rapidly spread across the egg, propagating the cortical granule reaction.

The zygote

After the sperm enters the egg, most of the body of the sperm is broken down, and only the nucleus and centriole survive (a few mitochondria may also survive, but there may be hundreds of thousands of egg-derived mitochondria). The nucleus decondenses as the sperm-specific histones and/or protamines are replaced by histones from the egg cytoplasm, and becomes known as the **male pronucleus**. The egg nucleus is the **female pronucleus**.

In most species, the sperm centriole is required for pronuclear fusion and, later, cleavage. The male and female pronuclei are brought together by the formation of a microtubule aster using the sperm centriole as an organizing center and egg-derived tubulin subunits. The contraction the aster draws them together. In sea urchins, the egg is fertilized after meiosis, and the two nuclei fuse to form a diploid zygotic nucleus in the single-celled zygote. In mammals, the female pronucleus completes meiosis before fusion and ejects the second polar body. The male and female pronuclei undergo S-phase as they move towards each other, and then undergo mitosis using the same spindle apparatus. The first diploid nuclei form after the first cleavage division.

E5 SEX DETERMINATION

Key Notes

Sexual phenotype and sex determination

Many species are sexually dimorphic, i.e. there are separate male and female individuals that produce the male and female gametes (sperm and eggs), respectively. In other species, hermaphrodites produce both types of gamete. The somatic sexual phenotype comprises primary and secondary sex characteristics. The primary sex characteristic is the sex of the gonad (testis or ovary) while secondary sex characteristics include e.g. genital structure, body size and markings. In most sexually dimorphic animals, determination of primary sex characteristics is controlled autonomously by sex chromosomes. Secondary sex characteristics may be determined in the same way, or alternatively by hormones released from the developing gonad.

Sex determination in mammals

In mammals, males are XY and females XX. Primary sex determination is controlled by the presence of the Y-chromosome, which directs male development. Secondary sex characteristics are determined by the hormones testosterone and Mullerian inhibiting substance secreted by the testis.

Sex determination in Drosophila

In *Drosophila*, males are XY and females XX. Primary sex characteristics are determined by the balance of X-chromosomes to autosomes, with males having a low X:A ratio of 0.5 and females a high X:A ratio of 1. Unlike mammals, the sex of each somatic cell in *Drosophila* is determined autonomously by the X:A ratio. It is therefore possible to produce mosaics called gynandromorphs with sectors of male and female tissue showing independent secondary sex characteristics.

Sex determination in C. elegans

In *C. elegans*, males are XO and hermaphrodites XX. Sex in *C. elegans* is also determined by the X:A balance, resulting in the activation of a gene called *XO-lethal* (*xol-1*) specifically in males. This encodes a repressor protein that initiates a downstream cascade of repressive interactions, culminating in the repression of the gene *transformer-1* (*tra-1*) whose function is to direct hermaphrodite-specific development. In hermaphrodites, *xol-1* is repressed by the high X:A ratio and *tra-1* is active. Secondary sex characteristics are controlled by the level of TRA-1 protein in somatic cells, but extensive cell–cell interactions are also involved.

Dosage compensation

Most of the genes carried on the X-chromosome are not involved in sexual differentiation, and the differential gene dosage between males and females could have disastrous consequences. Dosage compensation mechanisms are therefore used to make males and females equivalent in terms of X-linked gene expression. In mammals, this is achieved by inactivating one of the X-chromosomes randomly in each female cell. In *Drosophila*, there is male-specific transcriptional upregulation of X-linked genes, while in *C. elegans*, there is hermaphrodite-specific transcriptional repression of X-linked genes.

Environmental sex determination

Where sex chromosomes do not control sex-determination, the environment plays a significant role. For example, the sex of many reptiles is controlled by the temperature of the egg just after fertilization through the temperature-dependent regulation of enzymes that interconvert male and female hormones. Other animals may change their sex during their lifetime in response to signals from other members of the population.

Sex determination in plants

Unlike animals, most plants are hermaphrodites, producing flowers containing both male and female sex organs (stamens and carpels). In the minority of plants with independent sexes, male and female flowers may arise on the same plant (monoecy) or separate male and female plants (dioecy). Monoecy is often under hormonal control, so that the development of organs of one sex can be suppressed in particular regions of the plant. In dioecious species, the mechanisms of sex-determination appear striking similar to those in animals: both male determining Y-chromosomes and X:A balance mechanisms appear to operate in different species.

Related topics

Gene expression and regulation (C1)

Chromatin and DNA methylation (C2)

Gametogenesis (E3)

Plant *vs.* animal development (N1)

Flower development (N6)

Sexual phenotype and sex determination

Fertilization in all animals involves the fusion of **male and female gametes**. Male gametes are small and motile (sperm) whereas female gametes are large and immotile (eggs). This is the **germ cell sexual phenotype**, which becomes established during gametogenesis, as discussed in Topic E3. In many species, there are also *individuals* of two different **sexes**, males and females, each sex producing one type of gamete. In other species, all individuals are **hermaphrodite**, i.e. capable of producing both types of gamete. This is the **somatic sexual phenotype**, and where there are two sexes it is termed **sexual dimorphism**. In still other species, there are variations on this pattern: for example, *C. elegans* shows sexual dimorphism, but individuals are either male or hermaphrodite.

Sexual dimorphism can be described in terms of primary and secondary sex characteristics. **Primary sex characteristics** reflect the nature of the gonads. Males have **testes** that produce sperm, females have **ovaries** that produce eggs, and hermaphrodites have both types of gonad, or a bifunctional gonad allowing the production of both eggs and sperm. **Secondary sex characteristics** mark the sexes as phenotypically distinct but do not affect the nature of the gonad. Such characters include the structure of the genitals, body size, body markings, and the development of sex-specific structures such as the sex-combs in *Drosophila*.

In sexually dimorphic species, **sexual differentiation** involves deviation from a common developmental pathway, since early embryos of either sex are indistinguishable. Before differentiation occurs, there must be a mechanism of **sex determination** to choose the appropriate pathway. In most sexually dimorphic species, the primary sexual phenotype is determined genetically by **sex-chromo-**

somes (chromosomes that are distributed asymmetrically between the sexes and direct sex-specific developmental pathways due to the presence or dosage of particular genes). Secondary sex characteristics may also be determined using sex-chromosomes, or alternatively by hormonal signaling. The investigation of abnormal individuals that are mosaics or chimeras for cells with different sex-chromosome genotypes can help to distinguish between these alternatives.

Sex determination in mammals

In mammals, the primary sexual phenotype is dependent upon the sex-chromosome delivered by the sperm. Eggs normally contain one X-chromosome, and the sperm may deliver either an X-chromosome or a Y-chromosome. Embryos with two X-chromosomes (XX) usually develop as females, and those with a single X-chromosome plus a Y-chromosome (XY) usually develop as males. Maleness is conferred by the **presence of the Y-chromosome** not by the number of X-chromosomes (cf. *Drosophila* below). This is demonstrated by the primary sexual phenotype of individuals with abnormal numbers of sex chromosomes. Individuals with a single X chromosome are female, whereas XXY individuals are male, although in both cases these individuals show abnormal primary and secondary sex characteristics.

The sex-determining region of the human Y-chromosome contains a single gene, *SRY*, which is expressed in the male gonadal ridge and initiates male development. Rare XX males have been found to carry *SRY* as a translocation of the short arm of the Y-chromosome onto the X-chromosome. Similarly, some XY females appear to lack *SRY*. Genetically female mice transgenic for the equivalent mouse *Sry* gene also develop as males, although they are sterile because further Y-linked genes are required for sperm maturation. These results suggest that femaleness is a default state, and that expression of *SRY* is sufficient to switch to a pathway of male development. However, there are also genes on the X-chromosome that play an important role in ovarian differentiation, and in mutations these genes can feminize males with an active *SRY* gene. It therefore appears that both male and female primary sex determination pathways are actively promoted and neither is a true default state; the molecular factors involved are discussed below.

Unlike the primary sex characteristics, it appears that secondary sex characteristics in mammals are indeed female as a default state, and that maleness is conferred by hormonal signaling from the developing testis. Two signals, **testosterone** and **Müllerian inhibiting substance**, are released by the testis and direct male-specific development of the genitourinary system (*Fig. 1*). Testosterone also regulates the development of other male secondary sex characteristics, such as genital structure, and the male-specific changes that take place during puberty. In the absence of these hormones, the body develops female secondary sex characteristics regardless of the nature of the gonad and the sex-chromosome configuration of the somatic cells. Thus, genetic males develop phenotypically as females if the testis is removed (removing the source of hormones) or in **androgen-insensitivity syndrome,** where the nuclear receptor for testosterone is absent so that cells cannot respond to the presence of this hormone (see Topic C3 for discussion of steroid hormone signal transduction).

The molecular basis of mammalian sex determination involves the co-ordination of a number of signaling proteins and transcription factors. The SRY protein is a transcription factor. The expression profile of SRY suggests one of its roles may be to activate genes encoding other transcription factors, including SF-1 and SOX-9, which are also expressed in the male gonadal ridge, and in whose

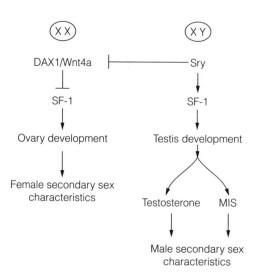

Fig. 1.   Molecular basis of sex-determination in mammals.

absence the XY male is feminized. SF-1 has several direct roles in male sexual development. This transcription factor regulates the gene encoding Müllerian inhibiting substance as well as genes involved in testosterone synthesis, and therefore controls synthesis of the hormones that direct the development of male secondary sex characteristics. However, SF-1 is produced in the gonadal ridge *before* SRY is expressed. In the testis, its expression is maintained, while in the ovary, its expression is shut off, perhaps under the influence of the signaling molecule Wnt4a. In genetically female mice lacking the *wnt4a* gene, the ovaries are poorly-developed and they continue to express SF-1, and therefore produce both MIS and testosterone. SRY may therefore maintain SF-1 expression by inhibiting Wnt4a in the bipotential gonad. Another gene with a predominant role in ovarian development is *DAX1*. This is found on the X-chromosome, and presumably in normal males its influence is repressed in the presence of SRY. In rare XY individuals carrying a tandem duplication of the *DAX1* gene, the gonads may develop as ovaries even if the *SRY* gene is expressed. XX individuals lacking the *DAX1* gene may also show male sexual differentiation. This suggests that while *SRY* acts as a switch from female to male development, the determination of primary sex characteristics actually reflects the balance of different transcription factors and signaling molecules that favor either male or female development.

Sex determination in *Drosophila*

Sex determination in *Drosophila* appears superficially similar to that in mammals because females possess two X-chromosomes and males possess a single X-chromosome plus a Y-chromosome. However, the mechanism used in flies differs from mammalian sex determination in two important respects.

Firstly, sex in *Drosophila* is determined by the number of X-chromosomes *relative to autosomes*, not by the presence or absence of the Y-chromosome. Females (XX) have an X:A ratio of 1, while males (XY) have an X:A ratio of 0.5. Thus, XXY individuals are female because the ratio is 1, whereas flies with a single X but no Y develop as males because the ratio is 0.5 (such male flies are sterile because Y-linked genes are required for male gametogenesis). This is a complete contrast to the situation in mammals described above, where XXY individuals are male and X individuals are female. Furthermore, triploid *Drosophila* are viable, so it is possible to generate individuals with X:A ratios that are less than 0.5, between 0.5 and 1, or greater than 1. These individuals display extreme or intermediate sexual phenotypes, and are known as **metamales, intersexes** and **metafemales**, respectively.

Secondly, whereas secondary sex characteristics in mammals are determined by hormonal signaling from the testis, secondary sex characteristics in *Drosophila* (which include structure of the genitals, size, wing length, body markings and the presence of sex combs in males) are determined genetically in a cell-autonomous fashion. Unlike mammals, where individual somatic cells outside the gonad have no particular intrinsic sex, each *Drosophila* cell is independently specified as male or female by its sex-chromosomes. Thus it is possible to create mosaic flies with sectors of male and female tissue, such individuals being known as **gynandromorphs**. Spectacular gynandromorphs can be generated in which the left side of the body is male and the right side is female. Such conditions do not usually occur in mammals because the secondary sex characteristics of XX and XY somatic tissue are dictated by signaling from the gonad.

The molecular basis of sex-determination in *Drosophila* depends on the balance of transcriptional repressors and activators acting on a single executive sex determining gene called *Sex-lethal (Sxl)*. The Sex lethal protein then initiates a cascade of regulatory activity that controls primary and secondary sex determination, and dosage compensation (see below). The X-chromosome carries a number of genes encoding transcription factors that activate *Sxl*, including *sis-a*, *sis-b* (*scute*), *sis-c* and *runt*. Conversely, the autosomal genes *deadpan* and *daughterless* encode transcriptional repressors. There is competition between these proteins for the regulation of *Sxl*: with two X-chromosomes, the balance is tipped in favor of *Sxl* activation, while with one it is tipped in favor of repression.

Sxl has two promoters, one for early expression and one for late expression. Early *Sxl* transcription is dependent on the competitive mechanism discussed above, and hence occurs only in females. Later in development, *Sxl* is transcribed from the second promoter in both males and females. The Sxl protein is a splicing factor that processes its own mRNA. In females, where early synthesis causes accumulation of Sxl protein, the late mRNA is productively spliced leading to further accumulation of the protein. In males, the lack of early Sxl means that the late mRNA is not productively spliced and a truncated, nonfunctional protein is produced. In females, the high level of active Sxl protein allows mRNA from a downstream gene, *transformer* (*tra*) to be productively spliced. Tra protein, which is also a splicing factor, cooperates with a second protein to splice the mRNA from the *doublesex* gene in a female-specific manner. In the default male pathway, where *tra* RNA is not productively spliced and there is no functional protein, different exons are included in the mature *doublesex* transcript. The male and female Doublesex proteins differentially regulate a battery of downstream genes. Therefore, due to the activity of Sex lethal, *doublesex* mRNA is spliced differently in males and females, initiating alternative pathways of primary sexual differentiation. This regulatory cascade, unusual in development because of the predominance of post-transcriptional regulation, is

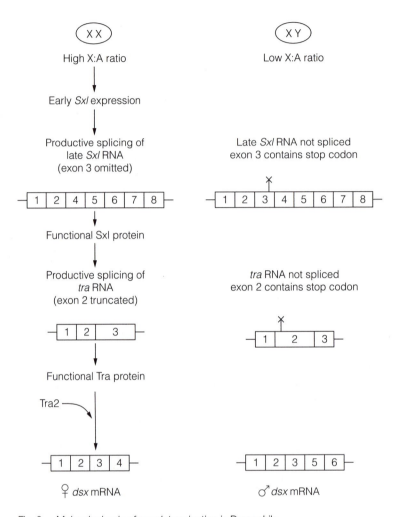

Fig. 2.   Molecular basis of sex determination in Drosophila.

shown in *Fig. 2*. The same pathway is used to control secondary sexual development, and involves not only the *doublesex* gene but also genes such as *dissatisfied*, *fruitless* and *hermaphrodite*.

Sex determination in *C. elegans*   *Caenorhabditis elegans* is unusual, even compared to other nematodes, in that sexual dimorphism produces males and hermaphrodites. In other nematode species there are separate male and female sexes as usual, and the sex-determination mechanism that gives rise to hermaphrodites in *C. elegans* gives rise to females in these other species. It is suggested that *C. elegans* hermaphrodites evolved from females, and in agreement with this, it is possible to isolate mutations that produce functional female worms (Topic E3).

As in *Drosophila*, the primary somatic sexual phenotype of *C. elegans* is determined by the number of X-chromosomes relative to autosomes. There is no Y-chromosome in *C. elegans*, so hermaphrodites are XX (X:A ratio = 1) and males are XO (X:A ratio = 0.5). As in *Drosophila*, it is possible to generate individuals with intermediate X:A ratios. However, unlike *Drosophila*, these individuals do not develop as intersexes, but show a sharp transition from male to female (an X:A ratio of 2:3 (0.67) promotes male development, whereas a slightly higher ratio of 3:4 (0.75) promotes hermaphrodite development). In terms of secondary sex characteristics, there is also extensive sexual dimorphism of the somatic cells, and while certain features of the general anatomy are similar there are also sex-specific somatic organs, such as the hermaphrodite vulva (whose development is discussed in Topic L1) and the larger and more elaborate tail of the male. Unlike the autonomy of sex determination described for *Drosophila*, there is extensive use of cell–cell signaling to control the sexual development of individual somatic cells in C. *elegans*.

The molecular basis of sex determination in C. *elegans* is, as in *Drosophila*, dependent on the balance of regulator proteins and their effect on a single executive sex determining locus, which in C. *elegans* is xol-1. The xol-1 gene is expressed at high levels when there is one X-chromosome and at low levels when there are two. The molecular nature of the X:A ratio signal elements that regulate xol-1 has not been fully elucidated, but two important components that have been identified are the genes sex-1 and fox-1 (both of which encode proteins that inhibit xol-1 in hermaphrodites). SEX-1 is a transcriptional repressor, while FOX-1 is a mRNA-binding protein, so it appears that xol-1 expression is regulated at least at two levels. The xol-1 transcript is alternatively spliced, and only one of the splice variants promotes XOL-1 activity, so FOX-1 is likely to regulate XOL-1 activity at this level.

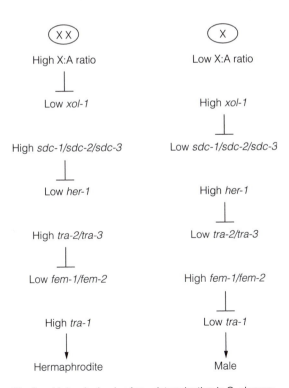

Fig. 3.   Molecular basis of sex determination in C. elegans.

Downstream of XOL-1, there is a cascade of repression (*Fig. 3*), resulting in either the activation or repression of the gene *transformer-1* (*tra-1*) which is the primary effector of the morphological changes involved in hermaphrodite development (equivalent to the *doublesex* gene in *Drosophila*). In hermaphrodites, the presence of two X-chromosomes represses *xol-1* expression, and the absence of XOL-1 results in high level expression of *tra-1*. Conversely, the single X-chromosome in males allows high level *xol-1* expression during gastrulation, leading to the inhibition of *tra-1*. Mutations in *tra-1* are epistatic to all other genes in the pathway, with loss-of-functional alleles causing male development even in XX worms, and gain-of-function alleles leading to hermaphrodite development even in XO worms. TRA-1 protein is a zinc finger transcription factor that regulates the downstream genes involved in sexual differentiation. An important immediate downstream target of XOL-1 is the *sdc-2* gene, which plays a pivotal role both in sex-determination and dosage compensation (see below). The SDC-2 protein binds to chromatin, and helps to assemble the repressive complex that inactivates *her-1* (see *Fig. 3*) and the regulatory complexes that reduce the transcriptional output of X-linked genes.

Dosage compensation

In sex determination mechanisms that involve the asymmetric distribution of sex chromosomes, there is an imbalance of gene dosage between males and females. In all three examples above, females (hermaphrodites in *C. elegans*) have twice the dosage of X-linked genes compared to males. Overall, only a small number of genes on the X-chromosome are specifically involved in sexual differentiation while the majority fulfil housekeeping functions or other roles that are equivalent in both sexes. Therefore males and females actually require equivalent amounts of most X-linked gene products.

The problem of sex-specific dosage is solved by **dosage compensation**, any of several genetic mechanisms that makes the two sexes equivalent in the levels of most X-linked gene products. In mammals this is achieved by the inactivation of one of the two X-chromosomes in every cell. The inactive X is sequestered into hypermethylated condensed chromatin within which most genes are transcriptionally repressed. The few genes that escape inactivation are those specifically required as a double dose in females. X-inactivation is controlled by a single locus, *Xist*, which is expressed specifically in the inactive X (X-inactive specific transcript). *Xist* does not encode a protein, but yields an untranslated RNA that is thought to coat the chromosome and in some manner induce chromatin condensation. More details of X-inactivation can be found in Topic C2.

In *Drosophila*, X chromosome dosage compensation occurs in the male, by upregulating the expression of all X-linked genes. The regulatory mechanism is closely associated with sex determination, as it is controlled by the splicing protein Sex lethal. The target of Sex lethal in dosage compensation is RNA from the gene *male sex lethal-2* (*msl-2*). In females, where Sex lethal is present, this RNA is spliced yielding a mature transcript that is not translated. Conversely, in males, where Sex lethal is absent, the RNA remains unspliced and yields a functional product. Four other proteins cooperate with MSL-2 in males: MSL-1, MSL-3, MOF and Maleless. The five proteins appear to form a chromatin-binding complex, which binds to many sites on the male X-chromosome and promotes high transcriptional activity by favorable chromatin remodeling and histone acetylation (Topic C2). Two nontranslated RNAs, products of the *rox1* and *rox2* genes (RNA on X), are also components of this complex. The loss of either RNA is tolerated, but in double mutants there is no dosage compensation, suggesting they are redundant. In *C. elegans*, dosage compensation is also mediated by a chromatin-binding complex, but in this species the complex is present only in hermaphrodites and assembles on both X-chromosomes. The complex includes the proteins SDC-2 and SDC-3, which are absent in males because the corresponding genes are repressed by XOL-1. The complex in this case represses gene expression by remodeling the chromatin in an unfavorable manner, reducing the transcriptional output by 50%. Many of the proteins in the dosage compensation repression complex are also found in the multiprotein complex that helps to condense chromatin during mitosis, so it is likely that these proteins have been recruited to a new function in dosage compensation.

Environmental sex determination

In some animals, the primary sexual phenotype is determined by extrinsic factors rather than sex chromosomes. The sex of crocodiles, some lizards and most turtles is determined by the temperature of the egg just after fertilization. In most cases, there is a range of temperatures over which both sexes can develop, with males becoming predominant at one extreme and females at the other. The molecular basis of this sex-determination mechanism may involve the temperature-dependent activation of enzymes that interconvert estrogen and testosterone. The environment in which the egg is laid to some extent controls the temperature of development, so females can influence the sex of their offspring by their egg-laying behavior.

Sex determination in plants

Plant and animal development show some important fundamental differences, and plant development is discussed separately in Section N. However, mechanisms of sex determination in plants are strikingly similar to those in animals, even though the vast majority of plants are hermaphrodite and only about 5–10% are unisexual (the converse applies in animals). Unisexuality in plants occurs either through the production of different male and female flowers on the same plant (**monoecy**) or separate male and female individuals (**dioecy**). As with animals, there are also deviations from this pattern, with some species producing males and hermaphrodites as in *C. elegans*, and others producing females and hermaphrodites.

Hermaphrodite flowers comprise concentric rings (or whorls) of organs, with the innermost whorls representing the male and female sex organs (the stamens and carpels, see Topic N6). In unisexual flowers, development of both sets of organs is often initiated, but the development of one set is aborted (usually early in development so that the male and female flowers are morphologically distinct) or the inappropriate organs are reabsorbed. However, in some species (e.g. asparagus) development is aborted at a very late stage so that male and female flowers are indistinguishable. In other species (e.g. *Humulus* spp.) the unwanted sex organs are not initiated at all. Monoecy is often a modification of hermaphroditism, with hormones released from a certain region of the plant suppressing the development of one set of sexual organs. However, in many dioecious species, sex is determined by differentially-distributed sex chromosomes. Both of the sex-determining mechanisms seen in animals are represented, i.e. the male-determining Y-chromosome and the X:A balance mechanism, as well as more complex systems involving multiple sex-chromosomes.

F1 CLEAVAGE

Key Notes

Properties of cleaving embryos	Cleavage involves the division of the fertilized egg into a number of smaller cells (blastomeres) and often generates a ball of cells surrounding a central cavity (the blastocoel). Early cleavage divisions in most species are rapid and synchronous because they are controlled by maternal gene-products and there is no net growth of the embryo.
The midblastula transition	Cleavage is often the stage of development at which the zygotic genome becomes active, because maternal gene-products become depleted. Typically, this midblastula transition is characterized by new RNA and protein synthesis, the slowing down of cell division and the loss of cleavage synchrony.
Types of cleavage	There are a number of distinct cleavage patterns in the animal kingdom, reflecting the order and orientation of successive cell divisions, whether the divisions are equal or asymmetrical, and the amount of yolk in the egg. Characteristic patterns include radial, spiral, rotational, bilateral and asymmetric.
The effect of egg yolk	Yolk inhibits cell division and therefore has a profound effect on the cleavage pattern in species with yolky eggs. In the moderately yolky eggs of amphibians, cleavage divisions are delayed. In species with very yolky eggs, such as birds, the cleavage divisions are incomplete, so that the blastomeres are initially continuous with the underlying yolk.
Formation of the blastocoel	In many species, cleavage results in the formation of an internal fluid filled cavity called the blastocoel. This initially forms as a result of loose packing between cells, but is often expanded by the architecture of the cleavage divisions and through increasing hydrostatic pressure brought about by the import of ions and water. The blastocoel is sealed off from the environment by tight junctions between cells, and this helps to maintain the integrity of the embryo.
Cleavage in sea urchin embryos	Sea urchins undergo simple radial cleavage, although some of the divisions are asymmetrical, producing blastomeres of different sizes. Radial divisions continue until the onset of gastrulation, producing a simple blastula comprising a single cell layer surrounding a large blastocoel.
Cleavage in *Drosophila*	In insects, the cleavage phase of development is replaced by a rapid series of nuclear divisions within a common cytoplasm (syncytial blastoderm). Cellularization occurs when the nuclei migrate to the periphery of the syncytial embryo, and become enclosed in cell membranes (cellular blastoderm). The large yolk prevents cellularization in the central region of the embryo.

Cleavage in zebrafish	After fertilization, the cytoplasm of the zebrafish egg separates from the yolk and forms a cap at the animal pole. Cleavage is restricted to this region, forming a raised blastoderm several layers thick, with the deep cells continuous with the yolk. The blastoderm differentiates to form an external enveloping layer, a deep layer, and a yolk syncytial layer formed by the fusion of the lower tier of deep cells with the large yolk cell.
Cleavage in *Xenopus*	The *Xenopus* egg has a yolky vegetal pole and a relatively yolk free animal pole. The egg undergoes radial cleavage, but the yolk inhibits cell division and displaces the mitotic spindle towards the animal pole, so that a large population of small animal cells is formed overlying a smaller population of larger yolky vegetal cells. The blastocoel is also restricted to the animal hemisphere of the embryo.
Cleavage in chicken embryos	The chicken egg is predominantly yolk, and cleavage is restricted to a small blastodisc at the animal pole. Initially the blastoderm has several layers, but this thins out to form a single-layer epiblast. The central region, overlying the subgerminal space, is the area pellucida. The peripheral region, in contact with the yolk, is the area opaca.
Cleavage, compaction, hatching and implantation in mammals	Cleavage in mammals is unique because of the slow, asynchronous divisions, the early activity of the zygotic genome, and the overt compaction stage where blastomeres flatten against each other and maximize their contacts. There is early differentiation of the mammalian morula to generate an outer trophoblast and an inner embryoblast, which becomes segregated to one end of the blastocoel to form the inner cell mass. As the blastocoel expands, the embryo hatches from the zona pellucida and implants into the uterine wall.

Related topics	Morphogenesis (A7)	Gastrulation in vertebrate embryos
	The cell division cycle (C4)	(F3)
	Gamete recognition, contact and	Early development of molluscs
	fertilization (E4)	and annelids (G4)
	Gastrulation in invertebrate	
	embryos (F2)	

Properties of cleaving embryos

Cleavage is the first stage of animal development, in which the zygote divides to form a number of smaller cells called **blastomeres**. The nature of the early cleavage divisions varies widely among different species as discussed below, but the result is generally a ball of cells called a **blastula** (or an equivalent term, see *Table 1*) often surrounding a fluid-filled cavity (the **blastocoel**).

In many animals (but not mammals), cleavage divisions are **rapid** and **synchronous**. The divisions are *rapid* because there is no net growth of the embryo – the cell cycle alternates between DNA replication and mitosis with no intervening gap phases. In the absence of growth, the cell number in the embryo increases while the cell size decreases. The divisions are *synchronous* because the zygotic genome is transcriptionally inactive, so cleavage depends predominantly on maternal gene-products distributed in the egg cytoplasm. Hence, the zygote can often cleave normally in the presence of drugs that inhibit transcription, and even in the absence of a nucleus! The maternal products determine the rate of cleavage (by regulating the cell cycle) and also the spatial arrangement and relative sizes of the blastomeres (by controlling the position of the mitotic spindles). The amount of yolk in the egg also affects the rate of cleavage and the size of blastomeres (see below).

In many species, the early cleavage divisions also depend on the provision of a centriole by the sperm, as this is required to organize the mitotic spindle. This is the only instance in which **paternal effect mutations** are found in the embryo. If mutations in the paternal genome affect centriole structure and function in the sperm, the mitotic spindle is not organized properly in early cleavage divisions, and such mutations cannot be rescued by a wild type egg.

Mammals are exceptional because the zygotic genome is activated early in development (in some species, as early as the two-cell stage). The cleavage divisions are controlled by the zygotic genome rather than the maternal genome, and are therefore slow (because the cell cycle includes gap phases) and asynchronous.

Table 1.   Nomenclature of cleaving embryos

Species	Name of cleaving embryo
Most invertebrates	Blastula
Insects	Syncytial blastoderm, then cellular blastoderm
Amphibians	Morula, then blastula
Fish, reptiles and birds	Blastodisc
Mammals	Morula, then blastocyst

In amphibians and mammals, the **morula** refers to the loosely-packed ball of blastomeres before the blastocoel arises, and the **blastula/blastocyst** refers to the embryo when the blastocoel has formed. Fish, reptiles and birds have very yolky eggs and cleavage is restricted to a flattened **blastodisc** at the periphery of the cell. A flat cell sheet in a cleaving embryo is described as **blastoderm** (e.g. in insects, and in birds).

The midblastula transition

In most animals used as developmental models (but not mammals), early development is controlled by maternal gene products stored in the egg. Eventually, these products become inactive or are used up, at which point developmental processes become dependent on zygotic transcription. In some species, the onset of zygotic gene expression is sudden and coincides with a dramatic slowing of cell division, the loss of synchrony, and changes in cell behavior (e.g. cells become motile). This landmark in early development is called the **midblastula transition (MBT)**, although it may occur at any time, from early cleavage to mid-gastrulation.

In zebrafish and *Xenopus*, the MBT occurs through the **titration** of a maternal gene product against the increasing amount of DNA. It appears that the egg stores a factor that inhibits zygotic gene expression, and that there is enough of this factor to repress transcription for nine nuclear divisions in zebrafish and 12 divisions in *Xenopus*. After this crucial final division, the zygotic genome is suddenly activated because all the available inhibitor is already bound to DNA. The timing of the transition can be controlled by artificially modifying the amount of DNA in the zygote, e.g. haploid zebrafish undergo the MBT one cycle later than diploid zebrafish. In other animals, the MBT occurs when maternal gene-products are **depleted**. In *Drosophila*, for example, the MBT occurs after 14 nuclear divisions, when a protein called **String** (a maternal regulator of the cell cycle) is exhausted, and cell division becomes dependent on zygotic *string* transcription (Topic C3).

Types of cleavage

There are a number of distinct **cleavage patterns** in the animal kingdom, reflecting both the influence of maternal gene products and the amount of yolk in the egg (*Fig. 1*). Essentially, all cleavage patterns are based on combinations of two types of division: vertical and horizontal. **Vertical cleavage divisions** occur parallel to the animal–vegetal axis of the egg and increase the number of blastomeres in a single tier, while **horizontal cleavage divisions** occur perpendicular to the animal–vegetal axis and generate additional tiers of blastomeres. The simplest cleavage pattern is **radial cleavage**, in which the first two cleavage divisions are vertical and perpendicular to each other, and the subsequent divisions alternate between horizontal and vertical. In the sea cucumber, the divisions are equal, producing a blastula of equally-sized blastomeres. The vertical divisions are **medial** and the horizontal divisions **equatorial**, in each case dividing the cell precisely in two. In other species, such as the sea urchin, some horizontal divisions are not equal, so that different tiers of cells may comprise different sized blastomeres (**micromeres**, **mesomeres** and **macromeres**).

A distinct strategy employed by annelids and molluscs is **spiral cleavage** (*Fig. 2*), where adjacent tiers of blastomeres are rotated slightly with respect to the other. The first two divisions are vertical, but the third horizontal division occurs at an oblique angle because the mitotic spindle is tilted away from the animal–vegetal axis of the egg and points either left or right, resulting in clockwise (dextral) or counterclockwise (sinistral) rotation of the upper tier. The role of spiral cleavage in the development of molluscs and annelids is discussed further in Topic G4. Other forms of cleavage include the **rotational cleavage** of mammalian embryos, in which the first cleavage plane is vertical but in the second division, one cell cleaves vertically and the other horizontally; and the **bilateral cleavage** of ascidians (Topic G3), where the first cleavage division is vertical and establishes the left and right sides of the embryo, and all further divisions on each side are mirror images of the other. Finally, in certain animals with a small cell number (e.g. *C. elegans*) the early cleavage divisions are asymmetrical and highly regulated. These divisions establish the principal axes of the body plan and the cell lineages that give rise to the germ line and major somatic tissues (Topic G2).

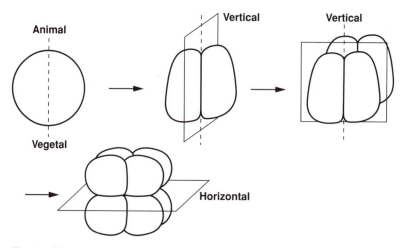

Fig. 1.   Orientations of cell division in radial cleavage.

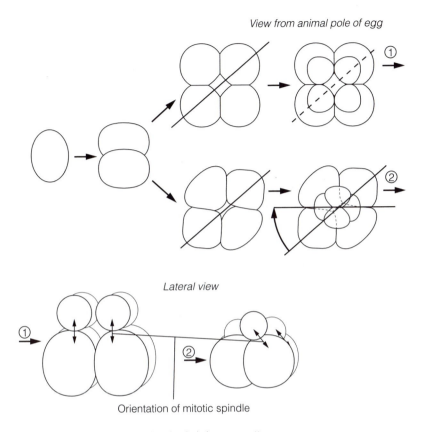

View from animal pole of egg

Lateral view

Orientation of mitotic spindle

Fig. 2.   Comparison of radial and spiral cleavage patterns.

The effect of egg yolk

The different cleavage patterns discussed above are controlled predominantly by maternal gene products stored in the egg (not in mammals). However, cleavage is also influenced by the amount of yolk in the egg, as yolk has an inhibitory effect on cell division. **Yolk** is required to nourish the developing embryo before it can feed independently, and is therefore not required in the eggs of those organisms that rapidly produce a feeding larva (e.g. sea urchins) nor in mammals, where the embryo is nourished through the placental blood supply. In such species, the zygote undergoes **holoblastic (complete) cleavage**. In eggs with a moderate amount of yolk (e.g. those of insects and amphibians) the yolk is not usually distributed uniformly, and cleavage is inhibited in the yolk-rich regions. In amphibian eggs, the yolk is concentrated at the **vegetal pole**, while the nucleus is displaced towards the opposite **animal pole**, which is relatively yolk free. Although amphibian zygotes undergo holoblastic cleavage, the divisions occur more rapidly in the animal hemisphere than in the vegetal hemisphere, with the result that the animal cells are much smaller and more numerous than the yolky vegetal cells. In *Drosophila* and other insects, the yolk is concentrated in the center of the egg, and cellularization is restricted to the peripheral region to form the superficial **cellular blastoderm**. In fish, reptiles and birds, the eggs contain such large amounts of yolk that they undergo **meroblastic** (incomplete) **cleavage**, and cleavage divisions are restricted to a peripheral region producing a small **blastodisc**.

Formation of the blastocoel

A characteristic of many cleaving embryos is the formation of an internal, fluid filled cavity termed the **blastocoel**. This may have evolved to set aside space inside the embryo for the complex cell movements of gastrulation, and may also serve to prevent premature contact between groups of cells that later take part in inductive interactions. The blastocoel originates as small gaps between cleaving cells, and then enlarges to form a distinct cavity. The gaps initially form passively, as a natural consequence of the loose packing of blastomeres. In some cases, the blastocoel enlarges simply due to the nature of cleavage divisions in the embryo. For example, in the sea urchin, the early radial cleavage divisions produce a blastula in which a central cavity is surrounded by a single layer of cells. All subsequent cleavage divisions occur in the plane of this single cell layer, and increase the cell number in the sheet without increasing cell volume. As a consequence, the cells become smaller and cannot 'fill' the internal space of the embryo, generating an internal cavity. In sea urchins and mammals, ions are also pumped into the cell spaces in the cleaving embryo, drawing in water by osmosis. This net movement of fluid increases the size of the blastocoel and maintains the integrity of the embryo by increasing its hydrostatic pressure.

Cleavage in sea urchin embryos

As discussed in Topic E4, part of the cortical granule reaction in the sea urchin egg involves the secretion of **hyalin**, a glycoprotein that forms a double layer around the egg. Microvilli on the egg surface extend and penetrate this **hyalin layer**, which is required for the integrity of cells during cleavage and plays an important role in gastrulation. The first two divisions are vertical and perpendicular to each other, and the next is horizontal and equatorial to generate eight equally sized blastomeres (*Fig. 3*). In the next divisions, the four blastomeres at the animal pole divide equatorially to generate eight equally sized **mesomeres**, while the four vegetal blastomeres divide asymmetrically to generate four **macromeres** and four **micromeres**. At this point, cleavage divisions become

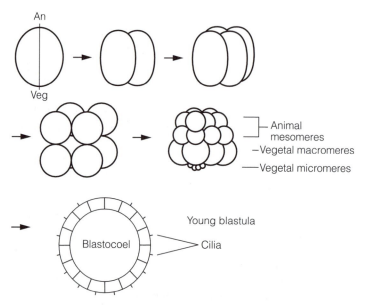

Fig. 3. Overview of cleavage in the sea urchin embryo.

asynchronous and a blastocoel begins to emerge, but all divisions occur within the plane of the emerging cell sheet so that while its size increases it remains one cell thick. The cells are polarized, with their apical surfaces covered in microvilli and in contact with the hyaline layer, while their basal surfaces secrete a basal lamina that lines the blastocoel cavity, and plays an important role in gastrulation. After about seven cleavage divisions, cilia begin to appear on the outside of the embryo. The embryo then hatches from the fertilization envelope and begins gastrulation (Topic F2).

Cleavage in Drosophila

As with most insects, 'cleavage' in *Drosophila* involves no cell division, but rather a rapid series of nuclear divisions to form a **multinucleate syncytium** (*Fig. 4*). After eight divisions, two to four nuclei in the posterior of the embryo migrate into the pole plasm and become enclosed in cell membranes to form pole cells, the *Drosophila* germ cells (see Topic E1). After another round of

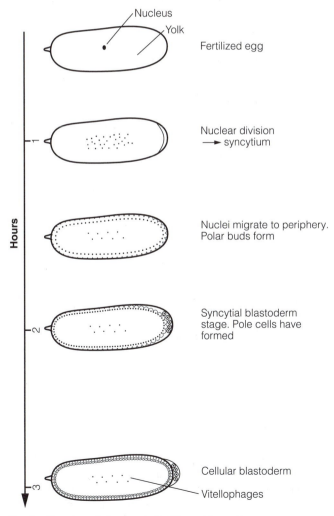

Fig. 4. Overview of cleavage in the Drosophila *embryo.*

division, most of the remaining nuclei migrate to the periphery of the embryo to form the **syncytial blastoderm**. After four more rounds of division, the nuclei are enclosed in cell membranes to form the **cellular blastoderm**, at which time the midblastula transition occurs and divisions become asynchronous. Nuclei remaining in the central yolky region of the embryo form structures termed **vitellophages** that are eventually enclosed in the gut lumen and excreted. From fertilization to the end of cleavage takes about 3 hours and generates 5000 blastoderm cells (as well as 32 pole cells). The embryo then undergoes gastrulation (Topic F2).

Cleavage in zebrafish

The yolk and cytoplasm of the zebrafish egg are initially evenly distributed. After fertilization, however, the cytoplasm streams to the animal pole and creates a raised cap above the yolk. All cleavage divisions occur in this region sitting on the yolk, to create a raised blastodisc (*Fig. 5*). The first five cleavage divisions are vertical and meroblastic, so that the 32 blastomeres are connected to each other and the large undivided yolk cell. After this, horizontal as well as vertical divisions occur, so that by the 64-cell stage the first independent blastomeres are produced. At around the 10th cleavage division, the midblastula transition occurs: zygotic transcription begins, and the cleavage divisions slow down and become asynchronous. By this stage, the outer layer of blastomeres has flattened and the inner layers have become rounded so that the blastula comprises two discrete populations – the outer **enveloping layer**, that eventually forms an extraembryonic structure called the **periderm**, and the **deep layer**, that undergoes gastrulation and forms the embryo proper. Some of the deep cells fuse with the yolk at approximately the 12th division to produce a **yolk syncytial layer** that later drives the epiboly of the blastoderm during gastrulation (Topic F3).

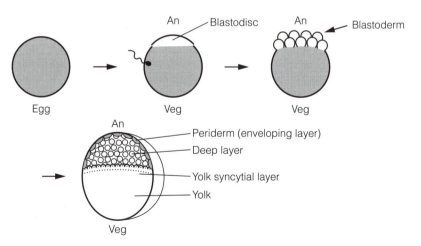

Fig. 5.   Overview of cleavage in the zebrafish embryo.

Cleavage in *Xenopus*

The unfertilized *Xenopus* egg has a pigmented animal hemisphere and a pale, yolky vegetal hemisphere, and is surrounded by a vitelline membrane and a jelly coat. The sperm enters the animal hemisphere. After fertilization, the egg completes meiosis, producing the second polar body, and then undergoes **cortical rotation**, where the cortical cytoplasm rotates relative to the inner cytoplasm towards the point of sperm entry. This establishes the Nieuwkoop center (Topic I2) which plays a predominant role in axis specification by inducing the organizer. The first two cleavage divisions are vertical and perpendicular, dividing the egg into four equal blastomeres (*Fig. 6*). However, while the cleavage furrow forms quickly in the animal hemisphere, it slows down dramatically as it passes through the vegetal hemisphere due to the high concentration of yolk. The next division is equatorial but the plane of cleavage is displaced towards the animal pole by the yolk so the animal blastomeres are smaller than the vegetal blastomeres. The animal blastomeres continue to divide rapidly and completely, while the vegetal blastomeres divide more slowly, so that the blastula stage embryo contains a large number of small animal cells and a smaller number of large yolky vegetal cells.

Until the 64-cell stage, the *Xenopus* embryo is termed a morula. The blastocoel emerges as a distinct cavity by the 128-cell stage, and the embryo is then termed a blastula. However, the blastocoel originates as a cleft between the first two blastomeres. This cleft is restricted to the animal hemisphere, and over subsequent divisions it becomes enlarged and sealed off from the external environment by tight junctions between the blastomeres. As the blastocoel becomes apparent it is also restricted to the animal hemisphere of the embryo. Gastrulation in *Xenopus* embryos is discussed in Topic F3.

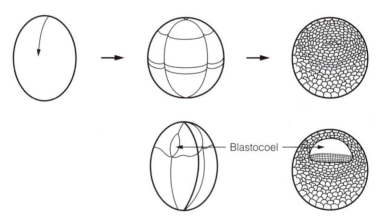

Fig. 6. Cleavage and blastocoel formation in the Xenopus *embryo. Top row shows external view. Bottom row shows sections at equivalent stages.*

Cleavage in chicken embryos

The chicken egg consists predominantly of yolk, and the nucleus is displaced to the animal pole where cleavage is restricted to a small area of cytoplasm just 2 mm in diameter, which becomes known as the **blastodisc**. As with zebrafish embryos, the initial cleavages are vertical and incomplete, so the blastomeres are continuous with the yolk. After several rounds of cleavage, horizontal divisions begin to occur and the first independent blastomeres are formed (*Fig. 7*). The **blastoderm** is eventually five to six cell layers thick, and secretes fluid into the underlying space adjacent to the large yolk cell to generate the **subgerminal cavity**. Unlike the other animals discussed in this Topic, cleavage in the chick embryo involves extensive cell death. This is required to thin out the blastoderm until it is a single cell layer. The delamination of cells from the blastoderm, and their migration into the subgerminal space where they die, occurs only in the central area of the blastodisc, to generate a translucent epithelial blastoderm

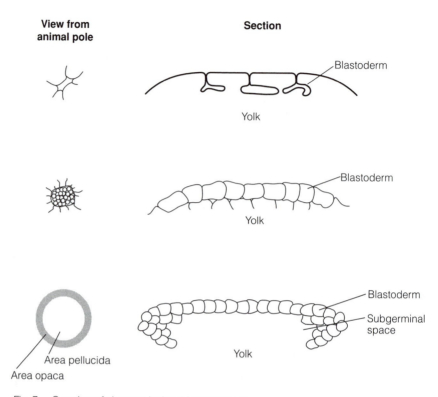

View from animal pole

Section

Fig. 7.   Overview of cleavage in the chicken embryo.

called the **area pellucida**. This is surrounded by a ring of thicker blastoderm in contact with the yolk, the **area opaca**. Gastrulation in chicken embryos is discussed in Topic F3.

Cleavage, compaction, hatching and implantation in mammals

As discussed above, cleavage in mammalian embryos is unusual for several reasons: the slow and asynchronous divisions, the early onset of zygotic gene expression, and the unique rotational cleavage pattern. Another important feature of mammalian cleavage is **compaction**, where the loosely associated blastomeres of the 8-cell embryo flatten against each other to maximize their contacts and form a tightly packed morula. Compaction occurs in many animals, including *Xenopus* and ascidians, but compaction in mammals is strikingly overt. The uncompacted blastomeres are rounded cells with uniformly distributed microvilli, and the cell adhesion molecule E-cadherin is found wherever cells contact each other. After compaction, the microvilli are restricted to the apical surface, E-cadherin is distributed over the basal and lateral surfaces, the cells form lateral tight junctions with their neighbors, and cytoskeletal elements are reorganized to form an apical band. In the 16-cell morula, it becomes possible to discriminate between two types of cell: external polarized cells and internal nonpolar cells. The surface blastomeres may divide in the plane of the outer cell layer, in which case two polarized cells are formed, or perpendicular to the layer, in which case division produces one external polarized cell and one internal nonpolar cell. As the population of nonpolar cells increases, the cells begin to communicate with each other through gap junctions.

The outer, polarized cells comprise the **trophoblast** or **trophectoderm**, which plays a major role in implantation and goes on to form the embryonic portion of the placenta (Topic F3). The internal, nonpolar cells comprise the **embryoblast**, which forms the embryo proper as well as other extraembryonic membranes (*Fig. 8*). As the blastocoel forms, these cells congregate at one end of the blastocoel to form the **inner cell mass (ICM)**. This polarized segregation defines the dorsoventral axis of the embryo (Topic I1). The decision as to which cells become trophoblast and which embryoblast appears to reflect their position in the morula. The mechanical reorganization of blastomeres and the transplantation of blastomeres from the inside of one morula to the outside of another has shown that cells are uncommitted until approximately the 64-cell stage, when they become determined as either trophoblast or embryoblast.

As discussed above, the blastocoel forms as sodium ions are pumped into the intracellular spaces of the embryoblast by trophoblast cells, and water is drawn in by osmosis. The size of the embryo therefore increases due to inflation of the blastocoel until it is pressed against the inside of the zona pellucida. At this point, the embryo hatches, a process achieved by the weakening of the zona by membrane bound proteases on the trophoblast cells, allowing the embryo within to squeeze out. **Hatching** is required for **implantation** into the uterine wall, since this involves the recognition of extracellular matrix proteins on uterine cells by integrins and heparan sulfate on the trophoblast cells. The timing of this process is critical, since early hatching can result in implantation in the oviduct or elsewhere, a dangerous condition known in humans as an **ectopic or tubal pregnancy**. Conversely, late hatching can result in failure to implant all together. Once the embryo has attached to the uterine wall, further proteases secreted by the trophoblast cells allow the blastocyst to embed itself into the uterine stroma as discussed in Topic F3.

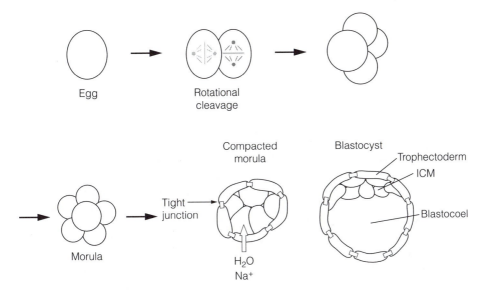

Fig. 8.   Cleavage in mammals.

F2 GASTRULATION IN INVERTEBRATE EMBRYOS

Key Notes

Gastrulation

Gastrulation follows cleavage and involves a complex series of coordinated cell movements to reorganize the simple blastula stage embryo, populating the interior with cells that will form mesoderm and endoderm, and covering the outside with ectoderm. The redistribution of cells involves many different types of cell movement, and often coincides with the specification of the three germ layers and the setting up of the embryonic axes.

Gastrulation in sea urchin embryos

Gastrulation in the sea urchin begins with primary mesenchyme cells migrating individually into the blastocoel. The vegetal region of the blastula then invaginates and involutes to form the archenteron. Convergent extension of endoderm cells lining the archenteron, together with tensile forces generated by secondary mesenchyme cells attached to the blastocoel through long filopodia, bring the archenteron and blastocoel into contact, to form the mouth.

Gastrulation in Drosophila

In *Drosophila*, gastrulation begins with the invagination of the blastoderm at the ventral furrow, under the control of the transcriptional regulators Twist and Snail. The invaginating cells form a tube of mesoderm and then delaminate to form a sheet. At the extremities of the embryo, presumptive endoderm cells also invaginate, and will meet to form the midgut. At the same time, the ectoderm on the surface of the embryo undergoes convergent extension, and the ventral blastoderm (germ band) extends, moving round to the dorsal side of the embryo following the curvature of the egg.

Related topics

Morphogenesis (A7)
Cleavage (F1)

Gastrulation in vertebrate embryos (F3)

Gastrulation

Gastrulation follows cleavage in animal development (Topic F1) and is the process by which the blastula, a simple ball or disc of cells (often enclosing a fluid filled cavity), undergoes extensive reorganization to generate a complex, three-dimensional body plan comprising three germ layers: ectoderm, mesoderm and endoderm. Specifically, gastrulation introduces cells that will become mesoderm and endoderm into the interior of the embryo, while covering its external surface with ectoderm.

Gastrulation involves a coordinated series of cell movements, which may include migration on the part of individual cells as well as the movement of cell sheets *en masse*. Gastrulation also involves the folding and reshaping of cell sheets, brought about by changes in cell shape and/or adhesion properties, or by forces generated elsewhere in the embryo that cause tension and distortion.

Some terms used to describe the different types of cell behavior observed in gastrulation are defined in *Table 2*, Topic A7. The complex cell behaviors that contribute to gastrulation have been studied in detail in several model organisms. Gastrulation in model invertebrates (sea urchin and *Drosophila*) is discussed in this Topic, while vertebrate gastrulation is discussed in Topic F3. Gastrulation in species with a small cell number, such as *C. elegans*, is considered in Section G.

Gastrulation in sea urchin embryos

The cell movements in sea urchin gastrulation are relatively simple, and this organism is discussed first. The movements can be summarized as follows (*Fig. 1*):

● Differentiation and ingression of primary mesenchyme cells.
● Invagination and involution of vegetal cells to form the archenteron.
● Differentiation and migration of secondary mesenchyme cells.
● Convergent extension of endoderm cells to lengthen the archenteron.
● Selective adhesion between the archenteron wall and the blastocoel wall to form the stomatodeum (mouth).

As discussed in Topic F1, cleavage of the fertilized sea urchin egg produces a roughly spherical blastula comprising a single layer of cells surrounding a large

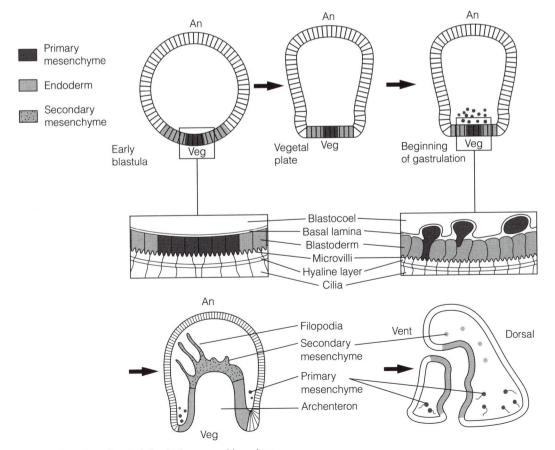

Fig. 1.   Overview of gastrulation in the sea urchin embryo.

blastocoel. The cells are polarized, with their apical surfaces in contact with the hyaline layer and their basal surfaces in contact with the basal lamina. The tiny cells at the vegetal pole of the blastula are termed **micromeres**, and these originate from the asymmetrical fourth cleavage division (Topic F1). After the blastula hatches from the vitelline envelope, the vegetal blastomeres thicken to form the **vegetal plate**. The micromeres differentiate into **primary mesenchyme cells (PMCs)** that detach from the vegetal plate and migrate individually into the blastocoel, where they collect at the vegetal surface. This involves loss of adhesion between the micromeres and the external hyaline layer of the embryo, but increased attraction to the basal lamina. A cadherin expressed by pre-migratory PMCs, **LvG-cadherin**, is rapidly internalized and forms intracellular aggregates as the cells begin to migrate. At the same time, several **integrins** are specifically downregulated in the PMCs. The downregulation and internalization of cell- and matrix-adhesion molecules therefore plays a major role in the delamination and migration of the PMCs. Once inside the blastocoel, further movement of the PMCs is directed by fibrils of extracellular matrix proteins aligned along the animal–vegetal axis. One protein that has been shown to have an important role in PMC guidance is **NG2**, which is related to the mammalian chondroitin sulfate proteoglycan core protein. The PMCs form skeletal rods that support the embryo.

The next stage of gastrulation involves the invagination of the lower tiers of vegetal cells to form a cavity called the **archenteron** (embryonic gut). The invaginating region of the gastrula-stage embryo is known as the **blastopore**. Invagination may be caused by a combination of factors, including changes in cell shape, and buckling caused by localized absorption of water. Macromere-derived cells in the vegetal plate are initially rounded, but become flask-shaped **bottle cells** as invagination commences. Ablating these cells, however, only blocks gastrulation temporarily, so they may play a role in the initiation of invagination, but are not critical for continued gastrulation. The vegetal plate may also secrete a proteoglycan into the inner hyaline layer, causing it to absorb water and expand. The outer hyaline layer remains the same size, and the forces generated cause the vegetal plate to buckle inwards. Invagination pulls the vegetal plate upwards, into the embryo, but is sufficient only to initiate the archenteron. Recent experiments indicate that cells in a ring around the blastopore then involute into the embryo, i.e. flow over the edge as a continuous sheet. This internalizes the upper tiers of vegetal cells, representing the prospective foregut, midgut and hindgut.

Cells representing the archenteron roof (the first cells to be internalized by invagination) then begin to differentiate as **secondary mesenchyme cells (SMCs)**. These cells extend long cytoplasmic processes, filopodia, which reach across the blastocoel and attach to the inner surface around the animal pole, although only at certain positions. Contraction of the filopodia provides tension that literally pulls the archenteron towards the blastocoel wall. Simultaneously, the endodermal cells of the archenteron undergo **convergent extension**, a process where cells change position to increase the length of a tissue at the expense of its thickness, without increasing cell number. The cells intercalate so that the number of cells forming the circumference of the archenteron is reduced from 30 to fewer than 10. The length of the archenteron increases as the thickness of the wall is reduced. Through these mechanisms, the tip of the archenteron eventually touches the blastocoel wall. The secondary mesenchyme cells disperse to form mesoderm-derived organs, and at the point of contact, the stomatodeum (mouth) is generated.

Gastrulation in
Drosophila

Drosophila gastrulation is also relatively simple. The movements can be summarized as follows (*Fig. 2*):

● Invagination of blastomeres at the ventral furrow to generate an internal sheet of mesoderm.
● Invagination of blastomeres at the anterior and posterior ends of the embryo to form endodermal pockets that eventually give rise to the gut.
● Convergent extension of remaining ectodermal cells form the germ band.

As discussed in Topic F1, cleavage in *Drosophila* embryos involves a series of rapid nuclear divisions giving rise to a syncytial blastoderm. The pole cells form after the 9th division, and the remaining nuclei then migrate to the periphery of the embryo and undergo several slower division cycles before cellularization occurs after the 13th division. As gastrulation begins, the *Drosophila* embryo is a cellular blastoderm surrounding a central yolk.

The first sign of gastrulation is the appearance of a **ventral furrow** along the ventral midline of the embryo, extending for about half its length. About 1000 cells in this region invaginate into the interior of the embryo, eventually forming a tube of presumptive mesoderm. The ventral furrow of the *Drosophila* embryo is therefore analogous to the blastopore of the sea urchin embryo, and cross sections of the two embryos are structurally very similar. Invagination at the

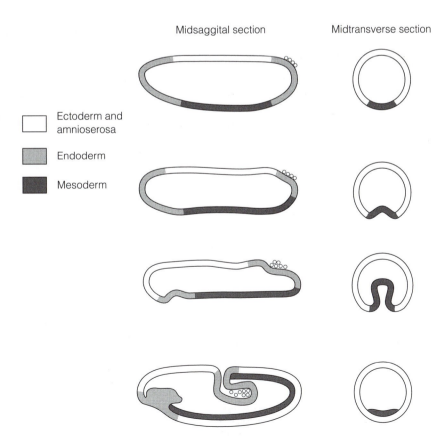

Midsaggital section   Midtransverse section

☐ Ectoderm and amnioserosa

▨ Endoderm

■ Mesoderm

Fig. 2. Gastrulation in Drosophila.

ventral furrow appears to be driven by apical constrictions of the ventral cells, a process that may be controlled by the transcription factors **Twist** and **Snail**, which are expressed specifically in this region (Topic H6). These transcription factors may regulate the expression of other genes, such as *concertina* and *folded gastrulation*, whose products in turn control the polymerization of actin filaments in the invaginating cells by acting on the GTPase **Rho**[1]. It is unclear how the anterior and posterior boundaries of the ventral furrow are established, but this depends on the terminal group transcription factor **Tailless** (Topic H3). Shortly after invagination, the cells lose adhesive contacts and dissociate from the tube forming a flat mesodermal sheet immediately adjacent to the ventral ectoderm.

Further invagination occurs at the anterior and posterior ends of the embryo to generate the tubes of endoderm that form the gut. These morphogenetic events are also under the control of **Tailless** (as well as other terminal transcription factors, such as **Huckebein** and **Forkhead**) and also require the activity of Rho. The anterior invagination occurs on the ventral side of the embryo while the posterior invagination occurs on the dorsal side. Eventually, the two tubes meet in the middle to form the anterior and posterior regions, respectively, of the midgut. The posterior invagination also internalizes the pole cells, bringing them to the midgut from where they migrate to the gonads (see Topic E2). Further ectodermal cells invaginating at the anterior and posterior ends of the embryo form the foregut and hindgut.

As the presumptive mesoderm and endoderm cells are invaginating into the embryo, the cells on the surface undergo convergent extension and migrate towards the ventral midline, to cover the embryo with ectoderm. There is little cell division during gastrulation in *Drosophila*, so invaginating cells are not replaced by the proliferation of ectoderm. The ventral blastoderm, or **germ band** as it is known, extends towards the posterior of the embryo, and constricted by the curvature of the egg case, curls round onto the dorsal surface. It is during this process of **germ-band extension** that the segments begin to appear (Topic H5).

---

[1]Rho is also known to be involved in cytoskeletal rearrangement in other developmental systems. For example, it is required for the migration of vertebrate neural crest cells (Topic J6).

F3 GASTRULATION IN VERTEBRATE EMBRYOS

Key Notes

Gastrulation in vertebrate embryos

Gastrulation in vertebrates is initiated by the organizer, which marks the future dorsal and posterior side of the embryo. Presumptive mesoderm and endoderm cells adopt their correct relative positions inside the embryo and cells converge and extend along the future anteroposterior axis to generate the notochord and other mesodermal and endodermal structures, thus establishing the embryo's principal axes. As cells move into the embryo, the remaining blastoderm (or epiblast in birds and mammals) spreads over the surface of the embryo to cover it with ectoderm. Gastrulation in amniotes is different to that in fish and frogs because the anteroposterior axis forms gradually rather than all at once, there is extensive cell proliferation, and because cells move individually into the embryo rather than involuting as a continuous sheet. Gastrulation in amniotes is also preceded by the differentiation of the blastoderm into two layers: epiblast and hypoblast.

Gastrulation in Xenopus

Xenopus gastrulation involves the following events: invagination of bottle cells to generate the blastopore (the blastopore forms gradually, beginning with the dorsal lip); involution of marginal zone cells over the lip of the blastopore; migration of leading mesoderm cells (future head mesoderm) across the blastocoel roof; convergent extension of the following mesoderm and endoderm cells along the anteroposterior axis (future notochord, somites, lateral plate mesoderm, gut); and epiboly of the ectoderm to cover the surface of the embryo.

Gastrulation in the zebrafish

Zebrafish gastrulation is very similar to that in *Xenopus*, although cells move into the blastula around its entire circumference, rather than initiating at the dorsal side. Gastrulation begins with the spreading of the blastoderm over the yolk by epiboly. Cells of the deep blastoderm layer then involute to form the equatorial germ ring, and involuting presumptive mesoderm and endoderm cells converge and extend through the embryonic shield along the anteroposterior axis.

Gastrulation in the chick

Chick gastrulation is preceded by the formation of the posterior marginal zone in the area opaca and the elaboration of the hypoblast. Gastrulation begins with the formation of the primitive streak and primitive node in the epiblast above the posterior marginal zone. The streak extends across the area pelludia, with the node at its anterior extremity. Migration of epiblast cells through the node forms the definitive endoderm, anterior notochord and prechordal mesoderm. Migration of epiblast cells through the remainder of the streak forms ventral and lateral mesoderm. The primitive streak then regresses, laying down the remainder of the notochord in its wake.

Gastrulation in mammals	Gastrulation in mammals is similar to that in chickens, beginning with the formation of a primitive streak in the future posterior region of the epiblast. Although the architecture of the epiblast differs in mice and humans, both processes involve similar cell movements. Most of the differences among mice, humans and other mammals reflect the formation of extraembryonic structures.
Related topics	Morphogenesis (A7) Cleavage (F1) Cleavage in invertebrate embryos (F2)

Gastrulation in vertebrate embryos

As in invertebrates, gastrulation in vertebrate embryos causes the reorganization of embryonic cells to populate the interior of the embryo with layers of mesoderm and endoderm, and covers the outside of the embryo with ectoderm. However, while gastrulation in sea urchins and *Drosophila* involves largely undifferentiated cells, a certain degree of cellular differentiation precedes vertebrate gastrulation, often concerning the formation of **extraembryonic structures**, such as the periderm in zebrafish, the yolk sac in birds and mammals, and uniquely in mammals, the embryonic portion of the placenta.

Gastrulation in vertebrate embryos is also closely associated with axis specification. Although the origins of the embryonic axes can be traced back to events before gastrulation, the point at which gastrulation initiates is usually the future dorsal side of the embryo, and the cells moving through this structure (the **embryonic shield** in zebrafish, the dorsal lip of the **blastopore** in *Xenopus*, and the **node** in birds and mammals) converge and extend to form the notochord and hence define the anteroposterior axis. This Topic is restricted to a description of the differentiation events preceding gastrulation and the cell movements involved in gastrulation in each of the model vertebrates. The molecular basis of axis specification and germ layer determination in vertebrates is discussed in Sections I, J and K.

Gastrulation in *Xenopus*

Among the model vertebrates discussed in this book, gastrulation is best characterized and most intensely studied in the frog *Xenopus laevis*, so it is this organism we consider first. *Xenopus* gastrulation involves several different forms of cell movement (invagination, involution, migration, convergent extension and epiboly, which are defined in *Table 2*, Topic A7). The movements can be summarized as follows (*Fig. 1*):

- Invagination of bottle cells to form the blastopore; the blastopore forms gradually, beginning with the dorsal lip.
- Involution of marginal zone cells (presumptive mesoderm and endoderm) over the lip of the blastopore.
- Migration of leading mesoderm cells (future head mesoderm) across the blastocoel roof.
- Convergent extension of the following mesoderm and endoderm cells along the anteroposterior axis (future notochord, somites, lateral plate mesoderm, gut).
- Epiboly of the ectoderm to cover the surface of the embryo.

As discussed in Topic F1, cleavage in *Xenopus* generates a blastula comprising numerous small pigmented cells at the animal pole and larger yolky cells at the

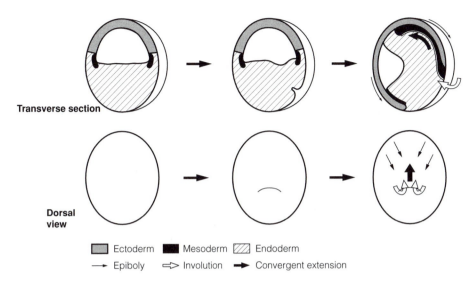

Transverse section

Dorsal view

☐ Ectoderm ■ Mesoderm ▧ Endoderm
→ Epiboly ⇨ Involution ➡ Convergent extension

Fig. 1.   Cell movements during gastrulation in Xenopus.

vegetal pole. The blastocoel is restricted to the animal hemisphere so that contact between animal and vegetal cells is limited to an equatorial ring around the equator of the embryo. As discussed in Topic K1, inductive interactions in this region specify a **marginal layer** of deep presumptive mesoderm cells and superficial presumptive endoderm cells (in other amphibians, there is no endoderm in the marginal layer, so both deep and superficial cells give rise to mesoderm). Gastrulation in *Xenopus* begins with the formation of a slit-shaped **blastopore** in the vegetal cells just below the marginal layer on the future dorsal side of the embryo. The position of the blastopore is specified by the Nieuwkoop center, which forms opposite the point of sperm entry following rotation of the cortical cytoplasm (Topic I1).

The formation of the blastopore is marked by the appearance of **bottle cells**, so called because of their unique shape, comprising a thick body and a long thin neck. The cells are organized with their bodies pointing inwards and their necks aligned at the surface. This change in shape and organization causes invagination, and forms the slit that marks the **dorsal lip of the blastopore**. This structure initially allows the presumptive mesodermal and endodermal cells of the marginal layer to involute over at the blastopore lip and extend along the inner surface of the ectodermal cell sheet, converting a single sheet of cells into three. However, the dorsal lip of the blastopore has a much more significant role in development, since transplantation to another embryo can initiate gastrulation and induce a secondary axis. This structure therefore contains all the signals needed to set up the embryonic body plan and is termed the **organizer**. The role of the organizer is discussed in Section I.

The first cells involuting into the embryo are destined to form mesodermal structures in the head. They lose their adhesive contacts and migrate individually along the inside of the blastocoel. The following presumptive mesoderm and overlying endoderm flows in as a continuous sheet. The involution reverses the order of the layers, so that the ectoderm is external, the endoderm internal and the mesoderm sandwiched in the middle. Also, as the endodermal sheet extends across the blastocoel roof, a new cavity is generated, the archenteron, which eventually becomes the lumen of the gut.

As gastrulation continues, more bottle cells form and the slit-like blastopore expands laterally eventually forming a ring. The cells of the marginal layer now involute around the entire circumference of the embryo, in much the same way as described for the zebrafish below. As cells move into the blastopore, they undergo convergent extension. They converge towards the dorsal side of the embryo and extend along the anteroposterior axis. The most dorsal mesoderm cells generate the notochord, and progressively more lateral cells converge to generate the somites and lateral plate mesoderm structures. Similarly, the endoderm converges to form the roof of the gut, while the gut floor is formed by the remaining vegetal cells. Meanwhile, the ectodermal cells covering the animal hemisphere of the embryo spread by epiboly to cover the entire surface. This involves the intercalation of layers of cells to form a thinner sheet, as well as changes in cell shape. Convergent extension also occurs in the ectodermal cells destined to form the neural plate (Section J).

Gastrulation in the zebrafish

Gastrulation in the zebrafish is very similar to that in *Xenopus* and can be broken into a similar series of stages. These are not sequential events that depend on the completion of the previous stage, but occur concurrently (*Fig. 2*).

- Spreading of the blastoderm over the yolk by epiboly.
- Involution of deep blastoderm cells to form the equatorial germ ring.
- Convergence and extension of involuting presumptive mesoderm and endoderm cells through the embryonic shield along the anteroposterior axis.

As discussed in Topic F1, cleavage in the zebrafish is restricted to the animal pole of the embryo, producing a mound of blastomeres sitting on top of the large yolk cell. This blastoderm comprises a single layer of flattened cells, the

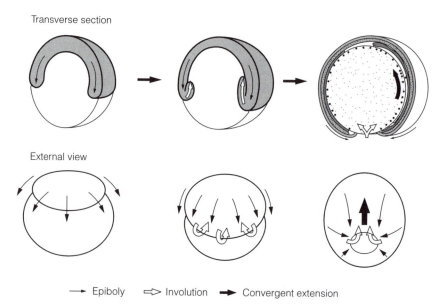

Fig. 2.   Cell movements during gastrulation in the zebrafish.

enveloping layer, and a deep layer of rounded cells in contact with the yolk through a yolk syncytial layer made up of deep cells that have fused with the yolk cell. After the midblastula transition, the blastomeres become motile and the blastoderm begins to spread over the yolk. This occurs around the entire circumference of the blastoderm and involves epiboly; it appears to be driven by expansion of the underlying yolk syncytial layer, which is closely associated with the blastoderm. These early stages of development are therefore classified as 30% epiboly, 40% epiboly etc.

Gastrulation proper begins at about 50% epiboly, i.e. when about half the yolk is covered by blastoderm. Around the equator of the embryo, deep cells at the leading edge of the advancing blastoderm begin to involute, forming a thickened ring around the equator of the embryo called the germ ring. The cells continue to involute at the blastoderm margin, even as the blastoderm itself continues to spread towards the vegetal pole of the embryo by epiboly. As the deep cells involute, they undergo convergence and extension at the dorsal midline of the embryo, initially forming a thickening called the embryonic shield and then extending to form the notochord, somites and other mesodermal derivatives. The embryonic shield can set up a secondary axis when transplanted to another embryo, and is therefore an organizer equivalent to the dorsal lip of the *Xenopus* blastopore. It thus appears that gastrulation in the zebrafish and *Xenopus* are very similar, but that while the *Xenopus* blastopore forms gradually starting at the dorsal lip, the whole zebrafish blastopore (the germ ring) is generated at once. In both cases, however, convergent extension is used to bring presumptive mesodermal and endodermal tissue involuting over the edge of the blastopore to extend along the dorsal midline to generate the notochord, somites, lateral plate mesoderm and gut. As the embryonic shield extends towards the animal pole of the embryo, it narrows and differentiates into the chordamesoderm that will form the notochord, and the adaxial cells that will form the somites. As in *Xenopus*, convergence and extension also occurs in the ectoderm cells on the outside of the embryo, generating a thickening at the dorsal midline that will form the neural tube (the neural tube forms as a solid cord in fish, rather than the neural plate that is characteristic of other vertebrates; see Topic J4).

Gastrulation in the chicken

Gastrulation in birds is quite different to that in frogs and fish because of the enormous amount of yolk in the egg, which restricts development to a small disc of peripheral cytoplasm at the animal pole. Furthermore, in common with mammals (see below), early development in birds is as much concerned with the production of extraembryonic structures as with the embryo itself. Such structures include the chorion, which encloses the entire developmental system and lies just beneath the egg shell; the amnion, which surrounds the delicate embryo within a fluid-filled sac and protects it from shock and desiccation; the yolk sac, which surrounds the yolk and connects the yolk to the embryonic gut; and the allantois, which facilitates gaseous exchange with the environment and stores waste products excreted from the embryo. The stages leading up to and including chicken gastrulation can be summarized as follows (*Fig. 3*):

- Formation of the posterior marginal zone in the area opaca and elaboration of the hypoblast.
- Formation of the primitive streak and primitive node in the epiblast above the posterior marginal zone.
- Extension of the streak across the area pelludia, with the node at its anterior extremity.

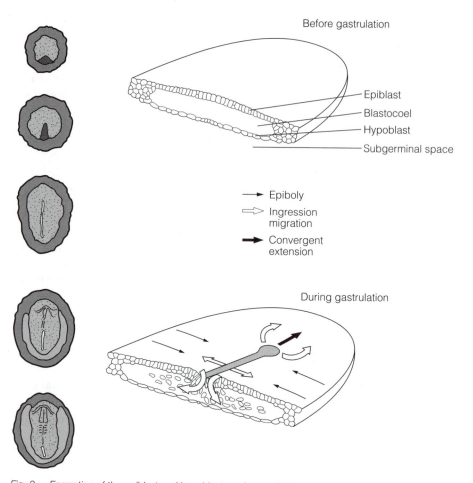

Fig. 3.   Formation of the epiblast and hypoblast, and gastrulation in the chicken embryo.

- Migration of epiblast cells through the node, forming the definitive endo-derm, anterior notochord and prechordal mesoderm.
- Migration of epiblast cells through the remainder of the streak, forming ventral and lateral mesoderm.
- Regression of the streak, laying down the remainder of the notochord in its wake.

As discussed in Topic F1, cleavage in the chick involves both cell division and cell death, resulting in a single epiblast layer comprising a central area pellucida (a clear area comprising a single cell layer separated from the yolk by a fluid filled subgerminal space) and a peripheral area opaca (a turbid ring comprising several layers of cells, with the bottom layer in direct contact with the yolk). Anteroposterior polarity in the embryo is first seen in the formation of the **posterior marginal zone**, a thickened block of blastoderm cells under the epiblast (see Topic I1). This is analogous to the Nieuwkoop center of the *Xenopus* embryo, since it induces the formation of the chick's organizer region in the overlying epiblast. Before this occurs, however, a second layer of cells called the **hypoblast** is formed beneath the epiblast. This involves a combination of delam-ination from the epiblast and migration of cells from the posterior marginal zone. The space between the two layers is then termed the **blastocoel**, and the hypoblast eventually forms part of the yolk sac.

Gastrulation begins with the formation of the **primitive streak**, a thickening of cells in the posterior epiblast caused by cells proliferating and moving in from the lateral regions. The streak forms in the epiblast above the posterior marginal zone and extends anteriorly until it reaches about three-quarters of the way across the area pellucida. At the anterior extremity of the streak is a swelling called **Hensen's node**. This is equivalent to the dorsal lip of the *Xenopus* blastopore (since it has organizer activity) while the streak is the equivalent of the rest of the blastopore. Cells migrate through a depression in the streak (and node) called the primitive groove (and primitive pit) into the blastocoel. Unlike *Xenopus* gastrulation, the epiblast cells moving through the streak do so not by involution, but by individual migration, as once in the streak, they lose adhesive contacts maintained by E-cadherin and delaminate inside the embryo. Another difference between chicks and *Xenopus* is that chick gastrulation involves rapid growth and cell proliferation, while there is little cell division in *Xenopus*.

The first cells moving through the node migrate anteriorly and displace the underlying hypoblast to form the foregut (these hypoblast cells form the germinal crescent, from where germ cells migrate into the vascular system; Topic E2). The following cells also migrate anteriorly and form head mesenchyme and the anterior portion of the notochord. Together with the overlying ectoderm, these cells comprise the **head process**. Cells moving through the rest of the streak first displace the underlying hypoblast to form the definitive endoderm, which contributes to the extraembryonic membranes and also forms the future gut (the displaced hypoblast cells also contribute to the extraembryonic membranes). The following cells then form a loose mesenchyme that eventually gives rise to the somites, intermediate mesoderm (kidneys and gonads) and lateral plate mesoderm (body wall and limbs). At this point the primitive streak begins to regress (retract towards the posterior of the embryo). As it does so, cells moving through the node are added to the notochord, so the notochord extends as the node regresses, and other cells contribute to the somites. Unlike *Xenopus*, therefore, the chick anteroposterior axis is established in gradient, with anterior structures forming before posterior structures. This carries through to later stages in development (neurulation and somitogenesis) so that the anterior of the embryo is more developmentally advanced than the posterior for several days. During gastrulation, the ectoderm proliferates and spreads by epiboly to over the exterior of the embryo. A remarkable feature of chick development is that the ectoderm continues to spread away from the embryo and eventually covers the entire yolk.

Gastrulation in mammals

Mammals have evolved so that development occurs within the mother's body. The embryo obtains nourishment from the maternal bloodstream through a specialized organ called a placenta, part of which is derived from the maternal reproductive system, and part from the zygote. The mammalian embryo does not therefore need a large amount of stored yolk, but requires a system of specialized extraembryonic membranes to protect it and facilitate development within another organism.

As discussed in Topic F1, early cleavage divisions in mammals produce a morula of equivalent blastomeres. By the 16 cell stage, however, the outer cells have begun to differentiate into flattened polarized **trophoblast cells** while the

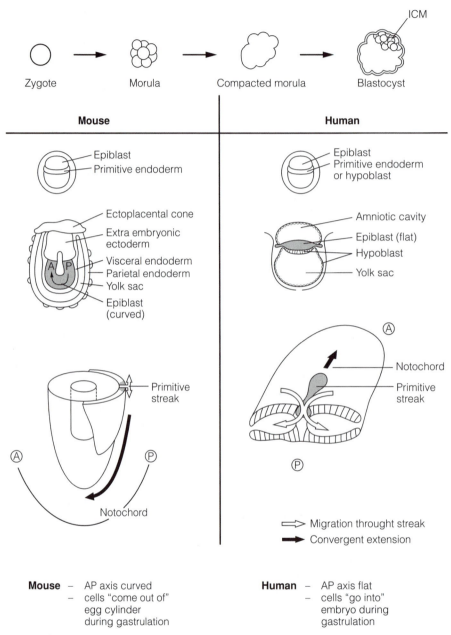

Fig. 4.   Comparison of gastrulation in mice and humans.

internal cells remain rounded and apolar **embryoblast cells**. The trophoblast is required for attachment and implantation, and develops into the embryonic side of the placenta. The embryoblast becomes segregated to one end of the blasto-coel as the amount of fluid increases, and is known as the **inner cell mass**. This develops into the embryo proper as well as producing further extraembryonic membranes. Up to this stage, the development of all placental mammals is much the same, and after gastrulation the body plan of all mammals is very similar. However, between these developmental time points there is considerable variation among species, both in the details of gastrulation and in the way that extra-embryonic membranes are formed. The mouse and human embryos are discussed below as a comparison (*Fig. 4*).

Differentiation and morphogenesis before gastrulation in mouse embryos

After differentiation of the trophoblast and embryoblast and the formation of the inner cell mass, the trophoblast is divided into the **polar trophectoderm**, a thickened region that overlies the inner cell mass; and the **mural trophectoderm**, a single cell layer delimiting the rest of the blastocoel. Upon implantation, the polar trophectoderm proliferates, while the mural trophectoderm undergoes multiple rounds of DNA replication without mitosis to generate **trophoblast giant cells**. Meanwhile, the ICM differentiates into the **epiblast (primitive ectoderm)** and **hypoblast (primitive endoderm)**. In mice, the epiblast is initially an inverted dome, and the hypoblast is tightly curved around its surface. At about 6.5 days after fertilization, cell death occurs within the epiblast to hollow out the proamniotic cavity. The embryo at this stage becomes known as the **egg cylinder**. The hypoblast cells proliferate and spread over the inner surface of the blastocoel to form the **yolk sac** (this structure is called a yolk sac even though there is no yolk, due to its homology with the yolk sac of birds and reptiles). The hypoblast lining the egg cylinder is the **visceral endoderm**, while that lining the blastocoel is termed the **parietal endoderm**. The polar trophectoderm continues to proliferate and generates a cap of extraembryonic ectoderm above the embryo, termed the **ectoplacental cone**. The polar trophectoderm also produces more giant cells that migrate around the embryo and become absorbed by the mural trophectoderm. The embryo is now ready to undergo gastrulation.

Differentiation and morphogenesis before gastrulation in human embryos

After differentiation of the trophoblast and embryoblast and the formation of the inner cell mass, the trophectoderm of primate embryos does not proliferate over the ICM to generate an ectoplacental cone as in mice. Rather, the trophoblast cells above the ICM begin to fuse together to generate a syncytial network, the **syncytiotrophoblast**. As the embryo embeds into the uterine wall, the syncytiotrophoblast spreads around the blastocyst, so that the ICM is surrounded by a cellular trophoblast layer (**cytotrophoblast**) and then the syncytiotrophoblast. As in the mouse embryo, the ICM differentiates into the epiblast and hypoblast. However, the epiblast is flat rather than dome shaped and the hypoblast likewise. Unlike the mouse, where the amniotic cavity is formed by cell death, the human amniotic cavity is generated by further differentiation within the epiblast, producing the **embryonic epiblast** (from which all cells in the embryo proper will develop) and a layer of **amniocytes**, which secrete **amniootic fluid** and generate a new cavity, the amnion, above the epiblast. As in the mouse embryo, the hypoblast proliferates and lines the blastocoel. In humans, this is called the **primary yolk sac**. The hypoblast then secretes a loose reticular matrix. The hypoblast is believed to be the source of **extraembryonic mesoderm**, which lines both the blastocoel and the primary yolk sac to generate two layers. The matrix between the layers of extraembryonic mesoderm breaks down and fills with fluid, and this becomes the **chorionic cavity (or exocoelom)**. The primary yolk sac then begins to constrict as a new wave of hypoblast migrates over the extraembryonic mesoderm and replaces the initial layer, forming the **secondary yolk sac**. The primary yolk sac then breaks up in the chorionic cavity. In primates, this is the point at which gastrulation begins. In mice, there appears to be no similar reorganization of the yolk sac and the extraembryonic mesoderm is formed during gastrulation. The mouse amnion and chorion form by the outgrowth of this extraembryonic mesoderm as well as some ectoderm across the proamniotic cavity, dividing it into two cavities.

Comparison of gastrulation in mouse and human embryos

The details of gastrulation in mammals are very similar to those in chickens (see summary above). Gastrulation begins with the formation of the primitive streak in the embryonic epiblast at the future posterior end of the embryo. Because the mouse epiblast is cup-shaped, the streak forms at one side of the egg cylinder, and extends around the base and up the other side, with the node at the anterior tip. The embryo thus develops with its anteroposterior axis curved around the cup. In humans and other primates, the epiblast is flat so the primitive streak forms on the surface and extends anteriorly, visually more similar to chickens than mice. Epiblast cells move through the streak as they do in chicken embryos and differentiate as mesoderm and endoderm. Cells migrating through the node form the anterior portion of the notochord as well as the mesoderm and endoderm of the head process, while cells migrating through the rest of the streak form mesodermal and endodermal structures of the body trunk. As the node regresses posteriorly, the remainder of the notochord is laid down, so the anteroposterior axis is formed in a temporal gradient starting at the anterior end as it does in chickens.

As in chickens, the first cells through the streak in mice and humans displace the hypoblast with definitive endoderm. The cells migrating though the node form the anterior portion of the notochord and the mesodermal and endodermal structures of the head process. However, unlike chickens, the cells that will eventually form the notochord first integrate into the endoderm and then roll up to form a tube, which is extended as the node regresses. In mice, as in the chicken, cells moving through the most posterior region of the streak form extraembryonic mesoderm, and contribute to the architecture of the extraembryonic membranes and cavities. Conversely, the extraembryonic mesoderm of the human embryo arises prior to gastrulation and all the mesoderm formed by cells migrating through the streak eventually contributes to tissues of the embryo.

G1 EARLY ANIMAL DEVELOPMENT BY SINGLE CELL SPECIFICATION

Key Notes

Single cell *vs.* group specification

In many animals, including vertebrates, the early embryo contains hundreds or thousands of cells, and cell fate is specified on a group basis by diffusible signaling molecules. Other animals such as nematodes, molluscs and ascidians have embryos with a small number of cells, and cell fate is allocated on an individual basis, often involving cytoplasmic determinants or local cell contacts.

Invariant cell lineage

One consequence of individual cell specification is a highly stereotyped developmental program, which can produce individuals with an invariant cell lineage during development and a conserved number of cells in the adult. Conversely, in organisms with a larger cell number there is a degree of random variation in the concentration and range of signaling molecules, resulting in a variable cell number and indeterminate lineage.

Mosaicism and regulation

Embryos with a small cell number often use the distribution of cytoplasmic determinants to allocate early cell fates. There are therefore good examples of mosaic developmental behavior in these embryos, revealed by blastomere isolation experiments and the results of redistributing certain cytoplasmic components. However, such experiments indicate that inductive interactions are also important in the development of these animals, showing that conditional specification by cell–cell signaling is required in addition to autonomous specification by the inheritance of cytoplasmic determinants.

Related topics

Mechanisms of developmental commitment (A3)
Mosaic and regulative development (A4)
Cell specification and patterning in *Caenorhabditis elegans* (G2)

Ascidian development (G3)
Early development of molluscs and annelids (G4)

Single cell *vs.* group specification

Most of the model organisms considered in this book (including sea urchins, *Drosophila* and all vertebrates) have a large cell number, and the embryo already comprises hundreds or thousands of cells when early cells fates are specified. Cell fate is acquired by groups of cells at a time, often in response to diffusible substances secreted by other cells (Topic A3). Similarly, positional information in such embryos is generated by diffusible substances, morphogens, emanating from a particular region of the embryo known as the organizer (Topic A6).

In contrast, the organisms discussed in this section (*Caenorhabditis elegans*, molluscs, annelids and ascidians) have a much lower cell number, and the embryo contains only a very few cells when initial patterning and specification of cell fates occurs. Rather than specifying groups of cells, cell fates in such embryos tend to be specified on an individual basis. Instead of diffusible substances, cell fates are often established by direct contact with other cells or the inheritance of asymmetrically-distributed cytoplasmic determinants upon cell division.

Invariant cell lineage

The acquisition of cell fate or positional identity on a group basis through the use of diffusible substances introduces a degree of randomness into development. Due to random fluctuations in the levels of a particular inductive signal or morphogen, signals in one embryo may not diffuse quite so far as in another. Also, since many developmental signals either stimulate or inhibit cell proliferation, the cell number in embryos using this strategy can vary considerably. Such randomness is characteristic of e.g. insect and vertebrate development, and introduces quantitative differences in phenotype that are described by the term **developmental noise**.

Conversely, the acquisition of cell fate or positional identity on an individual basis removes much of this variability. Consequently, the early development of many embryos that use single-cell specification is characterized by a **conserved cell number** (the number of cells in each individual is precisely the same at each stage of development, and often in the adult too) and an **invariant cell lineage** (all cell divisions are stereotyped and the fate of each cell entirely predictable). In species such as *C. elegans*, where the embryo is transparent, the lineage of each cell can be traced by simple microscopic observation using Normarski optics. In less amenable species, cells are externally labeled or injected with dyes. Such experiments have shown that adult ascidians, molluscs and nematodes are not all identical, although the variation may be limited and under genetic control. For example, variability between wild type *C. elegans* adults of the same sex is limited to 11 pairs of cells in postembryonic development. Specification on a cell-by-cell basis continues into postembryonic development, as shown by the cell–cell interactions in vulval cell specification (Topic L1).

Mosaicism and regulation

As discussed in Topic A3, there are two extreme modes of development described as mosaic and regulative. In mosaic development, the fate of each cell is specified autonomously, so an isolated cell develops into the part of the embryo it would normally give rise to *in situ*, while the remainder of the embryo develops normally but lacks the parts specified by the missing cell. Conversely, in regulative development, the fate of each cell is specified conditionally by its interactions with other cells, so an isolated cell often fails to develop according to its normal fate (because it lacks the appropriate interactions) while the remainder of the embryo can regulate for its missing parts and generate a whole embryo (because when cells are removed, the fates of other cells can be respecified).

The embryos discussed in this section are sometimes described as 'mosaic embryos' because there are some good examples of autonomous specification by the inheritance of cytoplasmic determinants. Such examples include determination of the germ cell lineage (P-cell lineage) in *C. elegans* by the inheritance of P-granules (Topic G2), determination of the muscle cell lineage in ascidians through inheritance of the cytoplasmic yellow crescent (Topic G3) and the specification of the D blastomere in certain molluscs by association with the polar lobe (Topic G4). Although there is evidence for mosaic behavior in these embryos, the results of experiments such as blastomere isolation or ablation, and cytoplasmic transfer or reorganization (e.g. by centrifugation) suggest that cell–cell interactions are also important in early development. Therefore, an invariant cell lineage does not necessarily correspond to mosaic development[1]. The following Topics illustrate examples of both autonomous and conditional specification in embryos with a small cell number.

[1]The developing root of *Arabidopsis thaliana* also undergoes stereotyped divisions and has a largely invariant cell lineage. However, as plant cells are totipotent (Topic A2) the fate of these cells must be determined by cell–cell interactions. In agreement with this, most laser ablation experiments do not perturb root development. See Topic N4 for further discussion.

G2 CELL SPECIFICATION AND PATTERNING IN *CAENORHABDITIS ELEGANS*

Key Notes

Properties of *Caenorhabditis elegans*	*Caenorhabditis elegans* is a species of predominantly hermaphrodite self-fertilizing nematode worms, which develop from egg to adult in about 2 days. Embryonic development is rapid, with hatching larvae produced in about 15 hours. The larvae undergo four molts to produce adult worms. *C. elegans* is remarkable because of its invariant somatic cell lineage, producing hermaphrodite individuals with exactly 959 somatic cells.
***C. elegans* embryonic development**	Eggs are fertilized by sperm stored in the gonad, and undergo cleavage to produce a ball of about 500 undifferentiated cells. Gastrulation begins when the embryo contains just 28 cells and is complete by the 3–400 cell stage. The first few cleavage divisions are asymmetrical, producing six founder cells that give rise to five somatic lineages and the germ line. Thereafter, cleavage divisions within each lineage are symmetrical and occur at a characteristic rate. After gastrulation there is extensive cell differentiation. The embryo elongates mainly by the stretching of cells in the hypodermis, and hatches when it has 558 cells.
Specification of the embryonic axes	Axis specification in *C. elegans* involves both cytoplasmic determinants and cell signaling. The anteroposterior axis is specified by anterior and posterior determinants, localized after fertilization in accordance with the point of sperm entry. This results in an asymmetrical first cleavage that produces a large anterior AB cell and a smaller posterior P_1 cell. The dorsoventral axis is specified at the next division, as the EMS blastomere arises on the future ventral side of the embryo. The left right axis is set up at the third division and reflects the orientation of two cells derived from the AB blastomere. The dorsoventral and left right axes can be reversed by manipulating individual cells, showing that cell–cell interactions are required for axis specification.
Mosaic *vs.* regulative development in *C. elegans*	Most of the determinants localized in the egg act autonomously to specify cell fates. The SKN-1 protein is such a determinant, and causes all blastomeres in which it is expressed to adopt EMS fates. The GLP-1 protein is exceptional in that it acts non-cell autonomously to specify cell fates in the AB lineage. The AB blastomere divides to produce equivalent anterior and posterior daughter cells, but only the posterior (ABp) daughter can interact with the P_1 cell, which displays the ligand for GLP-1. The differential fates of the daughters produced by the AB blastomere therefore depend on position rather than lineage.
Lineage mutants	The invariant cell lineage in *C. elegans* allows the identification of mutants that control the stereotyped cell division pattern. An interesting class of

so-called heterochronic mutants, e.g. *lin-14*, cause cells to behave in a manner more appropriate for an earlier or later developmental stage. The corresponding genes may act by establishing temporal gradients of regulatory proteins, acting in a manner conceptually very similar to morphogens, which establish spatial gradients to control pattern formation.

Related topics	Mechanisms of developmental commitment (A3)	Pattern formation and compartments (A6)
	Mosaic and regulative development (A4)	Early animal development by single cell specification (G1)

Properties of C. elegans

Caenorhabditis elegans is a soil-dwelling nematode that was chosen as a developmental model organism because of its amenability for genetic analysis (see Topic B1). The species is hermaphrodite, although males arise occasionally through nondisjunction of sex chromosomes (Topic E5). Hermaphrodites produce about 300 eggs per day and are self-fertilizing. Development is rapid, proceeding through embryogenesis and hatching to produce a first stage larva in about 15 hours at 25°C. Post-embryonic development comprises four larval stages, and adult worms emerge after about 45–50 hours. The entire somatic cell lineage during embryonic development is invariant, and postembryonic development is largely invariant, although a small number of cells can choose between alternative fates. All hermaphrodites have 959 somatic cells arranged in precisely the same way, while males have 1051 cells. The number of germ cells is indeterminate, but all germ cells arise from a single blastomere formed at the fourth cleavage division.

C. elegans embryonic development

As eggs pass through the gonad towards the vulva, they are fertilized by stored sperm, which may derive from the same (hermaphrodite) individual, or may be deposited there by a male. The sperm entry point defines the posterior of the egg. The egg nucleus then completes meiosis and a tough exterior coat is secreted. Before the fusion of pronuclei, the egg undergoes a process termed **pseudocleavage**, where cleavage furrows form but then regress. The pronuclei then fuse and cleavage proper begins.

Cleavage divisions in *C. elegans* are asymmetrical. They establish the principal embryonic axes and produce six **founder cells**. This lineage is shown in *Fig. 1*. The first cleavage generates an anterior **AB cell** and a posterior P_1 **cell**. Before cleavage occurs, uniformly distributed **P granules** migrate to the posterior and are incorporated into the P_1 cell. The P-cell lineage is like that of a stem cell. The P_1 cell undergoes three further asymmetrical divisions, in each case producing a further P cell and the somatic progenitor cells **EMS**, **C** and **D**. At each division, the P granules segregate into the P cell. The P_4 cell contains all the P granules and gives rise to all germ cells in the embryo (see Topic E1). The EMS cell also divides asymmetrically to produce the **E** and **MS** cells. The AB, C, D, E, MS and P_4 cells are the **founder cells**. Each founder cell then cleaves equally and synchronously at a unique rate, continuing to do so until zygotic transcription begins (when there are about 100 cells in the embryo).

Gastrulation begins when there are just 28 cells in the embryo, starting with the ingression of the two daughters of the E founder cell that go on to form the gut. The P_4 cell is next, followed by the MS-derived cells that go on to form body

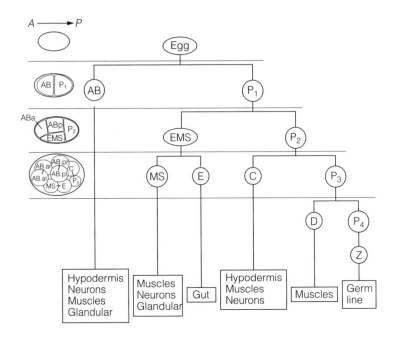

Fig. 1. Founder lineages in C. elegans.

muscle and neurons among other cell types. Some cells from the remaining lineages (AB, C and D) also ingress during gastrulation, which is completed after about four hours when there are 300–400 cells in the embryo. Apoptosis is an important component of the *C. elegans* developmental program (Topic C4), and the first programmed cell deaths occur during gastrulation. After gastrulation, the cells begin to differentiate and undergo morphogenesis to elongate the embryo. There is little cell division during this process, and most of the morphological reconstruction is achieved by changes in cell shape. The hatchling L1 larva contains only 558 cells. After hatching, the larva goes through four molts and quadruples in size to produce an adult. Although the cell number approximately doubles, much of the larval growth is achieved by changes in cell shape, since many of the new cells contribute to the gonad and the vulva (in hermaphrodites) or the tail (in males).

Specification of the embryonic axes

The three principal axes of the *C. elegans* embryo are specified during the first three cleavage divisions, and this involves a variety of processes including asymmetrical cell division, the segregation of cytoplasmic determinants and cell–cell signaling.

The anteroposterior axis is specified at the moment of fertilization by the point of sperm entry, which defines the posterior of the embryo. In some way, the sperm causes rearrangement of the microfilament cytoskeleton, resulting in (a) the asymmetric distribution of certain cytoplasmic determinants along the future anteroposterior axis and (b) an asymmetric first cleavage division. The involvement of microfilaments is demonstrated by treatment of eggs with cytochalasin D, which disrupts the early cleavage divisions and causes polarity defects in the embryo. The first division generates the large anterior AB founder cell and the smaller posterior P_1 cell. **Anterior determinants** such as GLP-1 and MEX-3 are segregated to the AB cell, while **posterior determinants** such as SKN-1 and PAL-1 are segregated to the P_1 cell (*Fig. 2*). The cytoplasm of the egg also contains uniformly distributed P granules, which are involved in germ line specification. Before the first cleavage division, the P granules sort to the posterior and are inherited by the P_1 cell. The P granules continue to segregate to the P cells in the next three cleavage divisions. They are associated with **germ-line determinants** such as PIE-1, which are discussed in Topic E1. If more than 25% of the cytoplasm is removed from the posterior of the egg after fertilization but *prior to* the redistribution of the P granules, the first cleavage division is symmetrical. This suggests that posterior cytoplasmic determinants first direct the asymmetrical cleavage division and then the recruitment of P granules to the P_1 cell.

The early cleavage divisions are disrupted in *par* (**partition defective**) **mutants**, and four *par* genes encode proteins that become asymmetrically localized in the egg prior to the first cleavage division. It is currently thought that the **PAR proteins**, all of which have domains consistent with a signaling role, are required to interpret the symmetry-breaking cue provided by the sperm and then localize downstream cytoplasmic determinants (*Fig. 2*). There is evidence that, in at least one case, this involves translational repression. The *glp-1* mRNA, distributed throughout the egg, is repressed in the posterior in a *par*-dependent mechanism so that the protein is synthesized only in the anterior. Translational repression is also used in establishing the anteroposterior axis in *Drosophila* (see Topic H2). The PAR proteins are not thought to act directly on determinants such as *glp-1*, but through a series of intermediates. A recently identified intermediate, MEX-5, appears to mediate the signal from PAR-1 to downstream determinants. In normal embryos, PAR-1 is restricted to the posterior and MEX-5 to the anterior. In *par-1* mutants, MEX-5 is distributed uniformly, indicating the PAR-1 helps to localize MEX-5 to the anterior. In *mex-5* mutants, the anterior determinant GLP-1 is absent while posterior determinants such as SKN-1 are uniformly distributed. The MEX-5 protein therefore acts upstream of these determinants and helps to localize them.

While the anteroposterior axis appears to be specified by the distribution of cytoplasmic determinants, dorsoventral and left right axis specification involve cell–cell interactions. The dorsoventral axis is specified at the second cleavage division, when the AB founder cell divides to produce two equal daughters, ABa and ABp (a and p meaning anterior and posterior, respectively). Shortly afterwards, the P_1 cell divides asymmetrically to produce the P_2 cell and the EMS cell. The relative positions of these four cells in the rhomboid stage four-

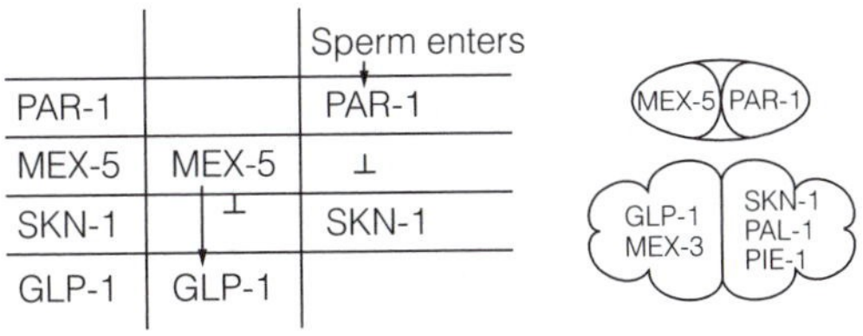

Fig. 2. Anteroposterior axis determinants in C. elegans.

cell embryo determine the dorsoventral axis, with the EMS cell located on the ventral side. However, if a blunt probe is used to prod the egg surface just before the AB cell divides, the positions of the ABa and ABp cells can be reversed relative to the P_1 cell. This also reverses the asymmetry of the P_1 cleavage division, so that the EMS arises on what is normally the dorsal side of the embryo, and the dorsoventral axis is reversed. The embryo develops normally but also has a reversed left–right axis, showing that neither the dorsoventral nor left–right axes are specified before the second cleavage division. Subsequent division of the ABa cell produces two equivalent blastomeres, ABal and ABar (l and r for left and right, respectively). The positions of these cells in the embryo specify the left–right axis, with the ABal cell protruding slightly more anteriorly than the ABar cell. Left–right axis reversal can be achieved by surgically manipulating the positions of these cells, but the dorsoventral axis is unaffected, showing that the left–right axis is specified last. Since the fate of many cells in the embryo can be influenced by moving individual cells, this shows that there must be communication between cells to set up the axes.

Mosaic vs. regulative development in C. elegans

Most of the determinants discussed above with respect to the anteroposterior axis act in a cell autonomous fashion to determine cell fates. This is shown by the effect of mutations that prevent such determinants becoming properly localized. In normal embryos, the SKN-1 protein is segregated to the posterior P_1 cell in the first cleavage division and then to the EMS cell in the second cleavage division. If the first phase of segregation is disrupted, as in *mex-1* mutants, SKN-1 is distributed to all somatic cells except those derived from the P_2 cell and all these cells adopt EMS fates. If the second phase of segregation is disrupted, as in *pie-1* mutants, then the P-cell lineage becomes an additional EMS lineage, resulting in the absence of a germ line and P-lineage derived somatic cells. If these mutants are combined in the same animal, all cells in the embryo adopt EMS fates by the four cell stage (*Fig. 3*).

The anterior determinant GLP-1 is an exception, in that it acts in a non-cell autonomous fashion. This protein is a member of the Notch family of transmembrane signaling proteins and mediates interactions with surrounding cells

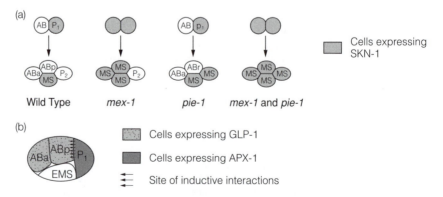

Fig. 3.   *Mosaic and regulative development in* C. elegans. *(a) Mutations that cause dislocalization of the SKN-1 protein show that it is an autonomous determinant of MS cell fate. (b) Specification of ABp cell shows that GLP-1 is a non-autonomous determinant, and must interact with the ligand APX-1 displayed by the adjacent P_1 cell.*

to establish cell fates in the AB lineage. The importance of inductive interactions in this lineage has been shown by blastomere isolation and laser ablation experiments. As discussed in Topic A3, if cell fate is controlled by an autonomous determinant, the cell should behave in the same manner in the embryo and in isolation. However, if cell–cell interactions are required, isolated blastomeres will not behave as they would *in situ* because they lack the interactions with other cells. Normally, the AB blastomere gives rise to a variety of cell types including muscle, hypodermis and neurons. However, the isolated AB blastomere fails to give rise to muscle, so cell–cell interactions are required for the specification of certain cell fates in the AB lineage.

The signaling events that regulate the AB lineage can be traced back to its first division, which produces the ABa and ABp cells. These cells have different fates: for example, only the ABa cell gives rise to anterior pharynx cells. However, the two cells are initially equivalent in potency – if the positions of the ABa and ABp cells are reversed by surgical manipulation, the embryo develops normally (this manipulation is carried out after the AB cell divides, and does not affect the dorsoventral axis, see above). Specification of the ABa and ABp cell fates requires interactions with the posterior P_2 cell, which synthesizes the ligand for GLP-1, called APX-1 (this is homologous to *Drosophila* Delta, the ligand for Notch). Both the ABa and ABp cells produce GLP-1, but only the ABp cell is exposed to APX-1 by virtue of its position adjacent to the P_2 cell (*Fig. 3*). Therefore, while the cell lineage in *C. elegans* is invariant, this is due not only to autonomous determinants but also stereotyped cell–cell interactions.

Lineage mutants The well-characterized and invariant cell lineage in *C. elegans*, together with its genetic amenability, make it possible to identify mutations that control the pattern of cell division. As discussed in Topic A3, animal cells make developmental decisions that are generally irreversible, reflecting an intrinsic developmental memory of past decisions directing them towards certain fates. An interesting class of mutants has been identified in *C. elegans* in which this developmental memory is disrupted, causing cells to behave in a manner more appropriate for an earlier or later stage of development, i.e. earlier or later in the cell lineage. These **heterochronic mutations** are conceptually equivalent to homeotic mutations (Topic A6), except that while homeotic mutations alter the positional value of a cell (the cell's awareness of where it is in relation to other cells), the heterochronic mutants affect a quality that could be called a **temporal value**, i.e. the cell's awareness of 'when' it is in relation to the developmental program.

The genes identified through heterochronic mutations encode proteins that govern cell behavior as a function of developmental time. The spatial patterning of cells is often controlled by gradients of certain molecules called morphogens, and altering the distribution or concentration of the morphogen in space causes the pattern to be disrupted. It appears that gradients may also be established over time, and can be disrupted in the same manner. The *lin-14* gene provides an excellent example of this mechanism. Mutations in this gene affect many lineages but the **T.ap lineage** is often used as an example (*Fig. 4*). In wild type worms, the T.ap cell undergoes a series of stereotyped divisions to produce 10 cells of various types (11 cells are actually produced, but one undergoes apoptosis). These divisions span two larval stages. In loss of function *lin-14* mutants, the T.ap cell skips the early divisions, and produces just four cells that are characteristic of the second larval stage. Conversely, in gain of function *LIN-14*

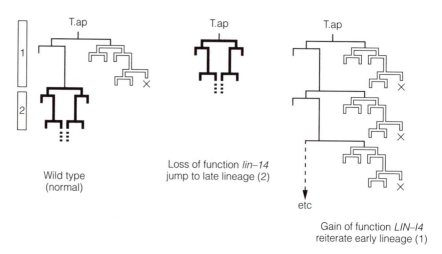

Fig. 4.   Heterochronic mutants of the lin-14 *gene in the T.ap lineage of* C. elegans.

mutants, the early divisions are repeated generating many more cells than usual, reminiscent of the first larval stage. In normal development, the concentration of LIN-14 protein drops gradually, providing evidence for a temporal gradient controlling cell behavior. In loss of function mutants, the cells behave precociously because the absence of LIN-14 protein provides them with the wrong temporal information, suggesting they are at a late developmental stage. In gain of function mutants, the LIN-14 protein is maintained at a high level, and the cells behave as they would in early development.

G3 ASCIDIAN DEVELOPMENT

Key Notes

Properties of ascidian embryos	Ascidians are chordates with a remarkably simple body plan. The fate map of the ascidian embryo is similar to that of *Xenopus*, and ascidian larva are morphologically very similar to vertebrate embryos. However, ascidian development also has certain properties in common with *C. elegans*, i.e. it is rapid, with a small cell number and invariant lineage. Single-cell specification occurs in the cleaving embryo and involves both cytoplasmic determinants and inductive interactions between cells. A striking feature of ascidian development is the presence of colorful regions of cytoplasm that correspond to the localization of cytoplasmic determinants.
Overview of ascidian development	Ascidian eggs comprise a number of distinct regions of cytoplasm. These are rearranged after fertilization through the reorganization of the egg cytoskeleton. Ascidian embryos undergo bilateral cleavage, i.e. the first division produces blastomeres that independently give rise to the left- and right-hand sides of the embryo, and all subsequent cleavage divisions on one side are mirror images of the other. Gastrulation begins at the 110-cell stage with presumptive endoderm, notochord, mesenchyme and muscle moving into the embryo while the ectoderm spreads over its surface by epiboly. Following gastrulation, the embryo elongates to form the swimming tadpole.
Mosaic *vs.* regulative development in ascidians – specification of muscle cells	Blastomere isolation and ablation experiments show that the ascidian embryo uses cytoplasmic determinants to establish certain cell fates. Muscle specification, for example, corresponds to the inheritance of a specific region of cytoplasm, the myoplasm, which in ascidians such as *Styela* is associated with yellow pigment granules (the yellow crescent). This is segregated into the B4.1 blastomeres during cleavage, and derivatives of this cell give rise to primary muscle cells. However, there is also evidence for cell–cell interactions in ascidian development. Isolated blastomeres never give rise to neural tissue, so neurons must arise through inductive interactions. There is also a requirement for inductive interactions in the formation of secondary muscle cells.
Related topics	Mechanisms of developmental commitment (A3) Early animal development by single-cell specification (G1) Mosaic and regulative development (A4)

Properties of ascidian embryos

Adult ascidians are sessile and superficially plant-like, but ascidians are chordates, and the larvae are similar to vertebrate embryos, possessing a dorsal notochord and hollow neural tube with a defined anteroposterior axis, and a tail[1]. Although the larvae resemble vertebrate embryos, they have a small cell number and thus a very simple body plan. Indeed, development to the larval stage is similar in certain respects to that of *C. elegans*, in that it is very rapid, the cell lineage is invariant, and the segregation of cytoplasmic factors initially distributed asymmetrically in the egg plays an important role in specifying cell fates. It is particularly striking that some ascidian eggs have differentially distributed pigment granules that mark different areas of cytoplasm like a natural fate map. The so-called **yellow crescent**, present in ascidian eggs of the genus *Styela*, is one such colorful determinant. This yellow cytoplasm segregates into the posterior vegetal (B4.1) blastomeres and is a determinant of myogenic fate, as discussed below.

Overview of ascidian development

Various ascidian species are used as developmental models, the most popular belonging to the genera *Styela* and *Halocynthia*, and more recently *Ciona*. Ascidians are hermaphrodites, but unlike *C. elegans* they usually cross-fertilize. The description below is of *Styela* development. The eggs complete meiosis after fertilization and the polar bodies segregate at the animal pole. Prior to fertilization, the egg comprises a number of distinct cytoplasmic regions: the cytoplasm in the animal hemisphere is clear and mainly taken up by the female pronucleus (germinal vesicle), while that in the vegetal hemisphere contains pigment granules that make it appear gray. Yellow pigment granules are evenly distributed in the cortical cytoplasm, but soon after fertilization these become localized at the vegetal pole and are drawn across to concentrate at what will be the posterior of the embryo, forming the yellow crescent.

The cleavage pattern in ascidian embryos is unique (*Fig. 1*). The first cleavage is vertical and medial, forming the **AB blastomeres** and separating the left and right halves of the embryo. Subsequent cleavage divisions are identical on both sides of the embryo and are mirror images. The blastomeres are given identical names, with those on the right hand side distinguished from those on the left by underlining (e.g. at the two cell stage left = AB, right = AB). The second cleavage division is vertical and separates anterior blastomeres (A3) and posterior blastomeres (B3). Anterior and posterior blastomere derivatives are distinguished by the letters A and B, respectively. The third division is horizontal and separates animal blastomeres (a4.2 and b4.2) and vegetal blastomeres (A4.1 and B4.1). Animal and vegetal blastomere derivatives are distinguished by lower and upper case letters respectively. Cell divisions become asynchronous by the 16-cell stage and gastrulation begins at the 110-cell stage. This involves

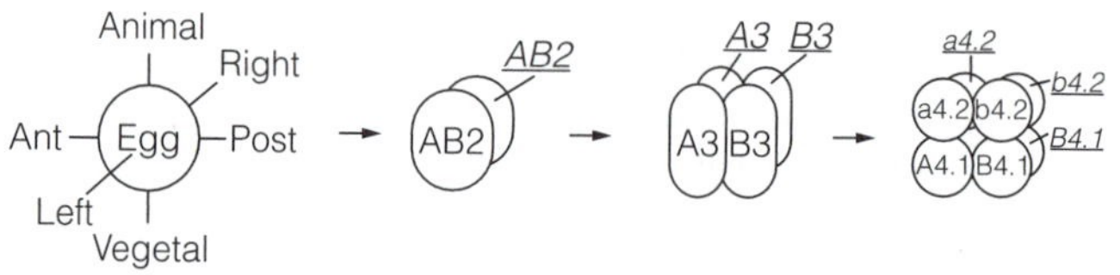

Fig. 1.   Nomenclature of ascidian cleavage divisions.

[1]Only some ascidians have tails. In the genus *Molgula*, the tadpoles of *M. oculata* have a tail, whereas those of *M. occulta* do not. The difference lies in the expression of a key regulatory gene, *Manx*, which is expressed in *M. oculata* but inhibited during tail-forming stages in *M. occulta*.

the invagination and involution of vegetal cells representing the prospective endoderm, notochord, mesenchyme and muscle, and epiboly of the animal cell layer, which forms the ectoderm and neural tube. Cell fate in the ascidian embryo appears to be determined as early as the 64-cell stage. As gastrulation continues, the embryo elongates to form the **tadpole**.

Mosaic *vs*. regulative development in ascidians – specification of muscle cells

Blastomere isolation experiments provided early evidence that ascidians undergo mosaic development. If, at the eight-cell stage, the four couples of blastomeres are separated and allowed to develop in isolation, they give rise to particular cell types commensurate with the ascidian fate map (*Fig. 2*). Both the a4.2 and b4.2 blastomeres give rise only to ectoderm, the A4.1 blastomere to notochord and endoderm, and the B4.1 blastomere to muscle, mesenchyme and endoderm. This mosaic behavior suggests that certain cell types are specified autonomously by the distribution of maternal determinants in the egg, and this has been confirmed for the epidermis, endoderm and some muscle cells. However, other cell types (e.g. notochord, mesenchyme, neurons) are specified conditionally, by inductive interactions. The specification of muscle cells shows examples of both autonomous and conditional specification, and is discussed below.

The muscle-forming region of the ascidian embryo is called the **myoplasm**, and corresponds to the yellow crescent in *Styela*, an area of pigmented cytoplasm in the posterior vegetal region of the egg. The myoplasm segregates to the B4.1 blastomeres during cleavage, and these cells give rise to muscle in isolation indicating that the myoplasm contains an autonomous **myogenic determinant**. Before fertilization, the yellow pigment granules are distributed evenly throughout the cortical cytoplasm of the egg. Soon after fertilization, **ooplasmic segregation** reorganizes the cytoplasm so that the myoplasm and other determinants are localized to particular regions. This is a two-phase process. The myoplasm is first localized to the vegetal pole and then translocated to the posterior of the egg following fertilization. Both microfilaments and microtubules have been implicated in this process. Drugs such as cytochalasin D that inhibit microfilament polymerization prevent the myoplasm from localizing at the vegetal pole, the first phase of reorganization. The second phase, where the myoplasm is localized to the posterior of the embryo, is inhibited by drugs such as colchicine that prevent the assembly of microtubules.

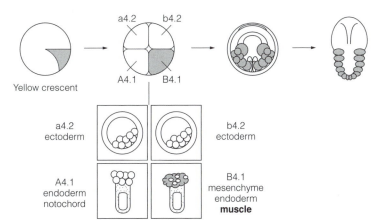

Fig. 2. Cell tracing and blastomere isolation experiments in Styela. Evidence for autonomous specification in the muscle lineage.

Detailed investigation of the behavior of isolated blastomeres and incomplete embryos suggests that cell–cell interactions are also involved in muscle-cell specification. In normal development, the descendants of B4.1 differentiate into a number of cell types including **primary muscle cells (B-line muscles)**, mesenchyme and endoderm. This is consistent with the myoplasm occupying a discrete sector of the B4.1 cell, and its segregation to only a proportion of the descendants. However, some of the cells that inherit the myoplasm also adopt non-muscle fates, indicating that the autonomously-specified fate of these cells can be changed by interactions with other cells in the embryo. Furthermore, there are additional **secondary muscle cells** in the normal embryo that are not derived from B4.1, but from A4.1 (**A-line muscles**) and b4.2 (**b-line muscles**). In isolation, and in embryos where the B4.1 blastomeres have been removed, neither of these blastomeres gives rise to muscle tissue, suggesting that cell–cell interactions are required to induce their myogenic fate, and that the appropriate signals originate from B4.1. Experiments in which myoplasm from B4.1 has been injected into blastomeres in different parts of the embryo surprisingly showed that, in most cases, the recipient cells did not produce muscle. Therefore, these cells may contain different determinants that commit them to alternative fates, and specific signaling pathways are required to respecify these cells.

The establishment of muscle-cell fates therefore involves both autonomous specification in the B4.1 lineage and conditional specification in the other lineages. The molecular nature of the myogenic determinant and the signals produced by B4.1 descendants to induce myogenic fates in other lineages remain unknown. Homologs of the vertebrate myogenic genes *MyoD1* and *MEF2* have been characterized, but these are not likely to be the maternal muscle determinants. *MyoD* is not expressed until the 64-cell stage, which is later than the onset of muscle-specific actin expression. MEF2 is a maternally-expressed gene, but the protein is distributed uniformly in the egg. The transcription factor **Snail** is expressed in b-line muscles, and its gene contains *cis*-acting sites that would be recognized by bHLH regulators, but the identity of these regulators is unknown. While the muscle determinants have remained elusive, experiments deigned to track them down have identified a number of genes, called *posterior end mark* (*pem*) genes, which are involved in setting up the anteroposterior axis. The reorganization of the myoplasm after fertilization is therefore required not only for the specification of different cell types, but also for the polarization of the embryo.

G4 EARLY DEVELOPMENT OF MOLLUSCS AND ANNELIDS

Key Notes

Properties of mollusc embryos

Mollusc embryos are renowned for their invariant cell lineage and characteristic spiral cleavage pattern, often involving asymmetrical early divisions to set up the body axes. The blastula undergoes gastrulation to form a larval stage known as a trochophore, which may develop into an adult directly, or via another larval stage called a veliger larva. Snails of the genera *Ilyanassa*, *Lymnaea* and *Nassarius* are used as developmental models.

Spiral cleavage

Molluscs undergo spiral cleavage, characterized by the clockwise (dextral) or counterclockwise (sinistral) rotation of successive tiers of blastomeres in the embryo, and the coiling of the shell in the adult. Spiral cleavage is directed by a single maternal gene product that controls the orientation of the mitotic spindle. Dextral cleavage is most common, while sinistral cleavage results from a loss-of-function maternal effect mutation.

Axis specification

In many molluscs and annelids, axis specification reflects asymmetrical early cleavage divisions to generate a four-cell embryo with one blastomere significantly larger than the others. In some molluscs, D-blastomere specification involves the formation of a cytoplasmic protuberance called the polar lobe, while in others there is asymmetric cell division by displacement of the mitotic spindle. In annelids, the D-blastomere is specified by the inheritance of a specialized cytoplasmic region called the teleoplasm.

Formation and organizer role of the polar lobe

In those molluscs that form a polar lobe, the position of the lobe may be specified by the distribution of cytoplasmic determinants in the egg. However, such components must be immobilized on the cytoskeleton because cytoplasmic reorganization does not affect polar lobe formation. When the polar lobe forms, its specialized properties are conferred by determinants in the cytoplasm, as redistribution of polar lobe cytoplasm has profound effects on development. Such experiments show that the polar lobe, and consequently the D blastomere, possesses organizer activity as well as the ability to induce specific cell types later in development.

Related topics

Mosaic and regulative development (A4)
Cleavage (F1)

Early animal development by single cell specification (G1)

Properties of mollusc embryos

Molluscs represent a diverse group of animals, which includes snails, limpets, squids, mussels and oysters. Adult molluscs vary widely in morphology, but embryonic development is highly conserved. In all molluscs except one order (the cephalopods) the fertilized egg undergoes **spiral cleavage**. The cell lineage is invariant, allowing accurate fate maps and lineage diagrams to be constructed. The early cleavage divisions are often asymmetrical, and in some species this reflects the extrusion of a large cytoplasmic bleb called the **polar lobe**, which retracts after cell division. After gastrulation, the mollusc embryo becomes a **trochophore** (larva) characterized by an equatorial ciliated band of cells, termed the **prototroch**. The trochophore may metamorphosize directly into the adult, or may pass through a second larval stage called the **veliger larva**. Snails of the genera *Ilyanassa*, *Lymnaea* and *Nassarius* are the most popular mollusc developmental model organisms. The early stages of development, up to the formation of the trochophore, are also similar in annelids such as the leech.

Spiral cleavage

Many invertebrates undergo spiral cleavage, but this is demonstrated most clearly in molluscs by the overt rotation of successive tiers of blastomeres in the cleaving embryo and the consequential spiral patterns of the shell. Spiral cleavage becomes readily apparent by the third division, which generates two tiers of blastomeres, vegetal macromeres and animal micromeres, one tier rotated slightly with respect to the other. The first two divisions are vertical, generating four blastomeres. The third division, which is horizontal in radially-cleaving embryos because the mitotic spindle is parallel to the animal–vegetal axis of the egg, is oblique in spirally-cleaving embryos because the spindle is tilted away from the animal–vegetal axis. Viewed from the animal pole, the spindles also point either clockwise or counterclockwise, resulting in either dextral cleavage (right handed) or sinistral cleavage (left handed). In dextral species, the third cleavage and all subsequent odd-numbered cleavage divisions are displaced clockwise, while in sinistral species the third cleavage and all subsequent odd-numbered cleavage divisions are displaced counterclockwise.

The direction of cleavage is controlled by a single maternal gene product stored in the egg. Most species of mollusc are dextral, although sinistral mutants can arise. In crosses, the 'dextral' allele is dominant to the 'sinistral' allele, suggesting that the maternal product directs clockwise rotation of the mitotic spindle at the third cleavage division, and in its absence, sinistral rotation occurs. Such crosses also show a classic maternal effect, i.e. individuals homozygous for the sinistral allele produce all sinistral offspring even if the sperm carries a dominant dextral allele and the embryos are consequently heterozygous – this is because the mutation affects oocyte maturation in the mother; the paternal nucleus is not activated until cleavage is already underway. Cytoplasm from dextral eggs can 'rescue' sinistral eggs resulting in the development of dextral embryos. The nature and mechanism of this maternal product, and how it controls both cleavage and the direction of shell coiling, are currently unknown. It is thought to be a regulator or structural component of the cytoskeleton, since it can direct the orientation of the mitotic spindle and the location of the contractile ring of microfilaments at cytokinesis.

Axis specification

Axis specification in many molluscs and annelids is related to early unequal cleavage divisions, such that at the four-cell stage, the embryo is divided into four **quadrants** with unique lineages (*Fig. 1*). This involves the formation of a

(a) Molluscs that use polar lobe

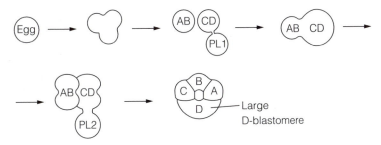

(b) Teleoplasm in annelids

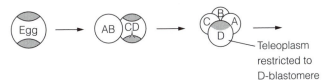

Fig. 1. Formation of the D blastomere in molluscs and annelids. (a) In many molluscs, the D blastomere is larger than the other three because of the extrusion of polar lobes (PL1, PL2) during the first two cleavage divisions. (b) In annelids, specification of the D blastomere involves the inheritance of a special region of cytoplasm, called the teleoplasm.

large **D blastomere** at the second cleavage division, which specifies the dorsal quadrant of the embryo. The formation of the D blastomere probably involves the segregation of cytoplasmic determinants in the egg (see below), but the function of the D blastomere is as an organizer, demonstrating the importance of cell–cell interactions in early mollusc development. In annelids, a large D blastomere is also formed by asymmetrical early cleavage divisions and this involves the inheritance of a specialized region of cytoplasm called the **teleoplasm**. The D blastomere gives rise to specialized population of cells called **teleoblasts** that produce ectoderm and mesoderm-derived structures in the adult. The following discussion relates to axis specification in molluscs, specifically in snails of the genus *Ilyanassa* where unequal division involves the extrusion of a polar lobe. It is applicable to other genera such as *Dentalium* and *Nassarius*.

In the first and second cleavage divisions of *Ilyanassa* spp., one of the blastomeres extrudes a large bleb of cytoplasm called the **polar lobe**. The first division divides the embryo into two cells, called AB and CD. The polar lobe is extruded from the vegetal pole of the egg and remains attached to the CD blastomere during cytokinesis through a narrow cytoplasmic bridge. The polar lobe is approximately the same size as the blastomere and resembles a third cell (the embryo is sometimes termed a **trefoil** for this reason). After cytokinesis, the polar lobe is retracted with the result that the CD cell is much larger than the AB cell. In the second cleavage division, the AB and CD blastomeres divide to form the A, B, C and D blastomeres, representing the four quadrants of the embryo. A second polar lobe is extruded, which remains attached to the D blastomere, so that this cell is much larger than the other three. In the subsequent divisions the descendants of the D blastomere remain larger than those of the other three blastomeres and go on to specify the mesodermal bands that give rise predominantly to larval muscle cells.

In other genera (e.g. *Spisula*), a large D blastomere arises through unequal cleavage but there is no visible polar lobe. Instead, cleavage involves displacement of the mitotic spindle away from the animal–vegetal axis of the egg and towards the point of sperm entry, in a similar fashion to the first cleavage division in *C. elegans*. In still other species, including the limpet *Patella vulgata*, the early cleavage divisions are equal and the A, B, C and D blastomeres are equivalent. In such species, differences arise later in development (often between the fifth and sixth cleavage divisions) due to inductive interactions between tiers of blastomeres.

Formation and organizer role of the polar lobe

In some molluscs (e.g. *Bythynaea*), the polar lobe cytoplasm contains distinct granules or **vegetal bodies** that appear to confer its specialized properties. The importance of the polar lobe cytoplasm has been demonstrated by various experimental approaches, including injecting lobe cytoplasm into other blastomeres and surgically removing the lobe. The redistribution of lobe cytoplasm at the two-cell stage can induce twinning, which suggests that it has organizer activity. If the lobe is removed at the four-cell stage, the D blastomere survives but is the same size as the other three blastomeres. It does not initiate the cell lineage that eventually produces the mesodermal bands of the trochophore, and the resultant larva has reduced muscle and mesenchyme tissue. Moreover, adults lack other structures (e.g. eyes) that are not derived from the D blastomere lineage in normal development, indicating that the D blastomere-derived cells are normally the source of inductive signals later in development.

Experiments with *Nassarius* and *Dentalium*, involving the redistribution of the cytoplasmic contents of the egg by centrifugation, do not prevent the polar lobe forming nor do they affect subsequent development. This indicates that although the polar lobe itself contains specialized cytoplasmic determinants, if these are already asymmetrically distributed in the egg, such determinants must be immobilized perhaps by attachment to the cytoskeleton. Therefore, cytoplasmic determinants may be involved in the formation of the D blastomere as they are in annelids, but its early organizer and later inductive roles in development are achieved through cell–cell signaling.

H1 MOLECULAR ASPECTS OF EMBRYONIC PATTERN FORMATION IN *DROSOPHILA*

Key Notes

Overview of patterning in early *Drosophila* development	Axis specification in *Drosophila* involves communication between somatic cells and the developing oocyte, followed by the asymmetric localization of maternal mRNAs. Early patterning occurs in the syncytial blastoderm through the localized expression of zygotic genes, dividing the embryo into domains representing future parasegments. When the cellular blastoderm forms, the parasegmental boundaries are established, thus dividing the embryo into a series of developmental compartments.
Classes of *Drosophila* patterning mutant	Screens for *Drosophila* patterning mutants have revealed a number of distinct classes of mutant phenotype, representing genes acting at different levels of the developmental hierarchy. First, maternal genes establish the embryonic axes by setting up signaling centers. Along the anteroposterior axis, sequential classes of zygotic genes are activated: gap genes, pair rule genes, segment polarity genes and homeotic selector genes. The segment polarity genes establish compartment boundaries, while the homeotic selector genes confer positional values on each compartment.
The role of maternal somatic cells	In insects, the developing oocyte is transcriptionally inactive, and surrounding somatic cells (follicle cells and nurse cells) play a major role in setting up the egg axes. Communication between the oocyte nucleus and the surrounding follicle cells causes them to become regionally specified. The follicle cells then signal back to the egg and activate local signaling pathways to establish egg polarity. This is reinforced by the import of maternal mRNAs from the nurse cells and their distribution to asymmetric sites within the egg.
The importance of syncytial development	Early patterning occurs in a syncytial environment. For this reason, most of the zygotic patterning genes encode transcription factors, which can diffuse freely to interact with zygotic nuclei. Early development is a cascade of regulatory interactions, whose timing depends on the rate of transcription, translation and mRNA/protein turnover for different genes. Intercellular signaling is not required until the cellular blastoderm stage, but by this time, most of the embryonic pattern is already established.
Long and short germ-band insects	*Drosophila* is a long germ-band insect in which the entire body plan is established at the syncytial stage and all segments form simultaneously. Many other insects have a short germ band, representing only the anterior-most segments – the posterior segments are added by growth after cellularization. Although early patterning mechanisms appear to be

conserved, there is a great diversity of insect species on the planet, and *Drosophila* represents a very polarized view of their developmental diversity.

| **Related topics** | Pattern formation and compartments (A6) | Cellular and microsurgical techniques (B4) |
| | Developmental mutants (B2) | |

Overview of patterning in early *Drosophila* development

Pattern formation in *Drosophila* begins in the developing egg, with the asymmetric localization of maternal mRNAs, and the localized activation of particular signaling pathways. By the time the egg is fertilized, the principal axes of the embryo (anteroposterior and dorsoventral) are already specified. Therefore, unlike organisms such as *C. elegans* and *Xenopus*, the site of sperm entry plays absolutely no instructive role in axis specification or patterning. However, fertilization is permissive for axis patterning because it triggers the translation of key maternal mRNAs, and this is instrumental in establishing **anterior and posterior signaling centers** that direct future development. In contrast to vertebrates, which have a pronounced left–right axis, *Drosophila* appears to be almost perfectly bilaterally symmetrical.

After the completion of meiosis and the fusion of the male and female pronuclei, cleavage commences. In insects this involves a series of rapid nuclear divisions without cell division, generating a syncytial embryo (more details in Topic F1). Most of the early events in pattern formation occur in the context of this syncytium, so that by the 14th division cycle, when cell membranes begin to form to generate the cellular blastoderm, the future segmental body plan is already established. After gastrulation, the *Drosophila* embryo undergoes **germ band extension**, where the ventral blastoderm (or **germ band**) extends towards the posterior, wraps around the curved inner surface of the chorion and extends along the dorsal surface of the egg bringing the anterior and posterior extremities of the embryo into close contact. During this process, the metameric organization of the embryo becomes apparent with the appearance of a series of ectodermal grooves. However, these do not correspond exactly to the future segments of the larva, but to developmental compartments called **parasegments** representing the anterior two thirds of one segment and the posterior third of the next. The segmental body plan is established when the germ band begins to retract, by which time the patterning of the embryo is complete.

Classes of *Drosophila* patterning mutant

The molecular basis of pattern formation during *Drosophila* embryogenesis was elucidated through the isolation and characterization of different **classes of patterning mutant**, many arising from a series of saturation mutagenesis screens carried out by Christine Nusslein-Volhard and Eric Weichaus in the 1980s. This revealed the existence of nine groups of genes with different roles in pattern formation, acting in a hierarchical manner (*Table 1; Fig. 1*).

The earliest patterning events depended on **maternal genes**, which can be divided into four major groups acting more or less independently. Three groups of genes (the **anterior class**, the **posterior class** and the **terminal class**) are involved in anteroposterior axis specification, while the fourth is required for dorsoventral axis specification. These maternal genes are named **egg polarity genes** because their products are localized or activated at specific positions in the egg to establish its polarity prior to fertilization (Topics H2 and H6).

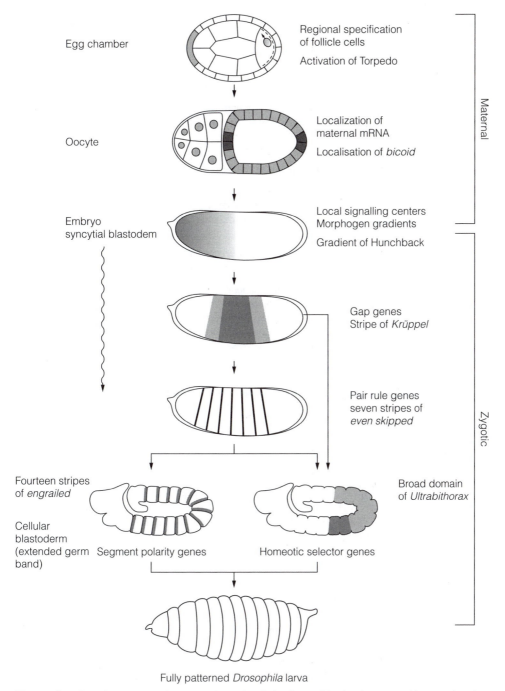

Fig. 1.   Overview of anteroposterior pattern formation during Drosophila *development, with examples of genes active at each stage.*

The function of the maternal genes is to establish **morphogen gradients** or **localized signaling centers** that control the expression of **zygotic genes** in the growing population of nuclei in the syncytial embryo. The first zygotic genes to be expressed along the anteroposterior axis are called **gap genes** because they

Table 1. *Properties of different classes of* Drosophila *patterning gene*

Gene class	Expression pattern	Mutant phenotype
Anteroposterior axis		
Maternal egg polarity (anterior)	Expressed in nurse cells (mRNAs imported into egg) or in follicle cells (regional signaling for egg polarization). See Topic H2 for details	Loss of head and thorax
Maternal egg polarity (posterior)		Loss of abdomen
Maternal egg polarity (terminal)		Loss of specialized terminal structures (acron and telson)
Zygotic gap genes	Broad transverse stripes, non-repetitive, in syncytial blastoderm (Topic H3)	Loss of contiguous segments
Zygotic pair rule genes	Seven narrow transverse stripes in syncytial and cellular blastoderm (Topic H4)	Loss of alternating segments
Zygotic segment polarity genes	Fourteen narrow transverse stripes in cellular blastoderm (Topic H4)	Loss of part of every segment, replaced with mirror image duplication
Zygotic homeotic selector genes	Broad transverse stripes, non-repetitive, in cellular blastoderm (Topic H5)	Homeotic transformation
Dorsoventral axis		
Maternal egg polarity (dorsoventral)	See Topic H6	Dorsalization or ventralization of embryo
Zygotic genes	Broad lateral stripes, non-repetitive, in cellular blastoderm (Topic H6)	Loss of and/or mis-specification of different cell types along dorsoventral axis

are expressed in broad domains in the syncytial embryo corresponding to several contiguous segments in the larva. In gap gene mutants, these contiguous segments are lost. The morphogen gradients set up in the egg activate the gap genes in a concentration-dependent manner, and interactions between the gap proteins establish sharp boundaries of expression (Topic H3). Although not called gap genes, zygotic genes are also activated in a concentration dependent manner along the dorsoventral axis, and these specify longitudinal belts of different cell types (Topic H6).

The products of the gap genes activate the **pair rule genes**, so called because they are expressed in seven stripes in the syncytial embryo representing alternating parasegments in the larva, and mutants lack every even or every odd segment (Topic H4). As the pair rule genes are expressed, the embryo undergoes cellularization. The products of the pair rule genes cooperate to activate the **segment polarity genes**. These are expressed in 14 transverse strips of cells representing the same position within each parasegment. The segment polarity genes establish and maintain the **parasegment boundaries** so that each parasegment becomes a restricted **developmental compartment**. Furthermore, the segment polarity genes pattern the compartments, generating the characteristic arrangement of denticle bands and naked cuticle. Mutations in the segment polarity genes cause this pattern to be disrupted.

While the segment polarity genes establish the metameric organization of the embryo, a different class of genes is required to give the segments **positional values** along the axis. These **homeotic selector genes** are expressed in a series of overlapping domains along the anteroposterior axis, and mutations cause spectacular phenotypes in which one body part is replaced by the likeness of another. The homeotic genes are discussed in Topic H5.

The role of maternal somatic cells

As discussed in Topic E3, the *Drosophila* oocyte is completely transcriptionally inactive and relies on surrounding somatic cells for the accumulation of gene products required for fertilization and development. In the ovariole, the oocyte is adjacent to 15 nurse cells and surrounded by hundreds of follicle cells. Both types of cell are involved in axis specification.

The follicle cells are responsible for conferring both anteroposterior and dorsoventral polarity on the egg in response to signals from the oocyte nucleus. Patterning begins with the specification of follicle cells adjacent to the oocyte nucleus as 'posterior' and those overlying the route of the migrating nucleus as 'dorsal'. Signals from these regionalized follicle cells then establish the axes of the egg (Topic H2, Topic H6).

One of the most important functions of the follicle cells is to polarize the egg's cytoskeleton. This allows the localization of maternal mRNAs to the anterior and posterior poles of the egg and hence helps to establish the anterior and posterior signaling centers. The maternal mRNAs are synthesized in the nurse cells and imported into the egg. Once inside, they attach to the cytoskeleton and are targeted to the appropriate sites (Topic H2).

The importance of syncytial development

After signaling between the egg and the follicle cells establishes the embryonic axes, much of the early patterning of the *Drosophila* embryo occurs inside the syncytial blastoderm where many nuclei are suspended in a common cytoplasm. The absence of cells in the early embryo has a number of important consequences. First, there is *no need for intercellular signaling*, so in contrast to all the other animals discussed in this book, none of the genes involved in patterning the early *Drosophila* embryo *after the axes have been specified* encode secreted proteins, membrane spanning receptors or intracellular signal transduction components. Almost all of them encode transcription factors. Second, there is *free diffusion of gene products within the syncytium*. Many examples of morphogen gradients are discussed in this book, and in most cases the morphogen is a secreted protein. Secreted proteins may not establish true extracellular concentration gradients over a field of cells, but are likely to act by other mechanisms, as discussed in Topic A6. In contrast, the morphogen gradients established in the early *Drosophila* embryo are true concentration gradients, with limits controlled only by the rate of synthesis and the rate of degradation of the gene product. Furthermore, for the reasons discussed above, *Drosophila* provides the only example where transcription factors are used as morphogens. Third, the syncytial embryo provides a *dynamic interactive system*. The patterning genes encode transcription factors that regulate each other in complex feedback loops. The order and timing of gene activation in *Drosophila* development therefore reflects an intrinsic timing mechanism based on delays between transcription, translation and the establishment of stable protein gradients in a genetic hierarchy. For the same reason, the gene expression patterns are dynamic and transient, reflecting the sequential activation of genes, their translation to yield new transcription factors, their diffusion to establish domains of activity, and their effect on the expression of other genes. In the midst of this dynamic system, the embryo undergoes cellularization. The first set of genes expressed after cellularization are the segment polarity genes. As expected, these encode a variety of intercellular signaling proteins as well as transcription factors.

Long and short germ band insects

The information presented in this topic and all the following Section H topics relates to a single dipteran insect, *Drosophila melanogaster*. Developmental biologists study certain **model organisms**, which are chosen on the basis of their convenience to maintain in laboratory populations and their amenability to experimental manipulation and analysis (Topic B1). *Drosophila* has been adopted as the model insect species, and while it is fair to say that more is known about the molecular basis of development in *Drosophila* than any other animal, this does present a rather biased view of insect development in general. There are many millions of different insect species in the world, and their diverse developmental mechanisms have only just begun to be explored.

Dipterans are **long germ-band** insects because the ventral blastoderm (often called the germ band) corresponds to the entire future anteroposterior axis. Patterning along the whole axis occurs at the same time in the syncytial embryo, and all the segments form at once. Many other insects, however, are characterized by a **short germ-band**, which represents only a few of the anterior segments. These anterior segments are patterned at the syncytial stage, but the posterior segments are not added until after cellularization, by growth from a posterior **growth zone** (*Fig. 2*). At the stage of maximum germ band extension, the body patterns of long and short germ band insects are similar, and this represents the **phylotypic stage** of insect development (vertebrate embryos also have a phylotypic stage; see Topic I1). The flour beetle *Tribolium* is one of the few short germ band insects that have been studied at the molecular level. In *Tribolium*, patterning genes that are normally expressed in the central region of the *Drosophila* embryo are confined to the posterior blastoderm, and there are fewer pair rule gene stripes, representing the anterior segments only. This indicates that the same genes are used to pattern long and short germ band insects at the syncytial state, but confirms that the posterior segments of short germ band insects have yet to be specified at the syncytial stage of development.

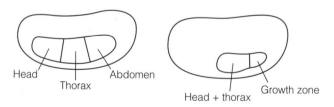

Fig. 2.   Fate maps of long and short germ band insects.

H2 ANTEROPOSTERIOR AXIS SPECIFICATION IN *DROSOPHILA*

Key Notes

The anteroposterior axis	The *Drosophila* larva comprises a series of segments defining a head, thorax and abdomen. Each segment has a unique cuticular pattern and internal structure, and gives rise to a particular set of appendages in the adult. The termini of the larva are represented by specialized structures called the acron and telson.
Three classes of maternal genes	The anteroposterior axis is specified by three classes of maternal egg polarity genes, which set up anterior, posterior and terminal signaling centers to control downstream patterning. This requires the prior regional specification of follicle cells, and results in the activation of a terminal signaling pathway as well as the import of maternal mRNAs and their localization to the anterior and posterior poles of the egg.
Anterior and posterior signaling centers – first coordinates	Four maternal genes, *bicoid*, *hunchback*, *nanos* and *caudal*, are critical for the establishment of anterior and posterior signaling centers. *bicoid* mRNA is localized at the anterior pole of the egg and Bicoid activates the transcription of *hunchback*. *nanos* mRNA is localized at the posterior pole of the egg and represses *hunchback* translation. Bicoid and Nanos together generate a gradient of Hunchback protein. Bicoid and Hunchback therefore represent the anterior signaling center, which drives head and thorax development. *caudal* mRNA is uniformly distributed but its translation is inhibited by Bicoid in the anterior of the egg. Caudal protein represents the posterior signaling center, which drives abdominal development. Bicoid, Hunchback and Caudal act as morphogens to activate downstream genes in a concentration-dependent manner.
The importance of signaling centers	Cytoplasmic transfer experiments show that a determinant located at the anterior pole of the egg is required and sufficient for head development. This determinant can induce ectopic head development and can rescue embryos from *bicoid* mutant mothers, suggesting it is the Bicoid protein. Increased amounts of Bicoid protein generate a steeper morphogen gradient, increasing the size of the head and thorax at the expense of the abdomen.
Localization of maternal mRNAs	The *bicoid* and *nanos* mRNAs are localized to the anterior and posterior poles of the egg by transport along polarized microtubules. Several gene products required for this localization process have been identified, and the 3'-UTR of the mRNAs is critical for correct targeting. The microtubules are polarized following the specification of posterior follicle cells by local signaling from the oocyte nucleus. The secreted protein Gurken and its receptor Torpedo are required for this process.

Control of translation	The translation of *bicoid* and *nanos* mRNAs is activated by fertilization, so that patterning occurs only in fertilized eggs. Both the Bicoid and Nanos proteins also act as translation regulators, Bicoid inhibiting *caudal* and Nanos inhibiting *hunchback*. Bicoid binds to *caudal* mRNA directly, while Nanos binds to *hunchback* mRNA in association with another protein called Pumilio.
Terminal signaling centers	The acron and telson are not specified by localized maternal mRNAs but by signaling restricted to the egg termini. Terminal specification involves the Ras-Raf-MAP kinase pathway downstream of the receptor tyrosine kinase Torso. Torso is ubiquitous, so terminal specificity is controlled by its ligand. The secreted protein Torsolike is produced only by terminal follicle cells and may interact with another protein, Trunk, to activate Torso signaling.
Related topics	Pattern formation and compartments (A6) Signal transduction in development (C3) Gene expression and regulation (C1)

The antero-posterior axis

The anteroposterior axis of the *Drosophila* larva is divided into a regular series of segments, defining a head, thorax and abdomen. The head is a highly special-ized structure comprising seven segments that are fused together. Conversely, the three thoracic segments and eight abdominal segments are marked by clear ectodermal grooves. Each segment has a unique cuticular pattern and internal structure. The three thoracic segments each give rise to a particular set of appendages in the adult. The larva also has specialized structures at the anterior and posterior termini, which are called the **acron** and **telson**, respectively.

Three classes of maternal genes

As discussed in Topic H1, patterning the anteroposterior axis involves the sequential and hierarchical activation of several different groups of genes. Maternal gene products are responsible for the first stage of this process, which is **axis specification** (*Table 1*). The maternal gene products establish **signaling centers** that pattern the zygotic nuclei during cleavage through the graded acti-vation of downstream zygotic genes. These signaling centers are conceptually no different from the **organizers** that pattern vertebrate organs.

Setting up the anteroposterior axis involves three different classes of maternal genes (sometimes called **egg polarity genes**), each controlling the development of a different part of the embryo by establishing a distinct signaling center. One class of genes establishes an **anterior signaling center** that controls head and thorax development. Another establishes a **posterior signaling center** that controls the development of the abdomen. In both cases, the principal genes are transcribed in the nurse cells and the products are deposited in the egg as **maternal mRNAs**, where they become asymmetrically localized. This requires pre-existing polarity in the egg, which is established by signaling between the oocyte and the surrounding follicle cells as discussed below. A third set of genes establishes **terminal signaling centers** that control the development of the acron and telson. This also involves signaling between the oocyte and follicle cells. Therefore, an important early component of axis specification in *Drosophila* is the **regional specification of follicle cells**. As discussed in Topic H6, regionally-specified follicle cells are also required to set up the dorsoventral axis.

Table 1.   Major maternal genes involved in axis specification in Drosophila

Gene	Product	Function
Polarizing the oocyte		
gurken	EGF homolog	⎫
torpedo	EGF receptor	⎬ Posteriorizes follicle cells
cornichon	Membrane-associated protein	⎭
Anterior signaling center		
bicoid	Transcription and translation factor	Anterior morphogen Activates *hunchback* transcription Represses *caudal* translation
hunchback	Transcription factor	Anterior morphogen
staufen	RNA-binding protein	Localization of *bicoid* mRNA
exuperantia	RNA-binding protein	Localization of *bicoid* mRNA
swallow	RNA-binding protein	Localization of *bicoid* mRNA
Posterior signaling center		
nanos	Translation factor	Represses *hunchback* translation
oskar		Localization of *nanos* mRNA Establishment of pole plasm
tudor, vasa, valios		Localization of *nanos* mRNA Establishment of pole plasm
staufen	RNA-binding protein	Transport of *oskar* mRNA
pumilio	RNA-binding protein	Required for inhibition of *hunchback* translation by Nanos
Terminal signaling centers		
Torso	Receptor tyrosine kinase	Receptor for terminal signal
torsolike, trunk	Secreted proteins	Candidates for terminal signal
polehole, rolled	Intracellular signaling proteins (Ras-Raf-MAP kinase pathway)	Transduction of terminal signal

Mutations in the egg polarity genes show a maternal effect because they function in the mother to pattern the early embryo, when the embryo's own genes are still switched off. Consequently, even if the mutant allele is recessive to wild type and the egg is fertilized by sperm from a wild type male, the developmental phenotype is not rescued.

Anterior and posterior signaling centers – first coordinates

There are about 50 maternal genes involved in anteroposterior axis specification but four genes are particularly important because their products form gradients along the axis and play a predominant role in patterning the nuclei of the syncytial blastoderm. These genes are *bicoid, hunchback, nanos* and *caudal* (*Fig. 1*).

The maternal *bicoid* mRNA is synthesized in the nurse cells and is localized at the anterior pole of the egg. After fertilization, the message is translated, producing a gradient of Bicoid protein along the anteroposterior axis, with the highest concentration at the anterior pole. Bicoid is a transcription factor that activates a number of downstream zygotic genes, including *hunchback*. The anteroposterior gradient of Bicoid therefore sets up a similar gradient of

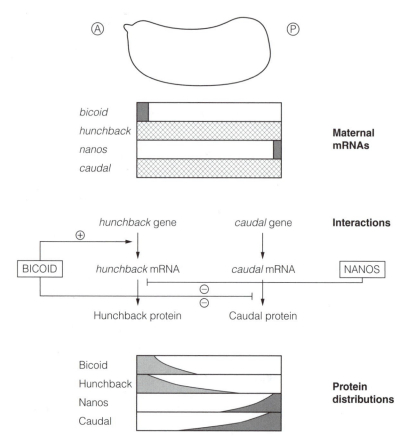

Fig. 1.   The Drosophila *anteroposterior axis is coordinated by four major genes:* bicoid, hunchback, nanos *and* caudal. *This figure shows distributions of the corresponding mRNAs and proteins, and interactions among them.*

Hunchback protein, also a transcription factor, with the highest concentration at the anterior pole of the egg. Both Bicoid and Hunchback regulate downstream genes that control anterior-specific developmental processes. However, there is also a low concentration of maternal *hunchback* mRNA in the egg, and this is distributed uniformly. The translation of maternal *hunchback* mRNA in the posterior of the embryo would yield enough Hunchback protein to initiate the development of head structures. The translation of *hunchback* mRNA is therefore blocked in the posterior region of the embryo, and this function is carried out by the Nanos protein. Maternal *nanos* mRNA is localized to the posterior pole of the egg and, like *bicoid* mRNA, is translated after fertilization to form a protein gradient (in this case running posterior to anterior). The gradient of Hunchback protein is therefore established by a dual mechanism: stimulation of zygotic *hunchback* transcription by Bicoid in the anterior of the embryo, and inhibition of maternal *hunchback* translation by Nanos in the posterior. The final gene in this quartet, *caudal*, encodes a transcription factor that controls the development of posterior structures. Like *hunchback*, maternal *caudal* mRNA is also distributed uniformly in the egg so its translation must be blocked in the anterior of the embryo. This function is carried out by Bicoid, therefore establishing a gradient of Caudal protein running from posterior to anterior. Bicoid is bifunctional: it acts as both a transcription factor to activate *hunchback* and a translational inhibitor to repress *caudal*. Conversely, while Nanos can act as a translational inhibitor, it does not act as a transcription factor. As a result of the interactions between these four genes, the syncytial embryo has gradients of Bicoid and Hunchback running from anterior to posterior (the anterior signaling center) and a gradient of Caudal running from posterior to anterior (the posterior signaling center). The three transcription factors act as morphogens to activate downstream gap genes in a concentration dependent manner. This is discussed in more detail in Topic H3. Note that Nanos is localized in the posterior of the embryo within the pole plasm that specifies the *Drosophila* germ line. Nanos also plays an important role in germ cell development, as discussed in Topic E1.

The importance of signaling centers

The importance of anterior and posterior signaling centers in *Drosophila* development was appreciated a long time before the *bicoid* and *nanos* genes were characterized, through cytoplasm redistribution experiments. For example, if anterior cytoplasm is drained from one egg and injected into the posterior of another, the recipient embryo develops with two heads (one at each end) and an interstitial thorax, while the donor embryo develops with abnormal anterior structures suggesting that the head and thorax have not been specified. The donor embryo is a close phenocopy of an embryo from a *bicoid* mutant mother, providing further evidence that maternal *bicoid* is required in the anterior cytoplasm to specify head and thoracic structures in the embryo. If anterior cytoplasm from a normal egg is injected into the anterior end of an egg from a *bicoid* mutant, anterior development can be rescued to a certain degree. Similar experiments can be carried out using posterior cytoplasm, and while the results are more complex, they establish a similar principle, i.e. that Nanos is required in the posterior cytoplasm for posterior specification. The role of Bicoid as a morphogen has been demonstrated by generating transgenic flies with increased dosage of the *bicoid* gene. This causes more *bicoid* mRNA to be imported into the oocyte and generates a steeper and longer *bicoid* gradient. These embryos develop with larger heads and thoraces than normal embryos, and severely reduced abdomens.

Localization of maternal mRNAs

The asymmetrical localization of *bicoid* and *nanos* mRNAs is critical for the establishment of anteroposterior polarity in the embryo. However, this localization process relies on a pre-existing asymmetry in the egg, which is established during oogenesis. This symmetry-breaking process involves a group of about 40 maternal genes.

The *Drosophila* ovariole initially comprises 16 identical and interlinked cells called cystocytes. One of these will form the oocyte, while the others develop as nurse cells. The posterior cystocyte always becomes the oocyte because of the way cytoplasmic bridges are established during the division of the original oogonium (this is discussed in more detail in Topic E3). The oocyte then interacts with the adjacent follicle cells and specifies them as **posterior follicle cells**. This involves the secretion of a short range signaling protein called **Gurken**[1] into the egg chamber. Gurken binds to a receptor called Torpedo on the surface of the adjacent follicle cells, initiating a signal transduction cascade that

---

[1] The *gurken-torpedo* pathway is used later by the oocyte nucleus to establish dorsoventral polarity – see Topic H6.

activates 'posterior' type genes. The *cornichon* gene is required for this signaling process. Posterior follicle cells then send an unknown signal back to the oocyte, causing reorganization of the microtubule cytoskeleton along the future antero-posterior axis (*Fig. 2*). This reorganization involves the replacement of a poste-rior microtubule organizing center (MTOC) with an anterior MTOC and the elaboration of a polarized microtubule array with the plus ends at the posterior. Recently, the *par-1* gene has been shown to be required for directing the micro-tubules. The homologous gene in C. *elegans* is also required for cytoskeletal reorganization in anteroposterior axis specification, although interestingly it is microfilaments that are involved in this species (Topic G2).

This polarized microtubule array is responsible for the intracellular transport and localization of *bicoid* and *oskar* mRNAs. Once *bicoid* mRNA has been imported into the oocyte, its 3' untranslated region (3'-UTR) is attached to microtubules and the message is transported to the anterior pole. This requires the products of the genes *staufen, exuperantia* and *swallow*. The mRNA for *oskar* is also attached to microtubules by its 3'-UTR, but this message is transported to the posterior pole, a process dependent on the products of the genes *staufen, cappuccino* and *spire*. The 3' UTR sequences are critical in this differential local-ization process. Recombinant mRNA with an *oskar* coding region but a *bicoid* 3'-UTR is localized to the anterior pole. Conversely, recombinant mRNA with a *bicoid* coding region and an *oskar* 3'-UTR is localized to the posterior pole, gener-ating embryos with two heads just as surely as the cytoplasmic transfer experi-ment discussed above.

Oskar protein is the key component of the posterior germ plasm, which speci-fies the *Drosophila* germ cells (Topic E1). It probably acts as a central scaffold to recruit other gene products required for germ cells development, and this involves the *vasa, valois* and *tudor* gene products. One of the maternal transcripts localized to the Oskar germ plasm complex is *nanos*, as Nanos protein plays a crucial role in germ cell migration as well as posterior specification. The *nanos* mRNA is then tethered to the posterior cytoskeleton of the egg.

Control of translation

Since there is no transcription in the *Drosophila* oocyte, certain early develop-mental events rely on the control of translation. In the egg, the *bicoid* and *nanos* mRNAs are tightly localized to the anterior and posterior poles, respectively. Neither transcript is translated until after fertilization. In the case of *bicoid*, this is

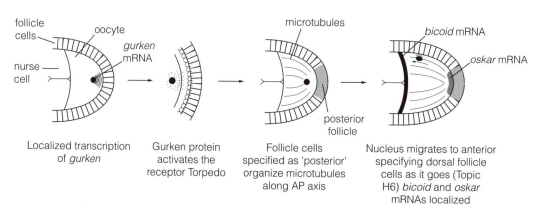

Fig. 2.   Polarization of the anteroposterior axis in the Drosophila *egg.*

because the polyadenylate tail is too short. After fertilization, it is elongated through a process requiring the products of the *cortex, grauzone* and *staufen* genes. In the case of *nanos*, this is because its translation is inhibited by a protein called Smaug. Oskar protein is required for the translation of *nanos* mRNA.

Not only are the *bicoid* and *nanos* mRNAs regulated at the level of translation, but the Bicoid and Nanos proteins both act as translational regulators themselves. In the case of Bicoid, the protein's homeodomain binds to *caudal* mRNA and represses its translation. The homeodomain can therefore function as both a DNA-binding and an RNA-binding module. Nanos does not bind to *hunchback* mRNA directly, but first associates with the product of another posterior class gene, *pumilio*.

Terminal signaling centers

The genes of the terminal group specify the **acron** and **telson**, structures that form at the anterior and posterior extremities of the embryo (*Fig. 3*). The most important gene in this group is *Torso*, which encodes a receptor tyrosine kinase. Dominant, constitutively active *Torso* mutants generate embryos with a large acron and telson, and no head, thorax or abdomen in between, indicating that singling through the **receptor Torso** is sufficient to establish acron and telson fates. *Torso* mRNA is distributed uniformly in the egg cytoplasm, and the receptor itself (which faces out into the perivitelline space) is found throughout the egg plasma membrane. The terminal specificity of signaling must therefore be generated by the ligand. One candidate for the Torso ligand is **Trunk**, because it is found in the perivitelline space and is related to the growth factor family of secreted proteins that are known to activate receptor tyrosine kinases. However, Trunk is distributed uniformly. Another candidate is **Torsolike**, which is secreted only by the anterior and posterior follicle cells. In this case the localization is correct, but Torsolike is not a growth factor-related protein.

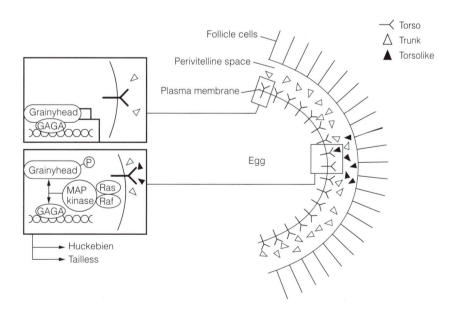

Fig. 3. Activation of huckebein *and* tailless *to establish the terminal signaling centers in the* Drosophila *embryo.*

Terminal activation of the receptor in normal development initiates a signaling cascade involving Ras, Raf and MAP kinase (see Topic C3) resulting in the activation of terminal gap genes at the poles of the egg (see Topic H3). The default developmental fate of the embryonic termini is the telson. Bicoid protein is required in addition to the terminal gene products to generate the acron.

H3 GAP GENES

Key Notes

Gap genes	The gap genes are the first zygotic genes to be expressed along the anteroposterior axis. They are expressed in broad domains corresponding to several contiguous segments, which are lost in gap gene mutants. The gap genes regulate each other as well as downstream patterning genes.
Establishing central gap gene expression	Gap genes in the central segmented region of the embryo, such as *Krüppel*, *knirps* and *giant*, respond to morphogen gradients represented by the transcription factors Bicoid, Hunchback and Caudal. Each gap gene is activated in a concentration window defined by one or more of these morphogens, resulting in a stripe of expression at a defined position along the axis. Manipulating the gradients of the upstream transcription factors causes the gap gene stripes to be displaced.
Refinement of gap gene expression	Once the gap genes have been activated by the pre-existing morphogen gradients in the syncytial blastoderm, their products (all transcription factors) interact with each other to establish complex regulatory loops. Most of the interactions among the gap genes are negative, serving to establish sharp boundaries between expression domains.
Complexities of gap gene regulation	The transcription factors encoded by the gap genes regulate downstream genes by binding to *cis*-acting control elements. Such interactions are complex, because the effect of each transcription factor depends upon its concentration, the arrangement of response elements in the target gene, and the other transcription factors that are present. The gap gene transcription factors can act alone or as heterodimers, in each case regulating alternative targets. In this way, the same gap transcription factor can have different effects on the same target gene in different parts of the embryo.
Gap genes in the head	The *Drosophila* head comprises seven segments. Individual segments are defined by the overlapping expression domains of central gap genes such as *hunchback* and *giant*, terminal gap genes such as *tailless* and *huckebein* and head-specific gap genes such as *orthodenticle*, *empty spiracles* and *buttonhead*.
Activation of terminal gap genes	The terminal gap genes *tailless* and *huckebein* are activated by the Ras-Raf-MAP kinase pathway, which is initiated by terminal signaling through the Torso receptor. Signal transduction culminates with the phosphorylation of a repressor called Grainyhead, which then detaches from the DNA and allows *tailless* and *huckebein* transcription. *tailless* and *huckebein* respond to different levels of signaling activity.

| **Related topics** | Pattern formation and compartments (A6) | Anteroposterior axis specification in *Drosophila* (H2) |
| | Gene expression and regulation (C1) | Molecular aspects of embryonic pattern formation in *Drosophila* (H1) |

Gap genes

The **gap genes** are the first zygotic genes to be expressed along the anteroposterior axis in the *Drosophila* embryo (*Fig. 1*). They are called gap genes because they are expressed in broad domains corresponding to several contiguous segments of the larva. The mutant phenotypes reflect the loss of these segments (i.e. leaving 'gaps' in the body plan) as well as other, more specific defects. The gap genes are expressed transiently in the syncytial blastoderm and all of them encode transcription factors (*Table 1*). The gap genes have at least three regulatory roles: (a) they regulate the expression of each other in complex feedback loops that establish sharp boundaries between expression domains; (b) they regulate the expression of downstream pair rule/segment polarity genes to establish the basic segmental body plan (Topic H4); (c) they regulate the downstream homeotic selector genes to give each segment a positional identity along the anteroposterior axis (Topic H5). Later in development, many of the gap genes are expressed again in restricted patterns to control the development of particular organs.

Establishing central gap gene expression

As discussed in Topic H2, interactions between the products of the maternal mRNAs for *bicoid*, *nanos*, *hunchback* and *caudal* just after fertilization establish gradients of Bicoid and Hunchback protein from the anterior signaling center, and gradients of Nanos and Caudal protein from the posterior signaling center.

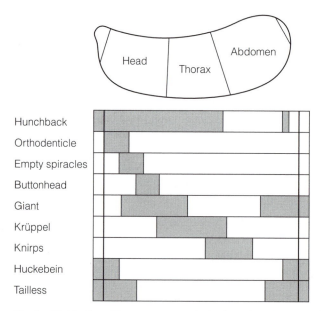

Fig. 1.   Distribution of the gap gene products along the anteroposteriorior axis of the syncytial Drosophila *embryo.*

Table 1.   Drosophila *gap genes*

Gene	Type of transcription factor	Summary of expression and function in the syncytial blastoderm
Major patterning genes of the anteroposterior axis		
hunchback	zinc finger	Anteroposterior morphogen gradient. Head specification and regulation of other gap genes in concert with Bicoid.
Krüppel	zinc finger	Broad stripe in middle of embryo. Trunk patterning.
knirps	steroid receptor	Two stripes, anterior and posterior. Trunk patterning.
giant	bZIP	Two stripes, anterior and posterior. Trunk patterning.
caudal	homeodomain	Posteroanterior morphogen gradient. Abdominal specification and regulation of other gap genes.
Organizers of embryonic termini		
huckebein	zinc finger	Terminal stripes. Pattern embryonic termini. Head patterning in concert with Bicoid.
tailless	steroid receptor	
Organizers of segmentation in the head		
buttonhead	zinc finger	Complex overlapping expression domains in the head. Specification and patterning of individual head segments.
cap'n'collar	bZIP	
knot	HLH	
crocodile	winged helix	
empty spiracles	homeodomain	
orthodenticle	homeodomain	
sloppy paired 1/2	winged helix	

Note that *hunchback* and *caudal* are classed here as (zygotic) gap genes, but they are also maternal egg polarity genes. In each case, the distribution of protein in the embryo is determined by the distribution of both maternal and zygotic mRNAs.

Bicoid, Hunchback and Caudal are all transcription factors. Their function is to activate downstream gap genes such as *Krüppel*, *knirps* and *giant* in the central area of the embryo. Initially, these genes are expressed in broad overlapping domains. They then become resolved into sharp stripes. In the early syncytial blastoderm, *Krüppel* is expressed as a central stripe, flanked on both sides by stripes of *knirps* and *giant*. These expression domains specify contiguous segments that are lost from the body plan when these genes are mutated.

In some way then, the **gradients** of transcription factors emanating from the anterior and posterior signaling centers are being used to establish **stripes** of gap gene expression. The striped expression pattern is specified because the transcription factors act as **morphogens**, activating and repressing the downstream gap genes at different concentrations. For example, the position of the central *Krüppel* stripe is predominantly controlled by the concentration of

Hunchback protein. *Krüppel* transcription is activated in the presence of Bicoid and low levels of Hunchback, but high concentrations of Hunchback repress *Krüppel* transcription. The *Krüppel* stripe is therefore positioned between two thresholds of Hunchback concentration, one so high that it inhibits the gene and the other too low to activate it (*Fig. 2*). In embryos from *bicoid* mutant mothers, the stripe of *Krüppel* expression is broadened, extending all the way to the anterior pole. This is because zygotic *hunchback* transcription is not activated in the embryos of these mutants, and the concentration of Hunchback protein in the anterior of the embryo remains low enough to permit *Krüppel* expression. If the dosage of *bicoid* in the maternal genome is increased, the Hunchback morphogen gradient is steeper and extends further into the posterior half of the embryo. In this case, the stripe of *Krüppel* expression is also shifted posteriorly, responding to the same concentration thresholds. The initial domains of *knirps* and *giant* are also established by concentration-dependent activation and repression, involving Bicoid, Hunchback and Caudal.

Refinement of gap gene expression

The conversion from gradients to stripes is initiated by the activation of gap genes within specific concentration windows, as discussed above for *Krüppel* regulation by Hunchback. However, the expression domains are stabilized and refined through interactions among the gap genes themselves. All the gap genes encode transcription factors that appear to act primarily in a negative way on

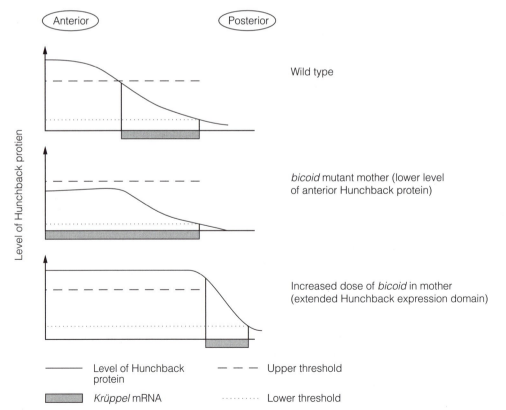

Fig. 2.   Repositioning the stripe of Krüppel *mRNA by altering the gradient of Hunchback protein. This can be achieved through modulation of the levels of maternal Bicoid protein by altering the number of copies of the* bicoid *gene.*

other gap genes. The initially broad domains of gap gene transcription are therefore resolved into sharp bands once the mRNA is translated, and the resulting transcription factors begin to act on surrounding nuclei. Each protein is rather unstable, and therefore does not diffuse far from its source.

For example, the *Krüppel* gene is initially expressed in a broad central band that tapers towards the edges, but this becomes resolved into a sharp central stripe. Repression is mediated by the products of the gap genes *knirps*, *giant* and *tailless*, which are expressed in adjacent domains (*Fig. 3*). If any of these genes is ectopically expressed throughout the embryo, the expression of *Krüppel* is reduced or even abolished. Knirps protein inhibits *Krüppel* expression by acting as a competitive inhibitor of Bicoid. The binding sites for Bicoid and Knirps overlap on the *Krüppel* promoter, and high concentrations of Knirps can displace Bicoid from its binding site.

The *Krüppel*, *knirps* and *giant* genes set up a network of redundant negative regulation, each protein repressing the transcription of the other gap genes to establish sharp boundaries. The central gap genes are also inhibited by the terminal gap gene products Huckebein and Tailless, excluding their expression from the embryonic termini.

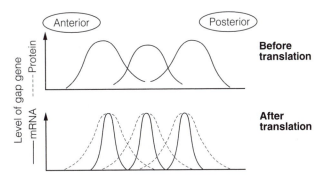

Fig. 3. Refining gap gene expression domains by negative regulation.

Complexities of gap gene regulation

The preceding sections paint a relatively simple picture in which gradients of transcription factors activate gap genes in a concentration dependent manner, and partially redundant negative interactions among the gap genes then establish sharp boundaries. In reality, however, the system is a lot more complex, because the products of the egg polarity genes and gap genes can act independently or cooperatively in various combinations to activate or repress different downstream genes. In some cases, this means that the same transcription factor can activate and repress the same target gene in different parts of the embryo by associating with a different partner.

For example, prior to segmentation, the terminal gap gene products Huckebein and Tailless cooperate with Knirps to inhibit *Krüppel* in the central domain of the embryo, thus establishing its posterior boundary of expression. Genetic analysis shows that Krüppel positively regulates *knirps*, but this is probably due to its negative regulation of *giant*, preventing the Giant protein from inhibiting *knirps*. Later in development, in the posterior region that forms the Malpighian tubules, Huckebein and Tailless activate *Krüppel*, because they associate with another transcription factor called Fork head. Krüppel itself can form homodimers, or heterodimers with Knirps and Hunchback. The effect of Krüppel acting as part of a heterodimer is different to its effect alone. Krüppel therefore activates different sets of downstream genes in different parts of the embryo, and at different developmental stages.

Another complication is that the gap gene expression patterns are transient and dynamic. As development proceeds, single stripes resolve into several, while other stripes disappear. The constantly shifting expression patterns reflect the complex physical and molecular interactions going on inside what is essentially still a single cell (Topic H1).

Gap genes in the head

The development of the head in *Drosophila* is a complex process that involves the specification of seven segments (six in most other insects) giving rise to the eyes, antennae and the various mouth parts. At least 10 of the gap genes are involved, including the trunk patterning genes *hunchback* and *giant*, the terminal gap genes *huckebein* and *tailless*, and others that are expressed specifically in the head.

Individual segments become defined through the overlapping expression patterns of these genes. The most anterior segments (including the acron) are established under the combined control of the anterior and terminal class maternal genes, acting through *huckebein*, *tailless* and *giant*. The definition of these and the other more posterior head segments is also controlled by head-specific gap genes such as *orthodenticle*, *empty spiracles* and *buttonhead*, whose expression is under the direct regulation of Bicoid. Mutations in these genes generate specific defects in particular regions of the head, in some cases resulting from the deletion or abnormal development of a single segment and in others covering several contiguous segments. The closely related genes *sloppy paired 1* and 2 behave as gap genes in the head, and are involved in the establishment of the antennal and intercalary segments. They are also expressed in the trunk, where their behavior is more like the pair rule and segment polarity genes described in Topic H4.

Activation of the terminal gap genes

The terminal gap genes *huckebein* and *tailless* are regulated independently of the gap genes that pattern the central region of the embryo. Expression of *tailless* results from the activation of the terminal Torso signaling pathway, probably by the ligand Torsolike, which is secreted by the terminal follicle cells. Torso is a receptor tyrosine kinase in the egg plasma membrane. Ligand-binding activates the Ras-Raf-MAP kinase pathway in the egg, resulting in the phosphorylation of **Grainyhead** (also called **NTF-1**) which binds to the *tailless* promoter as a co-repressor with **GAGA1**. Phosphorylation inactivates Grainyhead, allowing GAGA1 to activate *tailless* transcription (*Fig. 3*, Topic H2). The Torso pathway also appears to control *huckebein* expression, but there is a different response threshold. Mutations that abolish signaling through the Torso pathway prevent the expression of *huckebein* and *tailless*. However, hypomorphic mutations, which reduce the level of signaling through Torso, can generate embryos in which *tailless* is activated but *huckebein* is not.

H4 MOLECULAR CONTROL OF SEGMENTATION

Key Notes

Segments and parasegments	The *Drosophila* larva comprises a series of segments along the anteroposterior axis, but the initial developmental units that form in the embryo are parasegments, representing the posterior third of one segment and the anterior two thirds of the next. Each parasegment is a developmental compartment defined by boundaries of lineage restriction. The parasegments are specified by the expression of pair rule genes and the boundaries are established by segment polarity genes. The segmental organization of the embryo becomes apparent after germ band retraction.
Pair rule genes	The pair rule genes are expressed in the syncytial blastoderm, in most cases as a broad band resolving into seven evenly-distributed transverse stripes. These stripes represent alternating parasegments, and in pair rule mutants either all the even numbered segments or all the odd numbered segments are deleted.
Stripe positioning	The regular pattern of pair rule gene expression is controlled in two stages. First, the primary pair rule genes *even-skipped*, *hairy* and *runt* are activated by the gap gene transcription factors. Since the gap genes are expressed in non-repetitive domains along the axis, the different pair rule stripes must be independently positioned. Second, the products of the primary pair rule genes (all transcription factors) regulate the secondary pair rule genes such as *fushi tarazu* and *odd-skipped*. This establishes pair rule expression domains in complementary patterns.
Segment polarity genes	The segment polarity genes are expressed in the cellular blastoderm as 14 transverse stripes, under the combinatorial control of complementary pair rule gene products. They establish and maintain parasegment boundaries and generate the pattern of individual segments, which can be seen in the arrangement of denticle bands and naked cuticle. Mutations in the segment polarity genes are characterized by the disruption of this pattern.
Regulation of segment polarity gene expression	The pair rule genes are expressed in overlapping domains representing alternating parasegments, whereas the segment polarity genes are expressed in every parasegment, and in the same position relative to the parasegment boundaries. The even and odd stripes of segment polarity gene expression are therefore independently activated by combinations of pair rule transcription factors, defining subdomains in each parasegment.
Function of the segment polarity genes	The segment polarity genes have three functions. First, they establish an anterior and posterior compartment within each parasegment defined by the transcription factors Engrailed (anterior) and Cubitus interruptus (posterior). These differentially activate genes encoding cell adhesion

molecules leading to the generation of a compartmental boundary. Second, they maintain the boundary when pair rule gene expression stops. This is accomplished by a signaling loop in which the Engrailed and Cubitus interruptus maintain each other's expression in cells adjacent to the boundary through mutual signaling involving Hedgehog and Wingless. Finally, they pattern the cells of each parasegment. This involves the restricted diffusion of Wingless, which promotes naked cuticle development. Where Wingless is absent, denticle belts are generated.

Related topics	Pattern formation and compartments (A6)	Molecular aspects of embryonic pattern formation in *Drosophila* (H1)
	Gene expression and regulation (C1)	
	Signal transduction in development (C3)	Gap genes (H3)

Segments and parasegments

The anteroposterior axis of the *Drosophila* larva is divided into a repetitive series of morphologically similar **segments**. In the thorax and abdomen, these can be seen clearly because they are separated by distinct grooves. Segmentation is an important feature of *Drosophila* development, and as discussed below, a large number of zygotic genes is devoted to this process. However, the fundamental repetitive units established during development are not segments but **parasegments**. Each segment forms from the posterior two thirds of one parasegment and the anterior third of the next[1]. *Fig. 1* shows the relationship between segments and parasegments in *Drosophila*.

Parasegments first become apparent about 5–6 hours after fertilization (the stage of maximum germ band extension) when 15 grooves corresponding to parasegment boundaries appear in the ectoderm. Each parasegment is a **developmental compartment**, i.e. cells in one parasegment remain restricted therein and give rise to clones that never cross the boundary into the next parasegment (Topic A6). Although they do not become apparent until later, the parasegments are specified when the embryo is still a syncytial blastoderm. This involves the expression of **pair-rule genes** in overlapping patterns of **seven transverse stripes** to represent alternating parasegments. The pair rule gene products activate the expression of **segment polarity genes** in **14 transverse stripes**, and these genes establish and maintain the **parasegmental boundaries** (see summary in *Fig. 1*, Topic H1). As their name suggests, these genes also pattern the parasegments, resulting in the characteristic distribution of denticle belts on the larval segments. The initial parasegmental arrangement of the body changes to a segmental arrangement when the germ band retracts.

Pair rule genes

The position of the future parasegments is established in the syncytial *Drosophila* embryo by the expression of **pair-rule genes**, all of which encode transcription

---

[1]Insects are not the only animals in which the original developmental compartments are out of register with the final body plan. In vertebrates, the paraxial mesoderm is divided into segments called somites, which form the vertebrae of the backbone. However, each vertebra forms from the anterior half of one somite and the posterior half of the adjacent one (Topic K2).

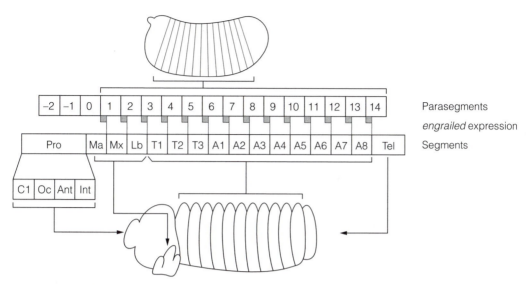

Fig. 1. Relationship between segments and parasegments in the syncytial embryo (top panel) and cellular blastoderm (bottom panel), with reference to early engrailed *expression in the trunk region. Pro = procehalon (occipital (Oc), antennal (Ant) and intercalary (Int) segments). Ma = mandlibles, Mx = maxillae, Lb = labium. T and A represent thoracic and abdominal segments respectively. Tel = telson.*

factors[2] (*Table 1*). Most of the pair rule genes are expressed initially as a series of **seven transverse stripes** along the anteroposterior axis. There are 14 parasegments, and the pair-rule genes are expressed in domains representing alternating parasegments. The *even-skipped* and *fushi tarazu* genes are expressed in complementary patterns that exactly correspond to the future parasegments (*even-skipped* expression represents the future odd-numbered parasegments,

Table 1. Drosophila *pair rule genes and their expression patterns*

Gene	Type of transcription factor	Expression
even-skipped	Homeodomain	Seven stripes representing odd-numbered parasegments
fushi tarasu	Homeodomain	Seven stripes representing even-numbered parasegments
hairy	bHLH	Seven stripes representing odd-numbered parasegments, although overlapping anterior border
runt	(novel)	Seven stripes representing even-numbered parasegments, although overlapping posterior border
odd paired	Zinc finger	Initially a broad central band, later resolving into 14 stripes. Classification as pair rule gene on the basis of its mutant phenotype not its expression pattern
sloppy paired 1 and *2*	Winged helix	Seven stripes resolving into 14
paired	Homeodomain	Seven stripes resolving into 14
odd skipped	Zinc finger	Seven stripes resolving into 14

[2]There is another pair rule gene called *tenascin major*, and in embryos with mutations in this gene, every odd-numbered body segment is deleted. However, *tenascin major* encodes a transmembrane protein not a transcription factor, and is so far the only example of pair rule gene that functions outside the cell.

while *fushi tarazu* represents the even-numbered parasegments). Other pair rule genes such as *hairy* and *runt* are expressed in stripes that are slightly offset and hence overlap future parasegmental boundaries. Further pair rule genes such as *paired* and *odd-skipped* are initially expressed as seven stripes, but these later resolve into 14, marking the position of every parasegment. As pair rule gene expression begins, the blastoderm undergoes cellularization. In this way, adjacent stripes of cells in each parasegment come to contain different combinations of the pair rule transcription factors, but these combinations show a two-parasegment periodicity along the axis (*Fig. 2*).

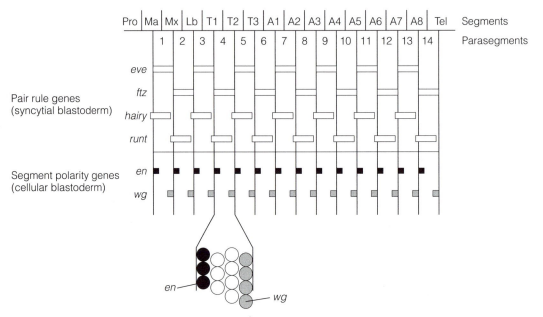

Fig. 2.   *Expression of key pair rule and segment polarity genes in the trunk.*

As is often the case, the convention of naming *Drosophila* genes according to the mutant phenotype causes confusion. With the pair rule genes, however, the confusion is doubled because the genes are named according to the effect of the mutation on the *segmental* body plan, while the gene expression patterns are generally discussed in terms of parasegmental organization. It is really not worth trying to work out a pair rule gene's expression pattern based on its name. For example, the *even-skipped* mutant lacks even numbered segments, but this is because the gene is expressed in odd-numbered parasegments.

Stripe positioning

The strikingly regular expression patterns of the pair rule genes are set up in the syncytial embryo in a two-stage process. First, there are three primary pair rule genes (*even-skipped*, *hairy* and *runt*) whose expression is controlled directly by the products of the gap genes (Topic H3). This control process is remarkable because different gap genes are expressed at restricted positions along the anteroposterior axis, but they cooperate to establish a repetitive pattern in which the same three pair rule genes are expressed in a series of stripes along the entire axis. This suggests that the different stripes of each primary pair rule gene are positioned independently by distinct regulatory mechanisms. Indeed, the promoters of the pair rule genes are highly modular, and experiments in which the promoters have been systematically mutated have shown that specific regions control the positioning of different stripes. The second stripe of the *even-skipped* gene has been studied in detail and its regulation is discussed below.

The second *even-skipped* stripe is positioned in the anterior region of the syncytial embryo between the expression boundaries of the gap genes *giant* and *Krüppel*. The *even-skipped* promoter contains several modules, each controlling an expression domain corresponding to one or more stripes (*Fig. 3*). The module controlling the positioning of stripe 2 contains binding sites for Bicoid and Hunchback, both of which are expressed in the anterior of the embryo and both of which activate *even-skipped* transcription. The module also contains binding sites for Giant and Krüppel, and both these proteins act as repressors of *even-skipped* transcription. It is interesting that across most of this regulatory region, the binding sites for the transcriptional activators and repressors are adjacent or overlapping, suggesting that *even-skipped* expression is regulated by **competition for binding sites**. Although there is a slight drop in the levels of Bicoid and Hunchback from anterior to posterior across the *even-skipped* stripe, these proteins are essentially present at a constant level. Therefore, the boundaries of *even-skipped* expression are determined by the changing levels of Giant and Krüppel. As Giant and Krüppel also inhibit each other's genes, there is a gap between the *giant* and *Krüppel* expression domains where both proteins are at a low concentration, and this is where the second stripe of *even-skipped* mRNA is positioned (*Fig. 3*).

In the second stage, the remaining **secondary pair rule genes** are expressed under the control of the products of the primary pair rule genes (the gap gene

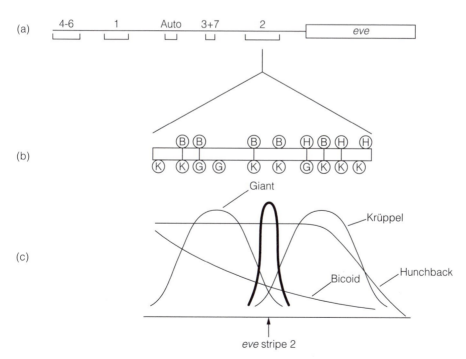

Fig. 3. *Positioning the second stripe of* even-skipped (eve) *gene expression. (a) The* eve *promoter, labeled to indicate the modular elements responsible for regulating different expression stripes (Auto is an autoregulatory region). (b) A detailed map of the second stripe region, showing binding sites for the activators Bicoid (B) and Hunchback (H) and the repressors Krüppel (K) and Giant (G). (c) Expression domains of the activators and repressors, showing how the second* eve *stripe is positioned.*

products play little or no role in their regulation). For example, the pair rule gene *fushi tarazu* is expressed specifically in even-numbered parasegments, in a precisely complementary pattern to *even-skipped*. This expression profile arises primarily because *fushi tarazu* is inhibited by Even-skipped, and transcription is therefore limited to the **interstripe regions** where *even skipped* is not expressed.

Segment polarity genes

Every segment of the *Drosophila* larva has a recognizable anteroposterior polarity. This is most noticeable in the ventral cuticle, where the anterior part of each segment has a pattern of hair-like projections called **denticles** and the posterior part is smooth. The function of the **segment polarity genes** is to establish the boundaries between parasegments and generate the pattern within each parasegment. Therefore, mutations in these genes generate a disturbance to the underlying patterning mechanism as shown by abnormal denticle patterns. Typically, it appears that one half of the segment has been deleted and replaced by a mirror image of the other half. Many of the segment polarity genes therefore have names reflecting the appearance of segments that lack denticle belts all together (e.g. *naked*, *smooth*, *fused*) or have denticle belts throughout (e.g. *hedgehog*, *shaggy*, *gooseberry*, *porcupine*).

The segment polarity genes are expressed in a series of 14 stripes, each stripe representing cells at the same position within each parasegment. Whereas the gap and pair rule genes are expressed in the syncytial blastoderm and all encode transcription factors, the segment polarity genes function in the cellular blastoderm and encode secreted proteins, receptors and intracellular signaling components as well as transcription factors (*Table 2*). Furthermore, whereas the gap and pair rule genes function to generate an initial pattern and are expressed

Table 2. Principal Drosophila *segment polarity genes grouped by function*

Gene	Protein	Function
engrailed (also *invected*)[a]	Transcription factor	Transcription factor activated in anterior cells in response to Wingless signaling. Defines anterior compartment of parasegment and establishes anterior cell sorting behavior. Activates *hedgehog*.
hedgehog	Secreted protein	Signal from anterior compartment, maintains posterior compartment. Regulates anteroposterior patterning.
patched, smoothened	Receptor complex	Receptor for Hedgehog in posterior cells.
fused, costal 2	Intracellular signaling proteins	Hedgehog signal transduction pathway in posterior cells.
cubitus interruptus	Transcription factor	Transcription factor activated in posterior cells in response to Hedgehog signaling. Establishes posterior cell sorting behavior. Activates *wingless*.
wingless	Secreted protein	Signal from posterior compartment, maintains anterior compartment. Responsible for anteroposterior patterning.
frizzled-2	Receptor	Receptor for Wingless in anterior cells.
dishevilled, shaggy, armadillo, pangolin	Intracellular signaling proteins	Wingless transduction pathway.

[a]The *invected* gene is closely related to *engrailed*, is activated by Engrailed and cooperates with *engrailed* to establish anterior compartment identity.

for only a few hours, the segment polarity genes are required to maintain that pattern, and some are expressed throughout the life of the fly. First we discuss how the 14-stripe expression pattern is established, and then we go on to discuss how the segment polarity genes establish and maintain the compartmental boundaries and how they pattern each segment.

Regulation of segment polarity gene expression

Most of the pair rule genes are initially expressed in seven transverse stripes corresponding to alternating parasegments. As the cellular blastoderm forms, the segment polarity genes are expressed in 14 stripes, each stripe representing a single parasegment. Each parasegment is initially four cells across, and can be thought of as comprising four transverse bands of cells. Different segment polarity genes are expressed in different bands of cells, but always the same band in each parasegment. For example, the *engrailed* gene is expressed in the first band of cells in each parasegment, marking the anterior border, while *wingless* is expressed in the fourth band of cells marking the posterior border (*Fig. 2*). These expression patterns suggest that the odd and even stripes of segment polarity gene expression are independently regulated by pair rule genes expressed in complementary patterns. Furthermore, as the segment polarity genes are restricted to certain cells within each parasegment, combinations of pair rule genes with overlapping expression patterns are likely to control where the segment polarity genes are switched on.

We will consider *engrailed* and *wingless* as examples, since they play the predominant role in establishing and maintaining compartment boundaries as discussed below. The *engrailed* gene is positively regulated by Even-skipped and Fushi tarazu, which are expressed in a complementary pattern of alternating parasegments. This dual control is necessary for *engrailed* to be expressed in every parasegment. Thus, in *fushi tarazu* mutants, *engrailed* expression is limited to the odd-numbered parasegments because there is nothing upstream to activate the gene in the even-numbered parasegments. Both Even-skipped and Fushi tarazu proteins have a graded distribution in each parasegment, with the peak at the anterior border. This alone could support the *engrailed* expression pattern if *engrailed* responded to a threshold concentration of these transcription factors. However, *engrailed* expression is disrupted in other pair rule mutants that do not affect the expression of *even-skipped* or *fushi tarazu*. This indicates that while Even-skipped and Fushi tarazu are instrumental in positioning the engrailed stripes, the expression pattern is fine-tuned by at least four other pair rule genes. For example, *engrailed* expression is regulated positively by Paired and negatively by Odd-skipped. The genes for both these regulators are initially expressed in seven stripes, but resolve to 14 so they can regulate *engrailed* expression in each parasegment.

Unlike *engrailed*, the *wingless* gene is repressed by both Even-skipped and Fushi tarazu. Recall from above that Even-skipped also represses Fushi tarazu, so there is a gap between their stripes. This gap corresponds to the fourth strip of cells in each parasegment where *wingless* is expressed. Like *engrailed*, *wingless* is also negatively regulated by Odd-skipped, and positively regulated by Paired. The result of this network of interactions is that *engrailed*-expressing cells are established at the anterior boundary of each parasegment and *wingless*-expressing cells are established at the immediately adjacent strip of cells at the posterior boundary. The *sloppy-paired* gene is necessary for the maintenance of both *engrailed* and *wingless*. It is initially expressed as seven stripes and then resolves to 14.

Function of the segment polarity genes

The segment polarity genes essentially carry out three functions:

- They establish an anterior and posterior compartment within each para-segment. Where anterior and posterior compartments meet, a boundary is generated.
- They maintain parasegment boundaries.
- They generate anteroposterior pattern within each parasegment.

Establishing compartments and boundaries

As discussed above, the initial expression domains of some key segment polarity genes such as *engrailed* and *wingless* are controlled by the pair rule genes. The function of these segment polarity genes is to establish anterior and posterior compartments within each presumptive parasegment. Where cells of the anterior and posterior compartments meet, they interact to form a compartment boundary (*Fig. 4*).

This process begins with the expression of *engrailed* in the anterior-most cells of each presumptive parasegment (this defines those cells as belonging to the anterior compartment, corresponding to the posterior compartment of each segment). Engrailed activates the *hedgehog* gene, which encodes a secreted protein. At the same time, Engrailed represses those genes encoding components of the Hedgehog signal transduction pathway, including the receptor Patched and the downstream transcription factor Cubitus interruptus. In this way, the anterior cells are prevented from responding to the Hedgehog signal they produce.

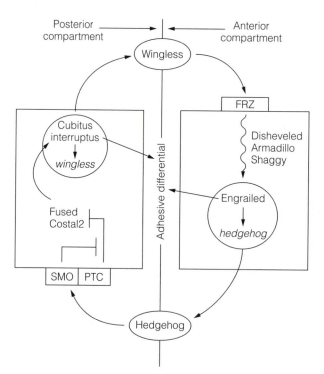

Fig. 4.   Establishment and maintenance of compartmental boundaries in Drosophila, *FRZ is Frizzled, the receptor for Wingless. SMO and PTC, the products of the* smoothened *and* patched *genes, form a receptor complex for Hedgehog.*

Conversely, both *patched* and *cubitus interruptus* are expressed in the adjacent Wingless-secreting cells corresponding to the posterior compartment of the next parasegment. These cells are therefore able to respond to the Hedgehog signal. Cubitus interruptus is present as a mixture of active and inactive forms in these cells. It is thought that Hedgehog signaling shifts the balance towards the active form, resulting in the induction of Cubitus interruptus-dependent genes.

A recently-proposed model suggests that Engrailed and Cubitus interruptus differentially activate genes for cell adhesion molecules. As discussed in Topic C5, the expression of different adhesion molecules results in the spontaneous sorting of cells into populations with similar adhesive properties and the formation of a boundary between them. However, experiments have shown that boundaries can arise when cells expressing low levels of either transcription factor are juxtaposed to cells expressing high levels of the same transcription factor. This indicates that Engrailed and Cubitus interruptus may be acting redundantly on the same cell adhesion molecule genes.

Maintaining parasegment boundaries
The expression of the segment polarity genes is induced by the pair rule gene products, but the pair rule genes are expressed only transiently. If the segment polarity genes relied on the pair rule transcription factors for their continued expression, they would soon become silent and the parasegment boundaries would eventually break down.

Maintenance of the parasegment boundaries requires a positive feedback loop involving the signaling proteins Wingless and Hedgehog. Wingless, secreted by cells in the posterior compartment of each parasegment, induces cells in the adjacent anterior compartment to express *engrailed*. Engrailed then activates *hedgehog*, and Hedgehog secreted by the anterior cells induces the adjacent posterior cells to produce Wingless by activating Cubitus interruptus. The components of this cyclical pathway are listed in *Table 2* and shown in *Fig. 4*.

Patterning within the parasegment
Hedgehog and Wingless are both secreted proteins produced at the boundary between anterior and posterior compartments. As well as acting on the immediately adjacent cell to maintain the signaling loop (see above) both proteins also diffuse away from their source, establishing gradients over the remaining cells of the parasegment. These gradients have the potential to act as morphogens, therefore patterning the parasegment and inducing the characteristic arrangement of denticle bands (*Fig. 5*). There is evidence that the receptors for Wingless and Hedgehog are differentially expressed among the central cells of the parasegment, indicating that these cells might vary in their competence to receive and respond to the secreted proteins. Furthermore, it appears that Wingless diffuses asymmetrically from its source, because it is in some way restricted from diffusing past the *engrailed*-expressing cells of the anterior compartment. This restriction of the Wingless signal is dependent on Hedgehog. Wingless induces the development of naked cuticle by suppressing the gene for the transcription factor Rhomboid. In the *engrailed*-expressing anterior compartment of the parasegment, which corresponds to the posterior region of the segment, Rhomboid is active and initiates denticle development through the EGF

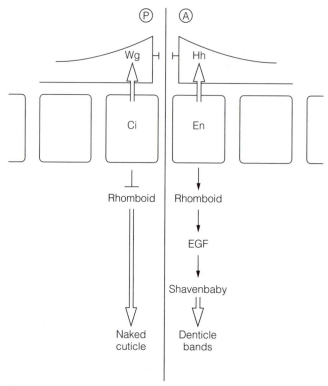

Fig. 5. Patterning the parasegments. Hedgehog (Hh) and Wingless (Wg) maintain anterior and posterior cells expressing the transcription factors Engrailed (En) and Cubitus interruptus (Ci). Only Engrailed activates the gene for Rhomboid, leading to denticle band development.

signaling pathway and the downstream gene *shavenbaby*. The asymmetric transport of Wingless therefore plays a critical role in establishing denticle belt and naked cuticle fates on either side of the segment boundary, thus generating the anteroposterior polarity of each segment (se Topic A6 for possible mechanisms of Wingless transport).

H5 HOMEOTIC SELECTOR GENES AND REGIONAL SPECIFICATION

Key Notes

Segmentation and positional identity	The process of segmentation (generating compartment boundaries, patterning) is controlled by the segment polarity genes but does not establish regional differences between segments. This requires a set of homeotic selector genes, which confer positional values on the segments.
Homeotic selector genes	The homeotic selector genes confer positional values on cells, and mutations result in homeotic transformations where one body part is replaced with another. This phenotype occurs because the cells have been given incorrect positional values and develop in a manner appropriate for a different position along the axis. Homeotic transformation may result from the selection of alternative batteries of downstream genes.
The homeotic complex (HOM-C)	The *Drosophila* homeotic selector genes map to two clusters (the Antennapedia and bithorax complexes), which together define the homeotic complex (HOM-C). There are eight homeotic genes in the complex, each of which encodes a homeodomain-class transcription factor. The HOM-C is homologous to the *Hox* complexes of other animals, suggesting that HOM/*Hox* genes play a fundamental role in anteroposterior patterning throughout the animal kingdom.
Colinearity and posterior dominance	The HOM-C genes are expressed in overlapping patterns along the anteroposterior axis, broadly corresponding to the order of the genes along the chromosome. This colinearity is conserved in other animals although its relevance is unclear. The posterior HOM-C genes negatively regulate those expressed in more anterior domains, a phenomenon termed posterior dominance.
How HOM-C genes confer positional identity	A simple model suggests that the HOM-C genes confer positional identity by establishing a combinatorial code of transcription factors, which would activate different sets of downstream genes and generate different regional structures. The analysis of mutations in the *Antennapedia* and *bithorax* complexes broadly supports this interpretation, although there are several variations in the mechanisms involved. In the head, pairs of genes can establish different fates using a simple on/off code. The *Antennapedia* gene shows how positional values can be generated by repressing a default developmental pathway. The *bithorax* complex most closely models the combinatorial code system, with different genes providing additive information to establish fates in the thorax and abdomen.

Initiation of HOM-C gene expression	The HOM-C genes are co-regulated by the gap and pair rule transcription factors. The role of the gap proteins is to establish the position of HOM-C gene expression along the anteroposterior axis, whereas the pair rule proteins serve to refine the expression domains to ensure they correspond to parasegment boundaries. The HOM-C genes are regulated by modular promoters and enhancers that specify different components of the full expression pattern. Mutations in such regulatory elements can generate specific types of homeotic transformation.
Maintenance of HOM-C gene expression	The gap and pair rule transcription factors are short lived, so HOM-C gene expression must be maintained by a different mechanism. Maintenance is achieved by proteins of the Polycomb and Trithorax groups, through the permanent modification of chromatin structure. This establishes open and repressed chromatin domains within the HOM-C, which differ along the anteroposterior axis. The remodelled chromatin is heritable through DNA replication and mitosis, ensuring that all descendants of a given cell express the same HOM-C genes
Molecular basis of HOM-C function	The link between regional specification and the development of regionally appropriate structures is the batteries of genes regulated by the HOM-C transcription factors. A surprisingly small number of such genes has been identified, some regulated by multiple HOM-C proteins and others on an individual basis.

Related topics	Pattern formation (A6)	Gap genes (H3)
	Gene expression and regulation (C1)	Patterning the anteroposterior neuraxis (J2)
	Molecular aspects of embryonic pattern formation in *Drosophila* (H1)	Somitogenesis and patterning (K2)
		Flower development (N6)

Segmentation and positional identity

Each segment of the *Drosophila* larva is unique, both in terms of its ventral denticle pattern and its internal organization. Furthermore, different segments give rise to different appendages in the adult fly. The unique properties of a given segment depend upon its position along the anteroposterior axis. The segment polarity genes (discussed in Topic H4) are expressed in the cellular blastoderm, and function to pattern the parasegments and generate parasegment boundaries. However, as the same genes are expressed in the same pattern in each parasegment, this does not provide a mechanism to make the segments different from each other. Further genes are required for **regional specification,** i.e. the process by which each segment is given a **positional value** so its cells can behave in the correct manner to generate regionally appropriate structures.

Homeotic selector genes

The genes that give cells their positional values along the anteroposterior axis of the *Drosophila* embryo are called **homeotic selector genes**. Mutations in these genes do not usually disrupt the segmental arrangement of the body, but they cause **homeosis**, the phenomenon in which one body part develops with the likeness of another. This suggests the cells are being given the 'wrong' positional information and behaving in a manner more appropriate for a different position along the axis. Although such **homeotic mutations** were first reported over 80 years ago, most were isolated and characterized in the 1980s through the use of saturation mutagenesis screens for developmental patterning phenotypes. Homeotic mutants show a diverse array of spectacular and bizarre phenotypes, e.g. the dominant *Antennapedia* mutant, in which legs grow in place of the antennae. In this mutant, it appears that the cells of the antennal imaginal disc have been given positional values appropriate for the thorax, where the legs would normally develop. It is thought that such mutations cause wholesale changes in the expression of batteries of downstream genes representing alternative developmental pathways, i.e. the homeotic genes are acting as 'selectors' of these pathways by controlling downstream gene expression.

The homeotic complex (HOM-C)

Most of the *Drosophila* homeotic selector genes are organized in two clusters on chromosome 3, called the **Antennapedia complex (ANT-C)** and the **bithorax complex (BX-C)** (*Fig. 1*). Although there are two clusters of homeotic genes in *Drosophila*, they correspond to the single **homeotic complex (HOM-C)** found in most other insects, and to the *Hox* complexes found in vertebrates and all other animals (see Topic K2 for a discussion of the vertebrate *Hox* complexes). Therefore, the ANT-C and BX-C are often discussed collectively as the *Drosophila* HOM-C, and the genes are called **HOM-C genes**. The *Drosophila* HOM-C contains a total of eight homeotic selector genes (*Table 1*) as well as the non-homeotic developmental genes *bicoid*, *fushi tarazu* and *zerknüllt*. There is also a complex array of regulatory elements in the HOM-C, located between genes and within introns.

Each HOM-C gene encodes a **homeodomain** transcription factor. Indeed, the homeodomain was named after these HOM-C gene products, in which it was first recognized. Genes encoding homeodomain transcription factors are called

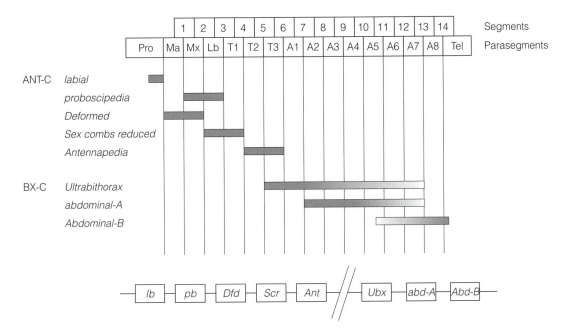

Fig. 1.   The Drosophila *HOM-C, and expression of the different HOM-C genes. Note the colinearity of gene order and expression.*

Table 1.   The Drosophila *HOM-C genes*

Gene	Expression	Function suggested by mutant phenotypes
ANT-C		
labial	Intercallary segment of head. This is not the region that gives rise to the labial mouth parts	Inhibits labial development in procephalon
proboscipedia	Maxillary and labial segments of head	Specification of proboscis and maxillary palps.
Deformed	Mandibular and maxillary segments of head.	Specification of mouth part identities
Sex combs reduced	Labial segment of head and first thoracic segment (later expression in CNS extends posteriorly to abdomen)	Specification of labial and first thoracic segments
Antennapedia	Anterior boundary parasegment 4	Specification of p4 (second thoracic segment)
BX-C		
Ultrabithorax	Anterior boundary parasegment 6 (anterior to posterior gradient)	Specification of p5 and p6
abdominal A	Anterior boundary parasegment 7 (anterior to posterior gradient)	Specification of p7-9
Abdominal B	Anterior boundary parasegment 10 (posterior to anterior gradient)	Specification of p 10-13 and p14

The expression domains of the head-specific genes *labial*, *proboscipedia*, *Deformed* and *Sex combs reduced* correspond to embryonic segments, while those of the other homeotic genes correspond to embryonic parasegments. Note on nomenclature: Some homeotic selector genes start with a capital letter and others do not. This is because some were identified as dominant mutants (capital letters, such as *Antennapedia*) and others as recessive mutants (lower case letters, such as *bithorax*). For more discussion on genetic nomenclature, see Topic B2.

homeobox (*Hox*) genes[1]. It should be stressed, however, that homeodomain transcription factors are not used solely for regional specification. Hundreds of homeodomain proteins have been identified with diverse roles in many species. The HOM-C/*Hox* complex genes are therefore a specialized subset of a large family of homeobox genes, and play a fundamental role in patterning the anteroposterior axis of all animals.

Colinearity and posterior dominance

The fact that the *Drosophila* homeotic selector genes are clustered together in the genome and all encode the same type of transcription factor is quite remarkable. However, even more surprising is the observation that the spatial and temporal order of HOM-C gene expression in the embryo corresponds almost exactly to the order of the genes along the chromosome (*Table 2*). The basis of this phenomenon, which has been termed **colinearity**, remains unknown. *Fig. 1* compares the genetic structure of the HOM-C with the expression patterns of the HOM-C genes in the blastoderm, in relation to the segmental and parasegmental organization of the embryo. The genes are expressed in a characteristic

---

[1]The region of a gene that encodes a particular conserved motif or domain of a protein is called a 'box'. Many vertebrate genes are named on the basis that they share a particular sequence box with a previously-cloned *Drosophila* gene. For example, Pax = Paired box, Krox = Krüppel-like box, Otx = Orthodenticle-like box.

Table 2. Properties of the HOM-C/Hox complex genes

Control regional specification along the anteroposterior axis in animals

Mutations cause homeotic transformations

Encode transcription factors of the homeodomain class

Map to a single gene complex (this is split in *Drosophila*). There are four copies in most vertebrates (seven in zebrafish!)

Expressed in a characteristic, posteriorly nested pattern, with each gene showing a sharp anterior boundary of expression

Colinearity between the order of genes on the chromosome and the spatial and temporal order of gene activation along the anteroposterior axis. Gene order is evolutionarily conserved, perhaps because the dense packing of regulatory elements prevents mixing

Regulation among the different genes establishes posterior dominance

Genes are regulated by retinoic acid

overlapping pattern, with the most posteriorly expressed genes showing nested expression domains. Many of the genes have sharp anterior boundaries of expression which dwindle in the posterior. The expression pattern arises in part due to the phenomenon of **posterior dominance**, in which the homeotic genes are negatively regulated by the homeotic gene products expressed in more posterior regions. In this way, the nested expression patterns are established but the strongest expression of each gene is found at the anterior margin.

How HOM-C genes confer positional identity

In Topic A6, we discussed a model in which overlapping nested patterns of homeotic gene expression could pattern an axis by generating a series of combinatorial codes. The analysis of mutant phenotypes and expression patterns has provided evidence that the HOM-C genes do indeed work in this manner, although with several variations, as discussed below.

Combinatorial specification in the head

Among the most anterior homeotic genes, *Deformed*, *proboscipedia* and *Sex combs reduced* have overlapping expression patterns in the head. The analysis of loss-of-function and ectopic expression mutants suggests that the products of *proboscipedia* and *Sex combs reduced* can specify four different segmental fates in the head and thorax (*Fig. 2*). In the absence of both proteins, the default developmental state is the antenna. In the presence of *proboscipedia* alone, the maxillary palps are specified. Where the domains overlap and both proteins are present, the proboscis is specified. Finally, *Sex combs reduced* expressed alone specifies the tarsus, the terminal portion of a leg. The experimental evidence follows. In the loss-of-function *proboscipedia* mutant, the proboscis is replaced by a pair of tarsi, while in a sublethal loss-of-function *sex combs reduced* mutant, it is converted into maxillary palps. In the double mutant, the proboscis is replaced by antennae. If *Sex combs reduced* is ectopically expressed throughout the head, the presumptive maxillary palps are converted into a proboscis. Similarly, if *proboscipedia* is ectopically expressed, the antennae are converted to maxillary palps. There is evidence that the transcription factors encoded by *Sex combs reduced* and *proboscipedia* form heterodimers where they are coexpressed. The heterodimers activate a unique set of genes, not the sum of the sets of genes activated by each individual transcription factor acting alone.

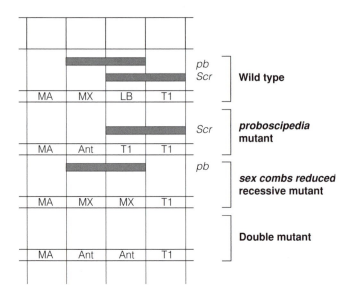

Fig. 2. Specification of head segment identities by the homeotic selector genes proboscipedia *and* Sex combs reduced, *MA = mandible, MX = maxilla, LB = labium; T1 = First thoracic segement; Ant = antenna.*

Antennapedia and labial – repression of default fates

The *Antennapedia* gene was identified in 1949 through a striking dominant mutation that converted the antennae into legs. In wild type flies, the gene is expressed most strongly in parasegment 4, roughly corresponding to the second thoracic segment where legs normally form. In the dominant *Antennapedia* mutant, the gene was found to be expressed in the head as well as in its normal site because an inversion had placed the *Antennapedia* coding region adjacent to a different promoter. This suggested that the *Antennapedia* gene actively promoted T2 fates by activating the genes for leg development. However, the recessive *antennapedia* mutant showed the opposite phenotype, where antennae grew in place of the second pair of legs. This argued against a direct role in leg specification, and suggested that the natural role of *Antennapedia* was to suppress antennal fates in the thorax. The target for repression is probably the transcription factor Salm, which is required for antennal development (see below). A similar example of negative control over cell specification is seen with the *labial* gene. This was named following the isolation of recessive mutants in which the labial segment of the head was deleted. However, the labial gene is not expressed in this compartment but in the more anterior intercalary segment. It is likely that the function of *labial* is to prevent labial development in the procephalon.

The bithorax complex – adding layers of information to default states

The BX-C controls regional specification in the abdomen in a manner that most closely fits the combinatorial code model discussed in Topic A6. The three genes of the BX-C are expressed in overlapping patterns in the posterior of the embryo, with the anterior boundary of *Ultrabithorax* expression at parasegment 5, the anterior boundary of *abdominal A* expression at parasegment 7, and the anterior boundary of *Abdominal B* expression at parasegment 10. Posterior dominance occurs within this group of genes, so that *Ultrabithorax* and *abdominal A* are inhibited by the product of *Abdominal B*, and *Ultrabithorax* is inhibited by the product of *abdominal A*.

The genes of the BX-C appear to positively impose new positional identities on a default state, which is **parasegment 4** (*Fig. 3*). Thus, in deletion mutants lacking the entire BX-C, each abdominal parasegment develops as parasegment 4. The role of the *Ultrabithorax* gene is to specify parasegments 5 and 6. Therefore, if *abdominal A* and *Abdominal B* are deleted, but *Ultrabithorax* is expressed, parasegments 5 and 6 form as normal, but all parasegments behind p6 are also specified as p6. If only *Abdominal B* is deleted, so that *Ultrabithorax* and *abdominal A* are expressed, parasegments 7–9 are specified as normal, but all parasegments behind p9 are also specified as p9. These results can be explained in terms of posterior dominance. In normal embryos, *Ultrabithorax* expression is inhibited posteriorly to parasegment 7 by the other BX-C genes. If these genes are absent, the expression domain of *Ultrabithorax* extends to the posterior end of the embryo. The function of *Ultrabithorax* is to specify parasegments 5 and 6, and this is what it does (ectopically) in the whole embryonic abdomen when its expression is not prevented. If the *Ultrabithorax* gene is deleted but the other BX-C genes continue to be expressed, parasegments 7–14 are specified as normal but parasegments 4–6 are all specified as parasegment 4. This is because the default state is parasegment 4, and in the absence of *Ultrabithorax*, parasegments 5 and 6 are not specified and hence remain in the default state. The development of more posterior segments is not grossly affected because *abdominal A* and *Abdominal B* are expressed as normal,

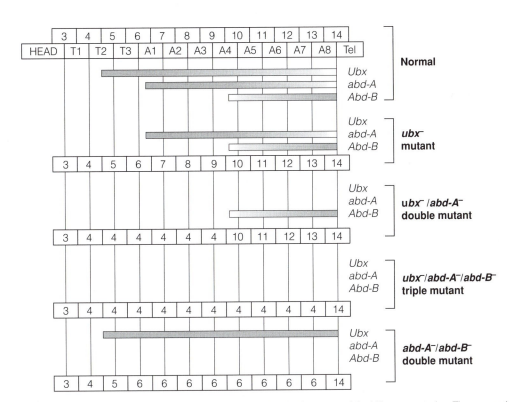

Fig. 3.   Simplified overview of parasegmental specification by the genes of the bithorax-complex. The expression of Ubx, abd-A and Abd-B are shown in a normal fly and in single and combinatorial loss of function ubx, abd-A and abd-B mutants, where the parasegments are respecified as shown.

and they restrict the expression of *Ultrabithorax* in the abdomen anyway, so its loss by mutation only complements their normal activity. However there are more subtle changes to the structures of parasegments 7–14 suggesting that *Ultrabithorax* does play a role in their specification. This indicates that positional identity in the abdomen requires a combinatorial code generated by all the genes of the BX-C, which closely resembles the model discussed in Topic A6.

Interestingly, mutational analysis of the BX-C initially revealed more than 10 genes, including the original *bithorax* locus after which the complex was named. Further studies, however, revealed that these 'genes' were in fact regulatory elements. The expression of the BX-C genes is driven in a highly modular way, such that mutating particular regulatory elements has highly specific effects on the gene expression patterns. While deletion of the entire *Ultrabithorax* gene converts parasegments 5 and 6 into parasegment 4 (a lethal effect), mutations in the *bithorax* locus specifically affect the posterior compartment of parasegment 5 (anterior compartment of the third thoracic segment) while those in the *posterobithorax* locus specifically affect the anterior compartment of parasegment 6 (posterior compartment of the third thoracic segment). These mutants are viable and produce adult flies with homeotic transformations of specific parts of each segment. A remarkable triple mutant (*anterobithorax*, *posterobithorax* and *bithoraxoid*) transforms the entire third thoracic segment into a second thoracic segment, producing a fly with four wings. It is speculated that the evolution of two-winged insects such as *Drosophila* may have been brought about by mutations in the BX-C.

Initiation of HOM-C gene expression

The basic body plan of the *Drosophila* embryo is complete when the positions of the parasegment boundaries are established and each parasegment has been given a positional value. The parasegment boundaries are established by the expression of the segment polarity genes in a repetitive series of stripes, while positional values are conferred by the ordered expression of the HOM-C genes along the anteroposterior axis. Coordination between these systems is critical so that positional values correspond to developmental compartments, and positional values change across compartment boundaries.

Such coordination is achieved by placing both the segment polarity genes and homeotic genes under the regulation of the upstream gap and pair rule gene products. The patterns of segment polarity gene expression are established by the pair rule transcription factors, which are themselves regulated in a modular fashion by the gap transcription factors to generate a repetitive series of stripes. The gap genes are expressed in a non-repetitive pattern along the anteroposterior axis, and this provides a simple mechanism to establish a similar distribution of HOM-C gene expression domains. The function of the pair rule genes in HOM-C regulation is to resolve the HOM-C expression patterns so they correspond to the compartment boundaries established in the syncytial blastoderm. The HOM-C genes therefore have modular promoters with multiple binding sites for gap and pair rule transcription factors. For example, the *bithorax* regulatory module of the *Ultrabithorax* gene discussed above contains binding sites for the gap transcription factors Hunchback and Tailless, as well as the pair rule transcription factor Fushi tarazu.

Maintenance of HOM-C gene expression

The gap and pair rule genes are expressed transiently in the syncytial embryo. Therefore, although they establish both the segment polarity and HOM-C gene expression patterns, these must be *maintained* by different mechanisms. As discussed in Topic H4, the segment polarity genes maintain their expression by establishing an **autoregulatory loop** involving the signaling proteins Hedgehog and Wingless. The HOM-C genes use an entirely different mechanism, based on the modification of **chromatin structure**.

The HOM-C gene expression patterns are dynamic, reflecting the cell movements in the embryo as well as complex regulatory networks in which the homeodomain transcription factors regulate each other's genes. After this initial dynamic phase, however, the expression patterns are stabilized so that each segment can differentiate appropriately. This expression pattern is fixed by the permanent modification of chromatin structure within the HOM-C, a process facilitated by the *trithorax* and *Polycomb* genes. These genes were first identified through their ability to mimic homeotic mutant phenotypes. Both these genes encode **chromatin-binding proteins** although they have opposite functions. The **Polycomb** protein binds to transcriptionally inactive DNA and remodels the chromatin structure so that the locus is permanently repressed. Conversely, **Trithorax** binds to transcriptionally active DNA and maintains the chromatin in an open state. Both proteins function as part of large multiprotein complexes. Indeed the Polycomb protein itself has no intrinsic DNA-binding activity and must assemble in a complex that has already bound to DNA. The modifications established and maintained by the Polycomb and Trithorax group gene products are heritable through successive rounds of DNA replication and cell division, so that all descendants of a given cell express the same group of HOM-C genes (mechanisms for this are discussed in Topic C2).

Molecular basis of HOM-C function

The final step in the developmental program is the link between positional information and morphogenesis, i.e. which genes the HOM-C transcription factors activate in order to activate the developmental sub-programs that generate regionally-appropriate structures such as legs, wings and antennae. A combination of approaches has been used to identify downstream genes including the analysis of gene expression patterns in HOM-C mutants, the use of enhancer traps (Topic B2) to identify genes with corresponding expression patterns and the analysis of gene sequences. Surprisingly, very few targets of the HOM-C transcription factors have been found, although the current availability of almost all of the *Drosophila* genome sequence allows genes to be screened for potential HOM-C gene response elements. Many of the genes that have been identified as potential targets have response elements to multiple HOM-C proteins. For example, *distal-less*, which plays an important role in limb development (Topic L2) and *apterous*, contain binding sites for Antennapedia and Ultrabithorax as well as other HOM-C proteins. Interestingly, the posterior dominance among the HOM-C genes appears to apply also to their regulation of downstream targets. Antennapedia activates *distal-less* and *apterous* whereas Ultrabithorax represses both genes. Where both regulators are present together, the repressive effect of Ultrabithorax prevails over the positive effects of Antennapedia. Some widely-expressed genes appear to be regulated by many of the HOM-C transcription factors, an example being *Serrate*, which is controlled by Deformed, Antennapedia, Ultrabithorax and Abdominal A. Mutational analysis suggests that other genes are controlled by individual HOM-C genes or a small number of them. For example, the BX-C genes regulate *unplugged* and *18-wheeler*, two genes expressed in the developing tracheal system. Antennapedia represses *salm*, which specifies antennal development and needs to be inactivated in the thorax. An important target of Sex combs reduced is *Creb-A*, which is normally expressed in the salivary glands but is expressed in other areas when Sex combs reduced is itself misexpressed. There are likely to be many more targets of the HOM-C genes awaiting discovery.

H6 DORSOVENTRAL AXIS SPECIFICATION AND PATTERNING

Key Notes

Dorsoventral polarity	The dorsoventral axis of the *Drosophila* embryo is divided into a series of spatial domains corresponding to different cell fates. The dorsal blastoderm gives rise to the extraembryonic amnioserosa and the dorsal ectoderm of the embryo. The ventral blastoderm gives rise to embryonic mesoderm and neurectoderm.
Dorsal and Cactus	Dorsoventral polarity is established by a gradient of the transcription factor Dorsal. The protein is uniformly distributed in the syncytial embryo, but is selectively taken up into ventral nuclei. Nuclear Dorsal is a morphogen, activating downstream zygotic genes in a concentration-dependent manner to establish broad domains of gene expression along the axis. The nuclear gradient is established by the selective degradation of Cactus, an inhibitory protein closely associated with Dorsal that prevents its nuclear uptake. The degradation of Cactus is triggered by a signal originating at the ventral side of the egg.
Spätzle and Toll	The ventral signal is provided by a secreted protein, Spätzle, which interacts with a receptor called Toll spanning the egg plasma membrane. Signal transduction through Toll results in the degradation of Cactus and the uptake of Dorsal protein into ventral nuclei.
Ventral activation of Spätzle	Spätzle is uniformly distributed in the perivitelline space as an inactive zymogen, but is activated specifically on the ventral side of the egg by a complex of serine proteases. These are also uniformly distributed, but selectively activated on the ventral side of the egg by proteins secreted from ventral follicle cells. These proteins, Neudel, Windbeutel and Pipe, are not secreted from dorsal follicle cells because the genes are repressed by a signal emanating from the oocyte nucleus.
Zygotic genes	The concentration of nuclear Dorsal is highest in the ventral nuclei, where it activates the genes *twist*, *snail* and *rhomboid*. Expression of the genes *decapentaplegic*, *tolloid* and *zerknüllt* is restricted to the dorsal region of the embryo because these genes are repressed by Dorsal. The initial expression domains are refined as the products of the zygotic genes (signaling proteins and transcription factors) begin to function, dividing the dorsoventral axis of the embryo into sharply-defined spatial domains of zygotic gene expression.

Related topics	Pattern formation (A6)	Anteroposterior axis specification
	Gene expression and regulation (C1)	in *Drosophila* (H2)
		The vertebrate organizer (I2)
	Signal transduction in development (C3)	Mesoderm induction and patterning (K1)
	Molecular aspects of embryonic pattern formation in *Drosophila* (H1)	

Dorsoventral polarity

The dorsoventral axis of the *Drosophila* embryo does not become divided into segments like the anteroposterior axis. However, it is divided into broad regions that give rise to distinct cell types (*Fig. 1*). The most dorsal blastoderm forms an extraembryonic membranous structure called the **amnioserosa**. Moving more ventrally, the cells give rise to ectoderm: **dorsal ectoderm**, **lateral ectoderm** and finally the ventral **neurogenic ectoderm** from which much of the nervous system arises. The most ventral region of the embryo gives rise to **mesoderm**, which invaginates into the embryo during gastrulation (the blastoderm that gives rise to endoderm is located at the anterior and posterior poles of the embryo, and invaginates separately to form the gut; gastrulation in *Drosophila* is discussed in more detail in Topic F2). The neurogenic ectoderm then comes to cover the ventral surface of the embryo, and constitutes the **germ band**. Cells within this region form the proneural clusters that give rise to the central nervous system (Topic J5). Specification of dorsoventral cell fates in *Drosophila* involves reciprocal signaling between the oocyte (later the embryo) and the overlying follicle cells. These signaling events are discussed below. The genes involved are summarized in *Table 1*.

Dorsal and Cactus

Dorsoventral polarity in the *Drosophila* embryo is controlled by a maternal transcription factor appropriately called **Dorsal**. Like Hunchback, this protein acts as a morphogen, activating downstream zygotic genes in a concentration-dependent manner, and dividing the dorsoventral axis into discrete spatial domains of zygotic gene expression (*Fig. 1*). These zygotic genes are analogous to the gap genes expressed along the anteroposterior axis (Topic H2). However, unlike Hunchback, both *dorsal* mRNA and Dorsal protein are distributed uniformly throughout the syncytial blastoderm. The morphogen gradient arises

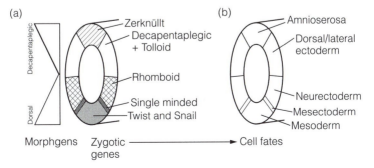

Fig. 1. *Establishment of dorsoventral polarity in* Drosophila. *(a) Two morphogens, Dorsal and Decapentaplegic, set up opposing gradients, resulting in the expression of transcription factors as regional bands along the axis. (b) The corresponding cell fates.*

Table 1.   Summary of major Drosophila genes involved in dorsoventral axis specification

Gene	Product	Maternal/zygotic and where required	Function
gurken	EGF homolog	Maternal required in oocyte	specifies dorsal follicle cells
torpedo	EGF receptor	Maternal required in follicle cells	specifies dorsal follicle cells
neudel	Intracellular signaling proteins	Maternal required in follicle cells	Expressed in ventral follicle cells. Activate ventralizing signal Spätzle
windbeutel			
pipe			
gastrulation defective	Serine proteases	Maternal required in embryo	Activate ventralizing signal Spätzle
easter			
snake			
spätzle	Secreted protein	Maternal required in embryo	Ventralizing signal
Toll	Receptor for Spätzle	Maternal required in embryo	Receptor for ventralizing signal
pelle	Intracellular signaling proteins	Maternal required in embryo	Transduction of ventralizing signal
tube			
dorsal	Transcription factor	Maternal required in embryo	Activates ventral genes
Cactus	Inhibitory protein	Maternal required in embryo	Prevents nuclear uptake of Dorsal
twist	Transcription factor	Zygotic	Specification of dorsal cell fates
snail	Transcription factor	Zygotic	
single minded	Transcription factor	Zygotic	
rhomboid	Transcription factor	Zygotic	
zerknüllt	Transcription factor	Zygotic	
toloid	BMP homolog	Zygotic	Specification of ventral cell fates
decapentaplegic	BMP homolog	Zygotic	
short gastrulation	Chordin homolog	Zygotic	

because the Dorsal protein is closely associated with a second protein called **Cactus**. Dorsal only functions as a transcription factor when translocated into the zygotic nuclei, but Cactus masks the **nuclear localization signal** of Dorsal, causing it to be retained in the cytoplasm. Dorsoventral polarity is established when a signal transduction pathway is initiated in the future ventral region of the blastoderm, culminating in the degradation of Cactus and translocation of Dorsal to the nucleus (*Fig. 2*). The amount of Dorsal translocated to the nucleus is proportional to the strength of the signal, so that a **gradient of nuclear Dorsal protein** is set up along the dorsoventral axis, with the highest concentration of nuclear Dorsal in the ventral nuclei and the lowest concentration in the dorsal nuclei. Therefore, like many other *Drosophila* genes, *dorsal* carries out the opposite function to that implied by its name. It is required in **ventral nuclei** to specify **ventral cell fates**. The gene is named *dorsal* because mutant mothers lacking this gene produce dorsalized embryos, as they are unable to specify ventral cell types.

Note that this mechanism of transcription factor regulation is not restricted to *Drosophila*. Dorsal is homologous to a vertebrate transcription factor NF-κB, which controls gene expression in B-lymphocytes. Cactus is homologous to its inhibitor, I-κB. As in *Drosophila*, NF-κB is retained in the cytoplasm by its association with I-κB, and external signals are required to trigger its degradation, followed by translocation of NF-κB to the nucleus. We consider such signals below.

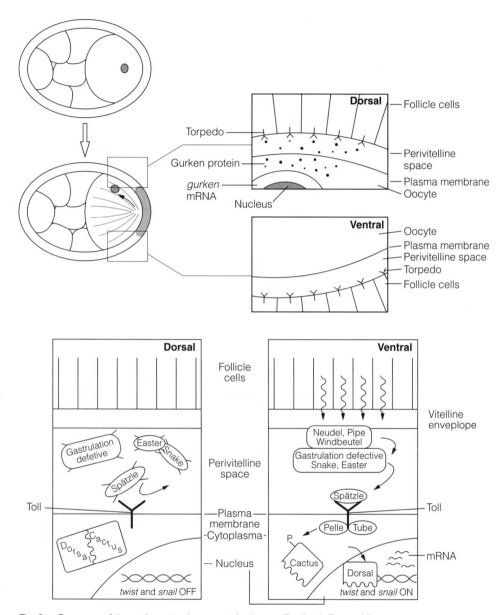

Fig. 2.   Summary of the pathway to dorsoventral axis specification in Drosophila.

Spätzle and Toll   The ventralizing signal that facilitates nuclear uptake of Dorsal is provided by the secreted protein **Spätzle** when it binds to the membrane-bound receptor **Toll**. Toll is tethered in the plasma membrane of the egg, but is uniformly distributed. The Spätzle protein is also uniformly distributed in the perivitelline space, however it is present as an inactive precursor that requires processing by **proteolytic cleavage**. As discussed below, this processing reaction occurs only on the ventral side of the embryo, resulting in the production of active Spätzle ligand specifically in the ventral region of the perivitelline space. The signal transduction cascade initiated when Spätzle binds to Toll involves the products of the genes *pelle* and *tube* (*Fig. 2*). **Pelle** is a **protein phosphorylase**, and this enzyme phosphorylates Cactus, resulting in its degradation and the nuclear uptake of Dorsal. The active Spätzle fragment is most concentrated on the ventral side of the embryo and establishes a concentration gradient in the perivitelline space along the dorsoventral axis. This results in progressively weaker signaling towards the dorsal side, and as less Toll receptors are activated, less Cactus is degraded and more Dorsal is retained in the cytoplasm.

Ventral activation of Spätzle

Spätzle is activated specifically on the ventral side of the embryo through a series of steps beginning in the oocyte (*Fig. 2*). Initially, the oocyte is symmetrical around the anteroposterior axis (i.e. there is no dorsoventral polarity) with the nucleus located at the posterior pole. As discussed in Topic H2, the oocyte nucleus specifies the posterior follicle cells and then moves towards the anterior pole through the cortical cytoplasm along an array of microtubules which follows the inside surface of the plasma membrane. The nucleus can follow any straight line to the anterior pole along the inside of the oocyte, and there appears to be no bias to the path taken. However, wherever the nucleus travels, follicle cells lining the egg plasma membrane along that path are specified as dorsal. This involves the local synthesis of *gurken* mRNA by the nucleus and its translation into protein. The protein can diffuse locally to the overlying follicle cells, but not across egg cytoplasm the other side. Follicle cells that do not receive the Gurken signal are specified as ventral. Mutations in the *gurken* gene therefore cause ventralization of the embryo. Gurken is homologous to mammalian epidermal growth factor (EGF). The receptor for Gurken, which is made in all follicle cells, is encoded by the gene *torpedo*. As well as becoming functionally distinct, the dorsal and ventral follicle cells also show morphological differentiation, with the dorsal cells becoming more columnar in appearance than the ventral cells.

Torpedo signal transduction in the dorsal follicle cells results in the repression of three key genes, *neudel*, *windbeutel* and *pipe*. In the ventral follicle cells, these genes are active, and the corresponding proteins are secreted where they form a complex tethered in the vitelline envelope. The function of this complex is to activate three **serine proteases** called **Gastrulation defective, Easter** and **Snake**, all of which are uniformly distributed as inactive precursors or **zymogens**. Activation occurs as a **proteolytic cascade**, in which the membrane complex activates Gastrulation defective, which in turn activates Snake, which in turn activates Easter. Finally, the Easter enzyme cleaves Spätzle, providing the active ligand for Toll.

Zygotic genes

Once a nuclear gradient of Dorsal protein is established, the syncytial blastoderm begins to undergo cellularization and transcription factors are no longer able to diffuse between nuclei. The nuclear gradient of Dorsal falls to zero approximately at the equator of the embryo, so that the dorsal half is more or less devoid of nuclear Dorsal protein. The Dorsal transcription factor acts as both an activator and repressor of different zygotic genes. It activates the genes *twist*, *snail*, *single minded* and *rhomboid*, and represses the genes *zerknüllt*, *tolloid* and *decapentaplegic*. The result is that *twist*, *snail*, *single minded* and *rhomboid* are expressed in the ventral half of the embryo, and *zerknüllt*, *tolloid* and *decapentaplegic* in the dorsal half (*Fig. 1*). The *twist* and *snail* genes are activated only at high concentrations of Dorsal protein, because their promoters contain low-affinity Dorsal-binding sites. Conversely, *rhomboid* is expressed at lower concentrations because its promoter contains high-affinity Dorsal-binding sites. However, the *snail* gene also encodes a transcription factor, which represses *rhomboid*. This establishes a sharp boundary between the *snail* and *rhomboid* expression domains. The *single minded* gene is repressed by both Snail and Rhomboid, and is hence expressed in a narrow stripe of cells between the *snail* and *rhomboid* expression domains, where the concentration of both proteins is lower. This divides the ventral region of the embryo into distinct spatial domains, which eventually form the ventral mesoderm and neurectoderm.

In the dorsal region of the embryo, there is no gradient of Dorsal protein to establish patterning. However, after cellularization a gradient is generated by the TGF-β signaling protein **Decapentaplegic**. The protein is distributed uniformly, but its activity is graded along the dorsoventral axis through the diffusion of an antagonist encoded by the *short gastrulation* gene. **Short gastrulation** is a ventral-specific secreted protein but its product forms a gradient that seeps into the dorsal half of the embryo. This results in strong inhibition of Decapentaplegic in the equatorial region, and progressively weaker inhibition towards the dorsal side of the embryo, as the concentration of Short gastrulation drops. At high concentrations of active Decapentaplegic, the *zerknüllt* gene is expressed. This results in a sharply delineated dorsal domain of *zerknüllt* gene expression, corresponding to the region of the embryo that will form the amnioserosa. The remainder of the dorsal side of the embryo is specified as ectoderm (future epidermis).

Remarkably, Decapentaplegic is homologous to the vertebrate signaling protein BMP4, while Short gastrulation is homologous to Chordin, a protein secreted by the vertebrate organizer and which inhibits BMP4 activity. The interaction between these proteins plays a crucial role in determining dorsoventral cell fates in the vertebrate embryo, as discussed in Sections I, J and K. It therefore appears that key genes in dorsoventral specification have been conserved for millions of years of animal evolution.

11 EARLY PATTERNING IN VERTEBRATES

Key Notes

Axes and germ layers	Vertebrates have a body plan based on three principal axes and three germ layers. The anteroposterior axis runs from head to tail, with the head at the anterior end. The dorsoventral axis runs from back to belly, with the mouth opening on the ventral surface. Certain internal organs are not symmetrically distributed, defining a third left-right axis.
Axis specification in vertebrates	Vertebrates use diverse mechanisms of axis specification, but the common result is a signaling center, known in amphibians as the Nieuwkoop center, that has the unique ability to specify the organizer. The eggs of all vertebrates are radially symmetrical and axis specification is therefore a symmetry-breaking process. This can involve either an external physical cue or, in amniotes, signals from the extraembryonic membranes.
Symmetry-breaking in amphibians	In *Xenopus* development, the symmetry-breaking cue for axis specification is provided by the sperm. Fertilization results in cortical rotation towards the site of sperm entry, establishing the Nieuwkoop center on the opposite side of the egg to the sperm. This becomes the dorsal side of the embryo.
Molecular basis of forming the Nieuwkoop center	Cortical rotation transfers maternal Disheveled protein, an inhibitor of GSK-3, to the dorsal side of the egg. This results in the stabilization of the transcription factor β-catenin, resulting in accumulation of this protein dorsally. β-catenin initiates a series of events that activates organizer-specific genes in the presumptive dorsal mesoderm.
Symmetry-breaking in birds (and reptiles)	The equivalent to the Nieuwkoop center in birds is the posterior marginal zone, which arises in the blastoderm in response to gravity as the egg is rotating on its way down the oviduct. The dorsoventral axis of the embryo is related to the animal–vegetal axis of the egg, and may be specified by the potential difference that arises between the albumen and the subgerminal cavity.
Symmetry-breaking in mammals	Mammalian eggs have no yolk and appear radially symmetrical. The blastocoel forms asymmetrically in the embryo, so that the inner cell mass is displaced to one end, and this defines the future dorsoventral axis. The anterior visceral endoderm plays an important role in polarizing the anteroposterior axis of the embryo, but the origins of initial anteroposterior polarity are unknown. There is evidence that signals from the uterus may be involved.

Related topics	Pattern formation and
	    compartments (A6)
	Gastrulation in vertebrate
	    embryos (F3)
		Molecular basis of embryonic
		    pattern formation in *Drosophila*
		    (H1)
		The vertebrate organizer (I2)

Axes and germ
layers

All vertebrates have a conserved body plan, based on the early specification of three **principal axes** and the diversification of three **germ layers**. The central nervous system with its enlarged brain and narrow spinal cord defines the **anteroposterior axis** (also called the craniocaudal or rostrocaudal axis), with the brain at the anterior end. The second principal axis is the **dorsoventral axis**, which runs from back to belly. The spinal cord is closest to the dorsal surface, and the gut closest to the ventral surface. The third axis is the **left–right axis**. Although the vertebrate body plan is superficially bilaterally symmetrical, certain internal organs are distributed asymmetrically on the left and right sides.

Axis specification
in vertebrates

The axes and germ layers of the vertebrate embryo are specified early in development. The mechanisms and timing of these processes differ greatly among the vertebrate classes, but despite these initial differences, gastrulation in all species produces an embryo with remarkably conserved features: an axial notochord and neural tube bracketed by paired somites, an enlarged head with segmented pharyngeal arches, two pairs of limb buds and a tail. This is known as the **phylotypic stage**, where all vertebrate embryos are similar in appearance. Later development involves the diversification of this body plan to produce features specific to each vertebrate class.

The eggs of all vertebrates except mammals have an overt **animal–vegetal axis**, with the yolk concentrated at the vegetal pole and the nucleus displaced towards the animal pole. The egg is **radially symmetrical** about this axis. In mammals, there is no yolk and the animal pole is defined as the site at which the polar bodies are extruded during meiosis. In fish and amphibians, the distribution of maternal determinants along the animal–vegetal axis of the egg is responsible for the initial specification of the germ layers, while the axes of the embryo are established at fertilization by an external physical cue. In amniotes, the axes are also established by physical cues, but the germ layers are specified by cell–cell interactions *after* the axes have been set up. It is important to differentiate between **axis formation**, which in all vertebrates occurs during gastrulation, and **axis specification**, which occurs beforehand to establish the site where gastrulation commences.

Axis specification in all vertebrates is a **symmetry-breaking process** that establishes a critical embryonic signaling center. In fish and amphibians, this is the **Nieuwkoop center**, while in reptiles and birds, the **posterior marginal zone** has been identified as the equivalent structure. This signaling center is required to induce the **organizer**. The organizer marks the future dorsal side of the embryo, and cells moving through the organizer during gastrulation converge and extend along the anteroposterior axis. In this Topic, we consider the diverse symmetry-breaking processes that initially polarize the vertebrate embryo. The formation and function of the vertebrate organizer is discussed in Topic I2. Finally in this section, Topic I3 covers the mechanisms of left–right axis specification in vertebrate embryos. The organizer is involved in this process also, and therefore plays a defining role in all three of the principal embryonic axes. The specification and further development of the germ layers is discussed in Sections J and K.

Symmetry-breaking in amphibians

The *Xenopus* egg shows obvious animal–vegetal polarity, with a darkly pigmented animal hemisphere, and a pale vegetal hemisphere rich in yolk platelets. Certain maternal gene products are asymmetrically distributed along this axis, and these help to establish the germ layers and the embryonic axes. However, while the germ layers can be mapped on to the animal–vegetal axis of the egg, the axes are initially specified by an external signal. The symmetry-breaking cue for axis specification is provided by the sperm at the moment of fertilization. The sperm can enter only the pigmented animal hemisphere of the egg. Fertilization triggers a series of events resulting in formation of the Nieuwkoop center opposite the **site of sperm entry**. As the Nieuwkoop center induces the organizer, which defines the dorsal side of the embryo, the sperm entry point marks the future ventral surface of the embryo.

The mechanism by which sperm entry polarizes the egg, resulting in the formation of the Nieuwkoop center, is not completely understood. Soon after fertilization, the egg plasma membrane and underlying cortex (the outer layer of cytoplasm, rich in actin filaments) rotate approximately 30°, with respect to the inner cytoplasm, towards the site of sperm entry (*Fig. 1*). The inner cytoplasm is laden with yolk and therefore held in position by gravity. The direction of this **cortical rotation** is controlled by an array of microtubules whose assembly is directed by the sperm-derived centriole. In some amphibians (but not *Xenopus*) the inner cytoplasm contains diffuse pigment granules while the cortex of the animal hemisphere is darkly pigmented. Cortical rotation reveals the lightly-pigmented inner cytoplasm as a **gray crescent** (*Fig. 1*). Cortical rotation can be disrupted with drugs that inhibit microtubule polymerization and by **UV-irradiation**, preventing the formation of the organizer and leading to the development of ventralized embryos. UV-irradiated embryos can be rescued by tilting the egg while it is supported in a solid matrix such as agarose, as this rotates the inner cytoplasm under the influence of gravity relative to the cortex and produces the same effect as normal cortical rotation. Interestingly, while tilting can rescue the dorsoventral and anteroposterior axes, it often results in disruption of the left–right axis. The basis of this phenomenon is discussed in Topic I3. If the tilting experiment is carried out on embryos that have already undergone

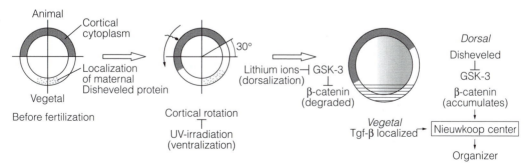

Fig. 1. Symmetry breaking in amphibians. Translocation of Disheveled protein inhibits glycogen synthase kinase 3 (GSK-3), allowing β-catenin to accumulate in the future dorsal side of the egg. The overlap of dorsal β-catenin and vegetal TGF-β signals specifies the Nieuwkoop center, which induces the organizer.

cortical rotation, a second Nieuwkoop center is established, resulting in the development of twin embryos. As discussed below, this suggests that the effects of cortical rotation are generated by the mixing of cytoplasm in the region of the egg that will become the Nieuwkoop center.

Molecular basis of forming the Nieuwkoop center

The key molecular component in the formation of the Nieuwkoop center is the transcription factor and cadherin-binding protein β-**catenin**. This protein can induce a second axis when its mRNA is injected into ventral blastomeres, and its role in endogenous axis specification is confirmed by experiments in which translation is blocked with antisense RNA, as this causes ventralization of the embryo. The mRNA for β-catenin is distributed uniformly throughout the egg, as is the mRNA for its inhibitor, **glycogen synthase kinase-3 (GSK-3)**. GSK-3 targets β-catenin for proteolytic degradation and therefore has the opposite effect of β-catenin in microinjection experiments: downregulation of its expression causes dorsalization of the embryo while overexpression in dorsal blastomeres prevents the normal axis forming, generating a ventralized embryo. Normally, β-catenin activity is stimulated by Wnt signaling, as this activates a repressor of GSK-3 called Disheveled. *Xenopus* embryos can be dorsalized by exposing them to **lithium ions**, and it is thought that lithium also exerts its effect by inhibiting GSK-3. In the unicellular environment of the egg, however, β-catenin must be activated independently of an external ligand, and in a manner that is dependent on cortical rotation. The maternal mRNA for *Xenopus disheveled* is localized to the vegetal pole of the egg, and a current model suggests that cortical rotation results in the transport of Disheveled protein to the future dorsal side of the embryo, stabilizing β-catenin by inhibiting GSK-3 and therefore enriching the dorsal side of the embryo for β-catenin protein. As discussed in Topics I2 and K1, β-catenin can then activate genes required for setting up the organizer, including the gene encoding a transcription factor called Siamois.

Symmetry-breaking in birds (and reptiles)

Bird and reptile eggs have pronounced animal–vegetal polarity caused by the massive yolk, and cleavage is restricted to a small disc of cytoplasm containing the nucleus, which defines the animal pole. This asymmetry plays a direct role in establishing the dorsoventral axis of the blastodisc, as the subgerminal cavity (beneath the blastodisc) is more acidic than the albumen, and this generates a **potential difference** across the blastoderm that is critical in axis specification. Reversing the electrical polarity across the blastoderm (e.g. by injecting acid into the albumen) results in the inversion of the dorsoventral axis so that the embryo develops upside down.

The anteroposterior axis is established by **gravity** as the egg moves down the oviduct. During this journey the egg rotates slowly, but the embryo and yolk tend to remain upright. Even so, there is a slight displacement of the blastoderm in the direction of rotation. In chickens, there is a critical period of about 2 hours approximately 20 hours after fertilization, during which the anteroposterior axis is specified. The thinning of the blastoderm, which precedes gastrulation, is initiated in the future posterior region of the blastodisc. In the anterior region, the blastoderm thins out so that the subgerminal cavity extends under the marginal zone, while in the posterior region, the marginal zone remains in contact with the yolk. This **posterior marginal zone (PMZ)**, which is the amniote equivalent of the *Xenopus* Nieuwkoop center, arises in the blastoderm that is nearest the top of the egg (*Fig. 2*). It is possible that gravity causes the

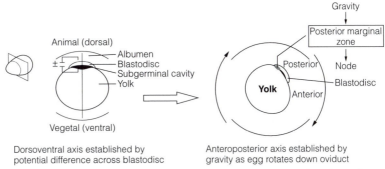

Fig. 2.   *Establishment of anteroposterior and dorsoventral axes in the avian embryo.*

displacement of certain cytoplasmic determinants in the rotating egg, resulting in the specification of the PMZ in the uppermost region of the blastodisc. Furthermore, the blastodisc also slides anteriorly along the yolk in a manner reminiscent of cortical rotation in amphibians. The PMZ induces the overlying epiblast cells to form **Hensen's node**, the avian equivalent of the *Xenopus* organizer (Topic I2).

<table>
<tr><td>

Symmetry-breaking in mammals

</td><td>

Unlike other vertebrates, mammalian embryos do not require yolk because they receive nutrients from secretions in the uterus, and later from the placenta. Mammalian eggs and early embryos therefore appear perfectly symmetrical. However, there is some recent evidence of molecular polarity at the oocyte stage, which persists through to the blastocyst. For example, leptin and a downstream kinase STAT-3 are localized in the oocyte and hence asymmetrically distributed to cells in the morula and early blastocyst. The significance of this is not clear, as until compaction, all cells of the mammalian morula are equivalent and totipotent, and can contribute to either the **ICM** or the **trophectoderm** if grafted to appropriate positions in another morula. The first sign of asymmetry is the segregation of the ICM, so that the **blastocoel** forms at one end of the blastocyst leaving the ICM attached to the trophectoderm at the other end. This may simply reflect the selection of ICM cells that happen to have the strongest adhesion to the trophectoderm. Whatever the mechanism, segregation establishes polarity in the conceptus, as the ICM is exposed to the trophectoderm on one side and the blastocoel on the other. This is the **embryonic–abembryonic axis**, which becomes the future dorsoventral axis of the embryo. The ICM cells nearest the trophectoderm differentiate in to the epiblast (future embryo and amnion) whereas those in contact with the blastocoel differentiate into the hypoblast or primitive endoderm (future lining of the yolk sac) (*Fig. 3*).

It is not known how the anteroposterior axis of the mammalian embryo is specified. Originally it was thought that, at the blastocyst stage, the mammalian embryo was radially symmetrical about the embyonic–abembryonic axis, and that the primitive streak was the first sign of anteroposterior polarity. However, careful analysis of early blastocysts reveals a definite bilateral symmetry. Furthermore, the blastocyst's bilateral axis is aligned with the future anteroposterior axis of the embryo. The precise relationship between the two is unknown because while the embryo's anteroposterior axis is *aligned* with the bilateral axis of the blastocyst, it can adopt either the same or the opposite polarity. There is some evidence that the maternal environment may play a role in axis specifica-

</td></tr>
</table>

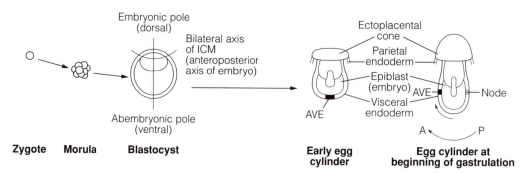

Fig. 3.   *Symmetry-breaking in the mouse embryo involves specification of the node and anterior visceral endoderm (AVE).*

tion, as the anteroposterior axis of the embryo is usually perpendicular to the long axis of the uterus. Whatever the initial cues for asymmetry, an important component of the polarization process is a small part of the visceral endoderm (the **anterior visceral endoderm, AVE**) which in mouse embryos is initially found at the tip of the egg cylinder aligned with the dorsoventral axis. The AVE rotates anteriorly just prior to gastrulation so that it occupies a position opposite the site where the primitive streak forms (*Fig. 3*). This rotation is prevented in mutants for the gene *Cripto*, which show anteroposterior polarity defects and the persistence of AVE markers in the original distal site of the AVE. Interestingly, grafted AVE from mammals can act as a head-inducer in chicken embryos, but the equivalent anterior hypoblast region of the chicken embryo cannot function in the same way, either in chickens or mammals. The recruitment of the AVE as a head inducer during early anteroposterior polarization may therefore be unique to mammals.

12 THE VERTEBRATE ORGANIZER

Key Notes

What is an organizer?

An organizer directs the development of surrounding tissues by acting as a source of inductive signals. Many organs have an organizer, but the most important in terms of vertebrate development is the primary embryonic organizer, which directs the development of the entire embryo by setting up and patterning the dorsoventral and anteroposterior axes. Organizer grafts induce a secondary axis resulting in the development of twin embryos.

The amphibian organizer

The amphibian organizer is the dorsal lip of the blastopore, whose remarkable inductive properties were discovered by Spemann and Mangold in 1924. This is equivalent to the teleost embryonic shield, and the amniote primitive node. Cross species grafts show that the organizer functions similarly in all vertebrates.

Early organizer-inducing centers

An early event in vertebrate development is the establishment of a signaling center, such as the amphibian Nieuwkoop center, whose function is to induce the organizer. This involves synergy between β-catenin and a TGF-β-dependent signal.

The organizer's role in development

The organizer's functions include initiating gastrulation, dorsalizing the mesoderm and neuralizing the endoderm, induction of the head, establishing the embryo's anteroposterior axis and forming the left–right axis. The organizer forms anterior endoderm and mesoderm, the notochord of the trunk, and the notochord and neural tube of the tail. A number of transcription factors and secreted proteins are expressed specifically in the organizer.

Late organizer inducing centers – organizer maintenance

The organizer's cell population is constantly changing, so it needs to be continuously redefined by signals from surrounding cells. In chickens, part of the primitive streak, the node-indicing center, acts to maintain the organizer during gastrulation.

Related topics

Mosaic and regulative development (A4)
Pattern formation and compartments (A6)
Gastrulation in vertebrate embryos (F3)
Early patterning in vertebrates (I1)

Mesoderm induction and patterning (K1)
The ectoderm: neural induction and epidermis (J1)
Patterning the anteroposterior neuraxis (J2)

What is an organizer?

An **organizer** is any region of an embryo that directs the development of surrounding tissues, usually by acting as a source of inductive signals. An organizer is responsible for generating the axes of a developing structure and thus initiating pattern formation. Consequently, removal of the organizer severely disrupts development, while the removal of any other part of the structure is tolerated because the remaining cells become re-specified under the control of the inductive signals from the organizer (this is the basis of regulative development, as discussed in Topic A4). Furthermore, grafting organizer tissue onto a structure that already contains its own organizer results in the setting up of a secondary axis, causing partial or full duplication of the developing structure.

Many developmental systems have their own organizer. Examples discussed in this book include the **zone of polarizing activity** in the posterior margin of the limb bud, which is the organizer for limb development (Topic M2) and the isthmus at the midbrain/hindbrain boundary (**mes/met boundary**), which is the organizer for midbrain development (Topic J2). However, when used in an unqualified sense, the term organizer usually refers to the **primary embryonic organizer**, known in the zebrafish as the **embryonic shield**, in amphibians as the **(Spemann) organizer**, in birds as **Hensen's node** and in mammals simply as the **node** (*Fig. 1*). This is responsible for forming and patterning the principal body axes, although the axes are initially specified by earlier-acting mechanisms (Topic I1). Organizer grafts have the unique property of being able to induce an entire secondary axis, resulting in the development of twin embryos.

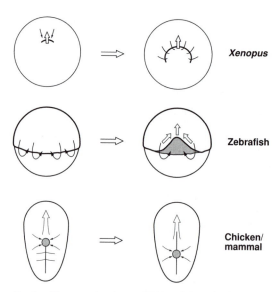

Fig. 1. *The organizer in amphibians, fish and amniotes during early gastrulation (left) and late gastrulation (right). The organizer is shaded in gray. Thin arrows show cells moving into the organizer, and thick arrows show convergent extension of cells that have already passed through the organizer.*

The amphibian organizer

In *Xenopus* and other amphibians, the primary embryonic organizer is the **dorsal lip of the blastopore** that forms at the beginning of gastrulation (Topic F3). This is a slit-like invagination through which cells move and give rise to (in the following order) anterior endoderm, prechordal plate and the notochord (as well as contributing to the somites, floor plate of the neural tube and dorsal endoderm). The importance of the organizer was discovered by Spemann and Mangold in their famous newt organizer graft experiments published in 1924. The grafting of organizer tissue to the ventral side of another embryo initiated gastrulation at the ectopic site and induced the formation of an entire secondary axis, producing twinned embryos. The secondary axis was produced from tissues of the host, not the graft, demonstrating that that the graft was respecifying the fates of the host cells surrounding it (*Fig. 2*).

The extraordinary inductive properties of the amphibian organizer demon-

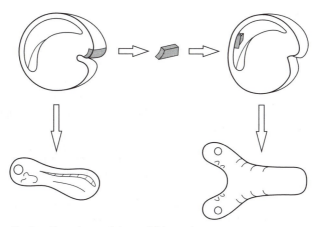

Fig. 2.   Organizer graft in amphibian embryos.

strated three important principles of organizer function. Firstly the organizer possesses the ability to *initiate the morphogenetic cell movements of gastrulation*. Secondly, the organizer has the ability to *confer dorsal fates on surrounding tissue*: the organizer itself forms the most dorsal mesodermal structures (notochord, prechordal plate) but induces adjacent ventral mesoderm to form dorsal structures such as somites, and also dorsalizes the adjacent ectoderm to form neural plate. The organizer therefore plays a critical role in patterning the embryo's dorsoventral axis. Finally, the organizer also has the ability to pattern the anteroposterior axis, as organizer grafts can produce secondary embryos with a head, trunk and tail. A more comprehensive list of the organizer's functions is provided below.

Grafting experiments have also confirmed the organizer activity of the equivalent structures in fish, birds and mammals. Furthermore, cross-class transplants have provided evidence that some of the molecular signals produced by the organizer are conserved among all vertebrates.

Early organizer-inducing centers

As discussed in Topic I1, an early event in the development of all vertebrates is the establishment of a signaling center whose role is to induce the organizer. In amphibians, the **Nieuwkoop center** is established opposite the site of sperm entry and induces the organizer in the overlying presumptive mesoderm. In birds and reptiles, the equivalent structure is thought to be the **PMZ**, a thickened region of blastoderm that induces the overlying epiblast to form the primitive streak and node. The similarities between the PMZ and the Nieuwkoop center occur at the structural, mechanistic and molecular levels. Both structures are capable of inducing organizer activity in overlying tissue. The formation of both structures involves gravity and rotation. Most importantly, certain key molecules are enriched specifically in these structures, and in the equivalent region of the zebrafish embryo, e.g. Vg1, β-catenin and other members of the TGF-β superfamily. The synergy between β-catenin-dependent and TGF-β-dependent signals is implicated in the induction of organizer mesoderm, as discussed in detail in Topic K1. Transplantation experiments in *Xenopus* have shown that blastomeres from the Nieuwkoop center grafted to the ventral side of another embryo can induce organizer activity in overlying mesoderm and form a secondary axis. Similarly, transplantation experiments in chick embryos have established that the

PMZ, when grafted to a lateral position in the blastoderm, can induce a secondary primitive streak in the overlying epiblast. However, this only occurs when the original primitive streak is removed, because the streak produces an inhibitor that prevents the development of ectopic streaks (see below).

The organizer's role in development

The organizer is a highly specialized structure that has multiple roles in axis specification, pattern formation and morphogenesis. The *Xenopus* organizer has been studied in great detail and is discussed below. The *Xenopus* organizer can be divided functionally and in terms of gene expression patterns into several domains, corresponding to a **head organizer**, a **trunk organizer** and a **tail organizer**. The head organizer is a specialized subpopulation of cells with head-inducing activity, which enters the embryo at the beginning of gastrulation and comes to lie at the anterior extremity of the axis. The trunk organizer gives rise to dorsal mesoderm and induces neural plate along the entire anteroposterior axis of the trunk. The tail organizer is what remains of the organizer after gastrulation. This produces the tail notochord and neural tube by secondary neurulation. The functions of the organizer are summarized below, with references to other Topics for further discussion. A number of genes encoding transcription factors and secreted proteins are expressed specifically in the organizer, and molecular correlates of organizer function are summarized in *Table 1*.

- The organizer initiates the cell movements of gastrulation (Topic F3).
- During gastrulation, the organizer predominantly forms anterior endoderm and dorsal mesoderm (notochord, prechordal plate), but also contributes to the somites, neural floor plate and dorsal endoderm of the gut (Topic K1, Topic J3).
- The organizer dorsalizes adjacent ventral/lateral mesoderm to generate a range of intermediate mesodermal cell types along the dorsoventral axis (Topic K1).
- The organizer dorsalizes the overlying ectoderm to induce the neural plate (Topic J1).
- The anterior endoderm and anterior dorsal mesoderm from the organizer induce the head (Topic J2). **Note:** in mammals, the anterior visceral endoderm is involved in head development (Topic I1).
- The organizer confers anteroposterior positional information on mesoderm and endoderm cells by controlling the pattern of *Hox* gene expression (this has been suggested on the basis of experiments carried out on the chick and mammalian organizers) (Topic K2).
- The organizer confers anteroposterior positional information on the neural plate, through the regional specificity of neural induction (Topic J2).
- The organizer-derived notochord and floor plate provide signals that pattern the dorsoventral axis of the neural tube and the somites (Topic J3).
- Asymmetric signaling and gene expression in the organizer plays an important role in left–right asymmetry (Topic I3).
- After gastrulation, the organizer forms the chordoneural hinge (caudal eminence) and lays down the notochord and neural tube of the tail (Topic J4).

Late organizer inducing centers – organizer maintenance

In chickens and other vertebrates, various experiments have shown that the organizer comprises a continually changing cell population. The inductive properties of the organizer cannot therefore be attributed to the specific characteristics of a dedicated population of organizer cells. Rather, it suggests that the

Table 1. Major classes of molecules expressed in the Xenopus organizer

Class	Examples	Function
Secreted proteins		
BMP antagonists	Chordin, Noggin, Follistatin, Cerberus, Xnr3	Dorsalization of mesoderm (Topic K1) Neuralization of ectoderm (Topic J1) Induction of head (in concert with Wnt antagonists) (Topic J2)
Wnt antagonists	Cerberus, Dickkopf-1, Frzb-1	Induction of head (in concert with BMP antagonists) (Topic J2) Establishment of anteroposterior neuraxis (perhaps in concert with FGF and retinoic acid) (Topic J2)
Antidorsalizing factors	Antidorsalizing morphogenetic protein (ADMP)	Inhibition of ectopic organizer induction (this Topic)
Transcription factors		
Early expression: establishing organizer functions	Siamois, Goosecoid	Siamois activates *goosecoid* gene in concert with a TGF-β-dependent signal (Topic K1). Goosecoid activates the genes for many other organizer proteins.
Later expression: maintaining organizer functions	XANF1, Pintallavis	Required for head and trunk organizer function.
	Otx2, Xlim-1	Required specifically for head organizer function and head morphogenesis. Knockouts of equivalent mouse genes prevent head development (Topic J2).
	Xnot (zebrafish homolog *floating head*)	Required for notochord differentiation as a separate process to its inductive abilities.

Note, Cerberus is both a BMP and Wnt antagonist and is the only protein capable of inducing ectopic head development alone

node is continuously redefined by signals emanating from elsewhere in the embryo. In chickens, this is supported by the remarkable demonstration that ablating the node even during the early stages of gastrulation does not inhibit normal development. Rather, the node regenerates and the new node shows the same characteristic patterns of gene expression as the original.

In chickens, it has been shown that the node is defined and (when necessary) regenerated in response to signals from the middle third of the primitive streak. This has been called the **node-inducing center (NIC).** It has the same molecular properties as the PMZ, but whereas the PMZ is fixed in the posterior of the embryo, the NIC is dynamic and follows the node as it moves anteriorly across the epiblast. The continuous redefinition of the node in this manner raises some important questions, such as how lateral cells surrounding the NIC are prevented from forming organizers. One possible explanation is the secretion of inhibitory signals by the node itself. One of the organizer-specific molecules expressed at this stage is **antidorsalizing morphogenetic protein (ADMP).** The role of an antidorsalizing signal in the organizer has been unclear, since the primary role of the organizer is to dorsalize the surrounding mesoderm and ectoderm. It is possible that this protein acts specifically to restrict the formation of additional organizers by inhibiting the NIC. This would also explain why the

NIC is not adjacent to the node, but maintains a constant distance from it. Cells posterior to the NIC are presumably prevented from forming organizers by other mechanisms, perhaps involving the ubiquitous antidorsalizing proteins, the BMPs.

13 LEFT–RIGHT ASYMMETRY IN VERTEBRATES

Key Notes

The left–right axis	Vertebrates appear bilaterally symmetrical but their internal organs are arranged asymmetrically. The body plan thus has distinct left and right sides. The constancy of this organization indicates that left and right must be specified during development.
Summary of left–right axis formation	The left–right axis is set up by a signaling center called the left–right coordinator, which is established in response to symmetry-breaking cues unique to the different vertebrate classes. This directs asymmetric signaling in the organizer, which in turn induces asymmetric gene expression in the lateral mesoderm, probably by relaying a signal through the paraxial mesoderm. In this way, organ primordia on each side of the body are exposed to different molecular environments.
The *iv* and *inv* genes, and the F-molecule hypothesis	Mutations in two mouse genes, *iv* and *inv*, show laterality defects. Such mutations indicate that the left–right axis is specified intrinsically, i.e. it does not rely on an external signal. The F-molecule hypothesis suggests that laterality could be generated by an asymmetrical molecule aligned with respect to the anteroposterior and dorsoventral axes, causing the accumulation of a particular determinant on one side of the embryo.
Symmetry breaking in mammals – motor proteins and nodal flow	The *iv* gene encodes a microtubule motor protein that appears to control the behavior of monocilia on ventral node cells. In normal embryos, the cilia cause the net flow of nodal fluid to the left, and this could cause a left–right axis determinant to accumulate on one side of the node. In *iv* mutants, the monocilia are inactive and there is no nodal flow.
Symmetry breaking in *Xenopus*	In *Xenopus*, left–right axis specification depends on the cortical rotation of the egg, resulting in the preferential activation of maternally-encoded Vg-1 protein on the left-hand side of the embryo. The injection of *Vg-1* mRNA into the right-hand side of the *Xenopus* blastula induces left–right axis reversal, whereas injection into the left-hand side has no effect.
Left–right asymmetry in the organizer and lateral plate mesoderm	The left–right coordinator may activate or inhibit certain genes on only one side of the organizer. Asymmetric gene expression is seen in the organizer at the early primitive streak stage in chicken embryos, with Sonic hedgehog restricted to the left side and activin restricted to the right. This causes the left-specific expression of the TGF-β proteins Nodal and Lefty2, which play pivotal roles in the development of laterality, and are highly conserved in all vertebrates. The role of other genes such as *lefty1*, expressed in the midline, may be to prevent the spread of laterality signals from one side to the other.
Downstream asymmetric gene expression	Lefty2 and Nodal activate the gene for the transcription factor Pitx2, specifically on the left-hand side of the embryo. The absence of these proteins on the right results in the activation of other transcription factors

such as Snail. These factors persist through to organogenesis, where they may directly regulate the laterality of developing organs.

| **Related topics** | Pattern formation (A6) | The vertebrate organizer (I2) |

The left–right axis

Vertebrates show **bilateral symmetry** for external characters but the internal organization of the body displays significant asymmetry, so that the **left and right sides** are fundamentally distinct. The most overt differences are seen in the structure and positioning of the heart and vasculature, the gut and its associated organs, and the lungs. Such asymmetry may have arisen during evolution to optimize the packaging and functional efficiency of these organs.

Summary of left–right axis formation

The conserved **left–right asymmetry** of the vertebrate body plan requires the establishment and polarization of a **left–right axis** during early development. However, left and right only have meaning in the context of existing antero-posterior and dorsoventral axes, so the left–right axis must be specified either at the same time as these axes or afterwards. The initial process of symmetry breaking appears to be achieved in different ways by different vertebrates, but this leads into a conserved molecular pathway beginning with the expression of particular nodal-related TGF-β proteins specifically in the left hand lateral plate mesoderm. Left–right axis formation has been studied in three principle vertebrate models: *Xenopus*, chick and mouse. The steps to left–right asymmetry can be summarized as shown below (*Fig. 1*).

- Initial symmetry breaking (establishment of the **left–right coordinator**)
- Asymmetric signaling initiated in the organizer
- Induction of asymmetric gene expression in the paraxial mesoderm
- Induction of asymmetric gene expression in the lateral plate mesoderm
- Establishment of midline block to prevent signals leaking into the contra-lateral side
- Asymmetric expression of transcription factors in developing organ primordia.

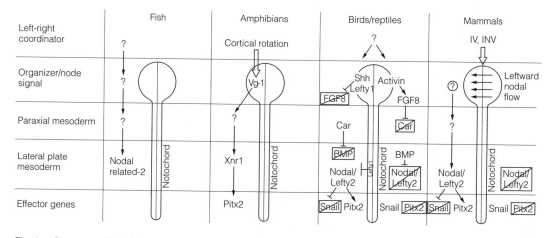

Fig. 1. Summary of left–right axis specification in different vertebrate classes. See text for details.

The *iv* and *inv* genes, and the F-molecule hypothesis

The first clues as to the origins of the left–right axis in vertebrates were found in humans. Approximately one in every 10 000 individuals has the condition **situs inversus**, where left–right asymmetry is abnormal. Individuals with **situs inversus totalis** have perfect reversed asymmetry and are usually healthy. However, in most cases, the axis reversal is incomplete so that there is a discordance of 'handedness' between individual organs (**heterotaxis**) or mirror image duplication (**isomerism**) with resulting cardiopulmonary defects. Some individuals also exhibit heterotaxis or isomerism for individual organs, while the rest of the body is normal.

Single gene mutations that disrupt global left–right polarity have been identified in the mouse, suggesting that initial polarization of the embryo is controlled intrinsically and by a small number of genes. Mice homozygous for the *iv* (*situs inversus viscerum*) mutation show heterotaxis. This suggests that the *iv* gene product sets a bias for polarization, and that in its absence this bias is lost so that the axis is polarized randomly and independently in different organs. Mice homozygous for the recessive *inv* (*inversion of embryonic turning*) mutation show situs inversus totalis, indicating that loss of *inv* function reverses the normal polarization.

A working model of left–right polarization was proposed by Brown and Wolpert. This has been called the **F-molecule hypothesis** because it involves the synthesis of an asymmetrical molecule (the letter F is asymmetrical) that would align with the anteroposterior and dorsoventral axes of the cell and point specifically in one direction. The important concept in this model is that axis specification is intrinsic, i.e. it requires no outside symmetry breaking process, and would thus explain the existence of mutants with left–right polarity defects. The function of the F-molecule could be to act as a pump to force the accumulation of a particular determinant on one side of the cell. However, the F-molecule itself could be expressed bilaterally. The alignment process would involve the cytoskeleton, which has been shown to generate polarized structures in alignment with the axes of a cell, e.g. the mitotic spindle. The *iv* gene could be envisaged as encoding the F-molecule, or at least one of its components. In homozygous *iv* mutant mice, the biasing mechanism would not work and axis polarization might depend on random fluctuations in the level of the determinant on each side of the embryo, which could differ in different parts resulting in heterotaxis. The role of the *inv* gene is less clear from this model, but could be required to attach the F-molecule to the cytoskeleton in the correct way. This model predicts that *iv* would be epistatic to *inv*, i.e. the double mutant would show the *iv* phenotype.

Symmetry breaking in mammals – motor proteins and nodal flow

Several recent breakthroughs, including the cloning of the *iv* gene, the characterization of its product and expression pattern, and ultrastructural functional analysis of the node, have provided exciting supporting data for the F-molecule hypothesis. A link between laterality defects and microtubule motor proteins has been appreciated for a long time because of the coincidence of situs inversus in humans and a respiratory disease called **Kartagener's syndrome** characterized by immotile cilia. The product of the *iv* gene is a microtubule motor protein (Topic C5), called **left–right dynein (LRD)**, and its loss is probably responsible for the immotile cilia syndrome. In early mouse development, LRD is expressed bilaterally in the node, and many cells in the node contain **monocilia**. The function of nodal monocilia has been a point of debate, but it was recently shown using minute fluorescent beads that the beating of the nodal cilia generates a net leftward flow of perinodal fluid, suggesting a mechanism whereby a determinant of the left–right axis could accumulate on one side of the node. In *iv* mutant mice, the nodal monocilia are immotile and there is no net flow of nodal fluid. In *inv* mutant mice, the nodal flow is not reversed as might have been expected, but is slowed. These results suggest that **nodal flow** breaks the initial bilateral symmetry of the mouse embryo, probably by causing the accumulation of a specific determinant on the left-hand side of the node, which would thus constitute the mammalian left–right coordinator. The nature of this determinant is unknown. This model is supported by the analysis of mice with targeted mutations in genes such as *hfh4* (which encodes a transcription factor), *kif3a* and *kif3b* (which encode kinesin motor proteins). In all these mutants, the monocilia of the node are absent, and there are strong laterality defects.

Symmetry breaking in *Xenopus*

In *Xenopus*, the left–right axis is thought to be specified along with the other axes at fertilization, by cortical rotation. As discussed in Topic I1, cortical rotation can be inhibited by UV-irradiation, but development can be rescued by tilting the embryo to mimic the relative rotation of the inner and cortical cytoplasm. Although gastrulation proceeds normally in these rescued embryos, many of the resulting tadpoles show heterotaxis. The left–right axis may be specified by the maternal protein Vg-1, an important functional component of the Nieuwkoop center (Topic I1). If mRNA for the mature form of Vg-1 is injected into the prospective right-hand side of the 16-cell *Xenopus* blastula, the left–right axis is reversed. No axis reversal occurs if the RNA is injected into the left-hand side, suggesting that Vg-1 is preferentially activated on the left in normal development. The left-hand side of the vegetal pole thus acts as the amphibian left–right coordinator. The link between cortical rotation and the specification of the left–right coordinator is unknown.

Left–right asymmetry in the organizer and lateral plate mesoderm

The function of the left–right coordinator is to initiate transient asymmetric signaling in the organizer, which eventually leads to permanent asymmetric gene expression in the lateral plate mesoderm. The role of the organizer in left–right axis formation has been studied extensively in the chicken. Although the initial symmetry-breaking mechanism in birds is unknown, many of the downstream signaling events have been elucidated. The first sign of molecular asymmetry in the chick embryo is the expression of activin and its receptor specifically on the right-hand side. Activin induces the expression of *fgf8* and suppresses the expression of *sonic hedgehog*. Therefore, FGF8 is produced specifically on the right-hand side of the node, while Sonic hedgehog is restricted to the left.

The downstream effect of this asymmetric signaling is the expression of two members of the TGF-β superfamily, ***nodal*** and ***lefty2***, specifically in the left-hand lateral plate mesoderm. In chickens this has been shown to be dependent on Sonic hedgehog (activator) and FGF8 (repressor). Blocking Sonic hedgehog signaling or the ectopic activation of activin signaling on the left abolishes the expression of these genes, while blocking activin signaling or activating Sonic hedgehog on the right results in bilateral expression. In both cases, the embryo shows heterotaxis. Other vertebrates have similar expression patterns for nodal-related genes. The mouse *nodal* and *lefty2* genes are expressed specifically in the left lateral-plate mesoderm, while in *Xenopus*, the nodal-related gene *Xnr1* is also left-specific. In both these species, the upstream signals remain to be characterized and may not be the same as those functioning in birds, as Sonic hedgehog appears to be expressed bilaterally. The *nodal* and *lefty2* genes play a pivotal role in the development of laterality as bilateral expression of either gene causes heterotaxis, whereas expression on the right-hand side alone causes axis reversal. The gene expression patterns are altered predictably by manipulating upstream genes such as *activin βb* and *sonic hedgehog* (chickens) or *iv* and *inv* (mice).

As well as its role in establishing asymmetric gene expression, the organizer also has a very important function in maintaining laterality. Removing the notochord causes downstream genes such as *nodal* and *lefty2* to be expressed symmetrically, resulting in heterotaxis. The notochord and neural floor plate appear to act as a barrier to prevent signals emanating from one side of the embryo diffusing across the midline to the other. The exact structural nature of this barrier is unknown, but many genes expressed in the notochord or floor plate affecting the structure of these tissues generate laterality defects when mutated. Some genes, such as *Hnf3β*, were originally thought to be involved in the specification of the left–right axis but are now thought to play a permissive role by maintaining the integrity of the midline. Other genes, such as *lefty1*, are expressed asymmetrically in the midline and have a specific function in its role as a left–right barrier.

Both Sonic hedgehog and FGF8 have a short range and appear to work indirectly, through a signal secreted by the paraxial mesoderm. One candidate for this signal is **Caronte**, a protein related to Cerberus that is expressed specifically in the left paraxial mesoderm of the chick embryo and directs the expression of *nodal* and *lefty2* through a mechanism involving the suppression of symmetrical BMP signaling. Interestingly, Caronte also maintains *lefty1* expression in the midline, also through the suppression of BMPs, and therefore plays a dual role in left–right axis formation, first by relaying an original laterality signal from the node to the lateral mesoderm, and then by establishing a midline block that prevents left-specific signals leaking into the right hand side of the embryo.

Downstream asymmetric gene expression

In all vertebrates, the function of nodal related proteins in the left-hand lateral plate mesoderm appears to be the activation of a left-specific transcription factor called **Pitx2**. In the absence of *nodal* and *lefty2* expression, right-specific transcription factors such as **Nkx3.2** and **Snail** are activated. While the signaling proteins are only transiently expressed, the transcription factors persist in e.g. the cardiogenic mesoderm and may directly influence the asymmetrical looping of the heart tube. This may involve the differential regulation of genes such as *flectin* and *fibrillin-2*, which are expressed on the left- and right-hand sides, respectively. Both genes encode components of the extracellular matrix.

J1 THE ECTODERM: NEURAL INDUCTION AND THE EPIDERMIS

Key Notes

Specification of the ectoderm

Ectoderm is the outer germ layer of the post-gastrulation vertebrate embryo. In amphibians, the ectoderm is specified in the developing oocyte, perhaps by the absence of mesodermal/endodermal determinants, but is not determined until after the midblastula transition. In birds and mammals, presumptive ectoderm lies anterior to the node and in the lateral epiblast. The ectoderm is not specified until gastrulation commences.

Fate of the ectoderm

The ectoderm gives rise to the neural plate (future central nervous system), epidermis and neural plate border (which forms ectodermal placodes and the neural crest). Neural plate and neural crest form from dorsal ectoderm, while epidermis forms from lateral and ventral ectoderm.

Neural induction

Neural induction is the process by which dorsal ectoderm becomes specified to form the neural plate. Neural induction depends on dorsalizing signals secreted by the organizer. The neurogenic ectoderm begins to express characteristic neural markers and differentiates to form tall columnar cells.

Grafting experiments

The ability of the organizer to induce the neural plate was established by organizer graft experiments in amphibians. Organizer tissue grafted to the ventral side of a host embryo induces the ventral (epidermal) ectoderm to form neural plate. The notochord and prechordal plate, derived from the organizer, can also induce neural plate, as can the neural plate itself (homeogenetic induction). More recent experiments have shown similar neural inducing activity in the organizers of other vertebrates.

Neural induction in *Xenopus* (neural default model)

Xenopus animal cap tissue forms epidermis in isolation but differentiates into neurons when the cells are dissociated. Neural development is therefore thought to be the default fate of the ectoderm, while epidermal differentiation is induced by BMPs secreted by the ectoderm itself. The BMPs are distributed uniformly in the embryo, but their effects are antagonized in the dorsal region by at least five anti-BMPs secreted by the organizer. This results in a morphogen gradient of BMP activity, which may be sufficient to specify the neural plate, neural plate border and epidermis.

Neural induction in birds and mammals

In birds and mammals, the expression patterns of the BMPs and anti-BMPs do not support a role in neural induction, although they may play a role in neural development after induction or in the specification of the neural plate border. The clear results from *Xenopus* may indicate that these earlier induction pathways are already activated in the explants used for induction assays, or that amphibians and amniotes use genuinely distinct processes for neural induction.

Specification of the epidermal ectoderm

In ventral and lateral ectoderm, the ventralizing BMP proteins are not inhibited by the organizer. BMPs inhibit the intrinsic tendency for ectoderm to form neural tissue and induce the epidermal developmental program, which involves the synthesis of transcription factors such as GATA-1 and Vent-1. These transcription factors activate downstream epidermis-specific genes.

Related topics

Mechanisms of developmental commitment (A3)

Signal transduction in development (C3)

Patterning the anteroposterior neuraxis (J2)

Mesoderm induction and patterning (K1)

Endoderm development (K6)

Specification of the ectoderm

The **ectoderm** is the outermost germ layer, which covers most of the metazoan embryo after gastrulation and gives rise primarily to epidermis, neural tissue and (in vertebrate embryos) the neural crest. The molecular basis of ectoderm specification in invertebrates is covered elsewhere (*Drosophila*, Topic H6; *C. elegans*, Topic G3) so the discussion below relates only to vertebrates. Of the three germ layers, least is known about the specification of the ectoderm. In amphibians, germ-layer fates are specified by the distribution of maternal determinants in the egg. Factors involved in the specification of both the mesoderm (Topic K1) and endoderm (Topic K6) have been identified through their ability to induce the expression of appropriate marker genes and/or induce the differentiation of appropriate cell types. The ectoderm may therefore represent a 'default state' where cells do not receive alternative instructions. As discussed below, a number of molecules have been isolated that influence early ectoderm development, but they all appear to control the choice between epidermal and neural fates once the ectoderm is already specified. Although the ectoderm may be **specified** early in development, ectodermal fate is not **determined** until the late blastula stage. Animal cap cells (presumptive ectoderm) grafted to the vegetal hemisphere of an early blastula host will become incorporated into endodermal tissues or give rise to mesoderm. Conversely, cells similarly transplanted from a late blastula stage embryo will form ectodermal tissue at the graft site.

In amniotes, fate-mapping shows that much of the epiblast anterior to the node, as well as much of the lateral epiblast, will give rise to ectoderm. However, the specification and determination of the germ layers occurs at a much later stage compared to *Xenopus*. Even at the early gastrula stage, cells from the ectodermal region of the fate map can form mesoderm or endoderm if moved nearer to the primitive streak, while individual epiblast cells in culture primarily give rise to muscle (see Topic K1 for more discussion on this experiment). This difference in timing between *Xenopus* and amniotes probably reflects the dependence of early *Xenopus* development on maternal factors distributed in the egg and the absence of cell proliferation until after gastrulation. In amniotes, zygotic gene transcription begins during early cleavage and there is extensive cell proliferation in the epiblast layer prior to and during gastrulation, which would allow for greater plasticity in patterning. The ability of different regions of the *Xenopus* blastula and amniote epiblast to be respecified by grafting reflects the remarkable regulative ability of the vertebrate embryo.

Fate of the ectoderm

The ectoderm gives rise to three major components of the vertebrate body: the **neural plate**, the **epidermis** and between them, the **neural plate border**. The neural plate forms from **dorsal ectoderm**, rolls up to form the **neural tube**, and gives rise to the **central nervous system** (brain and spinal cord). The epidermis forms from the **ventral and lateral ectoderm**, and forms the external layer of skin as well as cutaneous structures such as scales, feathers, hairs and claws. At the most anterior level of neural plate, the border region gives rise to **ectodermal placodes** that contribute to the development of certain sensory organs, as well as the amphibian **cement gland**. From the midbrain down, the border region forms the **neural crest**, a versatile tissue producing most of the **peripheral nervous system** as well as a diverse array of other cell types including cartilage, bone, connective tissue and smooth muscle. The neural crest is discussed in greater detail in Topic J6.

Neural induction

Neural induction is the process by which the **neural plate** arises from ectoderm on the dorsal surface of the vertebrate embryo. The neural plate later rolls up to form the neural tube (as discussed in Topic J4) and eventually develops into the brain and spinal cord. The ectoderm is induced to form the neural plate during gastrulation by the underlying dorsal mesoderm of the **organizer** (Topic I2). The organizer is said to **dorsalize** or **neuralize** the overlying ectoderm, both terms referring to the same inductive process.

Neural induction begins with the specification of **neural ectoderm (neurogenic ectoderm, neurectoderm)**. These cells express a number of **neural markers**, and undergo morphogenetic changes causing them to elongate to form a columnar epithelium. In this way, the neural ectoderm is raised above the surrounding epidermal ectoderm to form the neural plate. Early markers of neural commitment include genes encoding the transcription factors **Sox2** and **Sox3**. A marker expressed later, when the neural plate begins to differentiate, is the cell adhesion molecule **N-CAM**.

Grafting experiments

The ability of dorsal mesoderm to induce overlying ectoderm to form neural plate was established by the **organizer graft experiments** of Spemann and Mangold in 1924 (*Fig. 2*, Topic I2). When the amphibian organizer (the dorsal lip of the blastopore) is grafted onto the ventral side of a host embryo, a secondary axis is induced and the host ectoderm adjacent to the graft (originally specified as epidermal ectoderm) is converted to neural plate. Some tissues derived from the organizer, specifically the **notochord** and **prechordal plate**, also retain this neural-inducing activity. Interestingly, once the neural plate has formed, the neural plate itself is also able to induce ectoderm to form neural plate, a phenomenon termed **homeogenetic induction** (Topic A3).

Much more recently, grafting experiments with the zebrafish embryonic shield and Hensen's node of the chicken embryo have shown that these structures possess organizer activity similar to the Spemann organizer of *Xenopus*. In addition, cross-species transplantation of organizer tissue reveals that at least some of the molecular signals emanating from the organizer are conserved among all vertebrates. Experiments carried out in the 1960s showed that placing a permeable barrier between the organizer and the overlying ectoderm did not stop neural induction, even though cell contact was prevented. Neural induction therefore involves diffusible molecules secreted by the organizer. However, the identity of these molecules has only been established in the last decade.

Neural induction in *Xenopus* (neural default model)

When animal cap tissue from the *Xenopus* gastrula-stage embryo is explanted and cultured in isolation, it differentiates into epidermis. It was therefore thought originally that the default developmental pathway for ectoderm was epidermal, and that **instructive signals** from the organizer were required for neural induction. However, if the animal caps are dissociated in culture, the individual cells differentiate into neurons. This indicates that the default developmental pathway is in fact neural (the **neural default model**) and that epidermal fate arises from a **community effect** (Topic A6) in which molecules produced by the ectoderm itself suppresses neural development. Isolated cells do not produce these molecules in sufficient amounts to prevent neural differentiation, but a community of cells does. The role of the organizer is to inhibit these molecules and uncover the latent neural potential of the ectoderm (*Fig. 1*).

The neural differentiation of ectoderm cells is inhibited by members of the **bone morphogenetic protein (BMP)** family of signaling proteins. The key players in this system are **BMP2, BMP4** and **BMP7**. For example, isolated animal cap cells exposed to BMP4 differentiate into epidermis rather than neural tissue, and if BMP4 is forcibly expressed in dorsal blastomeres of the *Xenopus* embryo, gastrulation is prevented and the resulting embryo is ventralized (i.e. it lacks neural tissue as well as dorsal mesoderm structures such as the notochord and somites). Conversely, if the BMP signaling pathway is interrupted by expressing dominant negative BMPs or BMP receptors, or antisense *bmp* RNA (Topic B2), all ectodermal cells become neural and the whole embryo is dorsalized. BMPs play a fundamental role in specifying ventral cell fates in the *Xenopus* embryo in all three germ layers (see Topics I2, K1 and K6).

In normal *Xenopus* development, the effect of BMP signaling is opposed by dorsalizing signals from the organizer. The organizer secretes at least five anti-

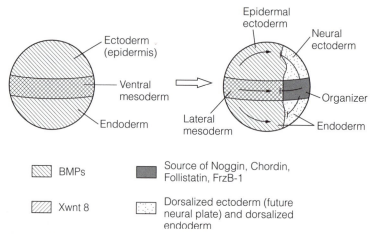

Fig. 1.   Neural induction in Xenopus *by molecules secreted from the organizer.*

BMP proteins with neural inducing activity: **Noggin, Chordin, Follistatin, Cerberus** and **Xnr3**. The first four of these proteins bind directly to multiple BMPs and do so with greater affinity than the corresponding BMP receptors, therefore competitively inhibiting BMP activity. Xnr3 (*Xenopus nodal*-related 3) is a signaling protein of the TGF-β superfamily, and is therefore related to the BMPs. Xnr3 is thought to inhibit BMP signaling by competing for the BMP receptors or shared intracellular signal transduction proteins (SMADs; see Topic C3). This competitive inhibition of BMPs by proteins secreted by the organizer results in a **gradient of BMP activity** from ventral to dorsal in the embryo. In the dorsal ectoderm where there is low BMP activity, the neural plate is induced. In the ventral and lateral ectoderm where there is high BMP activity, the epidermis forms. Between these extremes, where BMP activity is moderate, the **neural plate border** is specified (Topic J6). The BMPs therefore appear to act as **morphogens** in the early *Xenopus* embryo, specifying different cell fates at different concentrations (Topic A3). The BMPs are initially expressed throughout the ectoderm, consistent with their role as suppressors of neural differentiation. However, when the organizer forms, the BMPs are restricted to the ventral (non-neural) ectoderm. This is because BMP signaling is required to activate the *bmp* genes (positive feedback), so the inhibition of BMP activity by the organizer causes the domain of *bmp* gene expression to regress.

Neural induction in birds and mammals

In *Xenopus* (and zebrafish) the spatial and temporal expression patterns of the BMPs and their inhibitors fit well with their proposed roles in the neural default model of induction. However, in birds and mammals, the expression patterns are more complex and the molecules themselves do not appear to function in the same way as their amphibian counterparts.

In chickens, BMPs are not expressed in the ectoderm prior to neural induction, so there is no requirement for anti-BMPs in the organizer to inhibit them. Concordantly, while Noggin, Chordin and Follistatin are all expressed in the chicken node, they are not detected until after neural induction. Furthermore, none of these molecules shows neural inducing activity when expression is forced in non-neural ectoderm, and BMPs ectopically expressed in the neural plate do not induce the formation of epidermis. In mice, the genes encoding several of the BMPs and anti-BMPs have been knocked out, but the neural plate forms normally.

It therefore seems likely that other signals, which are not anti-BMPs, are required for initial neural induction in amniotes, while the BMPs and anti-BMPs may play a later role. For example, this is supported by experiments in chickens showing that Chordin can induce neural plate in ectoderm that has already been briefly exposed to the node. In *Xenopus*, there is conflicting evidence concerning the requirement of signals in addition to the anti-BMPs for neural induction. Firstly, it is thought that the animal cap tissue typically used for neural induction assays has already received some weak patterning signals, as dorsal and ventral explants show differing biases towards neural and epidermal fates. Secondly, if the FGF pathway is inhibited (e.g. by expressing a dominant negative FGF receptor) Noggin and Chordin are no longer able to neuralize animal cap tissue in culture. However, inhibiting the FGF signaling pathway in whole embryos does not prevent formation of the neural plate, and animal caps from UV-treated embryos (which never form an organizer) can still yield neural plate when treated with the anti-BMPs.

The success of cross-species neural induction by organizer transplants suggests that the fundamental mechanisms of neural induction *are* conserved among the different vertebrate classes. The major differences in the molecular basis of neural induction perceived between amphibians and amniotes, as well as the conflicting results from studies in *Xenopus*, may therefore represent artifacts of the assays used. In particular, when neural inducers are applied to ectoderm explants, it is difficult to determine whether the ectoderm is truly naïve or has already received some form of patterning information. Furthermore, the trauma caused by explanting tissue, or surgical manipulation *in vivo*, may result in the activation of unknown signaling pathways that have unpredictable effects.

An interesting suggestion is that the *in vivo* role of the BMPs and anti-BMPs may not be to induce the neural plate, but to stabilize the neural plate domain after induction and specify the position of the neural plate border, which gives rise to the neural crest. In amniotes, the BMPs are first expressed in the presumptive epidermal ectoderm adjacent to the neural plate, and changing the level of BMP activity affects the size of the neural plate and the position of the border. In *Xenopus*, it is possible that the entire neural ectoderm is initially specified as neural plate border and that the neural plate is specified later as *bmp* gene expression is eliminated from the dorsal region of the embryo. Specification of the neural plate border is discussed in more detail in Topic J6.

Specification of the epidermis

In the ventral and lateral ectoderm, where the organizer's secreted factors are absent, the ventralizing signals BMP2, BMP4 and BMP7 activate the epidermal development program (*Fig. 1*). In *Xenopus*, several components of this pathway have been identified. BMP signaling through type I and type II receptors is required to phosphorylate the type I receptor and activate the intracellular signaling molecule Smad1. This associates with Smad4 and the heterodimer translocates to the nucleus, where it activates a number of genes encoding **ventral fate transcription factors**, e.g. Vent-1 and GATA-1. These may in turn activate epidermal-specific genes in the ectoderm, such as the keratin genes. Keratin is a major component of the cornified outer layer of the epidermis, and of cutaneous structures such as hairs, nails and feathers.

J2 PATTERNING THE ANTEROPOSTERIOR NEURAXIS

Key Notes

The anteroposterior neuraxis	The anteroposterior polarity of the vertebrate central nervous system is apparent in the morphological division of the neural tube into a series of three anterior vesicles that form the forebrain, midbrain and hindbrain, and a narrow posterior tube that forms the spinal cord. There are also various marker genes expressed at different positions along the axis, which can be used to characterize neural plate tissue in culture.
Regional specificity of neural induction	The regional specificity of neural induction was first demonstrated by organizer graft experiments. Early/anterior dorsal mesoderm induces secondary heads, whereas late/posterior dorsal mesoderm induces secondary trunks. This indicates that different parts of the organizer induce regionally specific types of neural plate along the anteroposterior axis, providing evidence for multiple neural induction signals.
The activation–transformation model	In the activation–transformation model, it is proposed that the regional specificity of neural induction arises from just two signals from the organizer. The first (activation) signal is found throughout the organizer and induces neural plate of a default anterior character. The second (transformation) signal is restricted to the posterior region of the embryo and forms a concentration gradient that gradually posteriorizes the neural plate as its concentration rises. The anti-BMPs secreted by the organizer are good candidates for the activation signal. FGFs, Wnts and retinoic acid may all form components of the transformation signal.
Planar signals	Signals traveling in the plane of the ectoderm may also be involved in neural induction. If the organizer and overlying ectoderm are removed from an embryo and cultured as an elongated explant so that the ectoderm only contacts the mesoderm at one edge, neural plate is induced with a graded sequence of marker gene expression revealing an anteroposterior axis. Signals diffusing through the ectoderm from the organizer may therefore cooperate with those emanating vertically from the organizer-derived mesoderm to pattern the anteroposterior neuraxis.
Induction of the head	In amphibians, the head is induced by the most anterior organizer-derived tissue, which gives rise to pharyngeal endoderm, prechordal mesoderm and anterior notochord. This tissue secretes three potent head inducing molecules, Cerberus, Frzb-1 and Dickkopf-1, which antagonize Wnt signaling. These proteins can induce extra heads when expressed in combination with the anti-BMPs Noggin, Chordin or Follistatin, suggesting head development requires simultaneous inhibition of BMP and Wnt signal transduction pathways.

Molecular basis of activation and transformation	Anti-BMPs secreted by the organizer can neuralize the ectoderm, but produce neural plate of homogeneous anterior character. Thus, anti-BMPs probably represent the activation signal. The head organizer secretes anti-BMPs and anti-Wnts, both of which are required for head induction. The anti-Wnts may also diffuse away from the head organizer, establishing an anteroposterior gradient of an 'anti-transformation signal' to generate an opposite gradient of posteriorizing Wnt activity. FGFs can also posteriorize the neural plate and some FGFs are localized in the posterior mesoderm. The transformation signal may therefore comprise a number of cooperating signals, including FGFs and Wnts.
Specification and patterning of the midbrain	A number of transcription factors such as Otx2 and Lim1 are expressed specifically in the head organizer, and play important roles in head development by specifying the forebrain and midbrain. Deletion of the corresponding genes produces embryos with no heads. Positional identity in the forebrain and midbrain is controlled by the overlapping expression patterns of *Otx* and *Emx* homeobox genes, which help to establish the isthmic midbrain organizer at the mes/met boundary. This establishes a gradient of the transcription factor Engrailed, which is directly responsible for patterning in the midbrain-derived optic tectum.
Patterning the hindbrain	The hindbrain is transiently divided into a series of developmental compartments called rhombomeres. Boundaries form between odd- and even-numbered rhombomeres probably involving an adhesive differential. Many genes are involved in establishing and patterning the rhombomeres. Genes such as *Krox-20* and *kreisler* are expressed in specific domains of the pre-segmental hindbrain and may specify rhombomeres as 'odd' or 'even'. The compartment boundaries are stabilized by interactions between cells in alternating rhombomeres expressing Eph class receptor tyrosine kinases and their ligands.
Positional identity in the hindbrain and spinal cord	Regional specification in the hindbrain and spinal cord is conferred by overlapping, posteriorly-nested patterns of *Hox* gene expression. In the hindbrain it appears that the positional values of cells are fixed at the neural plate stage and that migrating neural crest cells with this positional information impose positional identities on the surrounding tissues. Conversely, positional identity in the spinal cord is imposed by signals from the surrounding paraxial mesoderm. The opposite roles of the neural ectoderm and paraxial mesoderm in the brain and spinal cord suggest fundamental differences in the development of the head and trunk.
Related topics	Pattern formation (A6) Homeotic selector genes and regional specification (H5) The ectoderm: neural induction and epidermis (J1)

Patterning the dorsoventral
  neuraxis (J3)
Neurogenesis (J5)
Mesoderm induction and
  patterning (K1)

The anteroposterior neuraxis

The anteroposterior polarity of the vertebrate nervous system is apparent from the earliest stages of neurulation due to the greatly expanded anterior region of the neural plate, which gives rise to the brain. After the closure of the neural tube, this anterior region becomes partitioned into a series of three **primary vesicles** that eventually form the forebrain, midbrain and hindbrain, while the posterior region forms a narrow tube that becomes the spinal cord (*Fig. 1*). Marker genes expressed at specific positions along the anteroposterior axis allow the regional identification of neural plate tissue in experimental studies. These markers include *Otx2* (forebrain) *engrailed-2* (midbrain), *Krox-20* (hindbrain) and the various *Hox* genes (hindbrain and spinal cord).

A crude anteroposterior neuraxis is established at the beginning of neurulation through the regional specificity of inductive signaling. The organizer can be divided into structural, functional and molecular domains corresponding to a head organizer, a trunk organizer and a tail organizer, each with specific inductive properties (Topic I2). As the nervous system develops, its anteroposterior pattern is progressively refined, although different regions of the neural tube utilize very different patterning mechanisms. A distinct signaling center called the **isthmic organizer** controls patterning at the midbrain/hindbrain boundary, whereas the hindbrain itself is patterned on the basis of its division into eight developmental compartments called **rhombomeres**. Positional identity in the neural tube is controlled by the overlapping expression patterns of *Otx* and *Emx* genes (forebrain, midbrain) and *Hox* genes (hindbrain, spinal cord) all of which encode homeodomain-class transcription factors. The positional identity of cells in the hindbrain is established early in development and appears to be an autonomous property of the neural tissue, which is imposed on the surrounding mesoderm and ectoderm by the emigrating neural crest cells (Topic J6). Conversely, the positional identity of cells in the posterior neural tube is established much later, and appears to be determined by signals emanating *from* the adjacent mesoderm.

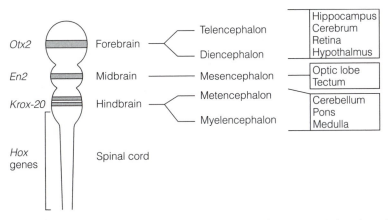

Fig. 1.   *Regional diversifacation of the neural tube along the anteroposterior axis, and associated marker genes.*

Regional specificity of neural induction

The organizer graft experiment published by Spemann and Mangold in 1924 showed that organizer tissue from an early newt gastrula could induce a complete secondary axis when grafted onto the ventral side of a host embryo. This demonstrated the ability of organizer tissue to recruit surrounding cells to form axial structures and to induce adjacent ectoderm to form neural plate. Further experiments carried out in the 1930s demonstrated the **regional specificity of neural induction** (*Fig. 2*). Organizer tissue from young salamander

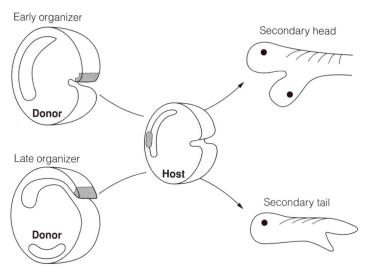

Fig. 2.   Grafting experiments show the regional specificity of neural induction.

gastrulae, which would normally give rise to the most anterior structures of the axis (anterior endoderm, prechordal plate), was shown to induce secondary heads in host embryos. Conversely, organizer tissue from older gastrulae, which would normally give rise to notochord, was shown to induce secondary trunks or tails. This regionally-specific inductive capability was also inherited by the axial mesoderm derived from the organizer. Grafts of dorsal mesoderm from different positions along the anteroposterior axis of a newt embryo that had just finished gastrulation were placed into the blastocoels of host embryos that were just starting gastrulation. Each of the grafts was able to induce neural plate, but the induced neural tissue was representative of the original position of the graft. The most anterior grafts induced head structures, while progressively more posterior grafts induced hindbrain, trunk and tail. More recently, the same types of experiment have been conducted in *Xenopus* and chickens with similar results.

The activation–transformation model

The experiments described above suggest that multiple signals are secreted by the organizer, each with the ability to induce a different region of the neural plate. The simplest interpretation of these results is the **activation–transformation model**. In this model, an **activation signal** is secreted throughout the dorsal mesoderm and induces ectoderm along the entire axis to form a default-state anterior-like neural plate tissue. A second **transformation signal** would then posteriorize the neural tissue. This signal would form a concentration gradient with the highest concentration in the posterior mesoderm (*Fig. 3a*).

In *Xenopus*, the secreted factors Noggin, Chordin and Follistatin are strong candidates for the activating signal, as they can neuralize cultured animal cap tissue and induce the expression of anterior neural markers. However, when injected into the ventral blastomeres of a blastula-stage *Xenopus* embryo, these three proteins are able to induce secondary trunks but not heads. This indicates that head induction requires additional signals, expressed specifically in the most anterior organizer-derived cells (see below).

There are several candidates for the transformation signal. **Fibroblast growth factors (FGFs)** and **Wnts** may play roles in posteriorizing the neural plate, since

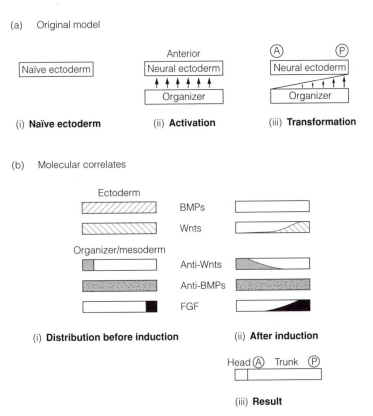

Fig. 3. (a) A simplified representation of the original 'activation-transformation' model of amphibian neural induction, and (b) molecular correlates in the organizer and surrounding tissue.

treating animal cap explants with various FGFs and Wnts (or β-catenin, which is activated by Wnt signaling) in addition to Noggin, Chordin or Follistatin, causes the induced neural tissue to express more posterior markers. It is not clear which molecules in particular are responsible for the transformation signal *in vivo*, although FGF4 and several Wnts are expressed in the posterior dorsal mesoderm of the *Xenopus* embryo at the appropriate stage. Wnts are particularly attractive candidates for the transformation signal because of the presence of anti-Wnt proteins in the most anterior region of the organizer (see below). Finally, **retinoic acid** can also posteriorize the neural plate, by altering the pattern of *Hox* gene expression. The role of *Hox* genes in anteroposterior patterning is discussed later in this Topic.

Planar signals   There is evidence that patterning the neuraxis may also involve signals travelling in the plane of the ectoderm. An explant containing organizer tissue and ectoderm can be removed from an early *Xenopus* gastrula and maintained in an elongated state so that the ectoderm remains in contact with the dorsal lip of the blastopore but never comes to lie adjacent to the mesoderm as it would in normal development (*Fig. 4*). The ectoderm cultured in this way differentiates into neural tissue and expresses positional neural markers in the correct order, with the most anterior markers furthest from the organizer. Furthermore, treatment with posteriorizing agents (FGF and Wnts) abolishes the expression of the

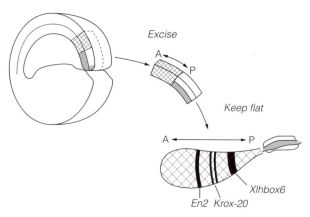

Fig. 4.   The role of planar signals from the organizer can be seen by the correct expression of anteroposterior neural markers in an elongated explant. See text for details.

anterior neural markers. This suggests that a signal is diffusing from the mesoderm through the plane of the ectodermal sheet and establishing a concentration gradient that controls positional identity, and that the signal responds to the same factors that control anteroposterior patterning in the whole embryo. It is therefore likely that both planar signals diffusing though the ectoderm from the point of contact with the blastopore lip, in combination with signals from the underlying mesoderm, act to pattern the anteroposterior neuraxis.

Induction of the head

As discussed above, treating explanted animal cap tissue with Noggin, Chordin and Follistatin induces neural tissue expressing anterior markers. However, the forced expression of these proteins in ventral blastomeres of the *Xenopus* blastula induces a second axis lacking a head. This indicates that head induction *in vivo* requires additional factors.

The head is induced by the first cells to migrate through the organizer, and these cells form the anterior endoderm and prechordal plate. In *Xenopus*, several proteins with potent head-inducing activity, including **Cerberus**, **Dickkopf-1** and **Frzb-1**, are secreted by these tissues. The *cerberus* gene is expressed in the anterior endoderm alone, *frzb-1* is expressed in the prechordal plate and anterior notochord, while *dickkopf-1* is expressed in all these tissues. All three molecules are potent inhibitors of Wnt signaling showing that head induction involves inhibition of the Wnt signal transduction pathway.

A clue to the role of Wnt signaling in patterning the anteroposterior neuraxis comes from manipulating the expression of Xwnt8, a Wnt protein expressed throughout the marginal layer (presumptive mesoderm) in the *Xenopus* gastrula. Overexpression of Xwnt8 generates headless embryos with no notochord but more muscle than usual, while dominant negative *Xwnt8* genes generate embryos with large heads and notochords, but little muscle. This confirms that Wnt proteins ventralize and posteriorize the embryonic axis, and that the specification of the most anterior dorsal structures requires Wnt signaling to be blocked[1].

[1]Note that the role of Wnts at this stage of development is to antagonize the activity of the organizer. However, earlier in development, the Wnt signal transduction pathway is required for organizer induction (Topic I1). The same signal transduction pathways are used over and over again in development, but have different (sometimes opposite) effects at different stages and in different cells.

However, while the inhibition of Wnt signaling can enlarge the head, it is not sufficient for the induction of extra heads. This requires the inhibition of both Wnt and BMP signal transduction pathways simultaneously. The overexpression of *dickkopf-1* or *frzb-1* in the *Xenopus* embryo causes the head to be bigger than normal, but overexpression of either of these genes in combination with an anti-BMP gene such as *noggin* or *chordin* leads to the growth of supernumerary heads. Remarkably, *cerberus* expression can accomplish head development in the absence of Noggin or Chordin, because the Cerberus protein inhibits both Wnts and BMPs. Homologs of *cerberus*, *dickkopf-1* and *frzb-1* have been identified in birds and mice and appear to have similar functions. In mammals, these molecules are present in the anterior visceral endoderm, which acts as a signaling center separate from the node to induce head development (Topic I1).

Molecular basis of activation and transformation

The molecular information discussed above can now be presented in the context of the activation–transformation model of neural induction (*Fig. 3b; Table 1*). In the early blastula, BMPs and Wnts are expressed widely. When overexpressed,

Table 1.   Neural induction: molecular components of the activation–transformation model

Xenopus gene	Function
Genes expressed in the organizer	
noggin	Activating signal Inhibits BMPs
chordin	Activating signal Inhibits BMPs
follistatin	Activating signal Inhibits BMPs
Xnr3	Activating signal Competes for BMP signal transduction components
cerberus	Head induction, anti-transformation signal Inhibits BMPs, Nodal and Xwnt8
dickkopf-1	Head induction, anti-transformation signal Inhibits Xwnt8
frzb-1	Head induction, anti-transformation signal Inhibits Xwnt8 and Xwnt1
Genes expressed outside the organizer	
bmp2	Specifies lateral and ventral cell fates. Inhibits neural induction.
bmp4	Specifies lateral and ventral cell fates. Inhibits neural induction.
bmp7	Specifies lateral and ventral cell fates. Inhibits neural induction. Acts upstream of *bmp4*
Xwnt8	Ventralizes and posteriorizes mesoderm. Inhibits neural induction and head induction. Probably a component of the transformation signal.
fgf4	Posteriorizes neural plate. Component of transformation signal?

both types of signaling molecule have been shown to prevent neural induction and generally ventralize and posteriorize the embryo, so the role of the organizer is to antagonize these molecules. The activation signal is represented by anti-BMP proteins synthesized in the organizer (Noggin, Chordin, Follistatin and Xnr3). Although not all of these proteins are made throughout the organizer, together they provide broad coverage, facilitating the inhibition of BMP signaling throughout the dorsal ectoderm to induce the neural plate. The neural plate induced by these proteins is characteristic of the hindbrain (anterior trunk). The transformation signal may be made up of several components, including Wnts and FGFs. Xwnt8 is a known posteriorizing agent that is initially active throughout the mesoderm, but the presence of anti-Wnts in the head organizer could establish a posterior to anterior gradient of Wnt activity that would posteriorize the trunk neural plate in a graded manner. Such a gradient could be established by diffusion of anti-Wnts through the mesoderm generating an 'anti-transformation signal'. A high concentration of anti-Wnts and anti-BMPs is required in the head organizer, because both signaling pathways must be blocked simultaneously to induce head neural plate (forebrain and midbrain).

Specification and patterning of the midbrain

A number of genes encoding transcription factors are expressed specifically in the anterior mesoderm, including *goosecoid*, *lim-1* and *Otx2*. In mouse embryos, if either *lim1* or *Otx2* is knocked out, the embryo lacks a forebrain and midbrain, while the rest of the nervous system is normal, indicating a specific role for these transcription factors in forebrain and midbrain specification. Following neural induction, these same transcription factors are expressed in the equivalent regions of the neural plate. It is possible that this transfer of expression may contribute to the ability of neural plate to induce more neural plate (homeogenetic induction; Topic A3).

We discuss below how positional identity in the hindbrain is controlled by the overlapping expression patterns of *Hox* genes, which are homologous to the HOM-C genes of *Drosophila* and are expressed in a remarkably similar pattern. While *Hox* genes are not expressed in the forebrain and midbrain, other homeobox genes of the *Otx* and *Emx* families are expressed in overlapping patterns and may play an equivalent role. The homologous *Drosophila* genes *orthodenticle* and *empty spiracles* are expressed in the most anterior segments of the developing fly larva and are required for head development (Topic H3). In *Drosophila*, the *orthodenticle* and HOM-C gene expression patterns overlap in the head. Conversely, in vertebrates, there is a gap between the *Otx/Emx* and *Hox* expression domains at the midbrain–hindbrain boundary, an important source of patterning signals.

The midbrain (or mesencephalon) gives rise to the optic lobes and tecta, while the adjacent metencephalon, which is derived from rhombomere 1 of the hindbrain, becomes the cerebellum (*Fig. 1*). At this stage of development, the **isthmus** that corresponds to the midbrain–hindbrain boundary (or **mes/met boundary**) is a specialized ring of cells with the properties of an organizer. If tissue from this boundary region is grafted to a more anterior part of the midbrain, it will induce tectal development with mirror-image symmetry to the true tectum. Similarly, if grafted into a more posterior part of the metencephalon, it will induce mirror image cerebellar development. This midbrain **isthmic organizer** does not correspond to a structural boundary in the brain, but forms at the posterior boundary of *Otx2* gene expression, which is also the anterior boundary of another homeobox gene, *Gbx2* (*Fig. 5*). There is mutual negative interaction between these genes to establish a sharp boundary of expression.

(a)   Homeobox gene expression in the brain

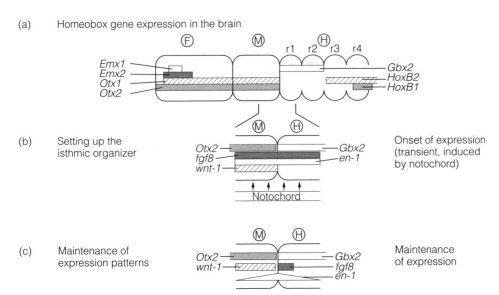

(b)   Setting up the
       isthmic organizer

(c)   Maintenance of
       expression patterns

Fig. 5.   Patterning the brain. (a) Expression patterns of homeobox genes in the brain. (b) Genes involves in establishing the isthmic organizer of the midbrain. (c) Genes involved in the reciprocal signaling loop that maintains midbrain patterning.

The establishment of the isthmic organizer requires a complex genetic cascade involving the transcription factors Engrailed-1 and Engrailed-2, as well as the signaling proteins FGF8 and Wnt1. The *engrailed* genes are initially activated in a broad domain spanning the mes/met boundary in response to inductive signals from the underlying mesoderm. The *fgf8* gene is activated by Engrailed and Gbx2 but repressed by Otx2, so that *fgf8* expression becomes restricted to a ring of cells on the hindbrain side of the boundary. FGF8 induces adjacent cells on the midbrain side of the boundary to express Wnt-1. The *engrailed*, *wnt-1*, *fgf8* and *Gbx2* genes maintain each other's expression through positive interactions, and this results in a gradient of Engrailed protein over both the mesencephalic and metencephalic vesicles with the highest concentration at the mes/met boundary. This gradient is instrumental in patterning the midbrain, and is probably directly responsible for the anteroposterior gradient of proteins such as ELF-1, which are used by axons projecting from the retina to generate a retino-tectal map (Topic J7).

Patterning the hindbrain

The hindbrain is the only part of the nervous system to show overt segmentation during development. The hindbrain becomes divided into eight segments, which are called **rhombomeres**, along the anteroposterior axis, each rhombomere corresponding to an individual developmental compartment (Topic A5). The overt segmentation of the rhombomeres disappears later in development, but certain adult structures derived from the hindbrain (e.g. the cranial nerves and peripheral nerve ganglia) retain a segmental arrangement. Furthermore, motor neurons and other cell types show metameric organization within this region of the brain.

Clonal analysis has confirmed that the rhombomeres are true developmental compartments. If neural plate cells are labeled with a fluorescent dye before the rhombomere boundaries become apparent, the descendants of each cell can span a boundary and populate two adjacent rhombomeres. However, if cells are labeled after the rhombomeres become apparent, the descendants of each labeled cell are restricted to one rhombomere. The compartmentalization of the hindbrain results from interactions between cells at the boundaries of odd and even-numbered rhombomeres. If explants are taken from one odd- and one even-numbered rhombomere and combined, a new boundary forms between them. Conversely, if explants are taken from two odd rhombomeres or two even rhombomeres, the cells mix together where the explants meet, and no boundary is generated. A number of genes have been identified whose products function in setting up rhombomere boundaries. One of the earliest acting genes is *Krox-20*, which is expressed in two stripes representing the future rhombomeres 3 and 5. If this gene is mutated, no boundaries form in this region of the hindbrain and rhombomeres 2–6 develop as a single, unsegmented structure. Other genes with similar mutant phenotypes have been identified, and these are expressed in alternative domains. For example, the gene *kreisler* is expressed in a domain representing future rhombomeres 5 and 6. The *kreisler* and *Krox-20* genes encode transcription factors. It is possible that overlapping patterns of these and other such genes specify cells in each region to behave as 'even' or 'odd' (*Fig. 6*).

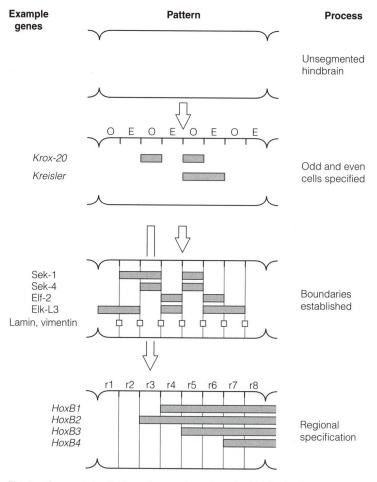

Fig. 6. Sequential activation of genes to pattern the hindbrain. O and E are odd and even rhombomeres.

The mechanism of segmentation itself is not understood although the morphogenetic changes probably involve the generation of an adhesive differential between odd- and even-numbered rhombomeres, as occurs during the formation of parasegment compartments in *Drosophila* (Topic H4). An important signaling pathway has been identified that may play a role in refining the segmental boundaries, and this involves receptor tyrosine kinases of the Eph family and their ligands, the ephrins. Several genes encoding the receptor tyrosine kinases are expressed in subsets of odd-numbered rhombomeres while the genes encoding the corresponding ligands are expressed in the even-numbered rhombomeres. The rhombomere boundaries are therefore stabilized by interactions between cells expressing the ligand and cells displaying the receptor (*Fig. 6*).

Positional identity in the hindbrain and spinal cord

Segmentation in the hindbrain generates eight structurally equivalent rhombomeres, but each has a unique function depending on its position. The cranial neural crest is a particularly useful model system because crest cells from each rhombomere migrate to different pharyngeal arches and give rise to a unique set of structures. In the chicken embryo, crest cells from rhombomeres 2, 4 and 6 migrate specifically to pharyngeal arches 1, 2 and 3/4, respectively. Most of the neural crest cells from rhombomeres 3 and 5 undergo apoptosis, while the remainder join the migrating crest cells from the adjacent, even-numbered rhombomeres (Topic J6). Grafting experiments in the chicken have shown that, by the time the rhombomeres have formed, the fate of the neural crest is already determined. Thus, if crest cells from rhombomere 2 are removed and replaced with crest cells from rhombomere 4, the grafted cells will migrate to pharyngeal arch 1 but form the structures appropriate for pharyngeal arch 2. Instead of forming the dermal bones of the jaw, the first pharyngeal arch gives rise to duplicated parts of the temporal and hyoid bones which are the normal skeletal derivatives of pharyngeal arch 2 (see Topic J6 for a detailed discussion of the neural crest in head development).

The positional identity of the cranial neural crest cells is probably established by the expression of *Hox* genes in the neural tube. Just before the rhombomeres become identifiable, *Hox* genes from the 3′ end of each *Hox* cluster (paralogous subgroups 1–3) begin to be expressed in the hindbrain (*Fig. 6*). As the rhombomeres take shape, the expression patterns of the *Hox* genes are refined, generating sharp anterior expression boundaries that correspond to the boundaries between rhombomeres. The *Hox* genes have characteristic overlapping expression patterns that are nested in the posterior, with successive paralogous subgroups showing two-rhombomere periodicity (*Fig. 6*). As discussed in Topic A6, such a nested pattern of *Hox* gene expression is sufficient to confer unique positional identities on each pair of rhombomeres. Similar *Hox* patterning is seen in the somites (Topic K3) the limb bud (Topic M2) and in *Drosophila* embryos (Topic H5).

Evidence that the *Hox* genes are indeed involved in establishing positional identify in the hindbrain has come from mouse gene knockout experiments. For example, the expression domain of *HoxA2* extends posteriorly from a sharp boundary at rhombomere 3, covering those neural crest cells that will populate pharyngeal arches 2, 3 and 4, but not those that will populate pharyngeal arch 1. Correspondingly, the mouse *hoxa2* knockout mutant shows skeletal defects in the structures derived from pharyngeal arches 2, 3 and 4, including evidence of a partial homeotic transformation resulting in the duplication of pharyngeal arch 1 derivatives. However, the phenotypes of other mutants are more complex and suggest that, unlike the situation in *Drosophila*, the *Hox* genes may also play a role in the actual process of segmentation itself, not just the patterning of segments that have already been set up. For example, knocking out the mouse *HoxA1* gene results in the deletion of rhombomere 5 and most of rhombomere 4 (as well as other more specific defects). This would not be expected if the role of the *Hox* genes was simply to confer positional identity on the rhombomeres. *Hox* gene expression in the hindbrain is controlled in part by genes such as *Krox-20* and *kreisler*. Krox-20 has been shown to directly regulate *HoxA2* and *HoxB2*, which are expressed strongly in rhombomeres r3 and r5. The *Hox* genes have complex modular regulatory elements and are therefore probably regulated by numerous different transcription factors.

There is no overt segmentation in the neural tube posterior to the hindbrain, but the distribution of cell types is not uniform, indicating the existence of an underlying pattern. For example, motor neurons along the anteroposterior axis arise in different positions with respect to the dorsoventral axis, and express different combinations of the LIM homeodomain transcription factors that control their axonal targeting specificity. Segmentation is also revealed by the repetitive organization of ventral motor nerves and dorsal root ganglia. Positional identity in the spinal cord is, like that in the hindbrain, thought to depend on the expression of *Hox* genes. *Hox* gene expression in the spinal cord is similar to that seen in the hindbrain, with the of *Hox* genes of paralogous subgroups 4–13 expressed in an overlapping posteriorly-nested pattern with sharp anterior boundaries of expression. This divides the neural tube into discrete regions marked by different combinations of *Hox* genes, and this may be sufficient to confer regional identity on the motor neurons and other cell types including the neural crest. However, a major distinction between the hindbrain and the spinal cord is the source of the positional information. In the hindbrain, positional identity is controlled by the neural tube itself, and this is imposed on surrounding structures such as the paraxial mesoderm and the ectoderm covering the pharyngeal arches. Neural crest cells arriving in each of the pharyngeal arches control the morphogenesis of the myoblasts derived from the adjacent somites and somitomeres. This is demonstrated by the neural crest graft experiment as described above, where ectopic structures are formed in the pharyngeal arches populated by neural crest cells grafted from a different position. In such experiments, the myoblasts also give rise to ectopic structures, and these structures are appropriate for the anteroposterior origin of the neural crest. Conversely, the *Hox* gene expression pattern in the spinal cord, and the resulting regional specification of motor neurons and neural crest derivatives, is imposed by the adjacent somites. The somites of the trunk express *Hox* genes in the same characteristic nested pattern seen in the hindbrain, and misexpression of *Hox* genes in the somites causes concurrent re-patterning of the neural tube. It is not clear why the head and trunk utilize such distinct methods to pattern the neural tissue, although it may reflect the lack of segmentation in the paraxial mesoderm flanking the hindbrain. The paraxial mesoderm is discussed in more detail in Topic K2.

J3 PATTERNING THE DORSOVENTRAL NEURAXIS

Key Notes

The dorsoventral neuraxis	The vertebrate neural tube shows dorsoventral polarity, with motor neurons restricted to the ventral region and other neural cell types such as association neurons and commissural neurons restricted to the dorsal region. Specialized non-neuronal structures called the roof plate and floor plate are found at the dorsal and ventral midlines, respectively.
Origins of the floor plate	The floor plate arises directly from the organizer and is inserted into the midline of the neural plate. Ectopic floor plate can be induced by grafted notochord tissue or beads soaked in Sonic hedgehog protein, but this may not be the major mechanism of floor plate formation in normal development.
Ventralizing and dorsalizing signals	Dorsoventral patterning is controlled predominantly by ventralizing signals secreted by the floor plate/notochord and dorsalizing signals originating in the ectoderm. The ventralizing signal is Sonic hedgehog, which is necessary and sufficient to establish ventral neuronal cell fates. The dorsalizing signal is a combination of bone morphogenetic proteins, which activate Dorsalin-1 expression in the dorsal neural tube. BMP-4 and Dorsalin-1 are required to specify dorsal neuronal cell fates.
Early molecular markers	The dorsalizing and ventralizing signals induce the expression of different transcription factors along the dorsoventral axis, and these can be used as markers of early regionalization. In particular, the expression of several different Pax, Gli and homeodomain transcription factors is activated in specific domains, and provides a code for neuronal differentiation.
Diversification of motor neurons	The dorsalizing and ventralizing signals act throughout the length of the neural tube to establish the basic subdivisions of neuronal cell types along the dorsoventral axis. These cell types are further diversified by more restricted patterns of transcription factor expression. This is particularly apparent in the motor neuron pool, generating various motor neuron subpopulations that innervate different muscle regions. These neuron subtypes express different combinations of LIM homeodomain transcription factors, which may be controlled both by dorsoventral patterning signals and local signals acting at restricted positions along the anteroposterior axis.
Related topics	Pattern formation (A6) Patterning the anteroposterior Dorsoventral axis specification neuraxis (J2) and patterning (H6) Neurogenesis (J5) The ectoderm: neural induction Mesoderm induction and and epidermis (J1) patterning (K1)

The dorsoventral neuraxis

The vertebrate nervous system shows dorsoventral as well as anteroposterior polarity. This is reflected in the regional specification of different cell types along the dorsoventral axis (*Fig. 1*). Dorsoventral polarity is established at the neural plate stage, when the dorsal-most extremity of the neural tube is represented by the lateral regions of the neural plate. At the lateral edges of the plate, cells are specified to become neural crest. These migrate away from the neural tube around the time of closure and differentiate to form a number of different cell types in various parts of the body, as discussed in Topic J6. The dorsal-most extremity of the closed neural tube (the **dorsal midline**) is represented by a specialized non-neuronal structure called the **roof plate**. This appears to be a source of dorsalizing signals that help pattern the dorsoventral neuraxis (see below). Another specialized non-neuronal structure is formed at the **ventral midline**, and is termed the **floor plate**. The floor plate has two important functions. First, in early development, it acts as a source of ventralizing signals that pattern the dorsoventral neuraxis (see below). Later, it is the source of chemo-attractants that guide commissural axons to the midline (Topic J7). Different neuronal subtypes arise between these two structures, symmetrically placed on each side of the neural tube. For example, various dorsal interneurons, such as **commissural neurons** and **association neurons** arise in the dorsal neural tube, while **motor neurons** and four classes of **ventral interneurons** (V0, V1, V2 and V3) arise in the ventral neural tube. The neuronal cell types are specified by different combinations of transcription factors, whose expression is established by competition between dorsalizing and ventralizing signals in the neural tube.

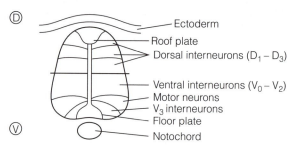

Fig. 1.   Distribution of cell types along the dorsoventral neuraxis.

Origins of the floor plate

Grafting pieces of notochord to a lateral region of the neural tube can induce the formation of an ectopic floor plate, suggesting that the notochord has an important role in the induction of the endogenous floor plate in normal development. However, cell tracing experiments in zebrafish, amphibians and chickens indicate that the floor plate is not induced but derived directly from the organizer itself. For example, the notochord, floor plate and midline dorsal endoderm distal to the regressing node all arise from quail cells when the chicken node is replaced with the quail node from an equivalent developmental stage. The notochord and floor plate are initially united as a population of **midline precursor cells**, which become inserted into the midline of the neural plate. As the node regresses, the notochord separates from the neural tube, leaving the floor plate behind. The ability of the notochord to induce floor-plate cells at ectopic sites in the neural tube may simply reflect the developmental plasticity of the neural tube at that stage of development. Floor-plate development is still possible in the absence of a notochord, although certain regions of the floor plate may fail to form when the notochord is ablated due to the absence of midline precursor cells.

Ventralizing and dorsalizing signals

Dorsoventral patterning of the neural tube is controlled by the combined effects of ventralizing and dorsalizing signals (*Fig. 2*). **Ventralizing signals** are released by the notochord, floor plate and somites to control the differentiation of **ventral neuronal cell types**. **Dorsalizing signals** are initially released by the **ectoderm**, which lies adjacent to the lateral edges of the neural plate (future neural crest and dorsal midline of the neural tube). This induces the formation of the **roof plate**, which along with some of the surrounding neural tissue also produces dorsalizing signals. The combined effect of signals from the ectoderm and dorsal neural tube controls the differentiation of **dorsal neuronal cell types**. The formation of different cell types along the dorsoventral neuraxis reflects **gradients** of these opposing signals and **competition** between them. Thus, in a similar manner to the dorsoventral axis of the early *Drosophila* embryo (see Topic H6), the dorsoventral axis of the neural tube appears to be set up through the interaction of two signaling centers producing factors that inhibit each other's activities.

The major ventralizing signal, the secreted protein **Sonic hedgehog,** is produced by the floor plate and notochord. Its expression is dependent on the winged helix transcription factor **HNF-3β**. The Sonic hedgehog protein acts as a morphogen, inducing different ventral fates by activating and repressing different transcription factors in a dose-dependent manner (see below). A bead soaked in Sonic hedgehog protein can also induce ectopic floor plate if implanted adjacent to the lateral neural tube, showing the basis of floor plate induction by grafted notochord. The ectopic floor plate also secretes Sonic hedgehog, so flanking neural cells then adopt ventral fates.

The dorsalizing signal from the ectoderm is composed of several **BMPs** with **BMP4** and **BMP7** playing principal roles. BMP signaling induces dorsal cell fates by opposing the ventralizing signals mediated by Sonic hedgehog. As the neural tube closes, BMP expression in the ectoderm is restricted to the dorsal midline, but by this time the BMPs are also synthesized by the dorsal neural tube, including the roof plate. One important effect of BMP signaling is the induction of another TGF-β signaling molecule, **Dorsalin-1,** in the dorsal neural tube. BMP4 and Dorsalin-1 establish a gradient that controls neuronal cell fate in the dorsal and lateral regions of the neural tube. Beads carrying BMP4 or BMP7 can replace the dorsalizing signal in embryos where the ectoderm above the neural tube has been removed. Therefore, Sonic hedgehog and BMPs have been shown to be **necessary and sufficient** as the ventralizing and dorsalizing signals, respectively. Competition between the ventralizing and dorsalizing signals is also revealed by such experiments. The induction of ectopic floor plate by notochord grafting causes local repression of Dorsalin-1. Conversely, where Shh signaling is abolished, the domain of Dorsalin-1 expression expands ventrally.

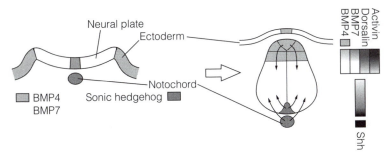

Fig. 2.   Dorsoventral signaling in the vertebrate CNS.

Early molecular markers

The gradients of active dorsalizing and ventralizing factors along the dorso-ventral axis of the neural tube act in turn to induce the expression of **transcription factors** that influence the phenotype of the differentiating neural cells (*Fig. 3*). The Sonic hedgehog signaling pathway activates members of the **Gli family** of transcription factors, related to the product of the *Drosophila* gene *cubitus interruptus*. After closure, three different Gli transcription factors are expressed in the ventricular zone in different regions of the neural tube. Gli1 is restricted to a small domain surrounding the floor plate, Gli2 is expressed in the remainder of the ventral half of the tube, and Gli3 is expressed in the dorsal half of the tube (*Fig. 3*). Interestingly, Gli3 exerts negative feedback on Sonic hedgehog and may therefore help to restrict its expression. Early specification also involves several members of the **Pax family** of transcription factors. Initially, the genes for **Pax3** and **Pax7** are expressed throughout the neural plate. One of the earliest effects of Sonic hedgehog signaling from the notochord and floor plate is the repression of *Pax3* and *Pax7* in the ventral neural tube. *Pax3* and *Pax7* are restricted to the neural folds, and later to the dorsal half of the closed neural tube. The gene for **Pax6** is first expressed at the neural fold stage. It is initially expressed throughout the neural plate, except for the floor plate itself. At closure, *Pax6* is also excluded from the most ventral cells but its domain of expression extends further into the ventral half of the neural tube than *Pax3* and *Pax7* (*Fig. 3*). Sonic hedgehog also regulates several different homeobox genes, resulting in the differential expression of various homeodomain transcription factors in the populations of neuronal cells destined to become motor neurons and ventral interneurons. These homeodomain transcription factors may provide a code that specifies the fate of ventral neuronal cells.

Other important genes are also regulated in the context of the dorsoventral axis. These include vertebrate homologs of the *Drosophila* proneural genes *achaete-scute* and *atonal*, and vertebrate homologs of the *Drosophila* neurogenic gene *Delta* (discussed in Topic J5). Together, the regionalized expression of these proteins may help to control the spatially restricted expression patterns of LIM homeobox genes, whose products appear to control neuronal subtype identity and axonal trajectory.

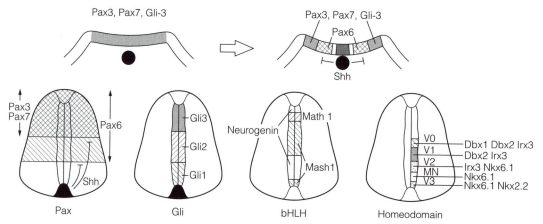

Fig. 3.   Early markers of dorsoventral patterning in the neural tube.

Diversification of motor neurons

Initial dorsoventral patterning generates a range of cell types, including floor-plate cells, motor neurons, ventral interneurons, commissural neurons and roof-plate cells. These cell types are present *throughout most of the anteroposterior axis,* although some are absent from parts of the specialized anterior region of the neural tube that forms the brain.

Later, there is further diversification within these neuronal cell types. This is best-characterized within the population of motor neurons, as different sub-types can be identified by the position of their cell bodies and their axonal projections (*Fig. 4*). The differentiation of motor neuron subtypes appears to be controlled in the same manner as earlier developmental decisions, i.e. by the expression of different classes of homeodomain-containing transcription factors. Initially, all motor neurons express the transcription factors **Islet-1** and **Islet-2**. As diversification occurs, different neuronal subtypes begin to express different combinations of these and other members of the **LIM homeodomain family** of transcription factors (*Fig. 4*). Importantly, this process of diversification is controlled by local interactions at *restricted positions along the anteroposterior axis,* as transplantation of motor neuron progenitor cells to the same position in a different anteroposterior segment results in respecification of the neuronal subtype and altered LIM homeodomain protein expression. Thus while signals from the notochord and ectoderm along the entire neural tube are instrumental in establishing the basic dorsoventral neuraxis, restricted signals probably originating from the paraxial mesoderm are involved in further specialization in the context of the anteroposterior axis (Topic J2).

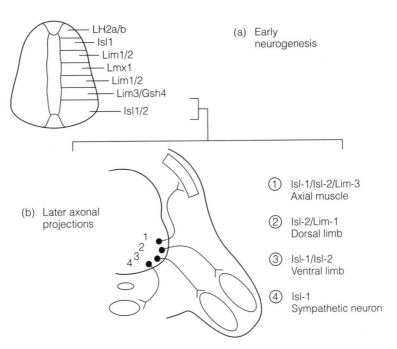

Fig. 4. (a) *Expression of LIM homeodomain proteins during early neurogenesis identifies populations of neurons with alternative dorsoventral fates.* (b) *Later, combinations of LIM homeodomain proteins divide the motor neuron pool into subpopulations with different innervation targets.*

J4 SHAPING THE NEURAL TUBE

Key Notes

Primary and secondary neurulation

Neurulation describes the formation of the neural tube. Primary neurulation begins with the induction of a flat neural plate, which rolls up along its anteroposterior axis to form the tube. In secondary neurulation, the organizer lays down a solid neural cord, which cavitates to form the tube. Most of the neural tube forms by primary neurulation (secondary neurulation is restricted to the tail) except in fishes, where the entire neural tube forms by secondary neurulation.

Cell behavior in primary neurulation

The rolling up of the neural plate during primary neurulation is driven by both intrinsic and extrinsic forces. Intrinsic forces include changes in cell shape at different positions in the neural plate and the anchoring of cells to surrounding tissues. In *Xenopus*, cells at the edge of the neural plate constrict, causing the edges of the plate to curl along its whole length. In amniotes, cells in the midline constrict causing the plate to fold at a hinge point, and this occurs progressively so that the neural folds appear to 'zip' together. The proliferating ectoderm provides extrinsic forces that push the neural folds towards the midline.

Molecular changes accompanying primary neurulation

A number of genes have been identified that affect neural tube closure, although there are also significant environmental influences. Birth defects such as spina bifida, which reflect the failure of neural tube closure, can therefore be avoided by dietary supplements of folic acid. When the neural folds make contact, they adhere to each other because they express common cell adhesion molecules. The surrounding ectoderm expresses different adhesion molecules, and this drives the separation of the neural tube and epidermal ectoderm.

Secondary neurulation

When gastrulation is complete, the movement of cells through the organizer and into the embryo ceases. The remainder of the organizer (the tail organizer) then moves posteriorly, extending the embryonic tail and laying down the solid notochord and neural cord. The neural cord cavitates around the neurenteric canal to generate the lumen of the tail spinal cord.

Related topics

Morphogenesis (A7)
Gastrulation in vertebrate
  embryos (F3)

The vertebrate organizer (I2)
Patterning the dorsoventral
  neuraxis (J3)

Primary and secondary neurulation

The formation and shaping of the neural tube in vertebrates occurs through a combination of primary and secondary neurulation. In all vertebrates except fishes, most of the neural tube forms by primary neurulation, with secondary neurulation restricted to the posterior of the embryo behind the future hind limbs (the extent of secondary neurulation varies in different vertebrate classes). In **primary neurulation**, the neural tube arises from the neural plate, which is induced by the head and trunk organizers (Topic J1). The plate narrows by convergent extension, driven by the intercalation of cell layers. The edges of the plate then become elevated to form **neural folds** that enclose a depressed **neural groove**. The neural folds continue to rise and move towards the dorsal midline until they meet and fuse, so creating the **neural tube**. The neural tube then detaches from the surrounding ectoderm and sinks into the embryo, while the remaining ectoderm differentiates as epidermis. In **secondary neurulation**, the neural tube forms directly from the tail organizer as a solid cord of neurecto-derm. The entire neural tube forms by secondary neurulation in fish embryos.

Cell behavior in primary neurulation

In primary neurulation, the formation of the neural tube is driven in part by changes in cell shape at key points in the neural plate, but the same mechanisms are not used in all vertebrates (*Fig. 1*). In *Xenopus*, the most important changes take place at the edges of the plate. The tall columnar cells in this region become constricted at their apical surfaces, causing the edges of the neural plate to curl, and this pulls the surrounding ectoderm towards the dorsal midline. Simultaneously the wedge-shaped cells extend along the inner surface of the ectoderm, which increases the pulling force on the innermost ectodermal cells. These changes take place throughout the whole of the anteroposterior axis of the neural plate, so that the entire neural tube rolls up at once. In chicks and mammals, neurulation occurs gradually. Neurulation commences in the midbrain region and the neural tube 'zips up' in both directions from this point.

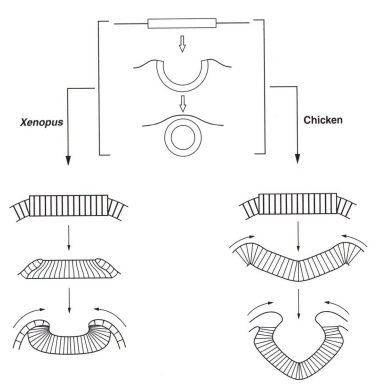

Fig. 1.   Neural tube formation in Xenopus *and chickens. In* Xenopus, *changes in cell shape at the boundary of the neural plate cause the neural tube to roll up. In chickens, important changes in cell shape occur at the medial and lateral hinge points.*

Unlike *Xenopus*, the important changes initially occur in the midline cells imme-diately overlying the notochord. These cells are in direct contact with the noto-chord, and in response to inductive interactions become squat and wedge-shaped, resulting in the formation of a definite **median hinge point**, so that the neural groove is initially V-shaped. Two **dorsolateral hinge points** then arise where the cells are in contact with the surface ectoderm. These cells also become wedge-shaped so that the neural folds are bent towards each other (*Fig. 1*). It appears that microtubule polymerization is required for the changes in cell height that accompany the formation of hinge points, while microfila-ments are necessary for the apical constrictions that cause bending.

As well as these intrinsic forces that change the shape of cells within the neural plate, there are also extrinsic forces generated by other tissues. In both *Xenopus* and amniotes, the epidermal ectoderm plays a major role in neural tube closure. The ectoderm at the edges of the neural plate proliferates and expands towards the midline, with the result that the neural folds are pushed together, becoming covered with ectoderm in the process.

In summary the folding of the neural plate to form the neural tube involves the following mechanisms acting in concert:

- Changes in cell shape within the neural plate, induced by surrounding meso-derm and epidermal ectoderm.
- Anchoring of specific regions of the neural plate to surrounding mesoderm and epidermal ectoderm.
- Proliferation of the epidermal ectoderm towards the midline.

Molecular changes accompanying primary neurulation

The molecular control of neural tube formation is not well understood. The mechanisms by which the notochord and epidermal ectoderm induce the changes in cell shape within the neural plate are unknown. A number of genes are known to play an important role in neural tube closure, including *sonic hedgehog* and a gene called *openbrain*, which is associated with a condition called **anencephaly**, where the anterior end of the neural tube (the **anterior neuropore**) fails to close. Failure of the **posterior neuropore** to close results in another condition called **spina bifida**. However, there are also significant environmental influences on neural tube closure. The occurrence of spina bifida can be dramati-cally reduced by a daily intake of about 0.5 mg of **folic acid** in the mother's diet.

Neural tube closure occurs upon contact between cells previously at opposite edges of the neural plate, as these cells express the same complement of cell adhesion molecules. Initially the neural plate expresses E-cadherin, but by the time closure occurs, this molecule has been replaced by the two principal neural cell adhesion molecules, N-CAM and N-cadherin. The surrounding ectoderm continues to express E-cadherin, therefore the two tissues no longer adhere to each other. The thermodynamic forces involved in this type of cell sorting play a significant role in the formation of the neural tube, as has been shown by forcing expression of the same cell adhesion molecules in both the neural tube and the ectoderm. This results in the failure of neural tube closure and prevents separa-tion of the neural tube and ectoderm, so that the neural tube does not sink into the embryo.

Secondary neurulation

Secondary neurulation in amphibians and amniotes is a continuation of gastru-lation, although instead of moving into the embryo and extending anteriorly, the cells extend posteriorly, increasing the length of the embryo and laying down the most posterior region of both the notochord and neural tube.

The process is best understood in *Xenopus*. Initially, organizer cells involute into the embryo and extend anteriorly to generate the notochord. The neural plate is induced in the overlying ectoderm. When gastrulation is complete, the involution of cells ceases and the remaining tail organizer region of the embryo (called the **chordoneural hinge**, equivalent to the **caudal eminence** of mammals) extends posteriorly. Both the mesodermal and ectodermal components of the hinge grow, the mesodermal component giving rise to the notochord and the ectodermal component to the neural cord. The lumen of the secondary neural tube is formed from the neurenteric canal, the last vestiges of the original blastocoel.

J5 NEUROGENESIS

Key Notes

Overview of neurogenesis

Neurogenesis describes the specification of neural cell fates. In insects and vertebrates, neurogenesis is a highly conserved, multi-step process involving a similar set of gene products.

Neurogenesis in the *Drosophila* CNS

The *Drosophila* CNS originates from small groups of cells called proneural clusters, specified by the expression of proneural genes of the *achaete-scute* complex. The clusters arise in a regular pattern, controlled by the overlapping expression domains of embryonic patterning genes. The proneural genes encode bHLH transcription factors that activate downstream neurogenic genes such as *Notch* and *Delta*. These genes encode signaling proteins that select a single neuronal progenitor (neuroblast) by lateral inhibition.

Neurogenesis in the *Drosophila* PNS

The *Drosophila* PNS arises from proneural clusters in the imaginal discs, which give rise to a variety of external and internal sensory organs. External sensory bristles are specified by the genes of the *achaete-scute* complex, while internal sensory organs are specified by another proneural gene, *atonal*. Both sets of proneural genes activate the same neurogenic genes, and a single neuronal progenitor (sensory mother cell) is selected by lateral inhibition.

Neurogenesis in the vertebrate CNS

Neural progenitors in the vertebrate CNS arise from proneural regions of the neural plate defined by the expression of proneural genes such as *neurogenin*, *MASH1* and *MATH1*. Neuronal progenitors (neuroblasts) are specified by lateral inhibition involving the Notch-Delta signaling pathway. Both the proneural and neurogenic levels of this process are represented by small multigene families, whose members are expressed in restricted domains. The use of different proneural and neurogenic genes may contribute to the differentiation of neuronal subtypes.

Asymmetric division of neural cells

In vertebrates and insects, the neural progenitor selected by lateral inhibition undergoes a series of asymmetric divisions. In the *Drosophila* and vertebrate CNS, asymmetric division maintains a population of neuroblast stem cells and produces a regular supply of committed neurons. In both cases, differentiation reflects the asymmetric distribution of certain proteins, one of which is the transmembrane receptor Numb. In the *Drosophila* PNS, the choice between different classes of sensory bristle is mediated by selector genes such as *paired box neuro* (*pox-neuro*). Each sensory mother cell undergoes two rounds of division to produce one neuron and three supportive cells. The cell fates in this lineage are determined in part by the asymmetric distribution of Numb and in part by lateral inhibition involving Notch and Delta.

Related topics	Cell fate and commitment (A2)	Signal transduction in
	Mechanisms of developmental	development (C3)
	commitment (A3)	Vulval specification in *C. elegans*
	Pattern formation and	(L1)
	compartments (A6)	Leaf development (N5)

Overview of neurogenesis

Insects and vertebrates use very different mechanisms to generate the central and peripheral elements of the nervous system. The central nervous system (CNS) in *Drosophila* is a **ventral nerve cord** formed by individual neuroblasts delaminating from the ventral ectoderm, while the vertebrate CNS is a dorsal **neural tube** formed predominantly by the orchestrated folding of the neural plate. In *Drosophila*, peripheral sensory organs are derived from neurogenic regions of the imaginal discs, while the vertebrate peripheral nervous system is derived mostly from neural crest cells originating at the neural plate border.

Despite such diversity in **neurulation** (the formation of the nervous system), the mechanism of **neurogenesis** (the commitment of cells to neuronal fates) is strikingly conserved. In both systems, neurogenesis involves the progressive restriction of cell fate (*Fig. 1*), and begins with the specification of **proneural regions** within which all cells have the potential to become neural progenitors. The proneural regions are specified by the expression of one or more members of a family of bHLH transcription factors, which are also conserved between *Drosophila* and vertebrates. Individual cells within the proneural regions are chosen as neural progenitors by **lateral inhibition**, through the expression of **neurogenic genes** such as *Notch* and *Delta*. The selected progenitor cells then undergo a series of asymmetric cell divisions to produce **neurons** and other supportive cell types, such as **glial cells**. This asymmetric division process involves another set of conserved genes, related to the *Drosophila* gene *numb*. There are multiple genes acting at the proneural, neurogenic and neuronal differentiation stages, which have different spatial expression patterns in the fly embryo and along the dorsoventral axis of the vertebrate central nervous

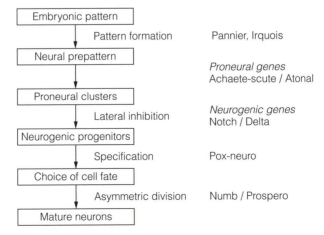

Fig. 1. *The genetic hierarchy of neurogenesis in* Drosophila, *with examples of gene products acting at each stage.*

system. These expression patterns may arise in response to anteroposterior and dorsoventral patterning signals, some of which are also conserved between *Drosophila* and vertebrates.

Neurogenesis in the *Drosophila* central nervous system (CNS)

As discussed in Topic H6, the dorsoventral axis of the syncytial *Drosophila* embryo is divided into a number of discrete regions, demarcated by the complementary expression patterns of the zygotic genes *twist*, *snail*, *single minded*, *rhomboid*, *zerknüllt*, *tolloid* and *decapentaplegic*. These genes are instrumental in the specification of different cell types, and the two broad longitudinal bands of cells lying either side of the ventral midline are specified as **neural ectoderm** by the transcription factor **Rhomboid**. During gastrulation, the band of presumptive mesoderm cells straddling the ventral midline invaginates into the interior of the embryo (Topic F2), so that the entire ventral surface of the embryo is then covered with neural ectoderm. This gives rise to the CNS (the **ventral nerve cord**) as well as epidermis.

Development of the CNS begins with the specification of regularly spaced **proneural clusters** in the neural ectoderm. A proneural cluster is a group of four to six cells, each cell having the potential to become a neuroblast. The proneural clusters arise in a precise pattern (eight clusters in each presumptive segment, arranged in two columns of four) and are characterized by the expression of **proneural genes**, all of which encode bHLH transcription factors. The most important proneural genes for CNS development are those of the *achaete-scute* **complex (AS-C),** as deletion of the AS-C generates larvae in which almost the entire CNS is absent. The AS-C contains four genes and a long complex regulatory region. The genes are expressed in overlapping patterns and co-operate to establish the precise pattern of proneural clusters. The AS-C genes are activated by upstream patterning genes expressed in orthogonal stripes along the anteroposterior and dorsoventral axes (e.g. *engrailed*, *wingless* and *decapentaplegic*). In some uncharacterized manner, these overlapping expression domains delimit a **neural prepattern**, defined in part by the expression of transcription factors such as **Irqouis** and **Pannier**.

Only one of the cells in each proneural cluster becomes a neuroblast, while the others form the ventral epidermis. The selection process involves **lateral inhibition** mediated by transmembrane signaling proteins encoded by the neurogenic genes *Notch* and *Delta*. Signaling though the **Notch** receptor represses proneural gene expression and therefore inhibits neuronal differentiation, but the transcription factors encoded by the proneural genes activate the expression of **Delta**, which is the ligand for Notch. In this way, individual proneural cells that transiently synthesize more Delta protein than their neighbors inhibit proneural gene expression in those neighboring cells. The neighboring cells produce less Delta protein and are less able to inhibit the 'stronger' cell. In this way, random fluctuations in the levels of Notch and Delta among individual cells in the proneural cluster results in a positive feedback loop in which one cell establishes dominance (*Fig. 1*). This cell suppresses neural differentiation in its neighbors and becomes the neuroblast. Lateral inhibition, mediated by Notch- and Delta-related proteins, is critical in several other developmental systems, including axis specification (Topic G2) and vulval development (Topic L1) in *C. elegans*.

Neurogenesis in the *Drosophila* PNS

Later in development, after the CNS has been specified, further proneural clusters arise in the imaginal discs to establish patterns of external bristles and internal sensory organs that make up the peripheral nervous system. There are four major types of external sensory organ in *Drosophila*: **mechanosensory bristles**, **campaniform sensilla** that monitor stretching at the base of appendages, **chemosensory bristles** around the mouthparts and **olfactory sensilla** on the antenna. Most of the internal sensory organs are **chordotonal organs** that span between cuticle cells and monitor stretching in limb joints. Therefore, neurogenesis in the PNS involves not only the specification of sensory organ precursors but also a choice between various types of organ.

Many of the sensory organs arise at invariant positions, reflecting the precise pattern of proneural clusters in the imaginal discs. As in the CNS, the position of each proneural cluster is specified by earlier acting patterning genes. Indeed many of the genes that pattern the embryonic axis are re-expressed in complex patterns within the imaginal discs. The proneural clusters that give rise to external sensory organs are specified by the same AS-C genes that function in the CNS. However, the chordotonal organs arise from proneural clusters specified by a different gene, *atonal*, which also encodes a bHLH transcription factor. Varying the expression levels of the AS-C and *atonal* genes results in changes in the relative numbers of bristles and chordotonal organs. Other genes that control the choice between bristles and chordotonal organs have also been identified. The *cut* gene, which encodes another transcription factor, is repressed by Atonal and is therefore expressed in external sensilla but not chordotonal organs. Expression of *cut* throughout the embryo using a heat-shock promoter produces more external sensilla, while *cut* mutations have the opposite effect.

Although the specification of proneural clusters is more complex in the PNS than in the CNS, the choice of the neuronal precursor is also mediated by lateral inhibition involving Notch and Delta.

Neurogenesis in the vertebrate CNS

In vertebrate embryos, the neurogenic ectoderm is induced by dorsalizing signals secreted by the organizer (Topic J1). As in *Drosophila*, neurons do not arise uniformly in the neural ectoderm but are restricted to specific regions demarcated by the expression of proneural genes. In *Xenopus*, the neural plate contains six longitudinal proneural stripes, three on each side of the midline. As the neural plate rolls up to form the tube, these give rise to dorsal, lateral and ventral neuronal populations.

The proneural regions are specified by bHLH proteins homologous to the proneural gene products of *Drosophila*. One such protein, **Neurogenin**, is expressed in the proneural regions of the *Xenopus* neurula and activates the expression of the *Xenopus Delta* gene. Ectopic expression of *neurogenin* leads to an expansion of the *Delta* expression domain and the differentiation of more neurons. Different proneural genes are expressed in complementary zones along the dorsoventral axis of the neural tube. *neurogenin* is predominantly expressed in the ventral region, while *MASH1* (mammalian *achaete-scute* homolog) and *MATH1* (mammalian *atonal* homolog) are expressed in the dorsal region. Notch is expressed throughout the neural tube, but there is regionalization for different ligands, Delta and Serrate. The regional differences in proneural and neurogenic gene expression may help to pattern the dorsoventral axis of the neural tube resulting in the differentiation of alternative neuronal subtypes. These overlapping expression patterns may arise in response to, or in concert with, the expression patterns of different Pax and Gli transcription factors (Topic J3). Downstream of Neurogenin, another bHLH transcription factor called **NeuroD** is synthesized. This can induce ectopic neuronal differentiation when expressed in non-neuronal ectoderm, and is probably a major determinant of neuronal cell fate.

Asymmetric division of neural cells

In both *Drosophila* and vertebrates, the neuroblast or neural progenitor selected by lateral inhibition undergoes several rounds of asymmetric cell division to give rise to neurons and other cell types. Once again, it appears that a highly conserved set of gene products is involved.

In the *Drosophila* CNS, each proneural cluster produces one neuroblast. This is a stem cell, which divides to produce two dissimilar daughters: a further neuroblast and a **ganglion mother cell**. The neuroblast continues to divide producing a string of ganglion mother cells, and each ganglion mother cell divides once to produce two neurons. This stem-cell behavior involves the asymmetric distribution of at least two proteins (*Fig. 2*). The **Numb** protein is thought to be a transmembrane receptor, while **Prospero** is a transcription factor. The proteins are uniformly distributed throughout interphase but co-localize at one of the spindle poles of the neuroblast during mitosis, so that they segregate into one of the daughter cells. The daughter cell that inherits these proteins becomes a ganglion mother cell. The localization mechanism is dependent on a specific domain shared by both proteins, and is mediated by another protein called **Inscuteable**.

The same proteins appear to control cell fates in sensory organs. Each proneural cluster produces a single neuronal precursor, which is called a **sensory mother cell**. The sensory mother cell undergoes two rounds of cell division, producing a neuron and three supportive cells. For example, to form an external sensory bristle, the sensory mother cell gives rise to one sensory neuron and a bristle cell, socket cell and sheath cell. At the first division, Numb and Prospero segregate to the daughter cell that will give rise to the sensory neuron and sheath cell. The choice between neuronal and sheath cell fates appears to involve further Notch-Delta signaling (*Fig. 3*). As discussed above, the choice between external sensilla and chordotonal organs involves the expression of different proneural genes. However, the various different types of external sensilla are all specified by the AS-C. The choice between alternative external

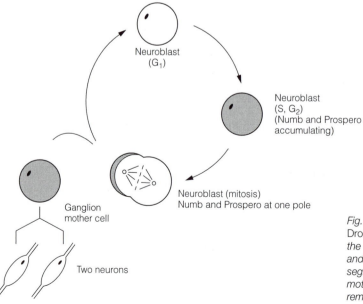

Fig. 2.   Stem cell behavior of
Drosophila *neuroblasts depends on
the asymmetric distribution of Numb
and Prospero (dark shading). These
segregate to the future ganglion
mother cell, while the other cell
remains as a neuroblast.*

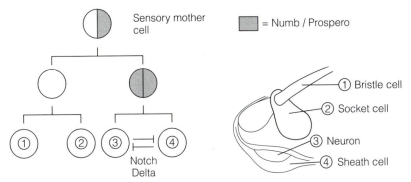

Fig. 3. Asymmetric cell division during sensory bristle development in Drosophila.

sensilla involves further selector genes such as **paired box-neuro** (*pox-neuro*) and **BarH** (Fig. 4) The expression of *pox-neuro* in the sensory mother cell results in the formation of a chemosensory bristle, whereas if this gene is not expressed, the cell can form a mechanosensory bristle or campaniform sensillum. The choice between the latter categories is controlled by the *BarH1* and *BarH2* genes. These genes are likely to be regulated by upstream patterning genes, such as *Krüppel*, *even-skipped* and *fushi tarazu*, expressed in the proneural clusters.

In vertebrates, the asymmetric distribution of Notch and Numb proteins is involved in neuronal differentiation. In the proliferative ventricular zone of the spinal cord (adjacent to the lumen) neuroblasts undergo repeated rounds of cell division. As in *Drosophila*, each neuroblast is a stem cell, capable of producing both further neuroblasts and neuronal precursors. Divisions in the plane of the ventricular zone generate two identical neuroblasts, each of which is a stem cell. However, divisions perpendicular to the ventricular zone generate a neuroblast and a neuronal precursor that will migrate out into the mantle zone. Before each division, the vertebrate Numb protein localizes to the apical pole of the dividing cell, whereas membrane-bound Notch collects at the basal pole. Therefore, Numb protein remains in the neuroblast whereas Notch is inherited solely by the differentiating neuronal precursor.

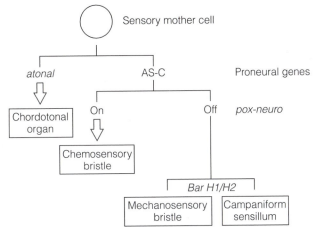

Fig. 4. Hierarchy of decisions to determine sensory organ fates in the Drosophila peripheral nervous system.

J6 THE NEURAL CREST

Key Notes

Properties of the neural crest

The neural crest is a population of cells found only in vertebrates, which arises at the neural plate border. As the neural tube closes, these cells undergo an epithelial to mesenchymal transition and migrate away from the midline to populate various parts of the body, giving rise to a diverse range of cell types. The fate of the cranial neural crest, which contributes to head and neck development, is already determined at the neural plate stage, and it imposes its positional identity on surrounding cells. Conversely, the fate of trunk neural crest (spinal cord) is determined much later by environmental cues on its migratory route. The positional identity of the trunk crest is controlled by the adjacent paraxial mesoderm.

Studying the neural crest

The neural crest is a stimulating model system because of the cells' extensive migration and complex lineage. A number of techniques have been particularly useful for studying the neural crest, including cell tracing, grafting and mutational analysis.

Specification of the neural plate border

In *Xenopus*, there is competition between ubiquitous BMPs and organizer-derived anti-BMPs to establish dorsoventral polarity. The inhibition of BMPs is sufficient to specify neural plate, while uninhibited BMP signaling specifies epidermis. The neural plate border may arise between these thresholds. BMPs may also be involved in specifying the neural plate border in amniotes, although there is increasing evidence that prior signaling involving FGFs and Wnts as well as BMPs may be required for neural border induction. The neural plate border expresses a range of specific markers, including the transcription factors Snail and Slug.

Role of cranial neural crest in head and neck development

The vertebrate head is a specialized developmental system, showing segmentation in all germ layers. The lateral mesoderm is divided into a series of pharyngeal arches, which become populated by cranial neural crest cells from specific rhombomeres. The cranial crest cells from each rhombomere have fixed positional values defined by their *Hox* gene expression profiles, and give rise to a stereotyped range of cartilaginous structures that form various bones of the face and jaw. Grafting experiments show that neural crest cells placed in an ectopic position will migrate to a different pharyngeal arch, but will give rise to cartilaginous structures appropriate for their original location. The neural crest cells also direct the morphogenesis of surrounding tissue, including the myoblasts derived from cranial somitomeres (which give rise to craniofacial muscle groups).

Fate of the trunk neural crest

The trunk neural crest migrates along either the ventral pathway (through the adjacent anterior somite) to establish populations of neurons and glia that form the peripheral nervous system, or along the dorsolateral pathway to populate the dermis and give rise to melanocytes. Trunk crest at the sacral and vagal levels gives rise to enteric ganglia. Several mutations have been identified that prevent the migration of this cell population, resulting in enteric disorders known in humans as Hirschsprung's disease.

Control of neural crest migration

The epithelial to mesenchymal transition that heralds neural crest migration is dependent on the Slug transcription factor. This may alter the pattern of cell adhesion molecules expressed in the neural plate border, allowing the neural crest to delaminate from the neural tube. Migration also depends on guidance cues in the extracellular matrix. Several matrix proteins that guide the cranial neural crest have been identified, but the trunk neural crest appears relatively unrestricted and may be guided predominantly by physical cues in the terrain over which it migrates.

Control of neural crest development

The diverse fates of the neural crest are controlled by both intrinsic and extrinsic factors. The fate of the cranial neural crest is already intrinsically determined at the neural plate stage of development. Conversely, while the trunk neural crest cells at different levels of the anteroposterior axis give rise to a restricted range of cell types, grafting experiments have shown that their potency is much greater, indicating their fates are controlled by the environment. A number of signals that affect trunk crest fate have been identified, including neuregulin, TGF-β and BMP4. These molecules favor different developmental pathways and appear to be expressed in the appropriate regions of the embryo to play an *in vivo* role in neural crest determination.

Related topics

Cell fate and commitment (A2)
Mechanisms of developmental
  commitment (A3)
Pattern formation and
  compartments (A6)
Morphogenesis (A7)

Germ cell migration (E2)
The ectoderm: neural induction
  and epidermis (J1)
Patterning the anteroposterior
  neuraxis (J2)
Somitogenesis and patterning (K2)

Properties of the neural crest

The **neural crest** is a population of cells found only in vertebrates that arises at the **neural plate border**, the boundary between the neural plate and epidermal ectoderm. As the neural plate rolls up, these cells are found at the crests of the neural folds, hence their name. At around the time of neural plate closure, the neural crest cell population undergoes an epithelial to mesenchymal transition and the cells migrate laterally, dispersing along a number of major routes to populate many different parts of the embryo. Once in place, they give rise to a diverse array of cell types including neurons and glia of the peripheral nervous system, craniofacial connective tissue, cartilage, melanocytes, muscle, and components of various glandular organs (*Table 1*).

There is a fundamental distinction between the **cranial neural crest** (located at the level of the mid- and hindbrain) and the **trunk neural crest**. In the trunk, the somites are the 'dominant' segmental units, giving rise to the axial skeletal and imposing their pattern on the neural crest derivatives. Conversely, in the hindbrain, the rhombomeres of the neural tube are the 'dominant' segmental units, the skeletal elements arise from the neural crest, and the crest cells impose their pattern on the (largely unsegmented) paraxial mesoderm. Only cranial

Table 1. *Migration routes and fates of cranial and trunk neural crest cells*

Origins	Major migration routes	Fates
Cranial neural crest	Craniofacial	Muscle and connective tissue of the eye Dermal bones in the skull Facial dermis Pia mater and arachnoid membranes Melanocytes
	Pharyngeal arches	Pharyngeal arch cartilages Odontoblasts in developing teeth Melanocytes Components of lachrymal, salivary, thymus, thyroid, parathyroid glands
	Cranial nerve ganglia	Cranial nerve ganglia Glial cells in peripheral nervous system
Cardiac crest	Ventral	Truncoconal septum of heart Endothelium of aortic arches
Trunk neural crest		
General trunk crest	Ventral (through anterior somite)	Dorsal root ganglia Chain ganglia Adrenal medulla Glial cells in peripheral nervous system
	Dorsolateral (around inside of epidermis)	Melanocytes
Vagal/sacral trunk crest	Ventral (through anterior somite)	Enteric ganglia

crest cells can differentiate into certain cell types, such as cartilage and aortic endothelium. The fate of cranial crest cells is already determined by the time they migrate, so if transplanted to a new position they give rise to ectopic structures (and also direct ectopic development of adjacent tissues). Conversely, the fate of trunk crest is not determined at this time, and grafted cells are respecified at their new position by signals from the paraxial mesoderm.

Studying the neural crest

The extensive migration of the neural crest population, together with its complex lineage and diverse fates, make it a particularly exciting and challenging topic for investigation. A range of experimental approaches has been used to study the neural crest and some of the major strategies are listed below:

- *Cell tracing*. Neural crest cells can be labeled with radioactivity or fluorescent dyes to trace their migration and characterize their lineage. This type of experiment has enabled migration pathways to be mapped and has shown that some crest cells are pluripotent while the potency of others is restricted. Cell tracing is greatly facilitated by the use of cross-species chimeras such as chick/quail or *Xenopus laevis/borealis*.
- *Explants, grafts and ablations*. Neural crest cells can be ablated from developing embryos, grafted to different sites or cultured in isolation. This type of experiment provides information concerning the fate, potency and state of commitment of neural crest cells from different axial positions, and has also established that depleted neural crest populations can regenerate. Explanted

neural crest cells in culture can be tested to see if they will migrate on a variety of different substrates.

● *Mutational analysis*. Mutations that affect neural crest development have been particularly useful in determining the factors controlling migration and differentiation. For example, the transcription factor **Pax3** (identified through the *Splotch* mutation) is expressed in the dorsal neural tube and is required for the correct migration and differentiation of cranial neural crest cells. The signaling molecule **Steel** and its receptor **c-Kit** (identified through the *White-spotting* mutation) are required for the proliferation of crest cells during migration, and play a similar role in germ cell migration (Topic E2). The signaling molecule **GDNF** and its receptor **c-Ret** as well as the growth factor **endothelin-3** and its receptor are required for crest cells to differentiate into enteric neurons. Many mutations that affect neural crest development cause 'spotting' phenotypes reflecting disorders of melanocyte migration/differentiation.

Specification of the neural plate border

In most vertebrates, the neural crest arises along the neural plate border at all positions caudal to the forebrain. At the level of the forebrain, the neural plate border gives rise predominantly to ectodermal placodes and (in *Xenopus*) to the cement gland, although some neural crest does form at this level in mice. Specification of the neural plate border is thought to involve inductive signals from the adjacent presumptive epidermal ectoderm. In *Xenopus*, specification of the neural plate border appears to involve the same secreted factors that mediate neural induction (Topic J1). Noggin, Chordin and Follistatin are secreted by the organizer and repress the activities of BMP2, BMP4 and BMP7. Near the organizer, the BMPs are strongly inhibited resulting in the specification of neural plate, while further away the inhibition is weaker and moderate levels of BMP signaling may specify the cells as neural plate border. Further away still, there is little or no inhibition, resulting in the specification of epidermis.

There is also evidence that BMPs are involved in the specification of the neural plate border in chicken and mouse embryos, as BMP4 applied to cultured neural plate can mimic the effects of epidermis and induce the expression of neural crest markers. However, recent *in vivo* experiments suggest the requirement for BMP4 may arise relatively late in neural crest specification, when the neural tube is closing. Thus, BMP4 may be required for the maintenance of the crest precursor cells, while other factors are required for initial specification. There is evidence that BMPs, FGFs and Wnts play a role in this initial specification process and signals may originate not only from the adjacent epidermal ectoderm, but also from the paraxial mesoderm. BMPs and Wnts continue to be expressed in the dorsal region of the closed neural tube, and this may be required for the maintenance and proliferation of the crest cells.

The result of the signaling processes discussed above is the establishment of a population of cells with the potential to become neural crest. These cells begin to express a series of transcription factors including Snail, Slug, Pax3, Pax7, several members of the Zic family (related to the *Drosophila* pair rule protein Odd-paired), Msx1 and Msx2. Many of the corresponding genes have been knocked out in mice, revealing mutant phenotypes with deficiencies in neural crest development. In the posterior of the neural tube, it appears that this population of cells is not determined as neural crest until the cells actually undergo the epithelial to mesenchymal transition and begin to migrate. Up to this point, they can be grafted to the neural tube and will contribute to the CNS. Similarly, cells from the ventral neural tube grafted to the neural crest can contribute to neural crest derived structures. These experiments suggest that the decision between neural plate and neural crest development is fixed at a relatively late stage, when differentiation has already begun.

Role of cranial neural crest in head and neck development

As in *Drosophila*, the head of the vertebrate embryo appears to be composed of modified segments that have become highly specialized. Segmentation in the vertebrate head is transiently apparent in all three germ layers, reflected in the rhombomeres of the neural tube, the somitomeres of the paraxial mesoderm and the pharyngeal arches (also called branchial arches) of the lateral mesoderm (these are covered with ectoderm and lined with pharyngeal endoderm). **Cranial neural crest** cells contribute to a variety of different structures in the head and neck, depending on their position along the anteroposterior axis (*Table 2*). The cranial crest cells have three major migration routes (*Fig. 1*). The most anterior crest cells migrate to various positions in the developing head and face, giving rise to **craniofacial mesenchyme** that will form dermal bones, connective tissue surrounding the eye, the eye's ciliary and pupillary muscles, and dermis of the face. Hindbrain neural crest migrates predominantly to the pharyngeal arches. The pharyngeal arches are populated by crest cells from the adjacent level of the neural tube, and myoblasts from the adjacent somitomeres. In most vertebrates, the first pair of arches gives rise to the jaws. The remainder give rise to the gills in fishes, and in terrestrial vertebrates to the cartilaginous

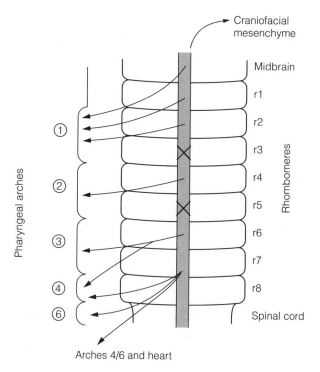

Fig. 1.   Migration routes of the cranial neural crest (gray). Note that neural crest from rhombomeres 3 and 5 undergoes extensive cell death.

Table 2. *Cranial neural crest: cranial ganglia and pharyngeal arch skeletal derivatives*

Neural crest origin	Cranial ganglia	Skeletal elements
forebrain, midbrain	Oculomotor (III)	Dermal bones of skull (derived from craniofacial mesenchyme) Some cells also migrate to pharyngeal arch 1 (see below)
r1, r2	Trigeminal (V)	(Pharyngeal arch 1) From maxillary cartilage: alisphenoid bone, incus of middle ear. From Meckel's cartilage: malleus of middle ear From dermal mesenchyme: jawbones, temporal bone.
r3	APOPTOSIS – or migrate to r2/r4	
r4	Facial (VII) Vestobulocochlear (VIII)	(Pharyngeal arch 2) From Reichert's cartilage: stapes of middle ear, styloid process, stylohyoid ligament, hyoid bone.
r5	APOPTOSIS – or migrate to r4/r6	
r6	Glossopharyngeal (IX) Vagus (X)	(Pharyngeal arches 3 and 4) From third-arch cartilage: Hyoid bone

Note that skeletal elements can derive from both cartilage and dermal mesenchyme, both of neural crest origin. Non-skeletal derivatives of the neural crest are not listed. Other cartilages, developing in pharyngeal arches 4 and 6 are derived from the lateral mesoderm, not the neural crest.

elements that form bones in the ear and neck, as well the odontoblasts of developing teeth and components of the thymus, thyroid and parathyroid glands. Some hindbrain crest also migrates laterally to form the **cranial ganglia**. A specialized population of **cardiac crest** is located in the posterior region of the hindbrain and has the unique ability to form endothelial tissue of the aortic arch and the truncoconal septum, which separates the aorta and the pulmonary artery. No other cranial or trunk crest can substitute for this population.

The arrangement of the hindbrain region of the head is similar in all vertebrates, although most experimental analysis has been carried out in chickens where chick/quail chimeras have been used to follow neural crest cells at different levels. The first three pharyngeal arches are aligned with pairs of rhombomeres in the hindbrain (*Fig. 1*). This correspondence is reflected in the pattern of motor neuron innervation, with motor nuclei of cranial nerve V projecting from r2 to the first arch, motor nuclei of cranial nerve VII projecting from r4 to the second arch, and motor nuclei from cranial nerve IX projecting from r6 to the third arch. The correspondence is also reflected in the migration of the neural crest cells. The midbrain neural crest contributes to the mesenchyme of the head and face but some cells also populate the first arch. Neural crest from r1 and r2 migrates to arch 1. Neural crest from r4 migrates to arch 2. Neural crest from r6 migrates to arch 3. The neural crest from r3 and r5 undergoes extensive cell death, and any remaining crest cells migrate along the neural tube and join the streams of cells migrating from the adjacent rhombomeres. In this way, neural crest cells from different rhombomeres give rise to unique cartilaginous structures in the head and neck (*Table 2; Fig. 1*).

The fate of the hindbrain neural crest cells is already determined when the rhombomeres first become apparent. This has been demonstrated by grafting experiments where neural crest cells are moved from one rhombomere to another prior to migration. For example, the crest cells from rhombomere 2 usually migrate to pharyngeal arch 1 where they form the jawbones and the incus and malleus bones of the middle ear. Crest cells from rhombomere 4 usually migrate to pharyngeal arch 2 and give rise to the stapes bone of the middle ear and part of the hyoid bone of the neck. If the crest cells are removed from rhombomere 4 and replaced with crest cells from rhombomere 2 of a donor embryo at the same developmental stage, the grafted cells will migrate to pharyngeal arch 2 but form structures appropriate for pharyngeal arch 1, i.e. a duplicated set of jaw and middle ear bones. Not only are the cranial neural crest cells determined, but they also appear to direct the development of the surrounding tissues. Myoblast cells migrating from the somitomeres follow the tracks left by neural crest cells and then differentiate *in situ* to form structures appropriate for the neural crest population. In the grafting experiment described above, the second pharyngeal arch is populated by myoblasts from the sixth cranial somitomere as usual, but instead of giving rise to facial muscles as they would normally, they give rise to the mastication muscles characteristic of pharyngeal arch 1. The positional identity of the neural crest cells is specified by the expression of *Hox* genes as discussed in Topic J2. This positional identity is imposed on other cell types in the pharyngeal arches, which begin to express the same set of *Hox* genes as the neural crest cells. The *Hox* genes expressed by the cranial mesoderm are altered when crest cells are grafted to ectopic positions.

Fate of the trunk neural crest

The trunk neural crest can be divided into two populations. Trunk crest cells along most of the anteroposterior axis form the dorsal root ganglia, sympathetic chain ganglia and melanocytes, as well as the adrenal medulla. However, crest cells at the vagal and sacral levels form the **enteric ganglia** of the alimentary canal. Several mutations that affect this population have been identified and are characterized by the failure of peristalsis in the digestive system, causing '**megacolon**'. In humans, this condition is called **Hirschsprung's disease**, and may result from disruption of GDNF-Ret signaling, or endothelin-3 signaling.

The trunk crest has two major migration pathways (*Fig. 2*). The **ventral pathway** passes directly through the sclerotome of the adjacent somites. All peripheral nerve ganglia, as well as the adrenal medulla, are formed by crest cells following this pathway. However, the crest cells migrate only through the *anterior* portion of each somite. Cells initially opposite the posterior portion of a somite change direction and migrate either anteriorly or posteriorly to join the adjacent streams of migrating cells. The **dorsolateral pathway** passes between the epidermis and the somites, around the periphery of the embryo and this is the route that melanocyte precursors use to reach the skin. Most early-migrating crest cells follow the ventral pathway, while late migrating cells follow the dorsolateral pathway.

Control of neural crest migration

The presumptive neural crest population in the neural plate is epithelial and unable to migrate. The ability to migrate arises following an **epithelial to mesenchymal transition**, which appears to be controlled by changes in cell

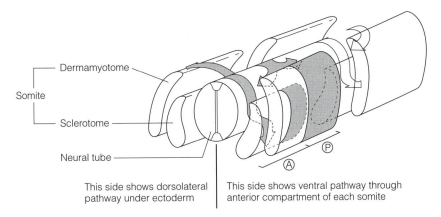

This side shows dorsolateral pathway under ectoderm | This side shows ventral pathway through anterior compartment of each somite

Fig. 2.   Migration routes of the trunk neural crest cells. Ventral pathway: crest cells adjacent to the anterior compartment of the somite may migrate through it, while those adjacent to the posterior compartment may not, and migrate either way along the neural tube to the next 'anterior' compartment. Dorsolateral pathway: crest cells destined to form melanocytes migrate between the somite and the ectoderm.

adhesion and cytoskeletal architecture. The transcription factor **Slug** is involved in this process because when expression of the *slug* gene is inhibited, neural crest cells are unable to migrate. The role of Slug may be to regulate certain genes encoding cell adhesion molecules. For example, the genes for **N-cadherin** and several other cadherins are expressed generally by cells of the dorsal neural tube, but are strongly downregulated in migratory crest cells (and expressed again when the cells reach their destination and begin to aggregate). Other cadherins are upregulated in migrating crest cells, and this may initiate migration by making the extracellular matrix a more attractive substrate than the neural tube. Cadherins may also be regulated at the protein level. Cadherins associate with many regulatory proteins including β-catenin, which is activated by Wnt signaling. Several Wnts are expressed in the dorsal neural tube at the time of neural crest migration and may play a role in modulating the adhesive properties of the cells. Another protein required for neural crest migration is **Rho**, an intracellular GTPase involved in remodeling the cytoskeleton. The role of the cytoskeleton in neural crest migration is not precisely understood, although changes in cell shape are likely to be an important requirement for delamination and migration.

Migration is also facilitated by changes in the extracellular matrix, which guides the cells towards their appropriate destination. These direction cues are region-specific since rotating or swapping the positions of blocks of mesoderm along the anteroposterior axis in chicken embryos causes the neural crest cells to migrate along different pathways. Inhibiting the binding of certain matrix components in chicken embryos has shown that cranial crest cells require fibronectin, tenascin and laminin for migration. The trunk neural crest cells appear to be less restricted by local guidance cues, although certain Eph family receptor tyrosine kinases have been identified that exclude migrating crest cells form particular regions of the embryo. Environmental signals are also important for the guidance of axons, as discussed in Topic J7.

Control of neural crest development

The fact that neural crest cells give rise to an incredibly diverse range of cell types raises a number of important questions. How is this diversification controlled? In particular, how much is the developmental potency of a neural crest cell determined by its lineage, and how much by interactions with other cells. These questions have been addressed using a combination of cell tracing and grafting experiments *in vivo* and studies of neural crest behavior *in vitro*.

As discussed above, there is a fundamental difference between cranial and trunk neural crest cells. Cell tracing in normal embryos has shown that the fate of cranial crest cells is determined before they migrate (*Fig. 3*). Cranial neural crest cells are pluripotent, but they give rise to the same predictable and restricted range of cell types whether in their normal position or when grafted to a different position along the axis. Trunk crest cells are also pluripotent, and also give rise to a restricted subset of the known crest derivatives at different positions along the anteroposterior axis. However, by grafting sections of trunk neural crest to different positions, investigators have shown that this is not due to any intrinsic restriction of potency, as the grafted cells can adopt new fates according to their new position (*Fig. 3*). Conversely, if trunk neural crest is grafted to the hindbrain region, the ectopic cells will not give rise to e.g. cartilage. This appears to be a fate reserved for cranial crest cells.

The studies described above involved labeling whole tissues. Therefore, while it is clear that crest *populations* are pluripotent, it is possible that each population is made up of individual cells whose fates are restricted by their lineage. The labeling of individual cells has confirmed this possibility: some cells give rise to many different cell types, while others give rise to only a single cell type. There is evidence that some crest cells are already restricted in terms of their developmental potency at the start of migration. However, it appears that crest cells become progressively restricted in developmental potency over time, as most late-migrating cells only give rise to melanocytes.

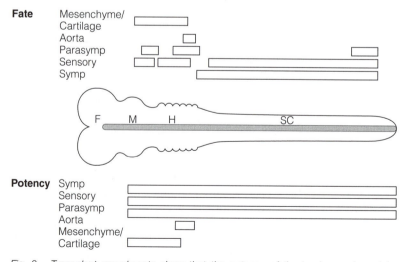

Fig. 3. Transplant experiments show that the potency of the trunk neural crest is much greater than its fate, since most trunk crest cells can give rise to parasympathetic, sympathetic and sensory neurons if moved to the appropriate position. Conversely, the fate of cranial neural crest cells is fixed before migration, and only these cells can give rise to mesenchyme, cartilage and aortic derivatives.

It thus appears that most neural crest cells are pluripotent at least at the start of migration. In the head, diversification is controlled by the positional identity of the crest cells, reflecting the expression of *Hox* genes in the neural tube (Topic J2). In the trunk, cell fate is not fixed, and diversification involves the perception of extracellular signals by the migrating cells. Many signals that influence the fate of neural crest cells have been identified, but three growth factors are particularly interesting because they have been shown to affect the differentiation of the same cultured neural crest cells and favor three different fates. The treatment of rat neural crest cells with **glial growth factor (neuregulin)** promotes differentiation into Schwaan cells, while **TGF-β** promotes differentiation into smooth muscle, and **BMP2** and **BMP4** promote differentiation into autonomic neurons. These molecules not only influence the fate of neural crest cells *in vitro*, but they are also expressed in the appropriate sites *in vivo* during development. The role of these signaling proteins is to induce the expression of transcription factors that orchestrate the changes in gene expression that underlie differentiation. A direct link has been established for BMP2, which has been shown to induce expression of a transcription factor called MASH1 (mammalian *achaete scute* homolog; Topic J5) that is essential for the differentiation of autonomic neurons. Many other transcription factors have been identified with specific roles in neural crest lineage determination, including the LIM domain proteins such as Islet-1 which are also involved in establishing neuronal fates in the CNS.

J7 NEURONAL CONNECTIONS

Key Notes

Specificity of neuronal connections	The functioning nervous system requires communication between neurons and other target cells, and this is established by the growth of axons. The specificity of neuronal connections is achieved in several stages. First, growth of axons towards specific targets; second, the selection of particular cells at the target site; third, refinement of initial neuronal connections through competition for neurotrophic factors (proteins that promote neuronal survival). There are initially many more neurons than required and such competition results in extensive neuronal cell death.
Axon guidance	Growing axons are guided to their targets by various guidance signals in the environment. Pioneer axons follow molecular guidance cues, often in a stepping-stone manner between intermediate targets to break a long journey into manageable sections. Later, growth cones can follow pre-existing bundles of axons (axon fascicles).
Molecular guidance	The growth cone responds to both long-range and contact-dependent molecular guidance cues, each of which may act positively or negatively. Long-range chemoattractants such as netrin and chemorepellants such as the semaphorins provide general directional guidance, while positive and negative contact-dependent signals map out available routes in the local environment. Several classes of molecule have been identified that mediate growth cone guidance, and surprisingly, the same type of molecule can act as both an attractant and a repellant, or a diffusible or contact-dependent signal. Growth cones may migrate up adhesive gradients by trading weak bonds for stronger ones.
Guidance of commissural axons	Commissural axons cross the midline of the CNS and provide examples of several types of guidance mechanism. The molecules involved are conserved between insects, nematodes and vertebrates. Netrins are diffusible chemoattractants found at the midline and they attract commissural axons bearing the Frazzled/DCC type netrin receptor. Other (ipsilateral) axons carry a different netrin receptor and these axons are repelled by netrins. At the midline, contact dependent signals are required to enable axons to cross. In *Drosophila*, crossing the midline is achieved by transiently blocking the ability of the growth cone to respond to repellant signals.
Fasciculation	The binding of axons to pre-existing axons considerably simplifies the task of finding appropriate targets. Fasciculation may be achieved by binding axons together with common cell adhesion molecules, or simply by making pre-existing axon fascicles more attractive substrates than the surrounding environment. Axons can selectively fasciculate and defasciculate, enabling them to chose different axon fascicles for different legs of their journey. This involves the local inhibition of axonal cell adhesion molecules, so that the surrounding terrain is a more attractive

substrate for migration. Several receptor tyrosine phosphatases that block axonal adhesion in this manner have been identified in *Drosophila*.

Target choice – the retino-tectal system

Axon growth from the retina to the optic tectum shows the remarkable phenomenon of topographical mapping, where neurons project their axons to a target in a precise array that maps the arrangement of the cell bodies. This involves target choice, which in the optic tectum is controlled by gradients of contact-dependent repulsion mediated by Ephrins and their receptors.

Refinement by competition for survival

Once an axon reaches its target it forms a synapse, which enables the neuron to communicate with its target cell. The synapse is also essential for neuronal survival, as functional synapses are required for neurons to take up neurotrophic factors (proteins that prevent neuronal cell death). Initially, many neurons may innervate the same target. Competition for neurotrophic factors is based on synaptic activity, with the most active synapses establishing dominance and forcing competitors to withdraw. The developing nervous system produces a gross excess of neurons, and competition among them for neurotrophic factors selects for those with correctly functioning synapses.

Requirement for neuronal activity – ocular dominance columns

The mammalian visual cortex initially receives overlapping axonal inputs from both eyes. During development, this system of connections is refined into ocular dominance columns, where adjacent groups of cells respond to the same position in the visual field but from separate eyes. Ocular dominance columns form by competition among axons for synaptic dominance. This in turn is dependent on neuronal activity, initially generated spontaneously in the developing brain and later reflecting visual experience in the newborn mammal.

Related topics

Morphogenesis (A7)
Signal transduction in
  development (C3)

The cell division cycle (C4)
Germ cell migration (E2)
The neural crest (J6)

Specificity of neuronal connections

The function of the nervous system is to detect and coordinate information concerning an animal's environment (internal and external) and initiate the appropriate responses. This requires communication between neurons, as well as between neurons and their target organs (muscles and glands). Communication is facilitated by an elaborate but precise network of hard-wired connections, established by the growth of axons. In both vertebrates and invertebrates, this intricate neuronal circuitry is generated *de novo* during development, predominantly by the *directed growth of axons* towards appropriate targets (in vertebrate embryos, forming the correct connections also depends in part on the migration of individual neurons – particularly those derived from the neural crest; Topic J6). The sheer number of neurons in a complex system such as the mammalian CNS makes this a formidable task, particularly as each neuron must find its target while millions of others are attempting the same feat. Once axons have reached their destination they make synaptic connections with target cells, a process that may involve *target selection*. In certain parts of the nervous system, there are initially many more synaptic connections than required. The pattern of connections is refined by competition among neurons for survival factors called neurotrophins, the outcome depending largely on the *differential activity of individual neurons*.

Axonal guidance

Axonal outgrowth from differentiating neurons is led by the **growth cone**. This is found at the tip of the axon and migrates by the extension of filopodia that attach to cell surface molecules and the extracellular matrix to pull the axon forward. At the same time, the length of the axon increases as more lipids are incorporated into the plasma membrane. The axon is therefore laid down behind the migrating growth cone.

Axons can grow over great distances, and do so in a highly stereotypical manner, following particular pathways and reaching their targets without error. The direct nature of axon growth reflects the ability of the growth cone to perceive and react to cellular and molecular **guidance signals** in the environment. Studies carried out in a number of model systems have shown that axons are guided to their targets by four major factors:

● Physical guidance in the form of grooves and channels. Physical guidance cues are uncommon, but one example is the channels between cells in the mouse retina, which are thought to guide retinal axons at least part of the way to the optic stalk.
● Molecular guidance by short range contact-mediated attraction or repulsion (involving cell-surface molecules or components of the extracellular matrix).
● Molecular guidance by long range diffusible chemoattractants and chemo-repellants.
● The fascicles of earlier 'pioneer' axons.

The ability of growing axons to bind to pre-existing axons and follow the same pathway (**fasciculation**) considerably simplifies the task of making accurate neuronal connections. As discussed in more detail below, axons have the ability to **selectively fasciculate**, so they can follow one axon fascicle for part of their journey, then defasciculate and swap to another. However, the first neurons to differentiate project their axons into virgin territory. In this case, the **pioneer axons** rely almost entirely on the molecular guidance cues in their environment. The long journey to their ultimate target may be simplified by its division into a series of short segments defined by groups of cells that act as **intermediate targets**. For example, in the insect limb, innervation is directed by small clusters of **guidepost cells**, while in vertebrate embryos, the intermediate targets are usually made up of larger groups of cells, an example being the midline cells of the CNS (discussed in more detail below). Pioneer axon growth is therefore built up of periods of linear growth between intermediate targets punctuated by changes of direction. As the axon grows, **secondary growth cones** may emerge from the elongating shaft and extend to other targets.

Molecular guidance

Axons respond to attractive and repulsive molecular guidance signals, each of which may be diffusible (acting over a long range) or contact-dependent (acting over a very short range). The growth cone is sensitive to the environment and carries the receptors that enable it to react to the appropriate signaling molecules. While the aim of molecular and biochemical studies has been to isolate individual signals, the growth of most axons is likely to be controlled by multiple guidance cues acting in concert (*Fig. 1*). The general direction of axon growth is specified by diffusible molecules acting over relatively long distances,

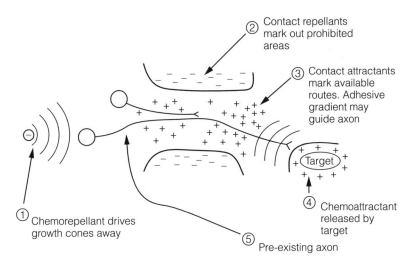

② Contact repellants mark out prohibited areas

③ Contact attractants mark available routes. Adhesive gradient may guide axon

① Chemorepellant drives growth cones away

④ Chemoattractant released by target

⑤ Pre-existing axon

Fig. 1. Summary of different molecular guidance cues to which growing axons may be exposed.

which either stimulate growth cone migration or cause the growth cone to collapse. At the same time, available routes in the local environment are marked by permissive, contact-dependent signals, while forbidden territories are marked by inhibitory contact-dependent signals. Axon growth through a permissive territory may involve **haptotaxis**, the movement of the growth cone up an adhesive gradient as the filopodia trade weak bonds for stronger ones. Furthermore, the receptors that enable axons to respond to environmental signals may be blocked or inhibited by other molecules, thus allowing axons to travel across inhibitory domains.

Genetic and biochemical studies have identified the ligands and receptors that provide axonal guidance signals and mediate growth cone responses. Remarkably, these ligand-receptor systems are highly conserved between vertebrates and invertebrates such as *Drosophila* and *C. elegans*, and can be grouped into a small number of families as shown in *Table 1*. Initially, different classes of molecules were thought to mediate different types of signal, e.g. netrins were regarded as attractive diffusible signals and semaphorins as repulsive diffusible signals. However, as more components have been identified it has emerged that the same class of molecule often carries out multiple signaling functions. For example, netrins can be repulsive as well as attractive, while both diffusible and membrane-spanning (contact-dependent) semaphorins have been identified. Different effects of the same signal are mediated by the expression of different types of receptors on the growing axons. The midline of the nervous system in insects and vertebrates provides an ideal system to study the roles of many of these molecules, and this is considered below.

Guidance of commissural axons

Although the CNS of vertebrates and invertebrates are organized very differently, each has bilateral symmetry with respect to the midline, and in both cases, commissural axons cross the midline to connect the two sides. Some of the molecules that play important roles in the guidance of commissural axons are conserved between vertebrates and invertebrates.

As stated above, **netrins** are diffusible molecules that can attract and repel

Table 1.   Classes of molecule that control axon growth

Class	Function
Cell adhesion molecules (CAMs)	Contact-dependent guidance. Involved in pathfinding and fasciculation. Examples include N-CAM, fasciculin and axonin. CAMs may show homophilic binding (allowing them to act as both ligands and receptors) or heterophilic binding (allowing them to act as ligands for other cell-surface or matrix-anchored receptors).
Components of the extracellular matrix (ECM)	Contact-dependent guidance. Molecules of the ECM map out permissive and nonpermissive territories for axon growth. Permissive molecules include laminin, while repellant components include tenascin. Receptors on growth cones (integrins, proteoglycans and certain receptors of the immunoglobulin superfamily) respond to these molecules.
Netrins	Long distance attraction and repulsion. Netrins are diffusible molecules related to the laminins in the extracellular matrix. A particular netrin can have attractive and repulsive effects on different types of axon, reflecting the existence of two types of receptor. The DCC/Frazzled/UNC-40 receptor family mediates attractive behavior, while the UNC-5 receptor family mediates repulsive behavior.
Receptor tyrosine kinases	Contact-dependent. Examples include FGF receptors (regulation of branching) and Ephrin receptors (mediate target selection – see section on retino-tectal maps in the main text)
Receptor tyrosine phosphatases	Contact-dependent. Control fasciculation and defasciculation (see main text)
Semaphorins	Long distance or contact-dependent repulsion (an attractive semaphorin has also been identified). A large family of proteins with diffusible and transmembrane members. Induce growth cone collapse.

different types of axons. In both vertebrates and invertebrates, one or more netrins is expressed by cells at the midline of the nervous system, helping to attract commissural axons and repel certain ipsilateral axons. In *Drosophila* two netrins, **Netrin-A** and **Netrin-B**, are expressed by midline cells, and attract commissural axons carrying the receptor, **Frazzled**. A similar function is mediated by the *C. elegans* netrin **UNC-6** and its receptor **UNC-40**. Two vertebrate netrins are also expressed in the midline. **Netrin-1** is restricted to the floor plate of the neural tube, while **Netrin-2** is expressed more widely in the ventral neural tube, but not in the floor plate. At the onset of outgrowth, dorsal commissural axons project ventrally along the lateral margin of the neural tube perhaps in response to the gradient of Netrin-2. However, as they enter the ventral half of the neural tube, they make a sudden sharp turn towards the floor plate. This abrupt change of direction is probably a response to Netrin-1. Experiments in which floor plate tissue is co-cultured with dorsal neural tube show that the axons grow towards the floor plate explant, and Netrin-1 has been purified from this tissue. The vertebrate homolog of Frazzled is called Deleted in Colorectal Cancer (DCC).[1] Both *netrin-1* and *dcc* genes have been knocked out in the mouse, and the mutant phenotype shows disruption of commissural axon projections.

[1]The human netrin receptor was called DCC (Deleted in Colorectal Cancer) because the gene is present in a chromosome region that is deleted in cases of genetically-determined colorectal cancer. However, the mouse *dcc* knockout provides no evidence that *DCC* is a tumor suppressor gene, so it is likely that another gene closely linked to *DCC* in humans is responsible for the clinical manifestation of colorectal cancer.

Once at the midline, contact-dependent interactions are important to allow the commissural axons to cross to the other side. This reflects the fact that the midline contains repulsive signals (probably mediated by one or more semaphorins) that cause growth cone collapse. In the chicken, interactions between the cell adhesion molecules **Axonin-1** on the growth cone and **NrCAM** expressed in the midline cells are required for the axons to cross the midline, since blocking such interactions causes the axons to stop growing or extend along the ipsilateral side of the midline. Blocking the response to a midline repellant may also be important for commissural axon growth in *Drosophila*. Two genes that are critical for the guidance of commissural axons are *roundabout* (*robo*) and *commissureless* (*comm*). In wild type flies, commissural axons cross the midline once and project longitudinally along the contralateral ventral nerve chain. In *robo* mutants, more axons than usual cross the midline, and criss-crossing from one side to the other occurs frequently. Conversely, in *comm* mutants, axons are prevented from crossing the midline and there are no commissures. These complementary phenotypes show that Comm is required for axonal growth across the midline while Robo has the opposite effect. Double mutants have the *robo* phenotype, suggesting that the function of Comm is to inhibit the activity of Robo. Comm is synthesized by midline cells, while Robo is synthesized by the neurons themselves. Axons expressing *robo* are unable to cross the midline, and it appears that Comm locally inactivates Robo in migrating neurons, enabling commissural axons to cross the midline. It therefore appears that Robo is the receptor for a repellant of commissural axons released by midline cells. This chemorepellant has recently been identified as the Slit protein. By inhibiting Robo, Comm enables the axons to cross through a region of high Slit concentration.

Fasciculation

The task of establishing neuronal connections is considerably simplified when growth cones can follow the paths laid down by pre-existing axons. Some of the different types of molecules involved in the contact-dependent guidance of pioneer axons are also involved in fasciculation (the binding of axons together in bundles). For example, axon fasciculation is promoted if there is no suitable substrate for growth except another axon, or if the surrounding territory is nonpermissive for growth. However, fasciculation is also actively directed by the expression of common cell adhesion molecules on adjacent axons, such as the *Drosophila* protein Fasciclin II which is related to N-CAM. A fascinating aspect of fasciculation is the ability of axons to selectively fasciculate and defasciculate, thus exploiting different fascicles for different stages of their journey. **Selective defasciculation** involves inhibiting the expression or activity of axonal cell adhesion molecules so the surrounding territory becomes more attractive for axon growth. In *Drosophila*, several genes have been identified that block the activity of Fasciculin II and thus promote defasciculation. Most encode **receptor tyrosine phosphatases**, whose signaling disrupts Fasciculin II binding activity.

Target choice – the retino-tectal system

In vertebrate embryos, the growth of axons from the retina to the brain is a well-studied system, which shows the remarkable phenomenon of **topographical mapping**. This is where neurons from one part of the nervous system project their axons to another in such a way that the spatial relationship between the neurons is mapped by the resulting synapses. Topographical mapping involves not only the guidance of axons to their target, but also an intricate mechanism of **target choice** once they arrive.

Impulses from the light-sensitive cells of the vertebrate retina are communicated to adjacent **retinal ganglion cells**. These neurons project axons along the optic nerve to a region of the brain called the **optic tectum** (or in mammals, the **lateral geniculate nucleus**). In turn, axons from the tectum connect to the **visual cortex** where the incoming visual stimuli are interpreted. Axons from the retinal ganglion cells project onto the optic tectum in a precisely ordered array, generating a topographical **retino-tectal map**. In mammals (with binocular vision) the left side of each retina projects to the right lateral geniculate nucleus and the right side of each retina projects to the left lateral geniculate nucleus. In lower vertebrates such as *Xenopus*, all the axons from the left retina map to the right tectum, while all the axons from the right retina map to the left tectum. As shown in *Fig. 2*, the tectal projections are reversed in polarity with respect to both the dorsoventral and anteroposterior (nasal-temporal) axes of the retina, because the ventral retinal axons project to the dorsal area of the tectum (and the dorsal retinal axons to the ventral area) while the anterior (nasal) retinal axons project to the posterior area of the tectum (and the posterior or temporal retinal axons to the anterior area).

The large number of axons extending through the optic nerve makes it unlikely that each is guided to its final destination by a qualitatively distinct

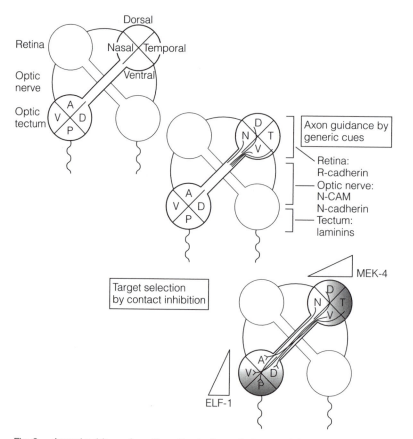

Fig. 2.   Axonal guidance from the retina to the optic tectum. Initial generic cues are used to bring retinal axons towards the tectum. Then target selection is carried out using gradients of repulsive interactions. See text for details.

signal. A number of guidance cues have been identified in the retina, optic nerve and brain, but these appear to act equivalently on all the axons. The cell adhesion molecules N-cadherin, R-cadherin and N-CAM are required to guide the axons across the retina and along the optic nerve, while in the brain itself, the axons grow along tracts of laminin until they reach the tectum. Once they arrive at the tectum, the axons spread across its surface and seek their appropriate targets. At this point, target selection appears to be mediated by a **gradient of contact-dependent repulsion**. Several molecules are distributed in an antero-posterior gradient across the tectum, including the ligand **ELF-1**, which binds to the receptor tyrosine kinase **Mek-4**. ELF-1 is expressed with its highest concentration in the posterior of the tectum, while the receptor Mek-4 is expressed on the growth cones of the incoming retinal axons, with its highest concentration on axons from the temporal (posterior) of the retina. Interactions between the receptor and ligand are repulsive, so temporal axons with the highest concentrations of Mek-4 are more likely to make contacts in the anterior tectum, where there is the least ELF-1. Conversely, nasal axons with the lowest Mek-4 levels are able to make contacts in areas of the tectum with high levels of ELF-1. Between these extremes is every possible combination in the relative levels of the two molecules. Therefore, incoming axons are sorted on the basis of optimal ELF-1/Mek-4 interactions, resulting in a precise map of neuronal connections on the tectum (*Fig. 2*). As discussed in Topic J2, the gradient of ELF-1 in the tectum is established by a pre-existing gradient of the transcription factor Engrailed-2, which is highest at the **isthmic midbrain organizer**.

Refinement by competition for survival

Once an axon has been guided to its destination and has selected its particular target, it forms a specialized junction with the target cell, called a **synapse.** The synapse enables the cells to communicate, as nerve impulses cannot jump directly from cell to cell. Instead, when an action potential reaches the end of an axon (the **pre-synaptic membrane**), chemicals called **neurotransmitters** are released from the synaptic terminal. The neurotransmitters diffuse across the gap and induce a new action potential in the **post-synaptic membrane** of the target cell.

The process of **synaptogenesis** has been studied extensively at the **neuro-muscular junctions** where axons contact muscle cells. An initial phase of non-specialized contact is followed by the clustering of receptors for **acetylcholine** on the muscle cell plasma membrane in the region of the growth cone (acetylcholine is the neurotransmitter utilized by motor neurons). The clustering of receptors is induced by the secretion of a molecule called **agrin** by the axon terminal, which binds to a receptor called **Musk** on the post-synaptic membrane. Furthermore, other proteins called **neuregulins** released by the axon terminal induce the acetylcholine receptor gene, resulting in the increased synthesis of the receptor under the synapse. The axon terminal then starts to accumulate neurotransmitter vesicles loaded with acetylcholine. At the same time, extracellular matrix accumulates in the space between the growth cone and the muscle cell, and the membranes of both cells are modified.

Synaptogenesis is essential for the correct function of the nervous system, but it also plays a major role in **neuronal survival**. A remarkable aspect of the developing vertebrate nervous system is that neurons are produced in great excess and must compete with each other for the right to make and keep synaptic connections with target cells. Cells that fail this challenge pay the ultimate price and undergo apoptosis. In extreme cases, such as in the mammalian retina, up to 80% of the initial population of neurons may die. Neuronal survival is controlled by the target tissue that is innervated. This is shown by experiments that modify the amount of available target tissue. For example, during innervation of the chick limb bud, it is normal for approximately half of the starting population of neurons to die. If the limb bud is removed, however, almost all the neurons die. Conversely, grafting extra muscle tissue into the limb bud causes a proportionately larger number of neurons to survive.

The target tissue controls neuronal survival by secreting limited amounts of **neurotrophic factors**. These proteins bind to **receptor tyrosine kinases** of the **Trk** family on the axon plasma membrane and initiate an intracellular signaling cascade that suppresses the apoptotic machinery of the neuronal cell. A number of different neurotrophic factors have been isolated, and appear to preferentially promote the survival of different types of neurons. **Nerve growth factor (NGF)**, for example, promotes the survival of sympathetic and sensory neurons, while **brain-derived neurotrophic factor (BDNF)** promotes the survival of motor neurons. Some neurons respond to several different neurotrophins, while for others, the requirement for particular neurotrophins changes during development. The function of neurotrophins in promoting the survival of different neuronal cell types has been supported by experiments in which the neurotrophic factors have been inactivated (e.g. with antibodies) and in mice with different neurotrophin genes knocked out. The correspondence between the amount of target tissue and the number of surviving neurons suggests that the target tissue secretes a limited amount of each neurotrophic factor, which is sufficient to promote the survival of a proportion of the neuronal population (the exact proportion varying from system to system, but particularly low for example in the mammalian retina).

Neurons must therefore compete for neurotrophic factors released by their target tissues and the competition occurs at two levels, reflecting the timing of neurotrophin release. Innervation of the limb bud is characterized by two phases of selection. In the initial phase, the first axons reaching the muscles get the lion's share of the available neurotrophins, reducing the amount available for those axons lagging behind. A wave of neuronal cell death therefore occurs as trailing axons fail to pick up neurotrophins and the neuronal cell bodies undergo apoptosis. Axons whose growth cones are near the target tissue converge on the first synaptic site to form, generating multiple inputs into the same muscle fiber. However, in the adult nervous system, there is only one synapse per muscle fiber, and this is achieved by further competition between the axons for dominance in terms of neuronal activity. The stimulation of a muscle cell through one synapse inhibits other synapses, and these axons withdraw from the muscle. Each axon may branch to innervate several muscle fibers and competition eventually sorts them so that each makes one connection. However, if a neuron loses all its competitions with other axons, all synapses are withdrawn and the neuron undergoes apoptosis.

This overproduction and selection approach to development may appear wasteful but it actually serves several useful purposes. Firstly, the system is perfect for matching the number of neurons to the number of available targets. Secondly, neuronal survival on the basis of dominant neuronal activity efficiently selects for correctly functioning synapses. Thirdly, selection on the basis of activity allows an initially coarse system of synaptic connections to be refined. This is particularly important in the developing visual system as discussed below.

Requirement for neuronal activity – ocular dominance columns

As discussed earlier in this Topic, the visual cortex in mammals must receive inputs from both eyes to enable binocular vision. The mammalian visual system is arranged so that axons from the left side of each retina project to the right lateral geniculate nucleus (LGN), and axons from the right side of each retina project to the left LGN. The connections from each eye are segregated to different layers of the LGN, but axons projected from the LGN to the visual cortex converge in one cell layer (*Fig. 3*). Initially, the visual fields of the left and right eyes overlap because axons from the LGN representing projections from the same region of the visual field of both eyes project to the same region of the visual cortex. However, later in development the cortex becomes divided into **ocular dominance columns**, where adjacent groups of cortical neurons respond to light from the same visual field, but one column responds to the left eye and one to the right eye.

The establishment of ocular dominance columns is dependent on neuronal activity. It is thought that adjacent cells receiving action potentials resulting

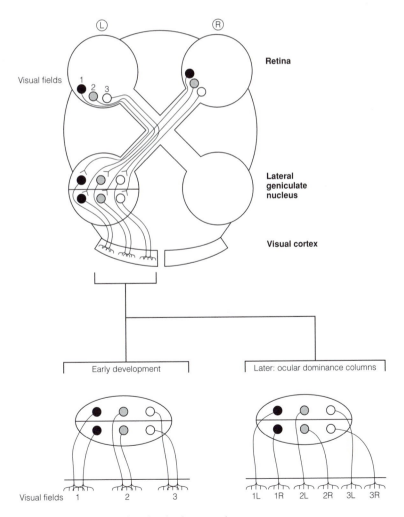

Fig. 3.   Development of ocular dominance columns.

from stimulation of the same eye initiate their own action potentials simultaneously, therefore cooperating to activate a given target neuron. This would help to establish the synapse at the expense of others in the same way as discussed above for neuromuscular junctions, so that axons representing the other eye would withdraw. The development of ocular dominance columns demonstrates an important principle of neural development: neuronal connections are initially established as part of the developmental program, but refinements are made on the basis of experience (in this case, the triggering of synapses by visual stimulation of the eye). Surprisingly, however, while synaptic activity is required for the establishment of ocular dominance columns, visual stimuli are only part of the process. In mammals, the ocular dominance columns are already formed at birth, even though the animal has had no exposure to light. It appears that spontaneous firing of synapses in the developing visual system is important for the establishment of the ocular dominance columns, while neural activity caused by visual experience takes over to refine the initial pattern.

K1 MESODERM INDUCTION AND PATTERNING

Key Notes

Mesoderm

The mesoderm of the vertebrate embryo is patterned along the dorsoventral axis. The most dorsal (axial) mesoderm forms the notochord and prechordal plate. The paraxial mesoderm becomes divided into somites, which form the axial skeleton, muscles and dermis. The intermediate mesoderm forms the gonadal ridge and excretory system. Finally, the lateral mesoderm forms the body wall and blood islands, as well as special organs such as the heart.

Mesoderm induction

In *Xenopus*, the mesoderm arises in the equatorial marginal layer after the midblastula transition in response to inductive signals from the vegetal blastomeres. Even at this early stage, there is a distinction between dorsal and ventral/lateral mesoderm, as dorsal and ventral explants give rise to different cell types. However, the full range of intermediate mesodermal cell types is not represented because further signaling within the mesoderm is required for complete dorsoventral patterning.

Animal cap assay

The animal cap assay involves removing the roof of the *Xenopus* blastocoel (animal cap) and recombining it with other tissues to test for inductive interactions. Animal caps combined with vegetal blastomeres can be used to assay for mesoderm induction.

Four-signal model

The four-signal model describes the molecular basis of mesoderm induction and patterning in *Xenopus*. The first two signals originate from the vegetal blastomeres and are required for mesoderm induction. Signal 1 is a general mesoderm inducer, while signal 2 is a dorsal modifier restricted to the Nieuwkoop center, which synergizes with the first signal to induce dorsal (organizer) mesoderm. The third and fourth signals are competitive. They originate in the ventral mesoderm and organizer, respectively, and establish dorsoventral pattern in the mesoderm.

Mesoderm inducing factors (signals 1 and 2)

It is thought that TGF-β signaling proteins secreted by vegetal blastomeres are responsible for general mesoderm induction (signal 1), although FGF signaling is also necessary to provide a permissive molecular environment. Four nodal-related proteins (Xnr1, Xnr2, Xnr4 and Derrière) have been identified as components of this signal, although two other TGF-β proteins, Vg-1 and activitin, may also be involved. The dorsal modifier (signal 2) is dependent on the presence of β-catenin, which is activated on the dorsal side of the egg by cortical rotation.

Mesoderm patterning factors (signals 3 and 4)

A number of secreted proteins with ventralizing activity are distributed in the *Xenopus* blastula, including several BMPs and Xwnt8. These molecules probably represent signal 3. Signal 4 comprises several anti-BMPs (Noggin, Chordin, Follistatin, Xnr3) and anti-Wnts (Cerberus,

Frzb-1, Dickkopf-1) secreted by the organizer. Opposing gradients of signals 3 and 4 provide the necessary range of molecular environments to specify a range of intermediate mesodermal cell types.

Related topics	Mechanisms of developmental commitment (A3)	Ectoderm: neural induction and epidermis (J1)
	Signal transduction in development (C3)	Patterning the anteroposterior neuraxis (J2)
		Endoderm development (K6)

Mesoderm

Mesoderm is the 'middle' germ layer of the post-gastrulation vertebrate embryo sandwiched between the outer covering of **ectoderm** and the inner **endoderm**. Mesoderm gives rise to a diverse variety of tissues, depending on its position along the embryo's **dorsoventral axis**. The most dorsal mesoderm arises directly from the **organizer** (Topic I2) and forms the **prechordal plate** and the **notochord**. Both of these tissues have important inductive functions early in development as discussed in Topics I2, J1 and J2. However, while the prechordal plate eventually contributes to the mesenchyme of the head, the notochord plays only a minor role in the adult organism, forming the nuclei pulposi of the intervertebral discs. Moving ventrally, the remaining mesoderm in the embryo is distributed bilaterally. First, the **paraxial mesoderm** becomes segmented to form the **somites** that flank the notochord and neural tube. These give rise to the cartilage and bone of the vertebral column, skeletal muscles and dermis (Topics K2 and K3). Next, the **intermediate mesoderm** gives rise to the urogenital system, which comprises gonads and associated ducts (Topic E5) and the embryonic and definitive kidneys (Topic K4). Finally, the **lateral** and **ventral mesoderm** gives rise to the remaining mesodermal derivatives of the body: the smooth muscle and mesenchyme of the gut and body wall, the limb buds, the pharyngeal arches and the blood. A very important organ originating from the lateral mesoderm is the heart, whose early development is discussed in Topic K5.

Mesoderm induction

In *Xenopus* and other amphibians, the ectoderm and endoderm are specified autonomously by maternal gene products localized in the egg. Therefore, animal caps cultured in isolation differentiate into epidermis, an ectodermal derivative, while vegetal tissue begins to express characteristic endodermal marker genes (Topics J1 and K6). Conversely, the mesoderm is specified conditionally by inductive interactions between animal and vegetal cells. Induction occurs at around the time of the midblastula transition, so equatorial explants taken before this stage give rise to epidermis, while those taken from later blastulae give rise to mesoderm. The equatorial ring of presumptive mesoderm is called the **marginal layer**. In *Xenopus*, but not other amphibians, the superficial cells in the marginal layer give rise to pharyngeal endoderm, while the deep cells give rise to mesoderm. The signals discussed below with respect to mesoderm induction are also required for the induction of this superficial endoderm, hence mesoderm induction is more properly termed **mesendoderm induction**.

The fate of explants from the marginal layer of the late blastula depends on the position from which the explant was taken (*Fig. 1*). Dorsal explants give rise primarily to notochord and somites, while ventral explants give rise to

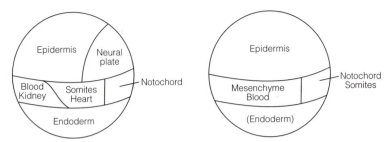

Fig. 1.   Fate map of Xenopus blastula (left), plus a specification map based on the behavior of explants (right). In some cases (e.g. endoderm) overt differentiation of explants does not take place, although appropriate markers are expressed.

mesenchyme and blood. This is because mesoderm induction involves at least two separate signals, one specifying dorsal mesoderm that will form the organizer, and one specifying ventral mesoderm. The range of cell types obtained from explants at this stage is not representative of the entire *Xenopus* fate map. For example, many intermediate and lateral mesodermal fates such as kidney and heart are not represented (*Fig. 1*). This is because further signaling between the dorsal and ventral mesoderm is required to establish the intermediate cell fates. These signals are discussed in more detail below.

Much less is known about mesoderm induction in amniotes. Fate mapping shows that dorsal mesoderm arises from the node, which is the amniote organizer, while epiblast lateral to and posterior to the node gives rise to other mesodermal cell types. However, the cells are not determined to form mesoderm until they migrate through the primitive streak, as reorganizing the epiblast leads to respecification of germ cell fates. Oddly enough, if undifferentiated epiblast cells are cultured in isolation they predominantly give rise to a mesodermal derivative, muscle. This innate myogenic tendency reflects low level expression of muscle determining transcription factors of the MyoD family (see Topic K3). It is possible that signals from other tissues repress myogenic activity in these cells during and after gastrulation, until they have become committed to other fates.

The early differentiation of the germ layers is revealed by the expression of particular marker genes, and early markers of the ectodermal and endodermal lineages are discussed in Topics J1 and K6. One of the earliest and most useful mesodermal markers is the gene encoding the transcription factor **Brachyury**. Like the transcription factors Mixer, Sox17α and Sox17β, which can convert presumptive ectoderm into endoderm (Topic K6), Brachyury can convert isolated ectoderm into mesoderm, and is thus a primary determinant of the mesodermal cell lineage. Although initially expressed throughout the mesoderm, *Brachyury* expression is later restricted to the notochord and posterior mesoderm. In mice, homozygous loss of the *Brachyury* gene is lethal because mesoderm fails to form, but if only one allele is lost, the phenotype predominantly reflects the function of Brachyury in the posterior mesoderm. Heterozygotes can be identified because of the reduced size of the tail. Indeed, the gene was first identified in mice on the basis of this phenotype, while loss of the *Brachyury* homolog in zebrafish generates a similar mutant, called *no tail*. Further mesodermal markers are used to investigate early patterning (*Table 1*). These markers are especially useful to monitor the effects of putative dorsalizing and ventralizing agents on explanted tissues.

Table 1.   Useful mesodermal marker genes

Gene or protein	Expression
Brachyury (*Xbra*)	Early pan-mesodermal, later restricted to posterior mesoderm
goosecoid	Dorsal anterior mesoderm (and neural plate at a later stage)
noggin	Dorsal anterior mesoderm
Cardiac actin	Dorsolateral mesoderm
Xwnt8	Initially throughout the marginal layer, then restricted to ventrolateral mesoderm

Animal cap assay

Mesoderm induction in *Xenopus* can be studied using the **animal cap assay** (*Fig. 2*). The roof of the blastocoel (the 'cap' of the animal hemisphere) is removed and placed adjacent to the vegetal cells. After a few days, the caps begin to express mesoderm marker genes and mesodermal cell types begin to differentiate. Endodermal markers are also expressed. Such experiments have shown, for example, that animal cap tissue is competent to respond to induction from about the 32-cell stage (3–4 hours after fertilization) and loses competence during gastrulation (12 hours after fertilization). Contact is not required throughout the cells' period of competence, just 4–5 hours of contact is sufficient for all mesodermal tissues to be induced. Such experiments have also shown that animal cap tissue will respond to mesoderm induction but individual blastomeres will not. Therefore, mesoderm induction involves a **community effect** (Topic A3).

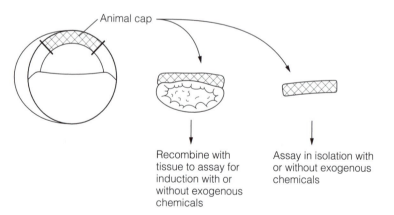

Recombine with tissue to assay for induction with or without exogenous chemicals

Assay in isolation with or without exogenous chemicals

Fig. 2.   The animal cap assay.

Four-signal model

As discussed above, dorsal and ventral explants from the marginal layer of the late *Xenopus* blastula give rise to different cell types, representative of dorsal and ventral mesoderm. Animal cap assays using early blastulae have provided evidence for two distinct signals emanating from the vegetal blastomeres, one specifying dorsal mesoderm and one specifying ventral mesoderm. Animal caps recombined with vegetal tissue from the dorsal side of the embryo (which includes the Nieuwkoop center; Topic I1) are induced to form dorsal mesoderm (notochord and muscle). Conversely, animal caps recombined with vegetal tissue from the ventral side of the embryo are induced to form ventral meso-derm (blood islands and mesenchyme). Since these represent the extreme dorsal and ventral cell types, further signals must pattern the mesoderm to generate intermediate mesodermal fates. This is the essence of the **four-signal model**, which can be summarized as follows (*Fig. 3*):

- **Signal 1**, a component of all vegetal cells, is the basic mesoderm inducer that causes all adjacent animal cap cells to adopt a ventral mesodermal fate. Signal 1 would induce the expression of pan-mesodermal markers like *Brachyury*.
- **Signal 2** emanates specifically from the dorsal vegetal cells (the Nieuwkoop center) and induces the formation of dorsal mesoderm, which has organizer activity. This signal would induce early organizer-specific genes such as *goosecoid*. The signal could be qualitatively or quantitatively different from signal 1.
- **Signal 3** is secreted by the organizer and acts on the adjacent ventral meso-derm, specifying intermediate mesodermal fates.
- **Signal 4** is present in the ventral mesoderm. It actively promotes ventral fates and competes with the dorsalizing signal emanating from the organizer.

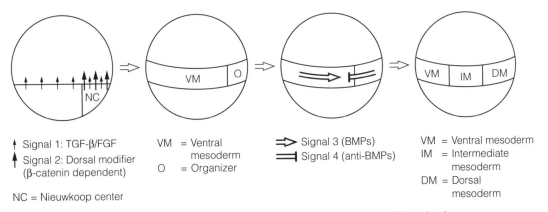

Fig. 3. The four-signal model of mesoderm induction and patterning, with some candidate signals.

Mesoderm-inducing factors (signals 1 and 2)

Barrier experiments have shown that mesoderm induction involves diffusible molecules rather than cell–cell contacts. Molecules of both the **TGF-β** and **FGF** families have been implicated in this inductive process, as both are present in the *Xenopus* embryo at the appropriate stage, and blocking each signaling pathway severely disrupts mesoderm induction. Several TGF-β proteins can induce different mesoderm cell types at different concentrations, with higher concentrations inducing more dorsal fates. FGF signaling is important for mesoderm induction but does not appear to affect dorsoventral patterning. This suggests that FGFs may play a permissive role to facilitate TGF-β signaling, and that more potent TGF-β signaling occurs in the dorsal region of the embryo, favoring the induction of dorsal mesoderm. The enhancement of TGF-β signaling on the dorsal side of the embryo is dependent on the transcription factor **β-catenin**, which accumulates on the dorsal side of the egg following cortical rotation (see Topic I1).

TGF-β signaling

At least four nodal-like members of the TGF-β superfamily (**Xnr1, Xnr2, Xnr4** and **Derrière**) are secreted from vegetal cells in the *Xenopus* midblastula and have direct roles in mesoderm induction. At the onset of zygotic gene expression, the genes for these proteins are activated by a maternally-encoded T-box transcription factor, **VegT**, which is localized to the vegetal hemisphere. In vegetal cells, VegT autonomously activates downstream transcription factors such as Bix4, resulting in endoderm specification (Topic K6). Mesoderm induction results from the non-autonomous activity of these signaling proteins on adjacent animal cells that lack VegT. The Xnr proteins are pan-mesodermal inducers while Derrière, as its name suggests, is required only for trunk and tail mesoderm development.

Two further TGF-β proteins, **Vg-1** and **activin**, are also likely to have roles in mesoderm induction. The treatment of isolated animal caps with either protein converts the ectoderm cells into mesoderm, with increasing doses inducing the expression of progressively more dorsal marker genes. High doses can induce organizer-specific markers such as *goosecoid* and *noggin*, and injection of *activin* or *Vg-1* mRNA into vegetal blastomeres of a *Xenopus* blastula can induce a secondary axis.

The role of β-catenin as a dorsal modifier of TGF-β signaling

As higher doses of TGF-β proteins generate progressively more dorsal mesoderm, a logical suggestion is that the role of the Nieuwkoop center is to enhance the level of TGF-β signals emanating from the vegetal blastomeres. As discussed in Topic I2, the defining feature of the Nieuwkoop center is the overlap between the dorsal zone of β-catenin accumulation and the vegetal zone of TGF-β signaling proteins, so it is likely that β-catenin and one or more of the TGF-β proteins synergize to induce dorsal (organizer) mesoderm. The mechanism is unclear, but may involve a transcription factor expressed in response to β-catenin, called Siamois. This has been shown to activate the *goosecoid* gene, which is a key component of organizer function, but only in the presence of TGF-β signals. The organizer may therefore form in presumptive mesoderm cells expressing both Siamois and one or more factors activated by TGF-β signaling from the vegetal hemisphere. In the absence of Siamois, the TGF-β-dependent factor would induce ventral mesoderm.

FGF signaling

FGF was initially thought to induce only ventral mesoderm, and initial models of mesoderm induction suggested that TGF-β and FGF signals acted separately as dorsal and ventral mesoderm inducers. However, blocking FGF signaling *in vivo* with dominant negative receptors prevents the formation of all mesoderm in the embryo except the prechordal plate. This suggests that FGFs play a more fundamental role in mesoderm induction than originally conceived. Furthermore, blocking FGF signaling *in vitro* inhibits the expression of mesodermal markers such as *Brachyury* in animal caps exposed to activin. This suggests FGF signaling may provide a permissive molecular environment for TGF-β signaling.

Mesoderm patterning factors (signals 3 and 4)

Xenopus embryos developing without the benefit of an organizer are ventralized, lacking dorsal mesoderm (notochord, somites) and dorsal ectoderm (neural plate). This reflects the ubiquitous expression of ventralizing signals, particularly those of the bone morphogenetic family of TGF-β signaling molecules (**BMP2, BMP4** and **BMP7**) and the Wnt family (**Xwnt8**). These signals establish ventral fates by activating transcription factors that in turn regulate ventral-mesoderm specific genes. For example, the secreted protein BMP4 appears to play an important role in specifying ventral mesoderm by inducing the transcription factors **Xvent-1**, **Xom** and **Vox**.

The primary role of the organizer is to protect itself from these ventralizing signals and secure its own development as dorsal mesoderm. The organizer therefore secretes a number of anti-BMPs (**Noggin, Chordin, Follistatin, Cerberus** and **Xnr3**) and anti-Wnts (**Cerberus, Frzb-1, Dickkopf**). These proteins function in different ways, some by direct binding to the ventralizing factors themselves and some by sequestering signal transduction components, to clear active BMPs and Wnts from the organizer. All the above are secreted proteins, so they diffuse from the organizer and establish gradients in all three germ layers. The dorsoventral axis of the mesoderm is set up as opposing gradients of ventralizing and dorsalizing factors, representing Signals 3 and 4 of the four-signal model. This results in the activation of different classes of transcription factors at different positions along the axis, and the subsequent differentiation of the full spectrum of mesodermal tissues.

The dorsalizing factors secreted by the organizer are important not only for the patterning of the mesoderm, they also play a major role in patterning the other germ layers. The wider role of the organizer is discussed in Topic I2. More details of the role of the organizer in neural induction and patterning can be found in Topics J1 and J2.

K2 SOMITOGENESIS AND PATTERNING

Key Notes

What are somites?

Somites are paired blocks of mesoderm bracketing the neural tube, which arise by segmentation of the paraxial mesoderm at all levels caudal to the anterior hindbrain. The somites are transient structures, but are critical for the segmental organization of the trunk (especially the axial skeleton).

Somitogenesis

Somite formation (somitogenesis) begins with the reorganization of paraxial mesoderm into whorls of cells called somitomeres. Along most of the anteroposterior axis, a mesenchymal to epithelial transition occurs that separates these somitomeres into somites. The number of somites formed is characteristic of each species, and the rate of somitogenesis represents an intrinsic measure of the progress of development. There is evidence that somitogenesis is regulated by an intrinsic molecular clock. Shortly after formation, each somite becomes divided into anterior and posterior compartments.

Regional diversification of somite derivatives

All somites are initially identical in appearance and they give rise to the same variety of cell types. However, each somite has a definitive positional value as shown by the generation of morphologically diverse structures along the axis, such as specific types of vertebrae. This positional identity is established very soon after the paraxial mesoderm is formed, as grafting blocks of paraxial mesoderm to a different position along the axis does not change the order of somite formation and results in the development of regionally inappropriate vertebrae.

Positional identity – *Hox* gene expression

Regional specification of somite identities is thought to be established by *Hox* gene expression. One piece of evidence is the characteristic overlapping and posteriorly nested expression patterns, which are similar to those seen for the homologous HOM-C genes in *Drosophila*.

The vertebrate *Hox* code

Further evidence for the role of *Hox* genes in the regional specification of somites is provided by the 'vertebrate *Hox* code'. Although different vertebrates have different numbers and types of vertebrae, their arrangement appears in each case to correspond to the expression domains of the *Hox* genes. For example, in all vertebrates, the transition from cervical to thoracic vertebrae and the position of the forelimbs corresponds to the anterior boundary of *HoxC6* expression.

Manipulation of positional identities

The most convincing evidence supporting the role of *Hox* genes in regional specification of the somites comes from experiments in which the *Hox* gene expression patterns are deliberately altered, either by mutation, forced overexpression, or treatment with retinoic acid. Such manipulations generate animals showing evidence of homeotic transformations among the vertebrae, indicating the positional identities of the somites have been changed. The effect of retinoic acid on *Hox* gene expression suggests one way in which the *Hox* gene expression pattern could be set up during gastrulation. The node is known to secrete retinoic acid as it regresses, and this could sequentially activate progressively more 3′ *Hox* genes as the node moves posteriorly.

Related topics

Pattern formation (A6)
Homeotic selector genes and
   regional specification (H5)
Patterning the anteroposterior
   neuraxis (J2)

The neural crest (J6)
Mesoderm induction and
   patterning (K1)
Somite differentiation (K3)

What are somites?

During gastrulation, presumptive dorsal mesoderm enters the embryo through the organizer (the dorsal lip of the blastopore in *Xenopus*, the node in amniotes) and establishes the **axial mesoderm** (notochord and prechordal plate). Presumptive mesoderm entering the embryo through the rest of the ingression zone (the lateral lips of the blastopore in *Xenopus*, the primitive streak in amniotes) becomes arranged as two broad lateral bands either side of the notochord. The mesoderm immediately adjacent to the notochord is called the **paraxial mesoderm**, and separates from the remaining lateral mesoderm to form a **segmental plate** or **presomitic plate** on each side of the notochord. The presomitic plate then undergoes a process of segmentation (**somitogenesis**) to form paired mesodermal blocks called **somites**.

Like the notochord, the somites are transient structures, but they are important for the segmental organization of the vertebrate body plan, giving rise to the axial skeleton (the vertebrae), skeletal muscles and contributing to the dermal blocks underlying the dorsal epidermis. In the trunk of the embryo, the somites also confer positional identity on the adjacent neural tube and neural crest (Topics J2 and J6) as well as guiding migrating neural crest cells and axons (Topics J6 and J7).

Somitogenesis

Somitogenesis begins with the formation of **somitomeres**, whorls of concentric mesodermal cells marking the position of future somites in the presomitic plate (*Fig. 1*). The conversion of a somitomere into a true somite involves a

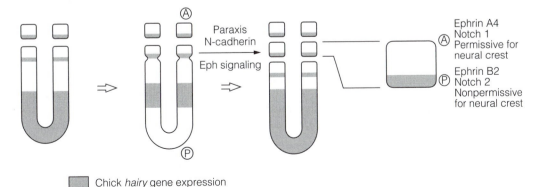

Chick *hairy* gene expression

Fig. 1. Somitogenesis in the chick embryo corresponds to cyclical patterns of hairy *gene expression. Each somite is divided into anterior and posterior compartments with different developmental properties and different patterns of gene expression.*

mesenchymal to epithelial transition, which requires the synthesis of the cell adhesion molecule **N-cadherin** possibly under the control of a transcription factor called **Paraxis**. Ephrins and Eph receptors have also been implicated in the process of separation. Most of the somitomeres undergo this transition, but the first seven (which flank the midbrain and hindbrain) remain as somitomeres. Myoblasts from these somitomeres migrate to the pharyngeal arches and give rise to skeletal muscles of the face and neck (Topic J6).

The somites form sequentially from anterior to posterior and at an even rate. However, the rate of somitogenesis and the total number of somites formed are particular for each species. For example, in humans, the first pair of somites form on day 20 of gestation and the last pair form approximately 10 days later. There are initially 44 pairs of somites although those arising in the transient embryonic tail do not persist. Somitogenesis can provide an objective control to determine the developmental stage of an embryo, as the actual rate of development is influenced by external factors such as temperature. Developmental biologists therefore talk of embryos 'at the four-somite stage' etc. The periodicity of somite formation has led to suggestions that an intrinsic molecular clock, perhaps linked to the cell cycle, might control somitogenesis. This is supported by the dynamic expression pattern of the chick gene *hairy1*, which oscillates in time with somite formation. It is currently unknown how this periodic expression pattern is regulated (*Fig. 1*).

Soon after formation, each somite becomes divided into anterior and posterior compartments. This division is first apparent based on the complementary expression domains of genes such as *Notch1*, *Notch2* and *Delta*, and indeed the anterior and posterior compartments are not specified properly when these genes are mutated. The anterior and posterior compartments have different properties (e.g. the anterior but not the posterior compartment is permissive for neural crest migration; Topic J6). The vertebrae are formed by the fusion of the anterior sclerotome from one somite to the posterior sclerotome of the next.

Regional diversification of somite derivatives

The presomitic plate gives rise to successive pairs of somites of identical appearance, which differentiate to form the same cell types. However, the structures formed by these cells vary according to their position along the anteroposterior axis. This is seen most clearly in the structure of the vertebrae. For example, the thoracic vertebrae include ribs, while the lumbar and sacral vertebrae do not. This indicates that the somites have unique positional identities along the anteroposterior axis.

Grafting experiments have demonstrated that the positional identity of somites is established very early in development. Thoracic somites grafted to another region of the embryo will produce vertebrae with ribs at the ectopic site. Moreover, the thoracic region of the unsegmented presomitic plate will also give rise to vertebrae with ribs at an ectopic site. Even the order of somitogenesis is determined, and is not affected by surgical manipulation. If the unsegmented region of paraxial mesoderm is inverted, the somites still form in the correct order in the inverted segment, proceeding posterior to anterior.

Positional identity – *Hox* gene expression

The positional identity of somites could be specified by the expression of *Hox* genes. As discussed in Topic A5, the vertebrate *Hox* genes are homologous to the *Drosophila* HOM-C genes, which are known to confer positional identity in the developing fly. In *Drosophila*, the HOM-C genes are expressed in overlapping expression patterns along the anteroposterior axis, conferring on each segment a unique combinatorial code of available transcription factors (Topic H5). This is thought to regulate particular combinations of downstream genes that control cell behavior, resulting in the development of regionally specific structures along the axis.

The *Hox* genes are expressed in the paraxial mesoderm as soon as it is formed, and the characteristic expression pattern suggests they could fulfil a similar patterning function. The expression pattern is posteriorly nested, with anterior boundaries of different paralogous *Hox* subgroups corresponding to individual somites. The dynamics of *Hox* gene expression are best-characterized in chick and mouse embryos, where the somites form in the wake of the regressing primitive node during gastrulation. As epiblast cells move through the streak into the embryo, they are specified as mesoderm and begin to express *Hox* genes. The first cells through the streak, which are destined to become anterior somites, express *HoxA1* and *HoxB1*, representing the most 3' paralogous subgroup. As the streak regresses, more posterior type *Hox* genes are switched on, until, in the most caudal regions of the embryo, all the *Hox* genes are expressed, including *HoxD12* and *HoxD13*. Not all of the *Hox* genes are expressed robustly throughout the embryo, indeed different *Hox* genes are expressed for different durations and their expression levels vary considerably. However, each has a precise anterior boundary of expression giving each somite a specific combinatorial expression pattern. The general picture of *Hox* gene expression that emerges from studies in the mouse is shown in *Fig. 2*. This figure also shows the structure of the four vertebrate *Hox* clusters compared to the *Drosophila* HOM-C. As described for the *Drosophila* HOM-C (Topic H5), the vertebrate *Hox* genes show colinearity and posterior dominance.

The vertebrate *Hox* code

The early onset of *Hox* gene expression in the paraxial mesoderm, and the characteristic overlapping expression patterns that arise during gastrulation, provide strong evidence that the *Hox* genes control positional identity in the trunk. Further evidence comes from comparing the anatomy and *Hox* gene expression patterns of different vertebrates. Different species have different numbers and different types of vertebrae, but in each case this appears to correlate with the *Hox* gene expression patterns (*Fig. 3*). For example, chickens have fewer ribs than mice because they have 14 cervical vertebrae and seven thoracic vertebrae, while mice have seven cervical vertebrae and 13 thoracic vertebrae. However, in both species the junction between cervical and thoracic vertebrae corresponds to the anterior boundary of *HoxC6* expression, which also marks the position of the forelimbs. Similar correlations exist at the thoracic/lumbar, lumbar/sacral and sacral/caudal boundaries. Interestingly, the absence of forelimbs in snakes may also reflect variations of *Hox* gene expression patterns. The python embryo has an unusually broad domain of anterior *Hox* gene expression, resulting in the specification of hundreds of identical thoracic vertebrae but no limb fields. The role of *Hox* genes in specifying the position of the limbs is discussed in Section M.

Manipulation of positional identities

Expression patterns and comparative anatomy provide only correlative evidence that *Hox* genes confer positional identity on the somites. Definitive proof requires the experimental manipulation of *Hox* gene expression resulting in specific and predictable homeotic transformations (the conversion of one body segment into the likeness of another). These types of experiments have been approached in four ways: (1) overexpression of *Hox* genes in transgenic

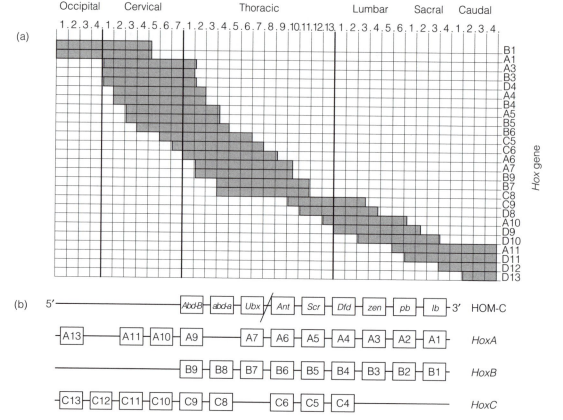

Fig. 2. (a) Hox gene expression in the mouse somitic mesoderm, in relation to final arrangement of vertebrae. (b) Hox gene clusters in the mouse compared to Drosophila HOM complex. Note colinearity of gene expression and order along the chromosome.

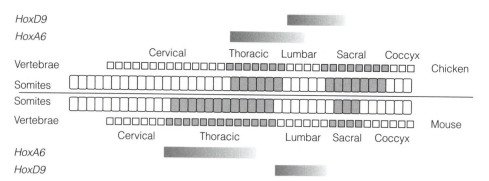

Fig. 3. The vertebrate Hox code. Comparison of vertebral pattern of chick and mouse embryos, with paraxial mesoderm expression patterns of HoxA6 and HoxD9, marking the cervical/thoracic and thoracic/lumbar transitions respectively.

mice; (2) *Hox* gene knockouts in mice; (3) knockouts of genes than regulate *Hox* gene expression; (4) retinoic acid treatment of embryos to alter *Hox* gene expression patterns.

Many of the *Hox* genes have been overexpressed or knocked out and in some cases this has revealed quite striking homeotic transformations. For example, the *HoxC8* gene is expressed in most of the thoracic vertebrae, but its expression drops off in more posterior regions of the embryo. Mice with targeted mutations in this gene develop an extra rib, as if the first lumbar vertebra had been converted into an additional 13th thoracic vertebra. More subtle changes occur in each rib from the seventh to the 13th, indicating that all the vertebrae in the normal domain of *HoxC8* expression had been anteriorized (*Fig. 4*). Other *Hox* knockouts show only minor phenotypes because the multiple paralogous genes expressed at each level appear to have partially overlapping functions. This problem has been addressed by generating multiple *Hox* knockouts to remove all active members of the same paralogous subgroup simultaneously.

Hox gene knockout and misexpression experiments demonstrate the principle that abolishing the expression of a *Hox* gene tends to shift positional identities towards the anterior, while expressing a *Hox* gene outside its normal expression domain (i.e. at a more anterior position) tends to shift positional identities towards the posterior, as would be expected from the expression patterns. However, some *Hox* genes show the opposite effect, while others cause simultaneous transformations in opposite directions at different positions along the axis. This indicates that the control of positional identify in the somites is probably not based on a simple combinatorial code, but involves interactions among the *Hox* genes and with other factors. Some genes controlling *Hox* gene expression have been identified, such as the mouse gene *Cdx1*, which is homologous to the *Drosophila* gene *caudal*. Posteriorization occurs in *cdx1* knockouts, and this has been correlated with a posterior shift in the expression patterns of all the *Hox* genes. Retinoic acid can induce both anterior and posterior transformations, depending on the dose and the timing of administration, and again this

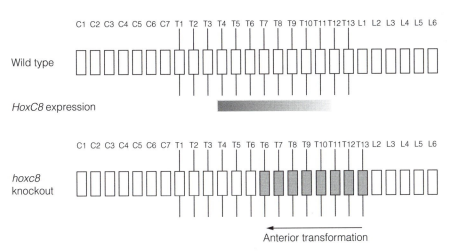

Fig. 4. Homeotic transformation following knockout of the HoxC8 gene in the mouse. There is an anterior transformation of vertebrae T7 to L1, resulting in duplication of T6 and Loss of L1. The conversion of L1 to T13 generates an extra set of ribs.

corresponds to shifting the expression domains of the *Hox* genes. Retinoic acid is probably involved in establishing *Hox* gene expression in the paraxial mesoderm during gastrulation. Retinoic acid can induce *Hox* gene expression in cultured mesoderm cells, and progressively more posterior type *Hox* genes are induced with increasing doses. Several of the *Hox* genes have retinoic acid response elements in their promoters, and mice with simultaneous targeted mutations in several retinoic acid receptor genes show anteriorization of the vertebrae. Furthermore, the node is a rich source of retinoic acid and could therefore induce *Hox* gene expression in mesoderm cells during gastrulation. The characteristic overlapping pattern of *Hox* genes could reflect a gradient of retinoic acid with its source at the node, a progressive increase in the level of retinoic acid produced by the node as it regresses, or the length of time cells in the node are exposed to retinoic acid.

K3 SOMITE DIFFERENTIATION

Key Notes

Fate of the somites

The somites differentiate to form sclerotome (future cartilage of the axial skeleton), two blocks of myotome (future muscles) and dermatome (cells that contribute to the dermis). These tissues arise in stages through sequential inductive interactions.

Signals and markers that specify somite fates

Early somitic cells are not patterned, because rotation has no effect on subsequent development. Somite patterning therefore depends on signals from other tissues. Sclerotome induction requires Sonic hedgehog from the notochord, while epaxial myotome induction requires both Sonic hedgehog from the notochord and Wnts from the dorsal neural tube. Hypaxial myotome is induced by the lateral mesoderm and ectoderm. Dermatome is induced by neurotrophin-3 from the dorsal neural tube. Following induction, these regions begin to express characteristic markers, representing transcription factors that promote tissue-specific differentiation.

Molecular basis of muscle differentiation

The myotome comprises a population of proliferating myoblasts, which must exit from the cell cycle and fuse together to form multinucleate myotubes. These myotubes accumulate contractile proteins to form mature myofibers. Skeletal muscle differentiation involves a small family of myogenic bHLH transcription factors. Initially MyoD1 or Myf5 commit somitic cells to the myoblast lineage. The expression of Myogenin is required to withdraw myoblasts from the cell cycle and permit fusion into myotubes. Then MRF4 is expressed and promotes further differentiation. Other myogenic transcription factors, such as MEF2, may cooperate with the myogenic bHLH transcription factors in skeletal muscle development, but play a more prominent role in the development of smooth and cardiac muscle, where the MyoD1 family is not expressed.

Related topics

Mechanisms of developmental commitment (A3)
Maintenance of differentiation (A5)

Signal transduction in development (C3)

Fate of the somites

The somites give rise to three major tissues: the cartilage and bone of the vertebral column, skeletal muscle and dermis. These tissues arise in several stages (*Fig. 1*). First, the medial region of each somite undergoes an epithelial to mesenchymal transition. The resulting mesenchymal cells establish the **sclerotome**, which migrates to surround both the neural tube and notochord, condenses and differentiates in **chondrocytes** (cartilage-secreting cells), and forms the rudiments of the vertebral bodies. The remaining epithelial component of the somite is called a **dermomyotome**. This splits into two populations, a **dermatome** (which contributes to the connective tissue of the dermis) and a **myotome** which gives rise to skeletal muscles. Both components undergo

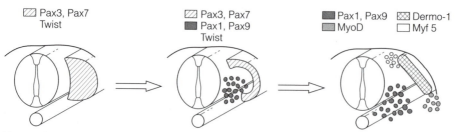

Fig. 1.   Somite differentiation, showing marker genes expressed in the sclerotome, myotome and dermatome.

epithelial to mesenchymal transitions to generate populations of migratory cells. The **dermal mesenchyme** migrates under the ectoderm and combines with mesenchyme from the lateral mesoderm to form the definitive dermis. The myotome splits into several subpopulations of **myoblasts** to produce different muscle groups. In the appropriate regions along the anteroposterior axis, myoblasts also migrate into the adjacent limb buds to generate the limb musculature.

Signals and markers that specify somite fates

When the somites form, the fate of individual cells is not specified as the somite can be rotated without affecting subsequent development. This indicates that cell specification requires interactions with the surrounding tissues, and it appears that the notochord, neural tube, lateral mesoderm and overlying ectoderm are all involved in this process (*Fig. 2*). After these signals have been received and interpreted, different parts of the somite become committed to different fates and begin to differentiate. This can be seen by the expression of characteristic markers of the alternative lineages (*Fig. 1*).

Sclerotome is specified in response to signals from the notochord and the floor plate of the neural tube. The major signal is thought to be **Sonic hedgehog**, as ectopically applied Sonic hedgehog protein can induce sclerotome differentiation in adjacent somite tissue, as can grafted pieces of notochord. Indeed, by grafting notochord or implanting a Sonic hedgehog bead dorsolaterally with respect to the neural tube, the entire somite can be converted into sclerotome. In response to Sonic hedgehog, the medial somitic cells undergo epithelial to mesenchymal transition and begin to express sclerotome-specific transcription factors such as **Pax1**, **Pax9** and the bHLH protein **Scleraxis**. These transcription factors activate chondrocyte-specific gene expression, resulting in the production of a cartilage model of the vertebral skeleton, which is later ossified.

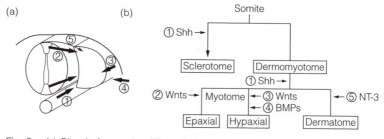

Fig. 2.   (a) Signals for somite differentiation come from the notochord and floor plate (1), dorsal neural tube (2, 5), ectoderm (3) and lateral plate mesoderm (4). (b) The nature of these signals and their roles in somite differentiation.

As the sclerotome cells migrate away, blocks of myotome arise at each end of the remaining dermomyotome, and this may involve two stages of signaling. The **epaxial myotome**, which gives rise to the muscles of the back, is induced by signals from the dorsal neural tube. At least three members of the Wnt family of signaling proteins are synthesized in this region (Wnt1, Wnt3 and Wnt4) and any combination of these could be required for myotome induction *in vivo*. However, *in vitro* studies have shown that proteins such as Wnt1 can only induce myotome in somites that have already been exposed to Sonic hedgehog. Therefore, it appears that sustained exposure to Sonic hedgehog is required for sclerotome differentiation, while a brief exposure to Sonic hedgehog, followed by one or more Wnts is required for myotome specification. The treatment of somites with Sonic hedgehog and Wnts results in the induction of muscle markers such as MyoD1, and the repression of the general somite marker Twist (this continues to be expressed in the sclerotome).

In contrast to the epaxial myotome, the hypaxial myotome (which gives rise to ventral body muscles and limb muscles) appears to be induced only by lateral tissues. The hypaxial muscles form normally in embryos where the neural tube has been removed, but do not differentiate if the somite is separated from adjacent lateral mesoderm and ectoderm. Wnt proteins in the ectoderm are thought to be involved in the specification of hypaxial myotomes, although there is evidence that BMP4 or FGF5 (secreted by the lateral mesoderm) are also required. The role of these proteins is inhibitory – they act to delay the onset of myogenic gene expression in order to allow the cells to migrate to their final destinations.

The dermatome is induced by signals from the dorsal region of the neural tube, including the neurotrophic factor neurotrophin-3. As with the sclerotome, this results in an epithelial to mesenchymal transition allowing the dermal mesenchyme to migrate under the epidermis. Early markers of dermatome specification include the transcription factor **Dermo-1**.

Molecular basis of muscle differentiation

The myotome initially comprises a population of proliferative cells called **myoblasts**. To differentiate into muscle, these cells must stop proliferating and fuse together to form long multinucleate syncytia, termed **myotubes**. The myotubes differentiate into mature **myofibers** by the accumulation of contractile proteins.

As discussed above, muscle differentiation begins with the induction of myogenic bHLH transcription factors, either MyoD1 or Myf5. In the mouse, Myf5 is expressed before MyoD1, whereas in the chicken, MyoD1 is activated first. The two transcription factors may initially be activated in different myotomes but their common function is to activate a third member of the so-called **MyoD family**, **myogenin**, as well as other muscle-specific genes. MyoD1 and Myf5 are required to commit somitic cells to the myoblast lineage. There appears to be a degree of redundancy in their function because knocking out either gene in the mouse does not prevent muscle development, while simultaneously knocking out both genes prevents the formation of all skeletal muscles. Intriguingly, if either *myoD1* or *myf5* is knocked out, the expression level of the remaining gene is elevated as if in compensation. This suggests that the two genes normally inhibit each other's expression or activity. The *myoD1*-knockout mouse has normal muscle development whereas in the *myf5*-knockout, muscle development is delayed. This phenotype probably reflects the brief period when Myf5 is active and MyoD1 has yet to be expressed. MyoD1 and Myf5 not only activate downstream genes responsible for muscle differentiation, but also activate their own genes. This self-sustaining feedback loop irreversibly commits each myoblast to the muscle cell lineage, and is one of several processes which form the molecular basis of determination (Topic A5).

The expression of *myogenin* is required for muscle differentiation, as *myogenin*-knockout mice form myoblasts that continue to proliferate and fail to form myofibers. Withdrawal from the cell cycle involves a combination of mechanisms including the activation of cell-cycle inhibitors and the downregulation of genes encoding growth factor receptors. The myoblasts are then unable to respond to growth factors in their environment.

Another MyoD family protein called MRF4 is expressed after myogenin and may co-operate with the other myogenic factors to activate genes required for differentiation into myofibers. The introduction of genes for any of the MyoD family proteins into non-muscle cells will convert them into muscle-like cells expressing muscle markers. The expression of one of the MyoD family proteins appears to induce the expression of the others, suggesting a comprehensive system of positive feedback loops to commit cells to the myogenic lineage. Another family of muscle-specific transcription factors, the MEF2 family, is also expressed during differentiation. These proteins encode transcription factors with a MADS box, which are also involved in patterning the flower during plant development (Topic N6). The MEF2 genes can also induce non-muscle cells to differentiate as muscles, and many muscle-specific genes have promoters and enhancers with binding sites for both the MyoD family and the MEF2 family. The MyoD and MEF2 family proteins are both expressed in skeletal muscle, but the MyoD family proteins are not present in the heart. The MEF2 proteins therefore play an important role in the differentiation of the cardiac muscle lineage (Topic K4).

K4 MAMMALIAN KIDNEY DEVELOPMENT

Key Notes

Three nephrogenic systems

During mammalian development, three sequential excretory systems arise in the intermediate mesoderm. The first is a vestigial system resembling the functional pronephric kidney of fish and amphibians, but this degenerates after a few days. The second mesonephric kidney is functional, and facilitates excretion while the final metanephric kidneys are developing. Signals from the ectoderm are required to induce the transient embryonic kidneys. Most of the mesonephric kidney regresses, but in males some caudal remnants persist as functional components of the reproductive system.

Metanephric kidney development

The metanephric kidney originates from metanephrogenic mesenchyme in the caudal intermediate mesoderm. Ureteric buds are induced to sprout from the mesonephric ducts and grow across until they penetrate the mesenchyme. Reciprocal inductive interactions then cause the mesenchyme to differentiate (first into proliferating stem cells and then into nephrogenic cell types) and the ureteric bud to branch and grow. Branching and nephrogenesis occur in a stereotyped pattern to generate the architecture of the adult kidneys.

Early inductive interactions

Early events in metanephric kidney development (formation of the metanephrogenic mesenchyme, initial induction of the ureteric buds) are not understood. Several signaling pathways have been shown to influence branching and growth of the ureteric bud when it penetrates the mesenchyme, and one of the most important components is glial derived neurotrophic factor (secreted by the mesenchyme) and its receptor Ret (localized on the growing tips of the ureteric branches). Signaling proteins of the BMP, Wnt and FGF families also play important roles in branching. Other molecules are required to avoid apoptosis in the mesenchyme and induce differentiation, including the transcription factors WT1 and Pax2.

Nephron development

The differentiation of nephrons involves a mesenchymal to epithelial transition, followed by morphogenesis in the resulting nephrogenic cyst to form an S-shaped tubule from which the structures of the mature nephron arise. Early events in the transition include dramatic changes to the composition of the extracellular matrix, and the expression of new cell adhesion molecules. Differentiation requires downregulation of the transcription factor Pax2, and signaling through the Trk nerve growth factor receptor.

Related topics

Mechanisms of developmental commitment (A3)
Signal transduction in development (C3)

Sex-determination (E5)
Mesoderm induction and patterning (K1)

Three nephrogenic systems

The mammalian kidneys develop from **intermediate mesoderm**, which is placed between the paraxial mesoderm (future somites) and lateral plate mesoderm (future viscera and body wall) on each side of the embryo during gastrulation (*Fig. 1*). The adult kidneys are derived from the most posterior section of this mesoderm on each side, but before the adult kidneys begin to emerge, two other transient excretory systems form in a craniocaudal sequence.

The first system is often termed the **pronephric kidney** because of its similarity to the embryonic kidneys of lower vertebrates. Mesenchymal cells of the intermediate mesoderm adjacent to the most anterior somites begin to condense into a solid rod of tissue, homologous to the **pronephric duct** of lower vertebrates. Signals from the ectoderm are essential for this process, and BMP is thought to be one of the signaling molecules involved, as beads soaked in BMP can compensate for the absence of ectoderm. The duct then induces the adjacent mesoderm to form hollow balls of epithelial cells termed **nephrotomes**. In fishes and amphibians, the pronephric ducts hollow out and the nephrotomes differentiate fully to form a functional excretory system. In mammals, however, the system does not differentiate and degenerates within a few days.

Before the pronephric kidney begins to degenerate, cells at the tip of the vestigial pronephric ducts begin to proliferate and migrate caudally through the intermediate mesoderm. Thus, as the pronephric kidney begins to regress at the cranial end, the vestigial pronephric ducts extend through the intermediate mesoderm to form the **mesonephric ducts**. As the mesonephric ducts extend,

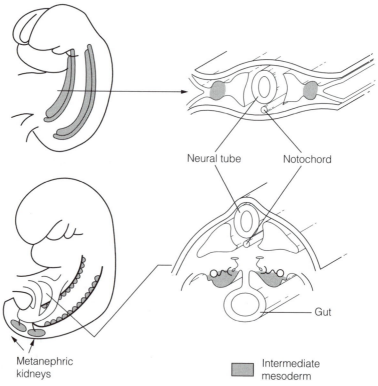

Neural tube   Notochord

Gut

Metanephric kidneys

Intermediate mesoderm

Fig. 1.   Distribution of intermediate mesoderm in the mammalian embryo (whole embryo left, transverse section right) and the positions of the mesonephric and metanephric kidneys.

they induce the condensation of adjacent mesenchyme to form mesonephric nephrotomes. However, unlike those of the pronephric kidney, each nephrotome or **mesonephros** differentiates into a fully functional excretory unit with a simple S-shaped tubule connecting the mesonephric duct to a Bowman's capsule surrounding a glomerulus derived from the dorsal aorta. The solid mesonephric ducts grow down through the intermediate mesoderm following an adhesion gradient. In the lumbar region of the embryo, the ducts grow out and away from the intermediate mesoderm and across to the endodermal wall of the cloaca. They fuse with the cloaca, and this induces the rods to hollow out, starting at the cloacal end and progressing cranially until the entire mesonephric ducts are hollow tubes. The mesonephroi form in a craniocaudal sequence. In humans, there are 40 in total although the most cranial mesonephroi degenerate before the most caudal ones have formed. The **mesonephric kidney** functions for a short time while the adult **metanephric kidney** is developing, and then undergoes regression. In females, this regression is complete, whereas in males the mesonephric duct and some of the most caudal nephrotomes persist to form functional elements of the reproductive system (Topic E5).

Metanephric kidney development

The **adult kidney (metanephric kidney)** forms in the caudal region of the intermediate mesoderm through a complex series of **reciprocal inductive interactions** (*Fig. 2*). The intermediate mesoderm in this region differentiates into a

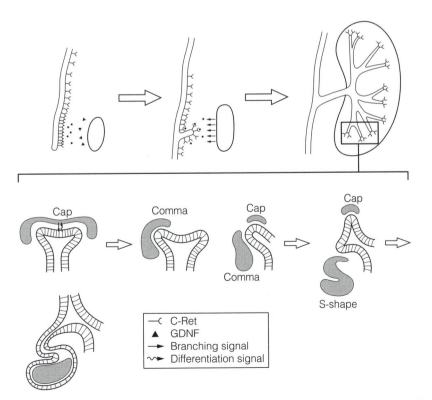

Fig. 2. Development of the metanephric kidney. Growth of the ureteric bud towards the metanephric balstema (top panel) and morphogenesis of branching (lower panel).

block of **metanephrogenic mesenchyme** called the **metanephric blastema**. This induces **ureteric buds** (future ureters) to sprout from the mesonephric ducts near the point at which each joins the cloaca, and grow across towards the blastema until they penetrate it. The signals involved in this process have not been identified. The ureteric buds first induce the blastema cells to differentiate into **proliferative stem cells**. Then groups of about 100 stem cells condense and form caps of epithelial tissue, which differentiate to form the functional nephrons of the adult kidney. Stem cells that do not differentiate into nephrons become the supportive **stromal cells** of the kidney. As the ureteric buds induce the metanephrogenic mesenchyme to differentiate, the mesenchyme induces the ureteric buds to branch and grow. Each branch induces the surrounding mesenchyme to differentiate into a nephron, so the number of nephrons in the kidney increases exponentially. The nature of branching differs in different mammals but follows a stereotyped pattern of bifurcations. Upon initial contact with the blastema, the ureteric bud expands at its tip to form the future renal pelvis. The first few generations of branching are followed by periods where the branches coalesce to form large chambers, the **major and minor calyces** of the kidney. Further branching occurs in the absence of coalescence, generating more than a million interconnected collecting ducts. Continued branching requires unknown factors secreted from stromal cells, but the ability of stromal cells to support continued branching is dependent on retinoic acid.

As the ureteric branches grow, they maintain their ability to induce the surrounding mesenchyme. The initial nephrons of each branch are therefore formed in direct contact with the ureteric bud, but as the branch elongates, newly arising nephrons are joined to the same collecting duct to form a long **arcade** (*Fig. 3*). The developing kidney therefore displays a gradient of developmental stages, with uninduced stem cells in the outermost cortex, and progressively more differentiated nephrogenic and stromal cells towards the center. The kidney also has an architecture where nephrons in the cortex are associated with the tips of the growing ureteric branches, while deeper nephrons are arranged in arcades associated with long ureteric branch bodies.

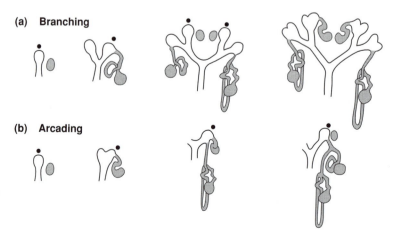

Fig. 3.   Nephron development during early kidney growth (branching) and later growth (arcading).

Early inductive interactions

Experiments in which ureteric bud tissue is cocultured with metanephric blastema tissue have shown that **reciprocal induction** occurs between the two tissues. The ureteric bud stimulates the metanephrogenic mesenchyme to differentiate, and in its absence, the mesenchyme cells undergo **apoptosis**. The mesenchyme cells induce the ureteric buds to grow and bifurcate, and ureteric bud tissue remains inert in the absence of metanephrogenic mesenchyme. Some of these signals are highly specific – only metanephrogenic mesenchyme can induce ureteric branching. Conversely, other tissues such as neural tube can induce differentiation in the metanephrogenic mesenchyme, but the responding tissue is only competent to differentiate into nephrotomes.

The early steps in this inductive pathway are unclear. It is unknown, for example, how the metanephrogenic mesenchyme arises from the intermediate mesoderm, and how it induces the initial sprouting of the ureteric buds. More is known about the reciprocal signaling pathways involved in mesenchyme differentiation and ureteric branching. One important signaling molecule, **glial-derived neurotrophic factor (GDNF)**, is secreted by the metanephrogenic mesenchyme and induces growth and branching in the ureteric buds. The receptor for GDNF, called **Ret**, is initially synthesized throughout the mesonephric ducts, and becomes progressively more localized to the zone of ureteric bud outgrowth, and finally to the very tip of the growing bud. Mice lacking either the *gdnf* or *c-ret* genes die due to **renal agenesis** (failure of kidney growth) caused by the absence of ureteric buds and consequently apoptosis of the mesenchyme cells. A number of signaling molecules and transcription factors have been identified with important roles in ureteric bud growth and branching, protection of the mesenchyme from cell death, mesenchymal to epithelial transition and the differentiation of condensed mesenchyme into nephrons. Some of these factors and their proposed functions are listed in *Table 1*.

Nephron development

Nephron development begins with the cap of condensed mesenchyme. This takes on epithelial characteristics and forms a hollow ball called a **nephrogenic cyst**. The cyst invaginates twice to form a **comma** and then an **S-shaped tubule**, which matures into a functional nephron. The mesenchyme to epithelial transition in nephrogenic mesenchyme involves changes in the nature of cell adhesion and extracellular matrix molecules. Initially, the cells begin to express two new adhesion molecules, **E-cadherin** and **Syndecan**, which appear to be required both for condensation of the mesenchyme and its proliferation. The mesenchyme cells initially secrete a loose reticular matrix composed of fibronectin and collagen types I and III. Conversely, the epithelial cells secrete **collagen type IV** and **laminin**, which form a basal lamina that is essential for differentiation. The transcription factor **Pax2** is expressed in response to induction and may orchestrate the changes that occur during the transition; condensation is prevented by blocking the expression of the *Pax2* gene. *Pax2* must be switched off, however, for the condensed cells to differentiate properly into the components of the nephron. Another important component is the **low affinity nerve growth factor receptor Trk**, whose absence blocks the differentiation of condensed cells into nephrons. There are more than 10 different cell types that differentiate in the nephron.

Table 1. Molecules with important roles in early kidney development

Molecule(s)	Function
(i) Signaling molecules and receptors	
GDNF/Ret	Signaling molecule/receptor tyrosine kinase. GDNF is expressed in the metanephrogenic mesenchyme, Ret in the tips of the ureteric bud and branches. This signaling pathway is responsible for continued ureteric bud growth and branching. Disruption of the pathway inhibits branching and also reduces levels of Wnt4 and Wnt11.
HGF/Met	Signaling molecule/receptor tyrosine kinase. HGF is expressed in metanephrogenic mesenchyme, Met in the ureteric bud. This signaling pathway is also involved in ureteric bud outgrowth, as disruption of the pathway inhibits branching.
Retinoic acid	The importance of retinoic acid in kidney development is shown by the small kidneys with few branches that arise in retinoic acid receptor mutants. The receptors are normally localized in the stromal cells which maintain Ret expression in the ureteric buds.
FGF2	Signaling molecule secreted by ureteric bud. Maintains blastema cells by maintaining WT1 expression (which prevents apoptosis); also required for differentiation.
LIF	Signaling molecule of the cytokine family secreted by ureteric bud. Receptor synthesized by metanephrogenic mesenchyme. Signal induces mesenchymal to epithelial transition.
BMP2	Signaling molecule. Inhibits branching, probably by increasing protein kinase A activity, which then acts on BMP7.
BMP4	Signaling molecule. Early expression in ectoderm induces nephric duct formation in intermediate mesoderm. Later expression in metanephric kidney inhibits branching.
BMP7	Signaling molecule synthesized in both ureteric bud and metanephrogenic mesenchyme. Prevents cell death in mesenchyme, and may have additional functions in differentiation. Low doses of BMP7 stimulate branching, but high doses inhibit branching.
Wnt4	Autocrine signaling molecule secreted by condensed metanephric mesenchyme. Required to complete mesenchyme to epithelial transition. Mesenchyme remains undifferentiated in *wnt4* mutants.
Wnt11	Signaling molecule expressed in ureteric bud. Expression induced by proteoglycans in mesenchyme after bud has penetrated blastema. Required for growth/branching within blastema.
(ii) Transcription factors	
WT1	Transcription factor expressed in gonadal ridge and metanephrogenic mesenchyme, and in nephrotomes throughout kidney development. Not required for initial differentiation of metanephrogenic mesenchyme as this tissue expresses the *gdnf* and Pax2 genes in *WT1* homozygous mutants, but protects metanephrogenic mesenchyme from cell death by activating *bcl-2* gene.
Eya1	Transcription factor with prominent role in eye and ear development and kidney development. In *eya1* mutants, metanephrogenic mesenchyme does not express GDNF, suggesting its role may be to regulate the *gdnf* gene.
Pax2	Transcription factor, required at many stages of kidney development. Early in development, required for specification of intermediate mesoderm during gastrulation, and also expressed in the early metanephrogenic mesenchyme. May regulate genes required for mesenchyme to epithelial transition, since blocking expression prevents condensation. Pax2 is switched off in epithelial tissue as it begins to differentiate (forced continued expression results in abnormal differentiation).

K5 HEART DEVELOPMENT

Key Notes

Overview of heart development

The vertebrate heart forms from lateral mesoderm in response to inductive signals from the underlying anterior endoderm. The primitive linear heart tube undergoes a complex series of looping and folding movements to generate the definitive heart with multiple chambers. In insects, the heart remains as a simple tube. Nevertheless, many of the genes required for insect heart development are conserved in vertebrates, although curiously they appear to act at a later stage of development to control vertebrate-specific looping and folding processes that do not occur in insects.

Molecular basis of heart specification in *Drosophila*

The most important molecule for heart development in *Drosophila* is the transcription factor Tinman, which is required for the development of all dorsal mesoderm structures. The *tinman* gene is induced by signals from the overlying ectoderm. Contact between the dorsal mesoderm and ectoderm is prevented by mutations in a gene called *heartless*, resulting in the failure of heart development.

Molecules involved in vertebrate heart development

There are several vertebrate homologs of Tinman, represented by transcription factors of the Nkx family. The expression patterns and mutant phenotypes of these genes suggest they are required for heart looping, not heart specification. Other genes, encoding the transcription factors dHAND and eHAND, also appear to be required for heart looping and morphogenesis.

Specification of the cardiac lineage

Some genes are required for heart morphogenesis while others are required for the differentiation of cardiac muscle cells. The MyoD family of transcription factors is not expressed in the heart, and cardiac muscle development is controlled by the MEF2 family of myogenic proteins. These proteins are also required for skeletal and smooth muscle development. GATA family transcription factors may also be required for muscle development, however mutant phenotypes show disruption to early heart morphogenesis. It is likely that morphogenesis and differentiation are intimately associated in vertebrate heart development.

Related topics

Maintenance of differentiation (A5)
Signal transduction in development (C3)

Mesoderm induction and patterning (K1)

Overview of heart development

In vertebrates, the heart develops from part of the lateral mesoderm called the **cardiogenic plate**, which in amniotes forms a horseshoe shape around the cranial end of the embryo. The cardiogenic region arises in response to inductive signals from the underlying endoderm and forms a pair of endocardial tubes that fuse together as the embryo folds to generate a single **primitive heart tube** at the midline (*Fig. 1*). The heart tube then undergoes **looping**. In vertebrates, the heart always loops to the right, and this is the first overt sign of left–right asymmetry. However, the factors that control the direction of looping are expressed throughout the lateral mesoderm, and have a global affect on the asymmetry of the embryo (Topic I3). After looping, the vertebrate heart undergoes a complex process of folding and septation to generate the four chambers that characterize the adult heart (the neural crest contributes to this process; see Topic J6; *Fig. 1*). Neither looping nor septation occur in the simpler hearts of invertebrates such as *Drosophila*, but surprisingly, many of the molecules involved in heart development are conserved.

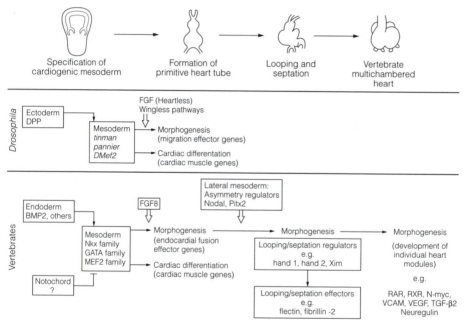

Fig. 1.   Overview of heart development in Drosophila *and vertebrates.*

Molecular basis of heart specification in *Drosophila*

Development of the *Drosophila* heart (and other dorsal mesoderm structures) requires a transcription factor encoded by the gene *tinman*. The gene is initially expressed throughout the mesoderm but is later restricted largely to the dorsal mesoderm, including the heart primordium. In *tinman* mutants, the heart, visceral mesoderm and dorsal muscles are entirely absent. The expression of *tinman* is regulated by the signaling molecules Decapentaplegic and Wingless, which are synthesized in orthogonal stripes in the overlying dorsal ectoderm. An FGF receptor, encoded by the gene *heartless*, must be activated for the embryonic mesoderm to migrate to the dorsal side of the embryo and make contact with the ectoderm.

Molecules involved in vertebrate heart development

Many of the molecules controlling heart development in *Drosophila* have counterparts in the vertebrate heart. Indeed due to a lack of heart cell culture based systems, most of the insights into vertebrate heart development have come directly from *Drosophila*.

A number of homeodomain-containing transcription factors related to Tinman have been identified in vertebrates, and are expressed during early heart development. The best-characterized is Nkx2.5, which is expressed in the cardiogenic plate and underlying endoderm. Expression of the *Nkx2.5* gene is induced by BMP2, BMP4 and BMP7, vertebrate homologs of Decapentaplegic, which are expressed in the anterior endoderm. The BMPs can induce *Nkx2.5* gene expression in cultured mesoderm cells and are attractive candidates for at least part of the cardiogenic plate-inducing signal. However, unlike the situation in *Drosophila*, *Nkx2.5* is not required for heart specification. In *nkx2.5*-knockout mice, the heart tube forms normally and begins to synthesize contractile proteins, but then fails to loop. Mutations in other genes also prevent looping, resulting in abnormal heart morphology. For example, two genes encoding bHLH transcription factors, *dHAND* and *eHAND*, are expressed in the cardiogenic plate and the heart tube; mouse knockouts for *dHAND* show a similar phenotype to *nkx2.5* knockouts.

Specification of cardiac lineage

In vertebrate embryos, the developing heart does not express any of the myogenic bHLH transcription factors, such as MyoD and myogenin, which are important for the specification and differentiation of muscle cells (Topic K3). However, the heart does express several genes of the **MEF2 family**, which encode MADS box transcription factors (these types of transcription factors are used to establish positional values in flower organ primordia; see Topic N6). In *Drosophila*, the *Dmef2* gene is required for the specification of all muscle types, and in mutants, the myoblasts fail to differentiate. Similarly, the vertebrate MEF2 family genes are expressed in all myogenic lineages, and mutations affect not only the heart, but also other muscle types.

Another class of transcription factors involved in the specification of the cardiac lineage is the **GATA family** of zinc-finger proteins. In vertebrates, GATA4, GATA5 and GATA6 are expressed in the heart and blocking the expression of GATA4 in cultured cells prevents cardiac myogenesis. However, *gata4*-knockout mice show normal cell differentiation although the heart tube fails to fuse at the midline. Therefore the GATA, MEF2, Nkx and HAND transcription factors are all likely to cooperate in the specification of the heart and the early myogenic lineage.

K6 ENDODERM DEVELOPMENT

Key Notes

Specification of the endoderm	In amphibians, the vegetal pole of the egg contains maternal determinants that specify endoderm development, although vegetal blastomeres do not become determined until gastrulation. The maternally-encoded transcription factor VegT is required, in addition to TGF-β signaling, for endoderm specification, and activates a series of endoderm-specific transcription factors such as Bix4, Mixer, Sox17α and Sox17β, which are necessary and sufficient to promote endoderm differentiation. Less is known about endoderm specification in amniotes, which occurs during early gastrulation.
Fate of the endoderm	The endoderm forms the gut tube and its associated organs, in close association with the visceral mesoderm. After the gut tube forms, it becomes regionalized into organ-forming domains. Budding then occurs to generate the organ primordia, and this is followed by differentiation to generate regionally appropriate endodermal cell types.
Anteroposterior pre-patterning	Before gut tube formation, the presumptive anterior and posterior endoderm are already distinct in terms of marker gene expression, suggesting that regionalization begins at an early developmental stage. The anterior and posterior endoderm are also morphologically and functionally distinct, with different rates of cell proliferation and different inductive properties on surrounding tissues.
Patterning the gut tube	As the somites form, the gut tube changes from a developmentally labile state to a more committed state characterized by its division into restricted organ-forming domains. Regionalization is controlled by signals from the visceral mesoderm, in combination with signals from other tissues, including the endoderm and notochord. Different regions of the gut tube become determined at different times – the liver-forming region of the foregut is determined shortly after the gut tube forms, while the specification of stomach and intestinal fates can be reversed after organogenesis has commenced.
Mechanisms of pattern formation	Regional specificity in the gut depends on signals from the visceral mesoderm, whose patterning is governed by *Hox* gene expression. However, the establishment of *Hox* gene expression in the visceral mesoderm appears to be dependent on the prior secretion of Sonic hedgehog by the hindgut. The positional information in the mesoderm induces the expression of gut transcription factors in particular zones, corresponding to organ-specific domains.
Pancreas development	Pancreas development begins with specification of a duodenal-pancreatic organ domain defined by the transcription factor Pdx1. Regional expression of Sonic hedgehog, constrained by inhibitory signals from the notochord, then establishes the pre-pancreatic endoderm. Development

of the dorsal and ventral pancreatic buds is controlled by a number of transcription factors that preferentially activate endocrine and exocrine-specific developmental pathways. Combinations of Nkx and Pax transcription factors act combinatorially to specify the different cell types in the pancreas.

Related topics   The ectoderm: epidermis and   Mesoderm induction and patterning
                       neural induction (J1)   (K1)

Specification of   The **endoderm** is the innermost germ layer of the animal embryo, which gives
the endoderm   rise to the gut tube and its associated organs. The specification of endoderm in
                       invertebrates such as *Drosophila* (Topic F2) and *C. elegans* (Topic G2) is covered
                       elsewhere, so the discussion in this Topic relates specifically to vertebrate
                       embryos.
                         In many amphibians, the endoderm originates entirely from the yolky vegetal
                       hemisphere of the blastula, reflecting the vegetal localization of maternal endo-
                       derm determinants in the egg. In *Xenopus*, however, the superficial cells of the
                       marginal layer also give rise to endoderm, and these are specified along with
                       the mesoderm by signals from the vegetal blastomeres (Topic K1). Vegetal
                       explants fail to differentiate fully in isolation, but nevertheless begin to express
                       characteristic endodermal markers such as **Endodermin**. This confirms that the
                       vegetal endoderm is *specified* autonomously by the distribution of maternal gene
                       products in the egg, but indicates that full differentiation requires interactions
                       with other cells. However, the fate of the endoderm is not *determined* until
                       gastrulation, as vegetal cells transplanted to the animal hemisphere of a host
                       blastula will give rise to ectoderm or mesoderm. Vegetal blastomeres are also
                       converted to mesoderm or ectoderm if TGF-β signaling is inhibited, suggesting
                       that TGF-β proteins present in the embryo at that time are required for endo-
                       derm specification.
                         Recently, many of the molecules in the pathway to endoderm specification in
                       Xenopus have been identified. The maternally-encoded T-box transcription
                       factor **VegT** is localized to the vegetal pole of the egg and experiments in which
                       this transcription factor is depleted have shown that it is required for both endo-
                       derm specification and mesoderm induction. The transcription factor regulates
                       (probably directly) a number of genes encoding nodal-related TGF-β signaling
                       proteins (**Xnr1, Xnr2, Xnr4** and **Dèrriere**) which have also been shown to be
                       required for endoderm and mesoderm development. The separation of endo-
                       derm and mesoderm fates may simply reflect the amount of VegT and its down-
                       stream effect on the TGF-β proteins, but this is probably not the case because
                       quantitative differences in the level of TGF-β signaling are implicated in the
                       diversification of dorsal and ventral mesoderm (Topic K1). A more likely model
                       is that VegT acts autonomously in concert with TGF-β proteins to specify the
                       endoderm, while the mesoderm arises only in cells lacking VegT but in response
                       to induction by TGF-β proteins secreted from VegT-expressing cells. In support
                       of this, VegT has been shown to directly activate a homeodomain-containing
                       transcription factor **Bix4** only in the vegetal cells where VegT is expressed, and
                       not in the marginal layer where the mesoderm forms. In embryos where VegT
                       has been depleted, the injection of mRNA for Bix4 can activate endoderm
                       marker genes but does not restore the ability of vegetal blastomeres to induce

mesoderm/endoderm in animal caps, confirming its specific role in autonomous endoderm development.

Downstream of the determinants and/or signaling events that specify the germ layers, transcription factors are activated that commit cells irreversibly to particular developmental pathways. These are pivotal molecules, and experimental manipulation shows that their expression is both necessary and sufficient for germ-layer development. As discussed above, grafting experiments show that *Xenopus* vegetal blastomeres are not determined to form endoderm until the beginning of gastrulation, so it is probably at this time that transcription factors irreversibly committing cells to the endoderm lineage become active. Brachyury serves this purpose in the mesoderm, while Neurogenin and Vent1/Vent2 may act at equivalent stages in the neural and epidermal lineages, respectively. Recently, three *Xenopus* transcription factors (**Mixer, Sox17α** and **Sox17β**) have been identified that induce endodermal development when expressed in animal cap explants. More significantly, the inhibition of these factors by expressing dominant negative DNA constructs prevents endoderm development *in vivo*. Overexpression of Mixer induces the expression of several proteins characteristic of the head-inducing leading endoderm of the organizer, including Cerberus and Dickkopf-1, while the Sox17 proteins only induce Cerberus. The genes encoding such proteins are probably downstream targets of Bix4.

So far, endoderm-determining factors have not been identified in amniotes, although the endoderm is likely to be specified by diffusible molecules secreted by the organizer. As discussed in Topic J1, the early activation of the zygotic genome and the extensive cell proliferation in the pre-gastrulation amniote embryo provides more plasticity in terms of early patterning. Consequently, epiblast cells can be respecified by grafting even during gastrulation. In amniotes, a layer of **primitive endoderm** (also called **hypoblast**) is formed prior to gastrulation, but this is displaced by the first epiblast cells to migrate through the primitive streak. The epiblast cells form the **definitive endoderm** of the embryo, while the primitive endoderm forms the lining of the **yolk sac**.

Fate of the endoderm

The endoderm gives rise to the gut tube and its associated organs, which are arranged in a particular sequence along the anteroposterior axis (*Table 1*; *Fig. 1*). The gut tube is initially divided into **foregut, midgut** and **hindgut** regions based on the morphological folding of the embryo. Later, these three regions are defined precisely on the basis of territories of vascularization, although these territories do not correspond to morphological boundaries between different organs in the gut or domains of gene expression. The development of the endoderm can be divided into a number of phases, beginning with the folding of the gut tube. This is followed by regional specification of different organ-forming regions, although this is based on an anteroposterior pre-pattern established at the blastula stage (amphibians) or just after gastrulation (amniotes). Patterning is followed by the appearance of buds, representing the primordia of the various endodermal organs. In some cases, these organs remain attached to the gut tube by a duct (e.g. liver, pancreas) while in others, the cells migrate away from the gut tube and develop independently (pharyngeal arch derivatives). The final stage of endoderm development is the differentiation of regionally-appropriate cell types. As well as forming a wide range of cells in different endodermal organs, the cytoarchitecture in the main gut tube also varies along the anteroposterior axis. Throughout development, the endo-

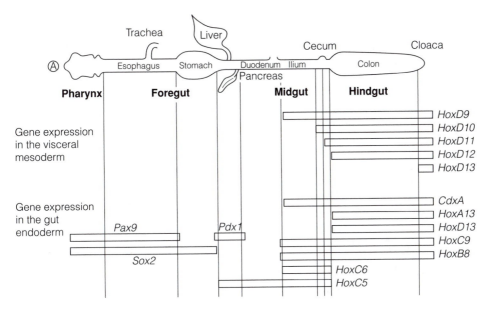

Fig. 1.   Molecular markers of anteroposterior polarity in the gut (visceral mesoderm and endoderm).

derm is intimately associated with **visceral mesoderm** (part of the lateral plate mesoderm, which forms the smooth muscle and connective tissue of the gut). As discussed below, this mesoderm plays an important role in endoderm development.

Table 1.   Regional diversity of the gut tube (in amniotes)

Region of the gut tube	Associated glands and organs
Foregut	
Pharynx	Tonsils, thymus, parathyroids, salivary glands, C cells of the thyroid.
Esophagus	Lungs.
Stomach	
Duodenum	Liver, gall bladder, bile duct, dorsal and ventral buds of pancreas.
Midgut	
Duodenum	
Jejunum	
Ilium	
Cecum	
Ascending colon and right two-thirds of transverse colon	
Hindgut	
Left one-third of transverse colon and descending colon	
Rectum	Bladder, ureter

Anteroposterior pre-patterning

The ability of the gut tube to give rise to various organs in a precise anteroposterior sequence indicates that the endoderm has anteroposterior polarity early in development. Much of this pattern arises after the gut tube has formed, in response to signaling from adjacent tissues. However, even at the blastula and gastrula stages of development, there is a fundamental distinction between anterior and posterior endoderm.

In *Xenopus*, vegetal explants reveal a pre-pattern in the presumptive endoderm even at the early blastula stage. Presumptive anterior vegetal cells express the pancreas-specific marker **Xlhbox8** while posterior cells express the intestine-specific marker **IFAP**. This evidence of an early pattern is supported by cell tracing studies showing that vegetal blastomeres are specified to contribute to particular regions of the gut tube as early as the 32-cell stage. Furthermore, the pharyngeal endoderm is formed specifically by the superficial cells of the marginal layer. These cells are the first to ingress during gastrulation and express a number of markers that are absent from posterior endoderm. The same differential marker gene expression is seen after gastrulation in mouse embryos, where the first cells migrating through the primitive node form pharyngeal endoderm, and later migrating cells form posterior endoderm. Anterior endoderm-specific markers include *cerberus* and *Otx1*, while posterior endoderm-specific markers include *Cdx2*. Note that *Otx1* and *Cdx2* are homologs of the *Drosophila* genes *orthodenticle* and *caudal*, which are expressed in similar relative positions in the fly embryo. The unique characteristics of the anterior endoderm are revealed by its specific inductive properties. Anterior endoderm alone can induce head development (Topic J2) and heart development (Topic K5). The anterior endoderm also proliferates more slowly that the posterior endoderm.

Patterning the gut tube

Before the somites form, the endoderm is developmentally labile and cell fates can be respecified by grafting to a different position along the axis. As the somites form however, the gut becomes regionalized into overlapping, organ-forming domains. Later, these broad domains are resolved into sharply defined zones corresponding to the morphological divisions of the gut.

Regional specification in the gut involves signaling from adjacent tissues, specifically the visceral mesoderm, ectoderm and the notochord. Liver cell fates are determined very early, before the organ buds form. The liver arises from the ventral foregut, which overlies the cardiogenic mesenchyme. Signals from the cardiogenic mesenchyme are required for **hepatocyte differentiation** and expression of the liver marker **albumin**, but these signals are permissive rather than instructive because if cardiogenic mesenchyme is placed adjacent to more posterior endoderm, it does not promote hepatocyte differentiation. However, if the posterior gut tube is separated from the overlying notochord, hepatocytes form spontaneously. This confirms the fundamental difference between anterior and posterior endoderm, as only the anterior endoderm can be induced by the cardiogenic mesenchyme to differentiate into hepatocytes *in the presence of the notochord*.

Other regions of the gut endoderm are determined much later. This is revealed by the ability of mesenchyme from different regions of the gut tube to respecify the fate of endoderm cells even after organogenesis has commenced. For example, non-glandular stomach endoderm can be converted into glandular stomach endoderm or intestinal endocrine tissue when combined with visceral mesoderm from the appropriate region.

Mechanisms of pattern formation

The experiments described above show that the visceral mesoderm plays an important role in the regional specification of the gut endoderm. Positional values in the visceral mesoderm are likely to be governed by the *Hox* genes, as these are expressed in overlapping, posteriorly nested domains as observed in the hindbrain (Topic J2), somites (Topic K2) and limb buds (Topic M2). Furthermore, the anterior boundaries of *Hox* gene expression in the visceral mesoderm correspond to morphological boundaries in the midgut and hindgut (*Fig. 1*). This indicates that the visceral mesoderm is providing the gut with positional information, although the nature of the signal(s) is currently unknown. It is clear that *Hox* gene expression patterns are not transferred from mesoderm to endoderm. This contrasts with the situation in other germ layers. For example, the pattern of *Hox* gene expression in the posterior neural tube is imposed by the *Hox* gene expression in the adjacent somites (Topic K2). However, only a small subset of *Hox* genes is expressed in the endoderm and the characteristic overlapping and nested pattern seen in the mesoderm is absent (*Fig. 1*). A number of transcription factors that mark organ-forming domains in the gut have been identified, including those encoded by *Hox*, *Cdx*, *Nkx*, *Pax* and *Pdx* type homeobox genes. The expression domains of some of these genes are shown in *Fig. 1*. Many of these genes play multiple roles in endoderm development, including regional specification, organogenesis and terminal differentiation. The development of the pancreas is considered as a model system below.

Pancreas development

The adult pancreas arises from dorsal and ventral pancreatic buds. The transcription factor **Pdx1** is initially expressed in a broad domain marking the entire **duodenal-pancreatic region**. The choice between rostral stomach, duodenal and pancreatic fates is controlled by **Sonic hedgehog**, which is initially restricted to the posterior of the gut tube but is later expressed throughout the endoderm, except for the **pre-pancreatic endoderm**. The expression of Sonic hedgehog in addition to Pdx1 promotes duodenal development, while the presence of Sonic hedgehog in the absence of Pdx1 promotes stomach development. The presence of Pdx1 in the absence of Sonic hedgehog is required for pancreas development, and the inhibition of Sonic hedgehog expression in the pre-pancreatic endoderm is dependent on the overlying notochord. If the notochord is removed, Sonic hedgehog expression is derepressed and pancreatic development is abolished. Conversely, if Sonic hedgehog activity outside the pre-pancreatic endoderm is inhibited with drugs, the stomach and duodenal zones begin to express dorsal pancreatic markers such as insulin. Activin and FGF2 are candidate proteins secreted by the notochord that could inhibit Sonic hedgehog expression in the pre-pancreatic endoderm.

Signals from the pancreatic mesenchyme are required to stimulate the differentiation of endocrine and exocrine cells. Both TGF-β and HGF (hepatocyte growth factor) signaling promote endocrine development, while Follistatin, which inhibits TGF-β signals, promotes exocrine development. Downstream of these signals, a number of transcription factors are activated. Uncommitted endocrine cells express Islet-1 and Pax6, and knocking out the gene for either of these transcription factors in mice prevents endocrine differentiation. Other transcription factors, such as HNF-6 and Mist1 are expressed specifically in exocrine cells. Combinations of Nkx and Pax transcription factors then cooperate to specify individual cell fates within the pancreas. The details of this developmental pathway are shown in *Fig. 2*.

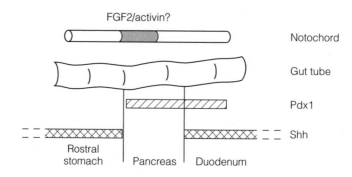

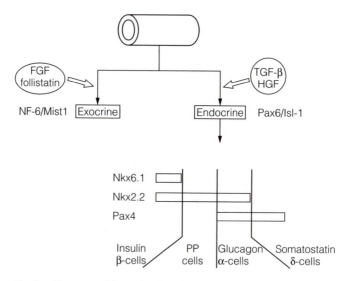

Fig. 2.   Summary of the molecular regulation of pancreas development.

L1 VULVAL SPECIFICATION IN C. ELEGANS

Key Notes

Vulval cell lineage

In hermaphrodite *Caenorhabditis elegans*, a vulva develops in the ventral body wall, through which eggs are delivered from the uterus. The vulva is a small organ comprising just 22 cells, and develops from three hypoblast cells (the vulval precursor cells p5, p6 and p7) under the influence of a fourth cell, the gonadal anchor cell. The p6 cell, directly under the anchor cell, enters the primary vulval lineage, while the flanking p5 and p7 cells enter the secondary vulval lineage. Initially, there are six competent vulval precursor cells, but only three normally form the vulva. The remaining cells (p3, p4 and p8) enter the non-vulval tertiary lineage and form hypodermis cells (the outer cell layer).

Specification of the anchor cell

Two adjacent cells in the gonad are competent to form the anchor cell. Mutual lateral inhibition occurs between them to inhibit anchor cell differentiation, involving the signaling molecule LAG-2 and its receptor LIN-12, which are initially synthesized by both cells (these are homologs of the *Drosophila* Delta and Notch proteins). Eventually one cell stops synthesizing LIN-12 and cannot respond to inhibition, and differentiates into the anchor cell, while the other differentiates to form the ventral uterine precursor. Random fluctuations that lead to an initial imbalance in the reciprocal signaling are reinforced by positive feedback, so that increased signaling through LIN-12 leads to an increase in *lin-12* gene expression in the presumptive uterine precursor, while reduced signaling leads to a decrease in *lin-12* gene expression in the presumptive anchor cell.

Induction of hypoblast cells

Once the anchor cell is specified, it secretes a signaling molecule, LIN-3, which is homologous to mammalian EGF. This induces the underlying hypoblast cells to differentiate into vulval cells. The hypoblast cells synthesize the receptor for LIN-3, called LIN-23, and other signal transduction components that are homologous to the components of the mammalian growth factor signaling pathway.

Specification of primary and secondary lineages

There are six vulval precursor cells and all are competent to differentiate into vulval cells under the inductive control of the anchor cell, while all will enter the tertiary (non-vulval) lineage in its absence. The decision to enter either the primary or secondary vulval lineage depends initially on the distance from the anchor cell, i.e. the vulval precursors respond to a concentration gradient of LIN-3. Initial differences in the levels of LIN-3 are reinforced by a second mechanism of lateral inhibition, whereby the central p6 cells produce a laterally-acting signal that prevents the flanking p5 and p7 cells adopting the primary vulval fate.

| **Related topics** | Mechanisms of developmental commitment (A3) | Early animal development by single cell specification (G1) |
| | Signal transduction in development (C3) | Cell specification and patterning in *Caenorhabditis elegans* (G2) |

Vulval cell lineage

The vulva of the nematode *Caenorhabditis elegans* is an external opening in the ventral hypodermis of the hermaphrodite, through which eggs are delivered from the uterus. This system is of particular interest to developmental biologists because the adult structure comprises just 22 cells, derived from three precursors over three rounds of cell division. The system provides examples of several types of inductive interaction occurring at the **single-cell level**. Initially, there are six hypoblast cells competent to enter the vulval cell lineage, the **vulval precursor cells (VPCs)** p3, p4, p5, p6, p7 and p8. The vulva is normally derived from cells p5, p6 and p7. The p6 cell enters the **primary vulval lineage** and undergoes three rounds of cell division to produce 12 **central vulval cells**. The flanking cells p5 and p7 enter the **secondary vulval lineage** and undergo three rounds of division to produce 11 **lateral vulval cells** (only 11 cells are produced because one of the second-generation cells does not divide a third time). The remaining VPCs do not enter the vulval lineage. They are said to adopt a **tertiary fate**; they divide once and differentiate as hypodermis. The vulval cell lineage is shown in *Fig. 1*.

The hypoblast cells are induced to form the vulva by a single cell in the overlying gonad, called the **gonadal anchor cell**. If this cell is destroyed by laser ablation, all hypoblast cells (including those that would normally enter vulval lineages) adopt tertiary fates and differentiate into hypodermis. If all cells in the gonad except the anchor cell are destroyed, the vulva forms as normal. If the

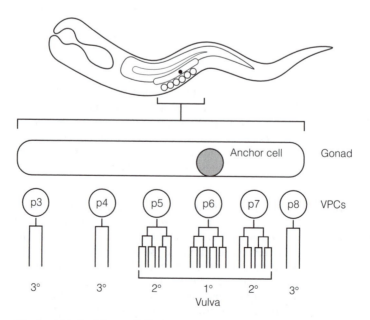

Fig. 1.   Vulval cell lineage in C. elegans.

anchor cell is moved mechanically, neighboring hypoblast cells that normally adopt tertiary fates can be induced to enter vulval lineages, but competence is restricted to the six VPCs.

The genetic amenability of *C. elegans* means that it is easy to screen for mutants in which vulval development is abnormal, particularly mutants with **vulvaless** or **multiple vulva** phenotypes. This has allowed the identification of over 40 genes with roles in various stages of vulval development. Some of the key regulators of this system are discussed below.

Specification of the anchor cell

Initially, two adjacent cells (Z1.ppp and Z4.aaa) are competent to become the anchor cell. The choice is made by **lateral inhibition** (Topic A3), involving the transmembrane signaling molecule LAG-2 (related to the *Drosophila* proteins Delta and Serrate) and its receptor LIN-12 (related to the *Drosophila* protein Notch), which are synthesized by both cells (*Fig. 2*). Signal transduction through the LAG-2/LIN-12 pathway inhibits anchor cell differentiation. In *lag-2* and *lin-12* mutants, both cells become anchor cells. Conversely, in dominant gain of function *lin-12* alleles where the receptor is constitutively active, neither cell becomes an anchor cell; both differentiate as **uterine precursor cells**, which divide to produce the cells of the uterus. In normal development, the anchor cell is thought to be specified by random fluctuations in the levels of these signaling molecules. Such fluctuations result in one cell producing more LAG-2 or LIN-12 than the other, resulting in increased signaling through LIN-12 in one of the cells. This initial imbalance is reinforced by positive feedback acting at the level of transcription – the level of *lin-12* mRNA has been shown to increase in the presumptive uterine precursor, and also decreases in the presumptive gonadal anchor cell (*Fig. 2*). A similar mechanism is thought to control the spacing of neuroblasts in *Drosophila*, and the Notch–Delta signal transduction pathway is involved in that system too (Topic J5).

Induction of hypoblast cells

The anchor cell induces the underlying hypoblast cells to adopt either primary or secondary vulval fates. Tertiary fate is the default outcome, and occurs in the absence of induction. The signal released by the anchor cell to induce the hypoblast cells is LIN-3, which is related to mammalian EGF. The receptor for this signal, LIN-23, is synthesized only by the six VPCs, accounting for the restriction of competence to these six cells. LIN-23 is homologous to the mammalian EGF receptor. Other genes have been identified, including *let-23*, *sem-5*, *let-60* and *lin-45*, which are required in the responding hypoblast cells and encode components of the signal transduction pathway. Each component has a homolog in the mammalian Raf-Ras-MAP kinase pathway (see Topic C3).

Specification of primary and secondary lineages

If all the VPCs except one are destroyed, the fate of the remaining cell (primary, secondary or tertiary) depends on its distance from the anchor cell. This suggests that the VPCs respond to LIN-3 in a concentration-dependent manner, i.e. LIN-3 acts as a morphogen to determine primary and secondary vulval fates, and cells that lie outside the influence of the signal adopt the tertiary fate (*Fig. 2*). This can be demonstrated in transgenic worms by placing the *lin-3* gene under the control of an inducible promoter. In such animals, the fate of individual VPCs remaining after laser ablation of the other five cells can be controlled by regulating the activity of the promoter.

However, it appears that a second mechanism is used to reinforce the morphogen. Once the primary fate of the p6 cell has been determined, this cell prevents the adjacent p5 and p7 cells from adopting the same fate. Once again, this process appears to involve lateral inhibition via the Notch/Delta signaling pathway. All the VPCs initially synthesize the Notch-related receptor LIN-12, but the *lin-12* gene is repressed in the p6 cell following induction by the anchor cell. It is possible that the VPCs initially use lateral inhibition to prevent each other from entering the primary vulval lineage, and that variations in level of LIN-12 result in positive feedback at the level of transcription eventually leading to determination. Unlike the specification of the anchor cell, which is due to random fluctuation in the levels of LIN-12 or LAG-2, the specification of primary vulval fate is biased by the inductive signal from the anchor cell causing the strongest repression of *lin-12* in the p6 cell. The p6 cell thus loses the ability to respond to the inhibitory signal from its neighbors, but retains its ability to inhibit them (*Fig. 2*).

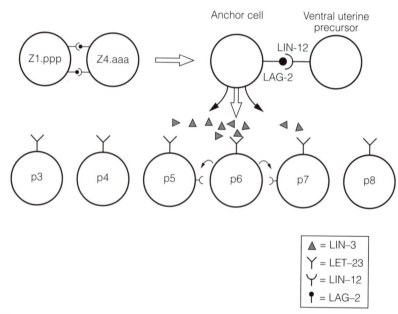

Fig. 2. Vulval specification in C. elegans. The anchor cell is specified by lateral inhibition between Z1.ppp and Z4.aaa. The anchor cell then specifies primary and secondary vulval cells using the LIN-3 protein as a morphogen. A lateral signal from the primary vulval precursor (p6) inhibits primary cell fates in its neighbors, through the receptor LIN-12.

L2 PATTERN FORMATION IN IMAGINAL DISCS

Key Notes

Imaginal discs	The appendages of the adult *Drosophila* are derived from sheets of epithelial tissue called imaginal discs that are formed by invagination of the cellular blastoderm during embryonic development. Initially comprising a small number of cells, the imaginal discs proliferate during larval development and then undergo profound morphogenetic changes to evert and form the mature structures during metamorphosis. Most of the information that patterns the appendages along all three axes is already established in the early imaginal discs.
Determination and identity of imaginal discs	Imaginal discs form in the *Drosophila* blastoderm where the domains of two signaling proteins, Wingless and Decapentaplegic, intersect. Cells in these intersection zones can express the *distal-less* gene and differentiate into imaginal precursors. Imaginal discs do not form at all intersection zones because *distal-less* gene expression is controlled by the homeotic selector (HOM-C) genes. For example, no imaginal discs arise in the abdomen because the *Hox* genes expressed there are non-permissive. The HOM-C genes also control the identity of the imaginal discs (i.e. which appendages they give rise to) evidenced by the fact that HOM-C gene mutations cause the discs to develop into inappropriate appendages. Such mutations also show that all discs use essentially the same mechanisms for patterning, but the results in terms of the final appendage generated differ according to the HOM-C genes that are expressed, probably because they activate specific master regulators of organogenesis such as the gene *eyeless* (for eye development) or *vestigial* (for wing development).
Anteroposterior and dorsoventral patterning	The leg, wing and haltere discs each span a parasegment boundary and thus comprise anterior and posterior developmental compartments, characterized by cells with different patterns of gene expression. Interactions at the border between adjacent cells from each compartment establish a strip of cells along the boundary as an organizer. These cells secrete a signaling protein that diffuses into both compartments, establishing a concentration gradient that acts as a primary patterning signal. The signaling proteins Hedgehog, Wingless and Decapentaplegic are involved in such patterning mechanisms. The mechanism can be demonstrated by placing cells from one compartment into the middle of the other: new boundaries are established around the ectopic cells and new organizers arise that cause abnormal patterning. Similar mechanisms described in the text are used to pattern the dorsoventral axis.
Proximodistal patterning	Proximodistal patterning is controlled by the *distal-less* gene. In the leg disc, *distal-less* gene expression requires both Decapentaplegic and Wingless signaling. As the imaginal disc grows, the domain of *distal-less*

expression is restricted to the center of the disc, where it defines the distal tip of the future appendage. The proximodistal axis is specified as a series of concentric circles radiating from this point, and as the disc everts from the body to form the structure, the cells nearest the center form the distal extremities of the appendage while those nearer the edge form more proximal structures.

Related topics	Pattern formation and compartments (A6)	Mesoderm induction and patterning (K1)
	Signal transduction in development (C3)	Vertebrate limb development (Section M)
	Homeotic selector genes and regional specification (H5)	

Imaginal discs

The *Drosophila* larva has no legs, wings, antennae or eyes. These, and other appendages of adult flies, are generated during metamorphosis and develop from epithelial structures termed **imaginal discs**. The imaginal discs are formed during embryonic development as invaginations of the cellular blastoderm. When the larvae hatch, the discs comprise less than 100 **epithelial cells** arranged in a flat sheet, as well as a few **adepithelial cells** that eventually give rise to the muscle cells and neurons of each appendage (*Fig. 1*). The epithelial cells proliferate rapidly during the instar stages of development and the sheet may fold up to accommodate the increase in size. Prior to metamorphosis, the largest imaginal disc (the wing disc) contains over 50 000 cells. When metamorphosis occurs, the imaginal discs undergo striking morphogenetic changes and evert from the body to form the mature appendages. This process is induced by the molting hormone **ecdysone**, and involves new mRNA and protein synthesis as well as changes in cell shape brought about by reorganization of the cytoskeleton. Remarkably, almost all of the patterning information that controls cellular phenotype along the anteroposterior, dorsoventral and proximodistal axes of these appendages is already established in the discs. This contrasts to vertebrate limb development (Section M) where proximodistal patterning occurs during limb growth.

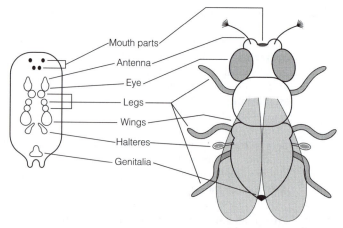

Fig. 1. Imaginal discs in the Drosophila *embryo, and the corresponding appendages in the adult fly.*

Determination and identity of imaginal discs

The positions of organ primordia in the *Drosophila* embryo are specified by the overlapping and intersecting patterns of gene expression along the antero-posterior and dorsoventral axes early in development (Section H). It has been shown that imaginal discs can originate where the expression domains of *decapentaplegic* (which is transcribed in a horizontal band in the dorsal ectoderm) and *wingless* (which is transcribed in vertical stripes representing each parasegment) intersect (*Fig. 2*). Cells in these overlap zones are potentially able to express the *distal-less* gene, and form the **imaginal precursors**. However, the expression of *distal-less* is repressed by the products of many of the homeotic selector genes (HOM-C genes), so initially only six pairs of discs and a single genital disc are formed. In the head, two pairs of discs specify mouthparts, and one pair of discs gives rise to the eyes and the antennae. In the thorax, there are three pairs of discs that give rise to three pairs of legs. Early in development, the discs in the second and third thoracic segments split into two, producing the wing disc and the haltere disc, respectively (*Fig. 2*). There are no discs formed in the abdomen because the *distal-less* gene is repressed throughout by the posterior HOM-C gene products.

The *Drosophila* imaginal discs are very similar at the cellular level, but they produce a diverse set of structures in the adult. Furthermore, patterning in the imaginal discs involves a common set of genes: *hedgehog*, *wingless* and *decapentaplegic*, indicating that broadly similar patterning mechanisms are used in all the discs, although with different outcomes. Indeed, each imaginal disc is competent to produce any of the various adult appendages. This is confirmed by the existence of homeotic mutants in which one appendage is transformed into another. One of the most striking homeotic mutations, *Antennapedia*, converts the antennae into legs, showing that the antennal imaginal discs are competent to produce leg structures. The universal nature of patterning information in the

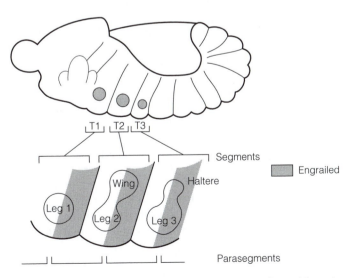

Fig. 2. Derivation of the thoracic imaginal discs in the Drosophila *embryo. The T2 and T3 discs split into two populations, to generate wing and leg precursors (T2) or haltere and leg precursors (T3). Note that each disc is partitioned into anterior and posterior compartments by a parasegment boundary, defined by the expression of* engrailed.

imaginal discs can be confirmed using mitotic recombination (Topic B2) to produce discs with small sectors of mutant tissue. For example, wild type antennal discs containing a small sector of *Antennapedia* mutant cells develop into antennae containing portions of leg tissue. However, the leg tissue develops with the appropriate positional identity along the proximodistal axis, i.e. if the mutant cells were in the center of the disc, leg tissue would develop at the distal tip of the appendage and would develop into a tarsus (foot), the structure that would normally appear at the equivalent distal position in the normal leg.

These experiments indicate that the HOM-C genes control not only the presence of imaginal discs, but also their normal segment-specific identity. The HOM-C genes may work by directly or indirectly controlling the expression of master regulators of organogenesis. Examples of such genes include *vestigial* and *eyeless*, which control wing and eye disc identity, respectively, and can induce ectopic wing or eye development when expressed in other discs.

Anteroposterior and dorsoventral patterning

The eye disc is a single developmental compartment because it originates from imaginal precursors formed in the middle of a parasegment in the head. Patterning of the eye disc is discussed in Topic L3. Conversely, the precursors of the thoracic imaginal discs (wing, leg and haltere discs) originate at the boundary of two parasegments. The discs themselves therefore span the boundary and possess anterior and posterior compartments across which no cell mixing occurs (developmental compartments are discussed in Topic A6). This boundary is critical for establishing an organizing center for anteroposterior patterning. However, as HOM-C gene expression domains in the thorax define parasegment boundaries, the division of each disc into two compartments also means that some homeotic mutations affect only one compartment, so that partial transformations occur. For example, in the *posterobithorax* mutant, only the posterior compartment of the haltere on thoracic segment 3 is converted into a wing, resulting in a hybrid appendage with a normal anterior haltere structure, and a mutant posterior wing structure (*Fig. 2*).

In the leg, wing and haltere discs, patterning along the anteroposterior axis involves signaling between cells on each side of the compartment boundary to establish an organizing center at the boundary. Initially, the transcription factor Engrailed is synthesized specifically in the posterior compartment of each imaginal disc, reflecting *engrailed* gene expression at the anterior margin of each parasegment in the embryo (Topic H4). Engrailed is a positive regulator of the *hedgehog* gene, so cells in the posterior compartment of each imaginal disc begin secreting Hedgehog protein.

In the wing, Hedgehog-secreting cells at the compartment boundary induce adjacent cells in the anterior compartment to synthesize another signaling protein, Decapentaplegic (*Fig. 3*). The *decapentaplegic* gene (*dpp*) is switched on in the anterior compartment in those cells within range of the Hedgehog signal, but *dpp* is not activated in the posterior compartment because the gene is repressed by Engrailed. Decapentaplegic protein diffuses into both the anterior and posterior compartments and is the primary patterning signal, acting on a number of downstream genes such as *spalt*. In the leg, Hedgehog-secreting cells at the compartment boundary also induce gene expression in adjacent anterior compartment cells, but the response differs in the dorsal and ventral halves of the imaginal disc (*Fig. 4*). In the dorsal region of the disc, the anterior boundary cells secrete Decapentaplegic as they do in the wing. In the ventral half of the disc, the anterior boundary cells secrete Wingless. Both these diffusible proteins

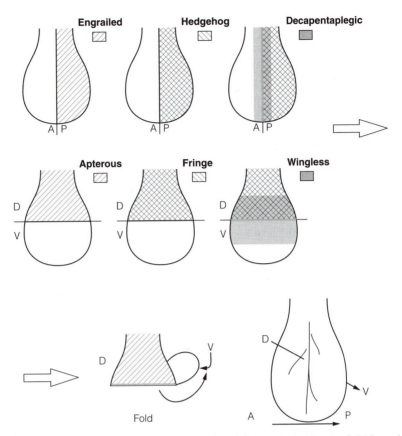

Fig. 3. Anteroposterior patterning (top row) and dorsoventral patterning (middle row) in the Drosophila *wing disc. In each case, signaling between two compartments establishes a midline organizer, represented by Decapentaplegic (anteroposterior axis) or Wingless (dorsoventral axis). Bottom row shows how the disc folds to generate the mature wing.*

then act as patterning signals. At the dorsoventral midline of the leg disc, the boundary of *wingless* and *dpp* expression is maintained by reciprocal repression between the two signaling proteins.

Dorsoventral patterning in the wing involves the initiation of a signaling center at a compartmental boundary, and is very similar in mechanism to anteroposterior patterning (*Fig. 3*). The dorsal compartment of the wing disc is defined by the expression of the *apterous* gene. In the same way that Engrailed induces *hedgehog* in posterior cells, Apterous induces *fringe* and *serrate* in the dorsal cells. Fringe-secreting cells then induce adjacent cells in the ventral compartment to express *wingless*, and the Wingless protein diffuses into both compartments and acts as the primary dorsoventral patterning signal.

Hedgehog, wingless and *dpp* are often expressed in adjacent developmental compartments, where they help to establish and maintain compartmental boundaries (see Topic H4 for a discussion of the role of these genes in segmentation in *Drosophila*, and see Topic M2 for discussion of the roles of related signaling molecules in vertebrate limb development). The function of these genes in imaginal disc compartmentalization and pattern formation can be demonstrated by ectopic expression experiments. For example, if *engrailed* or *hedgehog* is expressed in a clone of transgenic cells within the anterior compartment of the wing disc, the wild type cells surrounding it behave like the cells at the compartment border and begin to secrete Decapentaplegic. This sets up new signaling centers within the anterior compartment, resulting in abnormal, mirror image wing structures. Analogous phenotypes are generated by ectopically expressing *apterous* or *fringe* in sectors of the ventral wing disc.

Proximodistal patterning

Each thoracic appendage is generated by the eversion of a flat imaginal disc, so that the central part of the disc becomes the most distal elements of the appendage and the periphery of the disc represents the proximal elements of the appendage at the junction with the body. The proximodistal axis is therefore represented by a series of concentric circles of cells in the imaginal disc, with each smaller circle representing progressively more distal cell fates.

The proximodistal axis of the leg is established by the gene *distal-less*, which is expressed in the center of the leg disc to generate a signaling center. Mutants in *distal-less* lack distal elements of the leg, while ectopic expression of the gene generates extra distal elements. The expression of *distal-less* requires both Decapentaplegic and Wingless signaling. As these proteins mutually repress each other's synthesis and are restricted to the anteroposterior boundary in the dorsal and ventral halves of the leg, respectively, the only cells that can respond to both signals are those in the very center of the disc (*Fig. 4*).

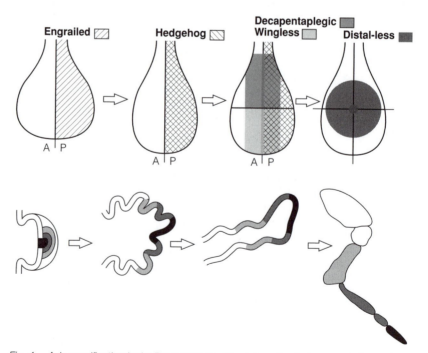

Fig. 4. Axis specification in the Drosophila leg disc. Interaction between anterior and posterior compartments establishes a midline signaling center. In the leg, Decapentaplegic is expressed in the dorsal half of the disc and Wingless in the ventral half. Interaction between cells expressing these two signaling proteins establishes a central zone of Distal-less expression. The concentration of Distal-less diminishes towards the periphery of the disc, and establishes and proximodistal axis as the disc everts.

L3 *DROSOPHILA* EYE DEVELOPMENT

Key Notes

The *Drosophila* compound eye

The *Drosophila* compound eye contains over 800 individual photoreceptor organs called ommatidia, arranged in a regular hexagonal array. Each ommatidium comprises 20 cells, including eight specialized neurons, photoreceptor cells, named R1–R8. The R1–R7 photoreceptors surround a central R8 cell.

Differentiation of ommatidia

The eye develops from the eye primordium of the eye/antennal imaginal disc during the third instar stage. Differentiation proceeds in a slow wave, the morphogenetic furrow, which sweeps from posterior to anterior. Behind the furrow, the R8 cells differentiate in a regularly spaced pattern, and the other photoreceptor cells differentiate around the R8 cell in a predictable sequence.

The morphogenetic furrow

The morphogenetic furrow is a dynamic band of non-dividing cells spanning the entire eye primordium, which can be defined by its production of the signaling protein Decapentaplegic. Differentiating cells in the posterior of the eye primordium secrete Hedgehog, and this protein diffuses into the adjacent rows of undifferentiated anterior cells and induces *dpp* expression, therefore generating the furrow. Decapentaplegic initiates R8 cell differentiation. As the furrow cells undergo differentiation they stop expressing *dpp* and start expressing *hedgehog*, and the Hedgehog protein diffuses away to induce *dpp* expression in the next row of adjacent undifferentiated cells. In this way, the morphogenetic furrow moves slowly across the eye primordium, completing its journey in approximately 2 days.

Spacing the ommatidia

Decapentaplegic protein in the morphogenetic furrow induces R8 cell differentiation. All undifferentiated cells are competent to differentiate, but R8 cells arise in a regular spaced pattern by lateral inhibition, involving the secreted protein Scabrous. Initially, all the undifferentiated cells secrete Scabrous and inhibit each other, but imbalances in the levels of Scabrous protein are amplified by feedback, so that individual cells establish dominance over their neighbors, and differentiate into R8 cells. Scabrous protein diffuses over 3–4 cell diameters, so R8 cells arise with 6–8 cells between them.

Cell signaling in differentiation

The photoreceptor cells are specified in a predictable series, involving the sequential expression of various transcription factors and signaling proteins. The pathway to R7 cell specification is understood in the most detail. A receptor called Sevenless in the membrane of the presumptive R7 cell interacts with a membrane bound ligand called Bride-of-sevenless (Boss) on the R8 cell. This initiates a signal transduction cascade homologous to the mammalian growth factor signaling pathway, resulting in the specification of the R7 cell. The other cells surrounding R8 also have a Sevenless receptor, but the signal transduction pathway in these cells is blocked by the products of other genes, including *seven-up* and *rough*.

Related topics
Mechanisms of developmental
    commitment (A3)
Pattern formation (A6)
Signal transduction in
    development (C3)

Homeotic selector genes and
    regional specification (H5)

The *Drosophila* compound eye
The compound eye of the fruit fly *Drosophila melanogaster* comprises 800 individual photoreceptor units called **ommatidia**, which are arranged in a regular pattern. Each ommatidium comprises 20 cells, eight of which are photoreceptors (R1–R8), and four of which are cone cells that secrete the material of the lens. The remaining cells are pigment cells that give the *Drosophila* eye its characteristic red color.

Hundreds of different mutations have been identified that alter eye color and morphology. Indeed the *white* eye color mutant was the very first *Drosophila* mutant to be isolated, and *white* mutants are often used to make transgenic flies so that the wild type eye color gene can be introduced along with the transgene as a marker (Topic B3). A very informative group of mutants characterize the cell–cell interactions occurring at the single-cell level that control the specification of the photoreceptor cell phenotypes. It appears that the cone cell is the default phenotype and that further inductive events are required to generate the photoreceptor cells.

Differentiation of ommatidia
The entire eye develops from the epithelial cell layer of the eye/antennal imaginal disc (imaginal discs are considered in more detail in Topic L2). During the third instar larval stage, the posterior cells proliferate so that the disc grows in size and produces the antennal primordium and eye primordium. At the same time, a wave of differentiation called the **morphogenetic furrow** sweeps slowly across the eye primordium from posterior to anterior, and in its wake the cells of the ommatidia begin to differentiate (*Fig. 1*). There are two important characteristics of this differentiation process: first, the ommatidia arise in a regularly spaced pattern, and second the photoreceptor cells differentiate in a predictable sequence starting with the central R8 cell. This indicates that specific patterning mechanisms are involved in the spacing of ommatidia, and that differentiation requires a progressive series of cell–cell interactions.

The morphogenetic furrow
As discussed in Topic L2, some imaginal discs have posterior and anterior compartments because they span two parasegments, and the compartments are characterized by the expression of the *hedgehog*, *wingless* and *dpp* genes encoding signaling proteins. There is no compartmentalization in the eye disc because it arises within a single parasegment, however, the posterior cells behind the furrow secrete Hedgehog, which induces the cells within the furrow to secrete Decapentaplegic. These cells then cease proliferating and begin to differentiate, switching off *dpp* and switching on *hedgehog*, which then induces the adjacent cells to become furrow cells. This feedback loop in which *hedgehog* stimulates *dpp*, which then induces differentiation, shutting down its own synthesis and

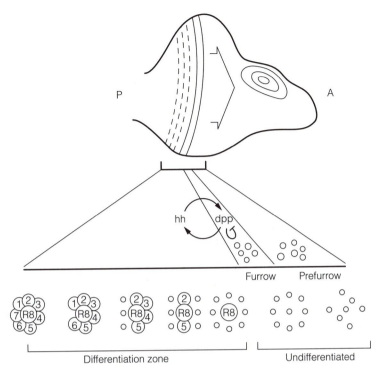

Fig. 1.   Eye development in Drosophila. *A morphogenetic furrow sweeps across the eye field, posterior to anterior (large arrow), and ommatidia arise in a regular pattern in its wake. See text for details.*

inducing *hedgehog*, is what drives the furrow across the eye (*Fig. 1*). The furrow initiates in the posterior of the eye because the Wingless protein is expressed at the lateral edges of the eye primordium and prevents the furrow initiating there. A number of important genes have been identified that have an executive role in initiating eye development, including *eyeless* and *eyes absent*. Mutations in either of these genes result in the absence of eyes, while ectopic expression of *eyeless* in other imaginal discs has the remarkable ability to initiate ectopic eye development, suggesting that it is capable of switching on all of the >1000 genes required for eye development. The *eyeless* and *eyes absent* genes have homologs in vertebrates (*Pax6* and *Eya1*) that appear to be equally critical in development of the eyes, ears and kidneys (Topic K4).

Spacing the ommatidia

As the morphogenetic furrow moves across the eye primordium, the ommatidia differentiate in rows parallel to the furrow, with each second row displaced slightly so that the ommatidia arise in hexagonal arrays. This regular spacing pattern is thought to be controlled by lateral inhibition (Topic A3) in the same way as the spacing of the neuroblasts in the *Drosophila* nervous system (Topic J5). As discussed above, cells in the morphogenetic furrow secrete **Decapentaplegic**. This protein induces differentiation of the epithelial cells into R8 photoreceptor cells. All cells in the pre-furrow epithelium have the ability to differentiate into R8 cells, but as part of the differentiation process they produce a secreted factor, **Scabrous**, which inhibits R8 differentiation in surrounding cells. Initial imbalances are probably amplified by positive feedback, as in other lateral inhibition systems, resulting in regularly spaced R8 cells separated by 6–8 undifferentiated cells. This suggests that the Scabrous protein can diffuse over 3–4 cells and establish a zone of inhibition. *scabrous* mutants show reduced distances between R8 cells and abnormal adult eye morphology, but not all cells become R8 cells. Other signaling molecules may also play a role in lateral inhibition, including those of the Notch/Delta pathway.

Cell signaling in differentiation

The photoreceptor cells are specified in a predictable series, starting with R8, then R2 and R5, then R4 and R6 and finally R7 (*Fig. 2*). The R8 cell first induces the two adjacent cells on the anteroposterior axis to become the functionally equivalent R2 and R5 cells. These cells then induce differentiation of the R3 and R4 cells, and then the R1 and R6 cells. Finally, the R8 cell induces the remaining undifferentiated cell to become the R7 cell. Although many of the signaling events involved in this series of inductive interactions are known, the pathway is most completely understood for the last event in the series, the specification of the R7 cell. Two genes, *sevenless* and *bride of sevenless* (*boss*) have been identified that pay important roles in R7 cell specification, and their products appear to act at the beginning of the signaling pathway. Mutations of either gene prevent R7 specification. Mosaic analysis has shown that the *sevenless* gene is required in the future R7 cell, while the *boss* gene is required in the inducing R8 cell. Boss and Sevenless are both transmembrane proteins, Boss a G-protein coupled receptor and Sevenless a receptor tyrosine kinase homologous to the mammalian epidermal growth factor receptor (Topic C3). Boss binds to Sevenless and initiates a signal transduction cascade in the future R7 cell, which is homologous to the Ras Raf MAP kinase pathway of mammals (Topic C3). The protein that links the receptor tyrosine kinase to Ras is called **Son-of-sevenless**, which is why homologous vertebrate proteins have the name SOS.

All the surrounding photoreceptor cells and the presumptive cone cells also express Sevenless, but only the R7 cell responds. There are number of different mechanisms that ensure the specificity of the R7 cell. Firstly, the Boss signal is restricted to the apical surface of the R8 cell, so not all Sevenless-expressing cells are exposed to the signal. Furthermore, transcription factors synthesized in the other photoreceptor cells as part of their differentiation specifically inhibit the Sevenless signaling pathway. One of these transcription factors is called **Seven-up**, and mutations in the *seven-up* gene result in the R1, R3, R4 and R6 cells adopting R7-like characteristics. Another transcription factor, **Rough**, is required in the R2 and R5 cells to induce *seven-up* expression in the other photoreceptor cells.

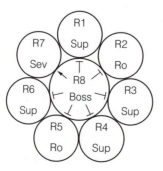

Fig. 2. Summary of molecular interactions in the specification of the R7 photoreceptor cell. The Sevenless receptor in the R7 cell (Sev) responds to the Bride-of-sevenless ligand displayed by the R8 cell (Boss). In the other photoreceptors, the Sevenless pathway is blocked by inhibitors such as Seven-up (Sup) or Rough (Ro).

M1 INITIATION AND MAINTENANCE OF LIMB GROWTH

Key Notes

The vertebrate limb as a model system

The vertebrate limb is a popular developmental system because it provides a useful model of many developmental processes, particularly pattern formation, yet is self-contained so that experimental manipulations do not affect the viability of the whole embryo. The results of experiments involving surgical manipulation of the limb are now being explained in terms of gene expression and cell–cell signaling.

Overview of limb development

The limb originates from a limb field in the lateral plate mesoderm of the body wall. The limb bud comprises a mesodermal core with an ectodermal cap and contains three key signaling centers, the apical ectodermal ridge (AER), the zone of polarizing activity (ZPA), and the dorsal ectoderm. These signaling centers interact to maintain each other's activities and coordinate limb outgrowth and patterning. The AER is the source of growth factors that maintain a population of proliferating mesenchyme cells, the progress zone, at the distal tip of the limb bud. Cells leaving the progress zone differentiate to form the skeletal elements of the limb, while immigrating myoblasts form the limb musculature.

The limb field

The limb field contains all the cells with the potential to form a limb. Only the central cells of the limb field are involved in limb development, but if removed, these can be replaced by more peripheral cells that normally do not contribute to the limb. If the entire limb field is removed, the limb does not form.

Limb field specification

The limbs always form at the same positions along the anteroposterior axis of the body, which are defined by the particular combination of *Hox* genes expressed in the paraxial mesoderm. Certain combination of HOX transcription factors are therefore likely to be permissive for limb development, perhaps by activating signaling pathways required for limb induction and morphogenesis.

Limb identity

The forelimbs and hindlimbs of the vertebrate embryo are distinct. The combinations of *Hox* genes permissive for limb development may also be instructive in terms of the type of limb that develops. Downstream of the *Hox* genes, *tbx5* and *tbx4* are expressed in the forelimbs and hindlimbs, respectively and are major determinants of limb identity.

Initiation of limb development

The initiation of limb development requires the establishment of the three signaling centers and the induction of cell proliferation in the mesenchyme. An important component of the inductive signal is FGF10, which can induce *fgf8* expression in the AER and *sonic hedgehog* expression in the ZPA, although not Wnt7a expression in the dorsal ectoderm. FGF10 is produced in the appropriate regions of the embryo prior to limb bud outgrowth.

Induction of the AER

The key molecular component of the AER is FGF8, which is required to maintain the proliferation of mesenchyme cells in the underlying progress zone. The expression of *fgf8* in the AER is induced in response to FGF10 signaling. The AER forms from cells at the boundary of dorsal and ventral ectoderm, which are marked by the expression of the gene *radical fringe*. The expression of *radical fringe* depends on signals from adjacent ventral cells in which this gene is repressed by the transcription factor Engrailed.

Induction of the ZPA

The key molecular component of the ZPA is the secreted protein Sonic hedgehog, which is required both for patterning the anteroposterior axis and the maintenance of the AER. FGF10 and Wnt7a combine to induce the expression of *sonic hedgehog* in the limb mesenchyme, but other genes such as *Alx4* and *Gli3* are required to restrict its expression to the posterior margin.

Limb outgrowth – roles of the AER and progress zone

The role of the AER is to maintain the proliferation of cells in the progress zone, but it does not determine the type of limb that forms. Conversely, the mesenchyme of the progress zone carries a positional identity, and grafted progress zones develop according to their original position. Non-limb mesenchyme will not support limb development, because only limb mesenchyme can maintain the AER.

Maintenance of limb outgrowth

The maintenance of limb outgrowth involves a self-sustaining signaling loop between the AER and ZPA. There is evidence that Sonic hedgehog induces the expression of *fgf4* in the AER, and that FGF4 may cooperate with other FGFs to maintain *sonic hedgehog* expression. In this way, the two organizing centers become mutually dependent.

Limbless vertebrates

Snakes and other limbless vertebrates are species in which defects in limb specification or outgrowth have become advantageous traits. In the python embryo, the forelimbs appear never to be specified because the correct combination of *Hox* genes is not expressed in the thoracic region of the embryo. Conversely, the hindlimbs are correctly specified and begin to form, but the AER collapses and limb growth ceases.

Related topics

Mechanisms of developmental commitment (A3)
Signal transduction in development (C3)
Homeotic selector genes and regional specification (H5)

Patterning the anteroposterior neuraxis (J2)
Somitogenesis and patterning (K2)
Patterning and morphogenesis in limb development (M2)
Limb regeneration (M3)

The vertebrate limb as a model system

For a number of reasons, the vertebrate limb is one of the most favored developmental model systems. First, the limb is polarized along three axes giving it a well-defined and easily recognizable anatomy (*Fig. 1*). This anatomy is especially clear in the cartilaginous elements, which can be visualized by staining so

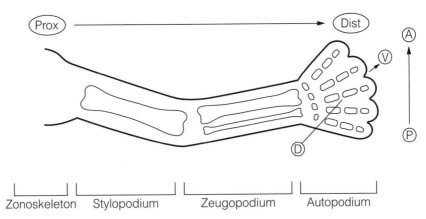

Zonoskeleton   Stylopodium   Zeugopodium   Autopodium

Fig. 1.   Anatomy of the vertebrate limb, showing the three principal axes.

that perturbations caused by mutation or experimental manipulation can be easily characterized. Second, the limb is a self-contained system, so that manipulations can be carried out without affecting other regions of the embryo, and thus without affecting the viability of the whole embryo. Third, the limb espouses many of the principles and processes of development that occur in the embryo as a whole, e.g. axis specification and patterning, region-specific morphogenetic behavior of similar cell types, cell migration, and axon guidance. Finally, many human birth defects are characterized by limb abnormalities, and the study of limb development has provided insights into the basis of such diseases.

Early studies of limb development generally involved monitoring the effects of surgical manipulation. More recently, the principles established by such experiments have been supported by molecular data generated by transgene misexpression and gene knockouts. The molecular basis of limb development is now fairly well understood. In this Topic, we begin with an overview of limb development, then discuss how limb growth is initiated and maintained. Mechanisms of axis specification, patterning and morphogenesis are discussed in Topic M2. Finally, we consider how certain vertebrates can regenerate limbs even as adults, and compare the mechanisms involved to those operating during limb development in the embryo (Topic M3).

Overview of limb development

Most vertebrates have four limbs, arranged as two symmetrical pairs. The term 'limb' covers a variety of structures – arms and legs in terrestrial vertebrates, wings in flying vertebrates, and flippers or fins in swimming ones. Some reptiles (snakes and certain lizards such as slow worms) lack limbs all together, and we consider some reasons for this at the end of the Topic.

Each vertebrate limb originates on the body wall from a circular area of the lateral plate mesoderm called the **limb field**. Cells at the center of the limb field proliferate, generating a protuberence called a **limb bud** comprising a mesenchyme core and a jacket of ectoderm. A critical event at the initiation of limb development is the establishment of three **limb organizing centers** (*Fig. 2*). First, ectodermal cells at the dorsal/ventral boundary of the limb bud form a specialized raised epithelium called the **apical ectodermal ridge (AER)**. FGFs secreted by this region maintain a population of rapidly proliferating

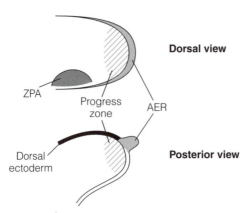

Fig. 2.   Dorsal and posterior aspects of the limb, showing the three limb organizing centers: the apical ectodermal ridge (AER), the zone of polarizing activity (ZPA) and the dorsal ectoderm. The progress zone is also shown.

mesenchyme cells at the distal tip of the limb bud (a region called the **progress zone**) and in this way control limb outgrowth along the proximodistal axis. Second, mesenchyme cells at the posterior of the limb bud form the **polarizing region** or **zone of polarizing activity (ZPA)**. The signaling protein Sonic hedgehog is synthesized in this region, and plays a major role in patterning the anteroposterior limb axis as well as maintaining the AER. Finally, the dorsal ectoderm is the source of **Wnt7a**, a secreted protein that patterns part of the dorsoventral axis. The three organizing centers interact to reinforce and maintain each other's activities. Therefore, the processes of limb outgrowth and patterning along all three axes are intimately linked.

Limb outgrowth depends entirely on the proliferation of cells in the progress zone, but the AER has a limited signaling range so the progress zone is only about 200 μm across. As growth continues, cells are steadily lost from the progress zone. These stop proliferating and differentiate to form the skeletal elements of the limb: cartilage (most of which later becomes bone), tendons and ligaments, as well as the vasculature. While the skeletal elements are derived from the lateral plate mesoderm, the musculature is derived from a population of myoblasts that migrate into the limb bud from the adjacent somite-derived myotomes. The muscles are later innervated by motor axons from the ventral neural tube. Neural crest cells (Topic J6) also migrate into the developing limb bud, and give rise to glial cells and melanocytes.

The limb skeleton is laid down in a progressive manner that establishes a **proximodistal axis** (running from body to extremity). The limb comprises four proximodistal elements: a zonoskeleton, stylopodium, zeugopodium and autopodium, each corresponding, for example, to a different bone or set of bones in the arm or leg (*Table 1*). As development proceeds it becomes clear that the limb is also polarized along two other axes. In adult humans, the **anteroposterior limb axis** runs from the thumb/big toe to the little finger/toe, and the digits are numbered consecutively from anterior to posterior (in the hand, the order is: thumb = I, index finger = II, middle finger = III, ring finger = IV, little finger = V). The same convention is followed in other vertebrates although the number of digits may vary (e.g. there are only three digits in the chicken wing, and four in the leg). The **dorsoventral limb axis** runs from the back of the hand to the palm.

Table 1.   *Proximodistal arrangement of skeletal elements in the vertebrate limb*

Element	Human arm	Human leg
Zonoskeleton	Clavicle	Hip girdle
Stylopodium	Humerus	Femur
Zeugopodium	Radius and ulna	Tibia and fibula
Autopodium	Bones of the wrist and hand	Bones of the ankle and foot

The limb field

Most vertebrates have the ability to regenerate limbs when limb buds are surgically removed from the embryo. Cell tracing studies have shown exactly which lateral plate mesoderm cells contribute to the limb, and even if all these cells are removed or ablated, a limb still forms albeit after a short delay. Such experiments reveal the presence of a **limb field**, i.e. a group of cells with the potential to form a limb, but from which only a proportion of the cells is used. When the limb bud is amputated, cells from the extremes of the limb field proliferate to close the wound, and the new central cells are then induced to form a new limb bud. If the entire limb field is surgically removed from an embryo, the limb fails to form. This shows that the limb field is endowed with unique and intrinsic limb-forming properties that cannot be emulated by cells outside the limb field.

Limb field specification

Limbed vertebrates possess two morphologically distinct pairs of limbs, called **forelimbs** and **hindlimbs** (or in fishes, pectoral and pelvic fins). These always form at the same positions in relation to the axial skeleton, suggesting that the limb fields are specified in response to positional values along the anteroposterior axis of the body. As discussed in Topic K2, regional specification in the paraxial mesoderm is controlled by *Hox* genes, which are expressed in characteristic overlapping patterns with nested anterior boundaries. Despite differences among vertebrates in the number and arrangement of different types of vertebrae, the limbs always form in the same *Hox*-defined region. The forelimbs, for example, always form at the junction between the cervical and thoracic vertebrae, which is marked by the anterior boundary of *HoxA6* expression. It is therefore likely that the specification of a limb field requires a certain combination of HOX transcription factors, which is permissive for the activation of the limb development pathway. The HOX proteins may act, for example, by activating genes encoding signaling molecules required for the induction of the AER and ZPA, and the stimulation of migratory behavior in the adjacent somites.

Limb identity

The forelimbs and hindlimbs of most vertebrates are morphologically distinct and specialized for different functions. As stated above, limb buds always form in the same positions along the anteroposterior axis, but moreover, the forelimbs and hindlimbs always develop in the right places. The combinations of HOX transcription factors that are permissive for limb development may therefore also be instructive in terms of the type of limb that forms. Genes from all four *Hox* clusters are expressed in the developing limb. As discussed in Topic M2, *HoxA* and *HoxD* genes are expressed in the distal parts of both limbs, and their function is to establish positional values along the anteroposterior and proximodistal axes. Conversely, *HoxB* and *HoxC* genes are expressed in the proximal region of the limb, and different genes are expressed in the forelimb and hindlimb. The *HoxC* genes may be instrumental in establishing limb identity – *HoxC4* and *HoxC5* are restricted to the forelimb, *HoxC6* and *HoxC8* are expressed in both, while *HoxC9*, *HoxC10* and *HoxC11* are restricted to the hindlimb. The *HoxB5* gene appears to play a role in positioning the forelimb. Mouse *hoxb5* knockouts show homeotic transformations in which the forelimbs are displaced anteriorly.

Differences in limb structure reflect underlying differences in cell behavior. Even at an early developmental stage, there are overt differences in the behavior of forelimb and hindlimb cells. For example, presumptive cartilage cells in the chicken forelimb and hindlimb buds respond differently to growth factors and other signaling molecules, and also differ in terms of the molecules they secrete into the extracellular matrix. The particular combination of HOX transcription factors present in each limb bud might activate genes that act as selectors for forelimb and hindlimb developmental programs. Four members of the **Tbx family** are expressed in the developing limb mesenchyme, and two are specific to particular limbs (**tbx5** and **tbx4** are expressed in mesenchyme of the forelimb and hindlimb, respectively). Ectopic limbs can be generated by implanting FGF-coated beads into the lateral plate mesoderm, and buds expressing *tbx5* develop as hindlimbs whereas those expressing *tbx4* develop as forelimbs, while those in between the expression domains develop as hybrid fore/hindlimbs. Recent experiments have shown that the overexpression of *tbx5* or *tbx4* throughout the flank mesoderm can induce the formation of forelimbs and hindlimbs respectively following the implantation of FGF beads. These *tbx* genes therefore appear to be determinants of forelimb and hindlimb identity.

Initiation of limb development

During most of the growth of the limb bud, the proliferating cells of the progress zone are maintained by signals from the AER. However, there is no AER at the beginning of limb development, so the signals that initiate limb growth must originate elsewhere. These signals must be able to set in motion a sequence of events resulting in the establishment of both the AER and the ZPA, since the AER is essential for continued growth and the maintenance of the AER depends on reciprocal signaling with the ZPA. The following evidence indicates that the intermediate mesoderm is required for the initiation of limb development, and that FGF10 is an important component of the inducing signal.

- Limb outgrowth is not initiated if an impermeable barrier is placed between the intermediate mesoderm (mesonephros) and the overlying limb field. However, if intermediate mesoderm underlying either of the limb fields is grafted to the inter-limb region of a host embryo, an ectopic limb bud is induced. *Conclusion: molecules from the intermediate mesoderm are necessary and sufficient for the initiation of limb development.*
- FGF10 is initially expressed throughout the intermediate mesoderm and the lateral plate mesoderm, but is downregulated in all areas except the lateral plate mesoderm of the limb fields as the limb buds begin to form. FGF10-expressing cells will induce limb development if placed under the lateral plate mesoderm in the inter-limb region. *fgf10*-knockout mice have truncated limbs resulting from the failure to induce either the AER or the ZPA. *Conclusion: FGF10 is expressed in the correct places to initiate and maintain limb development. FGF10 is necessary and sufficient for the initiation of limb development.* Note: the effects of FGF10 can be mimicked by other FGFs, but these are not expressed in the correct places in the developing embryo.
- FGF10 has been shown to induce the expression of *fgf8* in the AER and *sonic*

hedgehog in the ZPA, but not Wnt7a in dorsal ectoderm. *Conclusion: FGF10 can induce the expression of genes whose products represent the key functional components of the AER and ZPA but the dorsal organizer requires an additional, as yet unidentified signal.*

It is notable that both the AER and ZPA arise in restricted regions of the limb bud. Since FGF10 is not restricted in this manner, cells in different regions of the limb bud must be restricted in their competence to respond to the signal. We examine the basis of this spatial restriction below.

Induction of the AER

The AER forms at the boundary between dorsal and ventral ectoderm, and therefore positions each limb bud at the midline of the dorsoventral axis (*Fig. 3*). Even in ectopic limb buds induced by mesoderm grafts or FGF-containing beads, the AER still forms at the dorsal/ventral boundary. This suggests that the competence of ectoderm cells to respond to FGF10 signaling and form the AER is restricted to these cells.[1]

The restriction of competence to the dorsal/ventral boundary reflects the expression of a gene called *radical fringe* (related to the *Drosophila* gene *fringe*, which is involved in forming the wing margin; Topic L2). The *radical fringe* gene is initially expressed throughout the dorsal ectoderm. Its expression is induced by signals from the underlying progress zone, and although these signals have yet to be identified, they are dependent on a transcription factor called **Lhx2**. The ventral ectoderm does not express *radical fringe* because the gene is repressed by the transcription factor Engrailed. Although *radical fringe* is inhibited in the ventral ectoderm, maintenance of *radical fringe* expression in dorsal ectoderm is dependent on signals from the ventral cells. Therefore *radical fringe* becomes restricted to those dorsal ectoderm cells lying immediately adjacent to

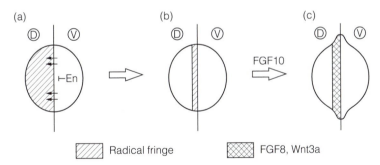

Fig. 3.   Distal view of the limb bud, showing establishment of the apical ectodermal ridge (AER). (a) Expression of the gene radical fringe is initially restricted to the dorsal limb because it is repressed by the ventral transcription factor Engrailed. (b) Later, radical fringe expression becomes dependent on signals from Engrailed-expressing cells, and is thus restricted to the dorsal/ventral boundary. (c) Following stimulation of limb outgrowth by FGF10, Wnt3a and FGF8 are expressed in the nascent AER.

[1] For a long time, it was thought that only cells at the dorsal/ventral boundary of the ectoderm were able to support AER development. However, recent experiments in chick embryos have shown that a strip of ectoderm along the dorsal midline is also competent to form an AER in response to FGF signaling. This lies within a uniform field of Radical fringe-expressing dorsal cells, indicating that a different mechanism is required to specify the competence of these responding cells. It is speculated that these cells might represent vestiges of the dorsal fin-field in fishes.

the dorsal/ventral boundary, and these cells become the AER. The juxtaposition of any dorsal and ventral limb ectoderm is sufficient to generate AER competence. If patches of dorsal ectoderm are grafted onto the ventral ectoderm or vice versa in the developing limb, an additional AER forms at the graft/host boundary, and a secondary limb bud grows out from the first.

The AER begins to secrete the growth factor FGF8, and this is sufficient to maintain the proliferating mesenchyme cells of the progress zone. The AER also secretes Wnt3a, a signaling molecule whose downstream target is the transcription factor β-catenin. As Wnt3a diffuses from the AER, β-catenin accumulates in the distal tip of the limb bud (AER and progress zone). The activity of the *fgf8* gene appears to be dependent on β-catenin.

Induction of the ZPA

The ZPA is formed in the early limb bud at the same time as the AER. FGF10 secreted by the underlying mesoderm, and Wnt7a secreted by the dorsal ectoderm, are thought to be responsible for inducing the ZPA. Neither the AER nor the ZPA are formed in *fgf10*-knockout mice, resulting in the truncation of all four limbs, while the ZPA does not form in *wnt7a* knockout mice, resulting in biventral limbs and the lack of posterior structures.

FGF10 can activate the *sonic hedgehog* gene in the limb field mesoderm, but this does not provide a mechanism for restricting *shh* expression to the posterior of the limb bud. A number of genes may be involved in this process, including *Alx4* (related to the *Drosophila* gene *arista-less*, which plays an important role in insect limb development; Topic L2) and *Gli3*, which encodes a repressor of Hedgehog signaling that is expressed before limb outgrowth begins (*Fig. 4*). The mouse mutant *Extra-toes* has a nonfunctional *Gli3* gene, resulting in the development of extra digits and the loss of digit identity.

Each limb bud is also divided into anterior and posterior sectors by an anterior boundary of *Hox* gene expression, and this may control the expression of *sonic hedgehog* and/or its repressors. The posterior half of both the mouse and chicken forelimbs expresses *HoxB8*. If this gene is forcibly expressed throughout the forelimb, *sonic hedgehog* is also expressed in the anterior mesenchyme, generating a phenotype similar to the *Extra-toes* mutant. As discussed in Topic M2, retinoic acid can also induce the formation of the ZPA, and this works by first activating *HoxB8*.

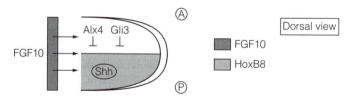

Fig. 4.   *Establishment of the ZPA. FGF10 induces* sonic hedgehog *expression throughout the limb bud, but its expression is restricted to the posterior by dependence on HoxB8 and inhibition by Alx4 and Gli3.*

Limb outgrowth – roles of the AER and progress zone

The major function of the AER in limb development is the regulated secretion of FGFs to maintain cell proliferation in the progress zone. If the AER is widened by grafting part of a donor AER onto the margin of the limb bud, the progress zone widens too, and the resulting limb is broader than usual and may contain supernumerary digits. Many cases of polydactyly in humans can be traced to an unusually broad AER. The AER is not responsible for any other properties of the limb, such as commitment to forelimb or hindlimb development. These characteristics of the limb depend on the mesenchyme, which expresses characteristic sets of *Hox* and *Tbx* genes as discussed above. The roles of the AER and mesenchyme can be derived from the results of reciprocal grafting experiments:

- If the AER from the chicken hindlimb is removed, limb growth ceases. Development can be rescued by implanting a bead soaked in various FGFs. If the AER grafted onto the side of a host limb bud, a secondary bud forms. *Conclusion: the AER is required for limb growth. It is a source of FGFs.*
- If the AER from the chicken hindlimb is replaced by AER from the chicken forelimb, a normal hindlimb forms. In the reciprocal graft, a normal forelimb is generated. *Conclusion: the AER supports limb development but the effect is general; FGF promotes cell proliferation but does not specify regional properties of the developing limb.*
- If the progress zone from the chicken hindlimb is grafted under the AER of the forelimb bud, a hindlimb is generated in the position of the forelimb. In the reciprocal graft (forelimb progress zone under hindlimb AER) a forelimb is generated in the position of the hindlimb. *Conclusion: the character of the limb depends on the mesenchyme, which has a positional value specified by Hox and Tbx genes.*
- If the progress zone from a chicken limb bud is replaced by non-limb mesenchyme, the AER collapses and limb growth ceases. *Conclusion: only mesenchyme from the limb bud can support limb development because only limb bud mesenchyme (expressing sonic hedgehog) can maintain the AER.*

Maintenance of limb outgrowth

As the limb extends away from the body wall, the expression of molecules that coordinate limb outgrowth and axis formation must be maintained, and this is achieved through a complex system of reciprocal signaling loops. FGF10 is required to initiate *sonic hedgehog* expression in the posterior mesenchyme (ZPA), and *fgf8* expression in the AER. FGF8 then maintains *fgf10* expression in the limb mesenchyme, which may explain why *fgf10* expression ceases in all intermediate and lateral plate mesoderm apart from the limb buds. This FGF10-FGF8 loop persists until the digits begin to separate. There is also a reciprocal signaling loop between the ZPA and AER. In chickens, there is evidence that Sonic hedgehog in the ZPA induces the expression of *fgf4* in the posterior AER. The FGF4 secreted by the AER then induces the expression of *sonic hedgehog*. In this way, Sonic hedgehog and FGF4 become mutually dependent (*Fig. 5*). Mice

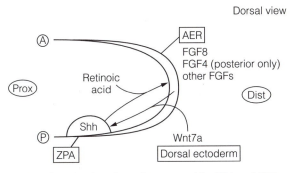

Fig. 5. Signaling loop for maintenance of the ZPA and AER.

lacking Sonic hedgehog or its downstream effector **Formin** do not maintain *fgf4* expression in the AER, although they do continue to express *fgf8* (probably reflecting its maintenance by FGF10). However, conditional *fgf4* knockout mice *can* maintain the ZPA and show normal *sonic hedgehog* expression. This suggests that no single FGF expressed in the AER is absolutely required for the maintenance of the ZPA. The exact nature of the signaling loop between the AER and ZPA remains to be worked out. Important downstream targets of Sonic hedgehog include the **BMPs** and the BMP antagonist **Gremlin**. The control of BMP levels in different parts of the limb is critical for maintaining a balance between cell proliferation and differentiation, so these proteins are likely to be subject to complex feedback controls.

Limbless vertebrates

We briefly return to the topic of snakes and similar limbless vertebrates, and ask why these animals do not develop limbs. There are a number of mutations in limbed vertebrates that result in the absence of limbs, including the chicken mutation *limbless* where a deficiency in the ectoderm results in the failure to form a dorsoventral boundary and hence the absence of an AER. Snakes might represent species where a similar defect has become the norm. In the python, for example, the hindlimbs are specified as in other vertebrates, but the AER collapses early in development and there is no further growth. Conversely, the forelimbs appear never to be specified in the first place. Analysis of the *Hox* gene expression pattern in the thoracic region of the python embryo reveals an unusually broad homogeneous expression domain, specifying hundreds of identical thoracic vertebrae and no limb fields. It remains to be seen whether the ectopic expression of *Hox* genes can induce leg development in snakes.

M2 PATTERNING AND MORPHOGENESIS IN LIMB DEVELOPMENT

Key Notes

Limb organizing centers	The limb has three organizing centers – the apical ectodermal ridge (AER), the zone of polarizing activity (ZPA) and the dorsal ectoderm – that help to pattern the proximodistal, anteroposterior and dorsoventral limb axes, respectively. As well as patterning the axes, these organizers also interact with each other to maintain limb outgrowth.
Specifying the proximodistal axis	The proximodistal limb axis emerges progressively as mesenchyme cells drop out of the progress zone and differentiate. The type of structure formed by these differentiating cells may depend on the amount of time they have spent in the progress zone, and hence the number of division cycles they have undergone in response to signaling from the AER.
Specifying the anteroposterior axis	The anteroposterior limb axis is specified by the zone of polarizing activity in the posterior mesenchyme, as grafting this region to the anterior side of the limb bud can induce duplication and mirror image reversal of the axis. The secreted protein Sonic hedgehog can substitute for the activity of the ZPA and appears to act in a dose-dependent manner to specify the fates of anteroposterior structures, such as the different digits of the hand.
Positional values along the proximodistal and anteroposterior axes	Along both the anteroposterior and proximodistal limb axes, the 5′ *HoxA* and *HoxD* genes are expressed in overlapping concentric patterns. These *Hox* genes play an important role in the regional specification of different skeletal elements along the two axes, as gene knockouts cause the deletion or respecification of particular limb structures.
Specifying the dorsoventral axis	The dorsoventral limb axis is specified by the secreted protein Wnt7a, which is synthesized in the dorsal ectoderm. This activates a transcription factor called Lmx1 in the dorsal mesenchyme. The inactivation of either gene generates biventral limbs, while overexpression throughout the limb generates bidorsal limbs.
Interaction and dependence in axis patterning	There is a significant interaction and interdependence between the three signaling pathways. The AER is maintained by Sonic hedgehog secreted by the ZPA. The maintenance of the ZPA is dependent on both FGFs secreted by the AER and Wnt7a secreted by the dorsal ectoderm. Furthermore, FGFs secreted by the AER are required in addition to Sonic hedgehog to establish the nested pattern of *HoxD* gene expression along the anteroposterior axis.

Other patterning influences	There is evidence that the limb is pre-patterned before the induction of the three signaling centers, as the absence of the AER and ZPA in certain mutants does not prevent the regionalized expression of asymmetrically expressed genes, such as those of the *HoxD* cluster. Furthermore, there is evidence that the limb has significant self-organizing ability, as shown by the development of recognizable digits in disaggregated and recombined limb buds.
Morphogenesis in the limb	The final structure of the limb depends largely on the regulation of cell death, controlled by the opposing activities of BMPs (which promote apoptosis) and BMP antagonists (which inhibit it). Complementary patterns of BMPs and their antagonists are seen in the limb, marking the interdigital and internal necrotic zones that separate the digits and the two bones of the zeugopodium, and the positions where joints will form.
Related topics	Pattern formation (A6) Morphogenesis (A7) Signal transduction in development (C3) The cell division cycle (C4) Homeotic selector genes and regional specification (H5) Patterning the anteroposterior neuraxis (J2) Mesoderm induction and patterning (K1) Somitogenesis and patterning (K2) Initiation and maintenance of limb growth (M1) Limb regeneration (M3)

Limb organizing centers

The limb is a self-contained system that demonstrates an extraordinary regulative ability. This reflects the existence of three organizing centers – the apical ectodermal ridge (AER), the zone of polarizing activity (ZPA) and the dorsal ectoderm – that cooperate to regulate outgrowth and control patterning along all three axes (*Table 1*). The role of these signaling centers in promoting and maintaining limb outgrowth was discussed in Topic M1. In this Topic, we consider how cells in the limb are given positional values, enabling them to form regionally-appropriate structures. We then consider the basis of morphogenesis in limb development, which produces the familiar pattern of bones and muscles in the mature limb.

Table 1.   Characteristics of limb organizing centers

Axis	Organizing center	Signals	Targets
Proximodistal	AER (ectoderm)	FGFs	5′ *HoxA* and *HoxD* genes
Anteroposterior	ZPA (posterior mesenchyme)	Sonic hedgehog	5′ *HoxD* genes
Dorsoventral	Dorsal ectoderm	Wnt7a	*lmx1*

Specifying the proximodistal axis

The anteroposterior and dorsoventral limb axes already exist in the limb field before limb development begins. Conversely, as the limb field is flat, there is no proximodistal axis until limb outgrowth commences. The proximodistal axis then emerges progressively, as cells drop out of the progress zone and differentiate.

Cells leaving the progress zone appear to have an intrinsic awareness of their position along the proximodistal axis, allowing them to form regionally appropriate structures. This determined state is demonstrated by the results of **heterochronic grafts**, where the progress zone of one limb bud is grafted under the AER of another at a different developmental stage (*Fig. 1*). When a young progress zone is grafted under the AER of an older limb (in which proximal elements have already formed), the young progress zone gives rise to a full proximodistal axis, resulting in a longer limb with duplicated proximal elements. Conversely, if an older progress zone is grafted under the AER of a young limb bud, distal elements are formed immediately and the limb lacks proximal elements.

One interpretation of the above results is that proximodistal patterning depends on the amount of time mesenchyme cells remain in the progress zone. The first cells leaving the progress zone have undergone only a few divisions before they differentiate, whereas the last cells will have undergone many rounds of division. In some way, the number of divisions assigns differentiating cells their positional values. This model is supported by experiments in which cell proliferation in the progress zone is blocked using mitotic inhibitors. The cells all differentiate into proximal structures. Conversely, if most of the progress zone cells are killed in the early limb bud, proximal structures fail to form. The few surviving cells repopulate the progress zone and then give rise to distal structures, perhaps because of the extra rounds of cell division they have undergone.

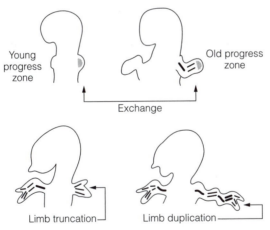

Fig. 1. *Heterochronic grafts of the entire distal tip of the limb bud show that the progress zone contains intrinsic positional information. A young progress zone will give rise to all limb elements even when grafted onto an older limb where some elements have already formed, whereas an older progress zone will give rise to distal limb elements even when grafted onto a younger limb bud lacking proximal elements.*

Specifying the anteroposterior axis

The anteroposterior limb axis, which in humans runs from thumb to little finger in the arm and from big toe to little toe in the leg, is specified by the **zone of polarizing activity (ZPA)** in the posterior mesenchyme. The role of the ZPA in anteroposterior patterning can be demonstrated by grafting a donor ZPA onto the anterior side of a host limb bud. In the chicken forelimb, which normally has three digits, such a graft results in the development of six digits. The extra digits have the same characteristic structure as the normal ones, but their polarity is reversed so that the limb is now symmetrical (*Fig. 2*). The extra digits come from

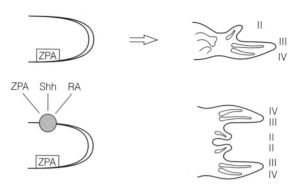

Fig. 2.   Grafting a second zone of polarising activity (ZPA) to the anterior margin of the chick
limb bud results in mirror image duplication of the digits. This effect can be mimicked by beads
coated in retinoic acid (RA) or Sonic hedgehog protein (Shh).

the tissue of the host limb bud, not the graft. This shows that signals from the
ZPA are reorganizing the limb mesenchyme.

A simple explanation of the above results is that the ZPA is the source of a
morphogen that forms a gradient across the limb bud and specifies posterior
digits at high concentrations and anterior digits at low concentrations. Support
for this model comes from experiments where the exposure of anterior
mesenchyme to the ZPA is limited, e.g. by grafting a very small part of the ZPA
or removing the graft after a short time. In such experiments, the mesenchyme
adjacent to the graft forms progressively more anterior structures as its
exposure to the ZPA is reduced.

As discussed in Topic M1, the secreted protein **Sonic hedgehog** is an important functional component of the ZPA and is required to maintain the AER. It
appears that Sonic hedgehog is also a key player in patterning the anteroposterior axis. Cells secreting Sonic hedgehog and plastic beads coated in Sonic
hedgehog protein can both mimic the effects of the ZPA when implanted in the
anterior mesenchyme. Furthermore, such experiments demonstrate the same
dose-dependent effects on digit fates as seen with ZPA grafts. Sonic hedgehog
appears sufficient to produce ZPA activity and is a secreted protein with the
potential to act as a morphogen. However, the distribution of Sonic hedgehog
protein in the limb bud is not consistent with this role. It is thought that Sonic
hedgehog may initiate a chain of cell–cell signaling events involving several
downstream components, with gradually diminishing efficacy across the limb.
There are several candidates for these downstream signaling proteins, one of
which is BMP2, which has been shown to be activated in response to Sonic
hedgehog signaling.

The vitamin A derivative **retinoic acid** can also mimic ZPA activity, as retinoic
acid-soaked beads implanted in the anterior mesenchyme generate the same
mirror image digit duplications seen in ZPA grafts. Initially, it was thought that
retinoic acid could be the morphogen controlling regional specification along the
anteroposterior axis, but it appears that the natural concentration of this molecule
in the developing limb is not high enough to sustain this role. Retinoic acid
mimics the ZPA by inducing *HoxB8* expression, resulting in the activation of *sonic
hedgehog*. The absence of retinoic acid results in limb patterning and outgrowth
defects and it is now thought that this molecule is required to sustain the interaction between the ZPA and AER during limb growth.

Positional values along the proximodistal and anteroposterior axes

The *Hox* genes are instrumental in patterning the anteroposterior axis of the vertebrate body, and they are expressed in all three germ layers (Topics J2, K2 and K6). There are four *Hox* clusters in birds and mammals, and as discussed in Topic M1, genes of the *HoxB* and *HoxC* clusters are expressed in limb-specific patterns consistent with their roles in limb positioning and identity. Conversely, genes of the *HoxA* and *HoxD* clusters are expressed in similar patterns in all the limbs, and their role is to confer positional values on cells in the limb bud. While the expression patterns of the genes from both clusters are distinct and dynamic, at different developmental stages they specify domains that generate positional values along both the proximodistal and anteroposterior axes.

As discussed above, cells leaving the progress zone differentiate into progressively more distal structures depending on the number of division cycles they have undergone. Although the mechanism is unknown, this in some way regulates the expression of the most 5' *HoxA* and *HoxD* genes (paralogous subgroups 9–13), so that they are expressed in a temporal sequence, generating characteristic overlapping expression patterns with nested boundaries corresponding to the subdivisions of the mature limb (*Fig. 3*). In the forelimb for example, cells leaving the progress zone to form proximal structures at the shoulder girdle express *HoxA9* and *HoxD9*. After several further rounds of division, cells leaving the progress zone differentiate to form the humerus, and at this time they express both *HoxA10* and *HoxD10* in addition to the *HoxAA9* and *HoxD9* genes. When the mesenchyme cells begin to form radius and ulna structures, they also express *HoxA11* and *HoxD11*. Wrist elements also express *HoxD12* (there is no *HoxA12* gene). Finally, in the hand itself, all 5' *HoxA* and *HoxD* genes are expressed, including *HoxA13* and *HoxD13*. The role of the *Hox* genes in conferring proximodistal positional values has been supported by the effects of gene knockouts, which result in the deletion of particular elements. For example, individual knockouts of *HoxA11* and *HoxD11* result in malformations of the zeugopodium (radius and ulna). However, double knockouts show almost complete deletion of these bones, suggesting that the role of *HoxA11* and *HoxD11* is to specify this region of the proximodistal axis.

As the zeugopodium forms, the expression pattern of the *HoxD* genes becomes nested along the anteroposterior axis (*Fig. 3*). The more 5' *HoxD* genes

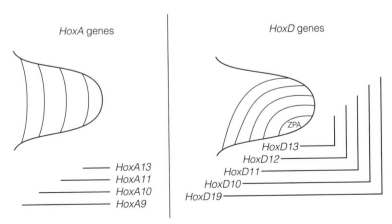

Fig. 3. At certain times during limb development, the HoxA genes mark out domains of regional specification along the limb's proximodistal axis, while the HoxD genes mark territories along both the proximodistal and anteroposterior axes.

are progressively restricted to the posterior mesenchyme, so that *HoxD9* is expressed along the entire axis, while *HoxD13* is restricted to the region immediately adjacent to and including the ZPA. This expression pattern is influenced by Sonic hedgehog expression, as ZPA grafts and Sonic hedgehog beads in the anterior mesenchyme induce a symmetrical concentric pattern of *HoxD* expression corresponding to the specification of extra digits in the reverse order (*Fig. 4*). It therefore appears that the *HoxD* genes are important downstream targets of the Sonic hedgehog signaling pathway in the limb, and that the ability of Sonic hedgehog to pattern the anteroposterior axis results from its ability to establish this characteristic pattern of *Hox* gene expression. A simple model suggests that domains of gene expression representing different combinations of *Hox* genes could specify different digits in a qualitative manner along the anteroposterior axis (Topic A6). Although certain gene knockouts in mice support this model by showing obvious homeotic transformations between digits, many show less clear phenotypes, affecting many different structures simultaneously. The general impression from the many knockout experiments that have now been carried out is that the HOX proteins act in a concentration-dependent manner to specify alternative cell fates, and there is no simple *Hox* gene code to define digit identities.

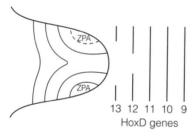

13 12 11 10 9
HoxD genes

Fig. 4.   The effect of a ZPA graft on HoxD gene expression in the limb.

Specifying the dorsoventral axis

Of the three limb axes, least is known about dorsoventral patterning. It is likely that most of the limb's dorsoventral axis is pre-patterned by molecules already expressed in the lateral plate mesoderm of the limb field. However, in the distal part of the limb, the ectoderm acts as a dorsoventral organizer to pattern the digits (*Fig. 5*). This is revealed by the effects of rotating the limb bud's jacket of ectoderm with respect to the inner core of mesenchyme. If the ectoderm is rotated such that the dorsal ectoderm overlies the ventral mesoderm and vice versa, the developing digits are inverted.

As discussed in Topic M1, ventral ectoderm expresses the transcription factor Engrailed-1, which represses the expression of *radical fringe* and leads to the formation of the AER. Another target for Engrailed repression is a gene encoding the signaling molecule **Wnt7a**, which is expressed specifically in dorsal ectoderm. Wnt7a appears to be the dorsalizing signal in limb development since mouse *wnt7a*-knockout mutants have biventral limbs, while the forced expression of *wnt7a* on the ventral side generates mice with bidorsal limbs. Mouse *engrailed-1* mutants also have bidorsal limbs because the effect of losing Engrailed in the ventral ectoderm is ubiquitous *wnt7a* expression. Ectodermal Wnt7a activates downstream transcription factors in the limb mesoderm promoting dorsal development. For example, the transcription factor

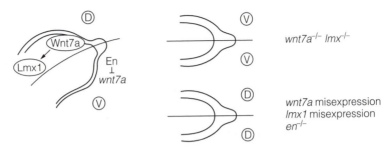

Fig. 5.   *Dorsoventral patterning in the distal limb bud is controlled by Wnt7a secreted by the dorsal ectoderm, which activates the transcription factor Lmx1 in the mesoderm. In the ventral ectoderm, wnt7a is repressed by Engrailed. Loss of function or transgenic overexpression mutants for these genes show predictable biventral or bidorsal phenotypes.*

Lmx1 is related to the product of the *Drosophila* gene *apterous*, which is also involved in dorsoventral patterning. Lmx1 is expressed in the dorsal mesoderm, and is both required and sufficient for dorsal specification. Knockout mice lacking the *lmx1* gene have biventral limbs, while forced overexpression in ventral mesoderm generates bidorsal limbs.

Interaction and dependence in axis patterning

The three axes of the limb are not patterned independently. There is extensive interaction between the molecules that control specification along each axis, and this ensures that limb patterning and outgrowth are coordinated.

Interactions between dorsoventral and anteroposterior axis patterning signals are revealed by the phenotypes of *wnt7a* and *lmx1* knockout mice. As discussed above, these genes are instrumental in establishing dorsoventral limb polarity and mutants have biventral distal elements. However, the limbs also lack the most posterior digits, indicating that Wnt7a signaling is necessary for antero-posterior patterning. The loss of posterior digits in these knockout mice reflects the lack of *sonic hedgehog* expression in the ZPA, and it has been shown that the ectopic expression of Wnt7a in these mutants is sufficient to restore *sonic hedgehog* expression and rescue anteroposterior patterning. Therefore it appears that the maintenance of the ZPA depends on both FGFs secreted from the AER and Wnt7a secreted from the dorsal ectoderm.

Interactions between the proximodistal and anteroposterior axis patterning signals are revealed by the effects of removing the AER. This not only prevents further limb outgrowth but also abolishes the expression of *HoxD12* and *HoxD13*, whose expression domains lie closest to the ZPA. It therefore appears that Sonic hedgehog alone is not sufficient to induce the expression of the most 5' *HoxD* genes, but acts in concert with factors secreted by the AER. FGF4, which is itself induced by Sonic hedgehog, is a likely candidate coactivator of *Hox* gene expression because FGF beads can activate *Hox* gene expression in limb buds lacking the AER.

A picture therefore emerges in which the three axes of the limb are patterned by different organizers secreting three different classes of signaling molecule representing the FGF, Wnt and Hedgehog families. However, all three signaling pathways are interdependent. FGF and Hedgehog pathways are mutually depen-dent, while the Hedgehog pathway also depends on Wnt signaling. Furthermore, the patterning functions of the three pathways are also interdependent.

Other patterning influences

Although the three signaling centers discussed above play a major role in patterning the limb, there is evidence that these merely reinforce a pre-pattern that already exists in the limb forming mesenchyme. This evidence comes from experiments in which the limb mesenchyme is dispersed and then reaggregated, but can still form recognizable limb elements, and from the analysis of mutant limbs that lack key regulators such as FGF and Sonic hedgehog, but which also form recognizable structures.

If a chicken limb bud is removed and separated into its ectodermal and mesenchymal components, the mesenchyme can be dispersed by treating it with chemicals that disrupt cell adhesion. The cells can then be mixed up, reaggregated and replaced inside the original ectoderm and grafted back onto the chicken. Initially, such reaggregated limb buds appear totally disorganized and give rise to proximal structures that do not form recognizable elements. However, the distal structures form recognizable digits, although these are not arranged in any particular anteroposterior pattern as in the normal autopodium. Nevertheless, it is remarkable that digits form at all as the ZPA has been dispersed, and cells expressing different combinations of *Hox* genes (and thus with different positional values) are randomly organized. These results suggest that the limb has some spontaneous pattern-forming ability enabling it to generate a repetitive sequence of equivalent digits that are given definitive regional identities by the *Hox* genes expressed along the anteroposterior axis. This pre-pattern of evenly spaced cartilaginous elements may explain why extra digits, rather than disorganized tissues, form when the AER is broader than usual, causing polydactyly.

There may also be a pre-patterned polarity in the limb bud, as shown by analysis of the chicken mutant *limbless*. In *limbless* embryos, *wnt7a* is expressed throughout the limb ectoderm, indicating that the ventral ectoderm has not been correctly specified. As the juxtaposition of dorsal and ventral ectoderm is required for the formation of the AER (Topic M1), these mutants lack an AER and therefore do not produce FGFs in this region. As *sonic hedgehog* is FGF-dependent, there is no *sonic hedgehog* expression in the limb buds and therefore no ZPA. Despite the lack of these key molecules, however, the 5' *HoxD* genes are expressed in the characteristic nested patterns seen in normal limbs, and other asymmetrically expressed gene products are also found in their normal locations.

It therefore appears that the limb field has a pre-pattern that helps to establish the limb organizer regions and may partially compensate for their functions, while the limb bud itself has a certain degree of self-organizing potential that generates a pre-pattern of the forming skeletal elements. The limb organizing centers are then activated to reinforce and refine this pre-pattern, resulting in the development of stereotyped limb elements.

Morphogenesis in the limb

Morphogenesis in limb development is required to form the final structures of the cartilaginous elements that give rise to the characteristically different bones arranged along the three limb axes. Cell death plays a major role in this process, defining bone shapes and forming joints between bones. Cell death also occurs in soft tissue, which is necessary to separate the digits. Cell death is restricted to certain areas, namely the **anterior and posterior necrotic zones** (which shape the distal limb elements), the **internal necrotic zone** (which separates the bones of the zeugopodium, e.g. the radius and ulna), and the **interdigital necrotic zones** (which separate the digits). Note that although these areas are called 'necrotic zones' this is misleading. The cells do not die through necrosis but by **programmed cell death** (apoptosis). The important distinctions between these modes of cell death are discussed in Topic C4.

The BMPs (BMP2, BMP4 and BMP7) are involved in specifying the areas that undergo apoptosis. This is shown by the effects of exposing developing limbs to high concentrations of BMP4 (more extensive cell death) and blocking BMP receptors (increased cell survival). BMP4 is expressed in the interdigital necrotic zones of the chicken hindlimb bud, and was originally thought to be absent from the corresponding areas in the hindlimb bud of the duck, which has webbed feet. However, more recent experiments with improved *in situ* hybridization techniques suggest that BMP is expressed in the duck interdigital zones, so the reason for webbing remains unclear. Anti-BMPs such as Noggin, Chordin and Follistatin are also expressed in the limb bud, and act as regulators of cell death by inhibiting BMP activity and restricting the expression of *bmp* genes. **Noggin** appears to play an important role in establishing joints, since *noggin* knockout mice have joint malformations and fused bones. Another important regulator is **Gremlin**, a BMP antagonist related to Cerberus. Gremlin is expressed in a complementary pattern to BMPs in the limb, and the expression of recombinant *gremlin* in the chicken limb bud inhibits cell death causing soft tissue syndactyly (the fusion of digits by connective tissue). Later, Gremlin expression is restricted to the forming skeletal elements and therefore appears to be an important regulator of chondrogenesis. While the digits of birds and mammals are separated by cell death, this is not true of all vertebrates. In *Xenopus*, for example, the digits resolve through differential growth rates of cells in the autopodium.

M3 LIMB REGENERATION

Key Notes

Regeneration

Regeneration is the ability to replace missing body parts after embryogenesis. There are two types of regeneration: morphallaxis, which involves the respecification of remaining tissues without further growth; and epimorphosis which involves the replacement of missing parts by the growth of new tissue. Limb regeneration occurs by epimorphosis.

Limb regeneration in amphibians

Urodele amphibians are unique among vertebrates in their ability to regenerate entire limbs as adults. This involves the proliferation of ectoderm to cover the wound and form an apical ectodermal cap, which induces the formation of a dedifferentiated tissue mass or blastema. The blastema proliferates and generates all the mesodermal components of the limb, progressively laying down the proximodistal axis in a manner very similar to embryonic limb development.

Requirements for regeneration

As well as the ectoderm cap, the regenerating limb transiently requires growth factors secreted by the severed axons at the wound site. If limbs are denervated prior to amputation, no regeneration occurs. The dense innervation of the urodele limb may explain, in part, its unique regenerative capacity.

Distal regeneration

Regeneration always precisely replaces the missing distal elements of the amputated limb, regardless of the position of the cut or the orientation of the limb's proximodistal axis. This shows that the cells of the limb retain their positional values and can generate a range of distal positional values in the blastema.

Molecular basis of regional specification

Regional specification in the blastema may be controlled by a gradient of retinoic acid, since this molecule is secreted by the wound ectoderm and establishes a distal to proximal gradient across the blastema. This is followed by the induction of *HoxA* gene expression, which could confer positional values on the blastema cells. Application of endogenous retinoic acid causes the positional identity of the blastema to be reset to a more proximal value. Different levels of retinoic acid produced at different amputation sites could therefore provide a mechanism for specifying the proximodistal location of the blastema cells.

Intercalary regeneration – polar coordinate model

As well as generating structures distal to a site of amputation, urodele limbs can also regenerate intermediate structures if, for example, an amputated hand is grafted onto shoulder stump. The ability of the regenerating limb to establish missing positional values between two extremes may reflect interactions between cells with nonadjacent positional identities, resulting in proliferation and redifferentiation until a complete set of positional values is re-established. Such interactions could result from a gradient of increasing cell adhesion along the limb, resulting in a significant adhesive differential when cells with distinct positional identities are brought into contact.

Related topics	Mechanisms of developmental	Initiation and maintenance of limb
	commitment (A3)	growth (M1)
	Maintenance of differentiation (A5)	Patterning and morphogenesis in
	Pattern formation (A6)	limb development (M2)
	Homeotic selector genes and	
	regional specification (H5)	

Regeneration

Regeneration can be defined as the ability to replace missing parts of the body. Two types of regeneration are often distinguished, termed morphallaxis and epimorphosis. **Morphallaxis** involves the reconstitution of missing body parts by the respecification of remaining tissues *in the absence of growth*. This is demonstrated by the coelenterate *Hydra*, which can generate new (smaller) individuals from pieces cut from an adult. **Epimorphosis**, by contrast, involves the progressive replacement of missing parts by cell proliferation and the growth of new tissue. Limb regeneration in amphibians and insects provide excellent examples of epimorphosis, although it is likely that in each case, growth is preceded by morphallaxis-type respecification. A related term, **regulation**, refers to the ability of an embryo to compensate for missing parts by respecifying the fates of remaining tissues. Regulation is therefore analogous to morphallaxis, the former term applied to embryos and the latter to the regeneration of body parts in adults.

Limb regeneration in amphibians

Most vertebrates have the ability to regenerate some parts of the limb post-embryonically, although this ability is often highly restricted. However, urodele amphibians such as salamanders and newts have the remarkable ability to regenerate entire limbs as adults. In such species, if a limb is amputated at the level of the zonoskeleton, an entire new limb can regenerate in 2–3 months.

Limb regeneration begins with the proliferation of ectodermal cells at the site of amputation. These cover the wound and then continue to proliferate to form a thickened **apical ectodermal cap (AEC)**, which is structurally and functionally analogous to the apical ectodermal ridge (AER) that forms in embryonic limb development. The AEC induces the underlying cells to form a mass of undifferentiated mesenchyme called the **regeneration blastema**. This is structurally and functionally analogous to the progress zone of the developing limb, and all new mesodermal structures formed during regeneration are derived from this cell population (*Fig. 1*).

The formation of the blastema indicates that the differentiated cells in the limb can **dedifferentiate**, proliferate and then **redifferentiate** into the various mesodermal components of the limb distal to the amputation site. Cell tracing studies have shown that the blastema is derived from all mesodermal cell types at the wound site – cartilage, muscle, dermis and connective tissue – indicating a

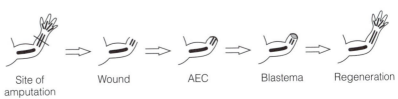

Site of   Wound   AEC   Blastema   Regeneration
amputation

Fig. 1.   Summary of limb regeneration in urodele amphibians. AEC = apical ectodermal cap.

surprising degree of developmental plasticity. This is particularly remarkable in the case of differentiated muscle cells, which are highly specialized multi-nucleate syncytia that have withdrawn from the cell cycle. Nevertheless, it has been shown that muscle, dermis and connective tissue cells can all dedifferentiate and then redifferentiate to give rise to any of the mesodermal components of the limb. Exceptionally, it has been shown that dedifferentiated cartilage gives rise only to new cartilage. The potency of muscle, dermis and connective tissue is therefore greater than that of cartilage in the context of the regenerating limb. None of the blastema cells give rise to epidermis, and the epidermis does not give rise to mesodermal derivatives.

Limb regeneration and limb development appear to share several structural and functional similarities. It is important to realize that the difference in size between the developing and regenerating limbs precludes the use of exactly the same molecular mechanisms for patterning. However, grafting experiments have shown that regenerating and developing limbs can interact to generate a normal pattern, indicating that key mechanisms may be conserved. Some of the major differences between limb development and regeneration are shown in *Table 1*.

Table 1.   Some key differences between limb development and limb regeneration

Context	Regenerating limb	Developing limb
Role of mesoderm	Blastema proliferates and produces all mesodermal structures *including muscles*.	Progress zone proliferates and produces skeletal derivatives. Muscles form from immigrating myoblasts.
Role of ectoderm	AEC induces dedifferentiation and maintains proliferation of blastema.	AER maintains proliferation of progress zone.
Role of nervous system	Required to initiate regeneration.	None.
Heterochronic mesenchyme grafts	Intercalary regeneration occurs when distal blastema grafted to proximal stump.	No intercallary growth occurs when older progress zone grafted to younger limb bud. Proximal elements missing from limb.
Proximodistal axis specification	Exposure of cells to retinoic acid specifies positional values? Downstream activation of *HoxA* genes.	Length of time cells remain in the progress zone specifies positional values. Downstream activation of *HoxA* genes.

Requirements for regeneration

As discussed above, the AEC of the regenerating limb plays a similar role to the AER in limb development. It has been shown that removal of the AEC prevents blastema proliferation, and stopping the epidermis from spreading over the wound prevents the blastema forming. The ectoderm therefore plays an essential role in limb regeneration by inducing and maintaining the blastema.

There is also a transient requirement for innervation. If nerves to the limb are cut *before* the limb is amputated, regeneration does not occur, but if they are cut *after* regeneration has begun, regeneration occurs normally, although at a slower rate.[1] This contrasts to the situation in embryonic limb development, where limb

---

[1]Although innervation is required for the regeneration of normal limbs, this dependence appears to have been acquired during development by the process of innervation itself. Limbs that have never been innervated (e.g. because the nerve was removed during development) are able to regenerate normally.

outgrowth and patterning occur well before the incipient muscle masses are innervated by incoming axons. It is thought that growth factors essential for limb regeneration are released by severed axons at the wound site. The identity of candidate growth factors has been suggested by the ability of certain exogenously applied molecules, such as FGF, neuregulin and transferrin, to rescue limb regeneration in denervated animals. Interestingly, FGF also plays an important role in the outgrowth of limbs during embryonic development.

The requirement for innervation provides a possible explanation as to why urodeles, but not other vertebrates, can regenerate their limbs. It appears that urodele limbs are more densely innervated than those of other vertebrates, suggesting that the amount of innervation may be an important factor in regeneration competence. In agreement with this model, the introduction of extra nerves into the limbs of anuran amphibians such as *Xenopus* can induce some regenerative potential. Furthermore, stepwise reduction of the number of nerves in the newt has revealed a 'threshold' of innervation required for regeneration.

Distal regeneration

Following limb amputation, regeneration always occurs in the distal direction, regardless of the position of the cut or the orientation of the limb. This is shown by the following experiments:

- If the limb is amputated at the wrist, only a new hand is regenerated, but if it is amputated at the shoulder, an entire new limb will form (*Fig. 2*).
- If the limb is amputated at the wrist, and then the cut surface of the stump is inserted into the flank of the body, a blood supply is established and the animal has a 'closed limb'. If this limb is then cut in the middle of the humerus, both cut surfaces produce a blastema and regenerate distal structures, even though the proximodistal axis of one of the limbs is inverted (*Fig. 3*).

These experiments indicate that the cells in the urodele limb retain a set of positional values along the proximodistal axis. Following limb amputation, the cells at the site of amputation use their own positional identity as a reference to generate a new set of positional values in the blastema, representing the more distal limb elements. Furthermore, the positional values established in the regeneration blastema are autonomous. If the blastema is removed from a regenerating limb and grafted to another region of the body that supports its growth, it will give rise to the same set of distal structures.

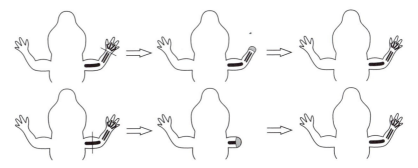

Fig. 2. Regeneration is always in the distal direction regardless of the level of amputation.

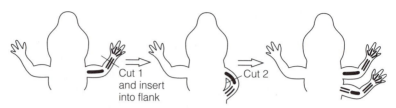

Fig. 3.   Regeneration is always in the distal direction, regardless of the orientation of the limb.

Molecular basis of regional specification

Regional specification in the blastema may involve **retinoic acid** and the expression of *HoxA* **genes**, indicating that despite their different sizes, the developing progress zone and the regenerating blastema use similar molecular mechanisms for proximodistal patterning. The wound epidermis is a rich source of retinoic acid, and this may establish a gradient in the blastema to generate a set of positional values along the future proximodistal axis. *Hox* genes are not expressed in the adult limb, but *HoxA* genes are activated in the regeneration blastema soon after it forms. The role of retinoic acid in regional specification is further suggested by the effects of exposing regenerating limbs to exogenous doses of this chemical. If the limb is amputated at the level of the wrist, it usually regenerates a hand. However, if the regenerating limb is treated with high doses of retinoic acid, the positional identity of the blastema is reset to its most proximal value, and an entire limb emerges from the wrist stump. The blastema shows a dose-dependent response to retinoic acid, so lower concentrations cause less extreme proximalization of the regenerating limb.

Following limb amputation at different levels, proximal and distal blastemas have been shown to contain different amounts of retinoic acid. It is therefore possible that the ectoderm at the level of amputation plays a significant role in setting the positional value of the blastema by secreting a regulated amount of retinoic acid, and that the gradient generated in the blastema then establishes the more distal values necessary to complete the proximodistal axis. The formation of a retinoic acid gradient with its source in the ectoderm would also explain the orientation-independent nature of distal regeneration. Regardless of the orientation of the amputated limb, the retinoic acid gradient would always form with its highest concentration at the distal extremity, in the same orientation with respect to the cut surface. The positional values established would therefore always run in the same direction.

Intercalary regeneration – polar coordinate model

The capacity of the urodele limb for distal regeneration shows that is has an inherent ability to fill in missing elements of its pattern. Remarkably, the limb can replace not only missing distal structures, but also missing intermediate structures by a process called **intercalary regeneration**. If the limb is amputated at the wrist, and the resulting blastema is grafted onto the stump of a host animal in which the limb has been amputated at the shoulder, a normal limb develops in which most of the proximal elements arise from the stump tissue and the distal elements from the blastema. This contrasts with embryonic development, where grafting the progress zone from an older limb bud onto a younger limb bud results in the omission of proximal structures (*Fig. 4*; cf. *Fig. 1*, Topic M2).

Both intercallary and distal regeneration can be explained in terms of a **polar coordinate model**, where cells are endowed with positional information in the

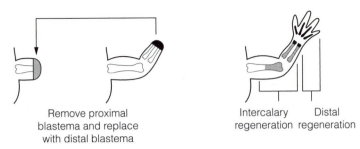

Remove proximal
blastema and replace
with distal blastema

Intercalary   Distal
regeneration regeneration

Fig. 4. Transfer of a distal blastema to a proximal wound site results in distal regeneration from the blastema tissue (black) and intercalary regeneration by tissue from the proximal wound site (gray).

form of polar coordinates on a three-dimensional map. In the normal limb, adjacent cells have adjacent coordinates and they remain quiescent. However, following amputation, cells with different positional values are brought together by the process of wound healing. This is thought to stimulate their proliferation and redifferentiation until a new complete set of polar coordinates is generated and the system once again comes to rest. There is evidence that differential cell adhesion may represent the molecular basis of this process.

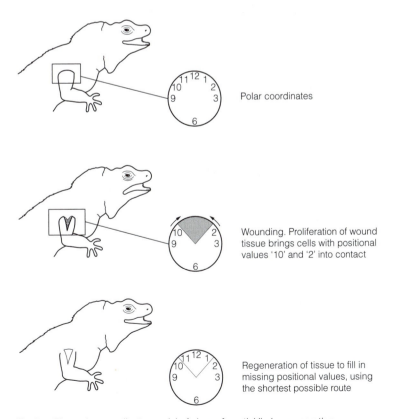

Polar coordinates

Wounding. Proliferation of wound tissue brings cells with positional values '10' and '2' into contact

Regeneration of tissue to fill in missing positional values, using the shortest possible route

Fig. 5. The polar coordinate model of circumferential limb regeneration.

Intercallary regeneration occurs not only in the context of the proximodistal limb axis, but also with respect to the circumference of the limb, representing the anteroposterior and dorsoventral axes. Therefore, if a sector of tissue is removed from the dorsal aspect of the limb, regeneration occurs to fill the gap with replacement dorsal tissues. The polar coordinate model treats the circumference of the limb as a clock face, with each number from 1 to 12 representing a different positional value. If dorsal tissue is removed (e.g. representing positional values 11, 12 and 1), wound healing brings cells with positional values 10 and 2 into juxtaposition. This stimulates the cells to divide and intercalate the missing positional values. Intercalation occurs by the shortest possible route (i.e. the regeneration of 11, 12 and 1 rather than the duplication of values 2 through to 10) (*Fig. 5*).

The polar coordinate model provides an explanation for some puzzling results of graft experiments. For example, if the right leg of a newt is amputated and then grafted back onto the same animal but upside down (so the palm faces upwards), the inverted limb becomes integrated at the new position but two additional limbs develop at the boundaries of the graft, with palms facing down. In terms of the polar coordinate model this can be explained by a confrontation between cells with extreme positional values (representing 6 and 12 on the clock face). The only logical choice for the regenerating limb is to intercalate all the missing values, resulting in the growth of two entire supernumerary limbs (*Fig. 6*).

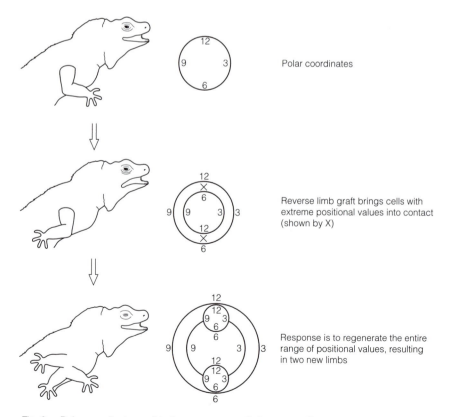

Polar coordinates

Reverse limb graft brings cells with extreme positional values into contact (shown by X)

Response is to regenerate the entire range of positional values, resulting in two new limbs

Fig. 6.   Polar coordinate model of supernumerary limb regeneration.

N1 PLANT VS. ANIMAL DEVELOPMENT

Key Notes

Plant developmental biology	The model organism *Arabidopsis thaliana* has similar properties in terms of genetic amenability to *Drosophila*, and a great deal of information has accumulated recently concerning the molecular pathways controlling development in this species. Many of the principles established for animal development can also be applied to plants, but there are some important fundamental differences.
The life cycle	Many animals set aside a germ line in the embryo, which is the sole source of gametes. Conversely, plants do not have a germ line in the embryo and can produce gametes from somatic cells during post-embryonic development. Fertilization in animals is simple, with one sperm fusing to one egg. In contrast, flowering plants undergo double fertilization, where one sperm cell fuses to the egg, and another fuses to the central cell of the embryo sac and generates the endosperm. In most animals, the gametes represent the haploid phase of the life cycle, whereas all vegetative growth occurs in the diploid phase. Many plants have more complex life cycles, with vegetative growth in both haploid and diploid phases. In flowering plants, the haploid gametophyte is reduced to a simple structure that is dependent on the diploid sporophyte, but it is still multicellular.
Embryonic and post-embryonic development	In animals, embryonic development establishes a fundamental body plan where the positions of most of the adult organs are represented. Post-embryonic development consists mainly of the refinement of this body plan and growth. Conversely, the plant embryo contains none of the organs of the adult plant. These are generated by post-embryonic development of proliferative cell populations called meristems.
Differentiation and cell potency	Animal cells become progressively restricted during development, so isolated differentiated cells can never give rise to an adult organism. Conversely, explants of vegetative tissue from many plants, and in some cases even single cells, can regenerate into whole plants. This can occur either by organogenesis (the direct growth of shoots and roots) or somatic embryogenesis (where even the embryonic stage of plant development is recapitulated). Differentiated plant cells, unlike animals cells, therefore remain totipotent.
Pattern formation	Pattern formation in plants appears to involve many of the molecular mechanisms seen in animals, including axis specification in the embryo, the formation of developmental compartments, localized induction, lateral inhibition and the use of homeotic selector genes to establish positional values in developing organ primordia. There are fundamental differences, however, in the establishment of the adult body plan. The

animal body plan is established in the embryo and determined by the developmental program. Conversely, the plant body plan arises from post-embryonic development and is strongly influenced by the environment in which the plant grows.

Morphogenesis	In animals, morphogenesis encompasses all manner of different types of cell behavior, many of which involve the movement of cells relative to other cells. In plants, this is not possible because cells are cemented tightly together. Morphogenesis in plants therefore involves a restricted set of processes, the most important of which are differential rates and planes of cell division, and changes in cell size.
Signaling pathways and transcription factors	In both animals and plants, the molecular control of developmental processes relies on transcription factors and signal transduction pathways. However, the diversity of animal cell–cell signaling pathways is much greater than that seen in plants, perhaps reflecting the smaller repertoire of cell types in plants and also the extensive networks of plasmodesmata, which connect cells together and allow communication by the flow of molecules through the common cytoplasm.

Related topics

Basic concepts in developmental
  biology (A1)
Mechanisms of developmental
  commitment (A3)

Maintenance of differentiation (A5)
Pattern formation and
  compartments (A6)
Morphogenesis (A7)

Plant developmental biology

Most of this book is devoted to animal development because the molecular basis of development has been studied for longer in animals than in plants, and more data have accumulated for animal systems. The foundations of plant development were laid with descriptive studies, later supplemented with experimental investigations based on surgical manipulation and the modulation of physiological and biochemical processes. Only in the last 10 years or so has the molecular basis of plant development come under the spotlight, and this is largely thanks to the emergence of *Arabidopsis thaliana* as a developmental model organism. *Arabidopsis* is a small, diploid flowering plant with a short life-cycle. It is self-fertilizing and produces thousands of seeds. It is therefore highly suited to genetic analysis and mutagenesis. It is easy to introduce DNA into the *Arabidopsis* genome, and the plants are also suitable for surgical manipulation allowing combined molecular and cellular approaches to the study of developmental processes. *Arabidopsis* and *Drosophila* are equivalent in status within the plant and animal development communities, respectively.

Throughout most of this section, *Arabidopsis* is used to illustrate the principles of plant development, although other species such as maize (*Zea mays*) and snapdragon (*Antirrhinum majus*) are introduced where appropriate. Many of the principles of animal development also apply to plants, and the conservation of processes such as differentiation, pattern formation and morphogenesis is strongly emphasized, with comparative examples cited wherever possible. However, to appreciate how these processes work in plants it is also necessary to understand the major differences between plant and animal development. These are discussed below and summarized in *Table 1*.

Table 1. Comparison of developmental processes in plants and animals

	Plants	Animals
Germ line	None. Germ cells arise from somatic cells during post-embryonic development.	A dedicated germ line often segregates from the somatic lineage in embryonic development.
Fertilization	Double fertilization. One sperm cell fuses with egg cell to generate diploid zygote, and the other to the diploid central cell to generate triploid endosperm.	Single fertilization. Male and female gametes fuse to generate diploid zygote.
Haploid phase of life cycle	Gametophyte is multicellular, although dependent on sporophyte in flowering plants.	Haploid phase of life cycle is restricted to gametes.
Ploidy	Many plants are polyploid.	Most animals are diploid.
Imprinting (see Topic C2)	Imprinting predominantly affects the endosperm, not the embryo.	In mammals, imprinting affects the embryo, making contributions from each parent necessary for development.
Differentiation	Relatively few cell types, many of which are general to all plant organs.	Hundreds of cell types, many organ-specific.
Potency of differentiated cells	Differentiated cells remain totipotent and can recapitulate development.	Cell fate progressively and (usually) irreversibly restricted during development.
Generating the body plan	Adult body plan not generated in the embryo, but established by post-embryonic development of meristems. Strongly influenced by the environment.	Adult body plan generated in the embryo. Predetermined by developmental program.
Pattern formation	Similar principles involved in animals and plants, including axis specification and patterning, the establishment of developmental compartments and the use of homeotic selector genes to confer positional identities.	
Morphogenesis	No relative cell movement or migration. Morphogenesis predominantly reflects the rate and plane of cell division, and changes in cell size.	Many processes involved including extensive relative cell movement (gastrulation is an example) and cell migration (neural crest, germ cells).

The life cycle   Some fundamental differences between plant and animal development reflect differences in reproductive biology. All eukaryotes have a **two-phase life cycle**, comprising a **haploid phase** and a **diploid phase**. In most animals, the haploid phase is represented by the single-celled **gametes**, which exist solely to reproduce the multicellular and vegetative diploid organism, and have no independent vegetative existence of their own. Conversely, some plants such as ferns and algae show a more complex life cycle, with alternating haploid and diploid generations, both comprising multicellular vegetative individuals. In such species, the diploid individual is a **sporophyte**, which produces reproductive cells called **spores** that undergo meiosis. The resulting haploid cells then develop into multicellular **gametophytes**, which produce gametes by mitosis. The gametes fuse to generate a zygote, which develops into a sporophyte in the third generation, and so on. In flowering plants, the complexity of the gametophyte is reduced, but not to the extent seen in animals. The male gametophyte is released as a **pollen grain** while the female gametophyte is represented by the **embryo sac**, which is retained within the ovule and is nutritionally dependent upon the sporophyte. Importantly, both the pollen grain and embryo sac are multicellular.

Fertilization in flowering plants is also more complex than in animals. In animals, the male gamete is motile and the female gamete is immotile; one sperm fertilizes one egg to generate one diploid zygote. In plants, neither male nor female gametes are motile. Fertilization begins when a pollen grain adheres to the exposed stigma of a flower. The pollen grain comprises a **generative cell** and a larger **vegetative cell**. The generative cell divides to produce two **sperm cells** (male gametes). A **pollen tube** then grows down through the stigma to the ovule, and the two sperm cells migrate down to the embryo sac housed within. The embryo sac comprises a number of cells including an **egg cell** (female gamete) and a **central cell** with two nuclei. Flowering plants undergo **double-fertilization**. One sperm cell fuses with the egg to generate a **diploid zygote**, while the other sperm cell fuses with the central cell to generate a **triploid primary endosperm cell**. The **endosperm** produced by the proliferation of this cell becomes a nutritive layer in the seed, which nourishes the embryo during development.

The origin of the gametes is also fundamentally different in plants and animals. As discussed in Topic E1, higher animals have a distinct **germ line**, which is set aside from the somatic cell lineage at an early stage of embryonic development. In plants there is no special germ line in the embryo, and germ cells arise from somatic cells during post-embryonic development, although they all appear to derive from a single cell layer, **L2** (see below).

Embryonic and post-embryonic development

In animals, embryonic development produces the basic body plan of the adult organism. In some animals, such as mammals, the embryo is essentially equivalent to the adult in miniature, and most post-embryonic development consists of growth. In other animals, the adult body does not emerge gradually but in discrete stages. Embryonic development produces a **larva**, and post-embryonic development consists of both growth and **metamorphosis**, sometimes through a series of molts, to produce the adult. In both cases, however, the embryo is representative of the adult body plan, and most of the adult organs are either present in the embryo or represented by dedicated cell populations such as insect imaginal discs (Topic L2).

In contrast to animals, the angiosperm embryo contains none of the organs of the adult. The plant embryo comprises two organ systems, the **embryonic axis** (which gives rise to the seedling) and either one or two **cotyledons** (which provide nourishment as the seedling grows). **Dicot plants** such as *Arabidopsis* have two cotyledons. **Monocot plants**, which include all the cereals, have one. The organs of the adult plant (shoots, roots, leaves, flowers) arise from two small populations of rapidly proliferating cells laid down in the embryo at either end of the axis, called the **shoot and root meristems**. Plant embryos and meristems can therefore be considered as separate developmental systems. Plant embryogenesis is discussed in Topic N2 and meristem development is discussed in Topic N4.

Differentiation and cell potency

In higher animals, development involves the progressive restriction of cell fates. Every time a cell divides, its descendants may have to choose between alternative developmental pathways, and once that choice is made, the decision is usually irreversible. Differentiated animal cells placed in isolation therefore cannot give rise to new individuals. There are examples where cells have a limited ability to dedifferentiate and recapitulate certain developmental processes, as seen in limb regeneration (Topic M3). However, in no case has it been possible to use a differentiated animal cell to recapitulate embryonic development.[1]

Conversely, plant cells show a remarkable capacity for regeneration. Many plants can regenerate new individuals from sections of vegetative tissue containing only differentiated cells. Under appropriate culture conditions, most plant tissues can be persuaded to dedifferentiate and produce a mass of undifferentiated cells called a **callus**. Such material can regenerate entire plants by two routes: **organogenesis** (the direct development of shoots and roots) or **somatic embryogenesis** (where the entire developmental pathway, including the embryonic stage, is recapitulated). Some species can even regenerate an entire plant from a single differentiated cell. This indicates that differentiated cells in plants are **totipotent**, and can (under appropriate conditions) form generative tissue resembling either the early zygote or post-embryonic meristems. This fundamental distinction between animal and plant cells reflects differences in concepts such as *determination* and the *maintenance of the differentiated cell state*. The molecular basis of such differences is discussed in more detail in Topic A6.

Both animals and plants contain numerous different cell types and, as discussed below, there is no evidence to suggest that any differences exist between animals and plants with respect to the fundamental processes underlying cell differentiation (i.e. the activity of different transcription factors and the expression of different sets of genes). However, while animals can have many hundreds of specialized cell types, plants have a more limited number, usually between 10 and 30. In part, this is because plants have relatively few specialized organs compared to animals, but furthermore, most plant organs are derived from the *same repertoire of cell types*. For example, many animal organs contain unique cells – hepatocytes in the liver, cardiac muscle in the heart, insulin β cells in the pancreas. Conversely, the stems, leaves and roots of plants all contain similar cell types derived from three fundamental cell layers established in the embryo: **L1**, which gives rise to the epidermis; **L2**, which gives rise predominantly to the cortical parenchyma cells and other tissue underlying the epidermis; and **L3**, which generates much of the vascular tissue and pith. Plant organs also contain specialized cells, but these are superimposed on a background organization that is similar throughout the plant, represented by L1, L2 and L3. Therefore, while organogenesis in animals may in some cases rely on the prior specification of a particular cell type, the relationship between

---

[1] While differentiated animal cells are not totipotent, their nuclei can be. See Topic A6 for a discussion of nuclear transfer experiments and animal cloning. Also note that in lower animals such as *Hydra* entire new individuals can regenerate from cut pieces of tissue, and new individuals can be generated by budding from the parent. Such primitive animals often show greater developmental plasticity than more advanced species and have a higher regenerative capacity, and are in this way much more similar to plants than higher animals.

organogenesis and cell differentiation in plants may be more complex. As discussed in later Topics, certain mutations have been identified that uncouple these two processes, suggesting they may be controlled independently.

Pattern formation

The concept of pattern formation was established in *Drosophila* and has been generally applied to all animals, reflecting the fact that the same genes appear to be used for patterning throughout the animal kingdom. As discussed in Topic A6, the first stage of patterning in animals is axis specification. This is followed by the allocation of positional values to cells along each axis, and then cells respond to their positional identities by differentiating and cooperating in morphogenetic processes in a regionally-appropriate manner. In many cases, pattern formation involves the division of an axis into broad regions defined by bands of gene expression, and the establishment of developmental compartments or segments. In this way, a body plan is generated that can be elaborated as development proceeds.

Pattern formation in plant development shows many mechanistic similarities to pattern formation in animals, but there are some fundamental differences. As in animals, one of the first patterning events in plants is axis specification in the embryo. In flowering plants, this often reflects asymmetric division of the zygote, generating an apical cell that gives rise to the **embryo** and a basal cell that forms a **suspensor** filament joining the embryo to the ovule. Asymmetric cell divisions are also important in polarizing certain animal embryos, such as *C. elegans* (Topic G2). In animals, axis specification may predominantly involve the distribution of maternal gene products (as in *Drosophila*) or may rely entirely on environmental cues (as in mammals) or in most cases a combination of both mechanisms (as in *Xenopus*). The situation in plants is not clear. Maternal influences are not essential for embryonic development, as somatic embryogenesis is possible. However, a small number of maternal genes which affect the pattern of the zygotic embryo have been identified in *Arabidopsis*, so maternal influences can play a role in embryonic polarity. It is likely that zygotic and somatic embryogenesis involve different early patterning mechanisms.

After axis specification, early pattern formation in animals establishes a basic body plan, which is representative of the adult. In plants, axis specification is also followed by the establishment of a body plan, but this does not correspond to the structure of the adult. The embryonic body plan corresponds to the structure of the seedling, but the adult body plan is determined by post-embryonic growth of the shoot and root meristems. As the adult body plan is not predetermined in the embryo, it is influenced greatly by the environment allowing the plant to use development in order to adapt to its surroundings.

In *Drosophila*, the anteroposterior and dorsoventral axes of the embryo are initially divided into broad regions by domains of gene expression, and mutations in these genes result in the deletion or misspecification of particular regions of the embryo (Section H). The *Arabidopsis* embryo is similarly divided into broad regions corresponding to different parts of the seedling, and similar patterning mutants have been identified (Topic N2). However, while the mechanisms used to establish these domains of gene expression are well characterized in *Drosophila*, it is not yet clear how they are set up in plants.

Later in development, pattern formation is used to elaborate the structure of emerging organs. In animals, many different mechanisms can be involved, including morphogen gradients, local induction, lateral inhibition, the establishment of compartment boundaries and the allocation of positional identities by homeotic selector genes. Some of these mechanisms have been identified in plants. For example, the spacing of leaves along the stem reflects the positions at which leaf primordia are initiated in the shoot apical meristem, and this appears to involve lateral inhibition. Lateral inhibition may also be involved in the spacing of stomata on the abaxial surface of the leaf. Leaf development and patterning is discussed in Topic N5. The developing flower is divided into a concentric series of compartments that generate different floral organs. These are conceptually similar to the segments of the *Drosophila* body plan. As in *Drosophila*, each compartment is given a positional value according to a combinatorial code of transcription factors established by overlapping expression patterns of homeotic selector genes. Homeotic mutations, which in *Drosophila* convert one body segment into the likeness of another, are also seen in the flower, resulting in the misspecification of floral organs (sepals, petals, stamens, carpels). Flower development and patterning is discussed in Topic N6.

Morphogenesis

As discussed in Topic A7, the generation of precise structures in a developing embryo (morphogenesis) reflects the control of cell behavior. In animals, this involves many different processes, but perhaps the most important of these is the ability of cells to move in relation to each other. This can be controlled by regulating the expression of genes encoding cell adhesion molecules and components of the extracellular matrix, and drives essential processes such as gastrulation, neurulation and the formation of the gut. Animal development also shows many examples of cell migration, the most dramatic and extensive of which include the migration of germ cells (Topic E2) and in vertebrates, the migration of neural crest cells (Topic J6).

In plants, cells are fixed in place by rigid cell walls, so there is no relative movement and no migration. Therefore, all morphogenetic processes in plant development reflect differential rates of cell division, the plane of cell divisions, and changes in cell size brought about by increasing the volume of the vacuole. As planes of cell division can be the most important factor behind the shaping of a particular tissue, two terms have been coined to describe them. **Anticlinal divisions** occur in the plane of a cell sheet, and therefore expand the sheet without increasing its thickness. Conversely, **periclinal divisions** occur at right angles to the plane of a cell sheet and result in its expansion into multiple layers. Switching from anticlinal to periclinal cell division is critical for certain morphogenetic processes in plants, such as the outgrowth of leaves.

Signaling pathways and transcription factors

Although there are fundamental differences between plants and animals in terms of developmental mechanisms, the underlying molecular processes in each case rely on transcription factors that regulate gene expression and signaling pathways that facilitate communication between cells. Therefore, many of the developmental genes identified in *Arabidopsis* and other plants encode transcription factors or components of signaling pathways, just as they do in animals. In this context, one major difference between animals and plants concerns the diversity of signaling pathways and networks. Plants appear to have a much smaller repertoire of signaling pathways than animals. This may reflect the smaller number of distinct cell types found in plants, but another factor is the interconnection of plant cells by cytoplasmic bridges termed **plasmodesmata**, allowing them to communicate by the transfer of small molecules and macromolecules directly from cell to cell. Animal cells display hundreds of different receptors on their surfaces, and respond to a diverse array of secreted proteins as well as other signaling molecules. Conversely, very few secreted proteins are used as signaling molecules in plants, and only a few major signaling pathways are recognized, responding to hormones such as auxins, cytokinins, abscisic acid, gibberellic acid, ethylene and jasmonic acid. The signaling pathways in animals are often promiscuous, redundant and inter-connected making the precise developmental roles of individual components difficult to establish (Topic C3).

N2 DEVELOPMENT OF THE PLANT EMBRYO

Key Notes

Properties of plant embryos

Plant embryos comprise two organ systems, the embryonic axis (which gives rise to the seedling) and either one or two cotyledons (which nourish the seedling during germination). At either end of the embryonic axis are meristem populations, which give rise to the structures of the adult plant during post-embryonic development.

Overview of *Arabidopsis* embryogenesis

Arabidopsis embryogenesis is typical of dicot flowering plants. A series of stereotyped cell divisions produces a globular embryo, which is patterned radially to produce three fundamental cell types, and axially to produce the shoot–root axis and cotyledons.

Specification of the apical–basal axis

The apical–basal axis of the *Arabidopsis* embryo is specified at the time of the first asymmetrical division in the zygote, and this process is thought to depend predominantly on zygotic genes. The *GNOM* gene is active at this stage, as the first division is approximately symmetrical in *gnom* mutants. Polarization is necessary to specify basal structures such as the root apical meristem.

Axial patterning

The proembryo is divided into horizontal tiers of cells corresponding to the different organs found along the apical–basal axis of the mature embryo. Patterning genes activated in the proembryo and early globular embryo, such as *MONOPTEROS*, *GURKE* and *FÄCKEL*, are expressed in different cell tiers, resulting in the specification of different organ-forming regions. In mutant embryos, particular organ-forming regions are either misspecified or not specified at all, resulting in the deletion or duplication of particular organs.

Radial patterning and cell differentiation

Mutants that affect axial patterning and organ formation do not appear to influence the radial organization of the embryo into L1, L2 and L3 layers. Other mutants, such as *knolle*, do affect radial organization, but do not appear to have an impact on cell differentiation. It therefore appears that the differentiation of the L1, L2 and L3 layers is controlled separately from their radial organization.

Somatic embryogenesis

The ability of plant cells to undergo somatic embryogenesis indicates that maternal influences are dispensable in early development. The structure of somatic and zygotic embryos is subtly different, but the embryos go through similar developmental stages. The molecular control of somatic embryogenesis is not understood, but a receptor kinase called SERK is expressed in somatic cells competent to undergo embryogenesis and may play a central role in this process.

Properties of plant embryos

Plant embryos, unlike those of animals, do not contain organ primordia representative of the adult body-plan. The mature embryo of a flowering plant such as *Arabidopsis thaliana* comprises only two organ systems, the **embryonic axis** and the **cotyledons** (*Fig. 1*). The embryonic axis contains the basic layout of the seedling, and has a population of meristem cells at each end which give rise to adult structures of the plant during post-embryonic development. The axis is therefore organized into regions representing the **shoot apical meristem (SAM)**, the future seedling shoot (**epicotyl** and **hypocotyl**), the embryonic root (**radicle**) and the **root apical meristem (RAM)** (*Fig. 1*). The cotyledons, which are attached to the hypocotyl, are storage organs that provide nourishment to the growing seedling. As well as being divided into organ-forming regions, the embryo is also organized into radial bands representing the three fundamental cell layers common to all flowering plants: **L1, L2** and **L3** (Topic N1).

In this Topic, we discuss three important aspects of embryonic development in flowering plants: specification of the apical–basal axis, patterning the axis to generate the organ-forming regions, and the control of cell differentiation. Finally, we discuss differences between normal **zygotic embryogenesis**, where the embryo is surrounded by maternal tissue, and **somatic embryogenesis**, where cultured plant cells can recapitulate the entire developmental pathway even when isolated from all maternal influences.

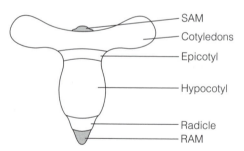

Fig. 1.   *Organ-forming regions of a dicot plant embryo.*

Overview of *Arabidopsis* embryogenesis

The *Arabidopsis* embryo is typical of dicot flowering plants, and we therefore begin with a brief overview of embryogenesis in this species (*Fig. 2*). There is considerable diversity among flowering plants in the details of embryonic development, but the principle stages seen in *Arabidopsis* have counterparts in all angiosperms.

Fertilization of the egg cell is followed by elongation of the zygote along the future **apical–basal axis** of the embryo, which corresponds to the future **shoot–root axis** of the seedling. The elongated zygote divides asymmetrically to produce a small, round apical cell and a longer basal cell. Descendants of the apical cell will give rise to most of the embryo, while descendants of the basal cell form the **suspensor** (a string of cells attaching the embryo to the ovule,

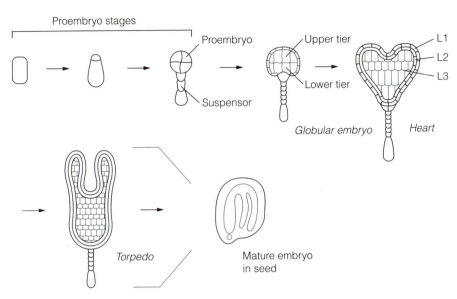

Proembryo stages

Proembryo

Upper tier

Lower tier

Suspensor

Globular embryo

L1
L2
L3

Heart

Torpedo

Mature embryo
in seed

Fig. 2.   Summary of embryogenesis in Arabidopsis, *which is representative of flowering plants.*

which is thought to transport nutrients to the embryo from the endosperm). The basal cell divides repeatedly in the same horizontal plane to form an elongated filament. Conversely, the apical cell undergoes a stereotyped sequence of horizontal and vertical divisions to generate a ball of cells. This is initially called a **proembryo** because no differentiated cell types can be recognized. However, by the 32-cell stage, the three fundamental cell layers present in all flowering plants are already clearly defined. Subsequently, the **globular embryo** is converted into an inverted **triangle** and then a **heart**, as the cotyledons begin to elevate. The **shoot and root meristems** form at the heart stage. The shoot apical meristem forms between the emerging cotyledons, while the root apical meristem forms at the basal extremity of the axis. In *Arabidopsis*, the root apical meristem is derived from both the proembryo and the apical cell of the suspensor. Further growth of the embryo causes the cotyledons to extend (**torpedo** stage) and then curve around the inner surface of the seed (**walking-stick** or **bent-cotyledon** stage) to generate the **mature embryo**.

Specification of the apical–basal axis

The apical–basal axis of the embryo is specified very early in development, and as it is aligned with the vertical axis of the ovule, this suggests maternal gene products may be involved in axis specification. However, very few maternal effect mutations with embryonic polarity phenotypes have been identified. This, together with the ability of *Arabidopsis* cells to undergo somatic embryogenesis suggests that, unlike the situation in *Drosophila*, early polarization of the plant embryo may be controlled predominantly by zygotic genes.

The first sign of apical–basal polarity in the *Arabidopsis* embryo is seen just after fertilization, when the zygote elongates along the future apical–basal axis and then divides asymmetrically to produce the small apical cell and long basal cell. The zygotic gene *GNOM* is involved at this stage because in *gnom* mutants, the zygote divides more-or-less symmetrically producing apical and basal cells of similar sizes. At the heart stage, *gnom* embryos are more rounded than wild-type embryos and fail to specify the RAM. This indicates that asymmetric division of the zygote is instrumental in specifying the future root-forming region of the embryo.

Axial patterning   The organ-forming regions of the *Arabidopsis* embryo are distributed along the apical–basal axis with the SAM at the apical pole, followed by the epicotyl and cotyledons, hypocotyl, radicle and the RAM at the basal pole. A number of mutants have been identified in which particular organ-forming regions are never specified (the corresponding organs are deleted), or misspecified (the corresponding organs are replaced by a different set of organs). This suggests that the appearance of organ-forming regions is a consequence of pattern formation along the apical–basal axis. Such mutations are conveniently divided into three groups: **apical mutants** (in which the apical region is deleted, corresponding to the SAM, epicotyl and cotyledons), **central mutants** in which the hypocotyl is deleted, and **basal mutants** in which the hypocotyl and the roots are deleted (*Fig. 3*).

Fate mapping has shown that the proembryo is divided into presumptive organ-forming regions at a very early stage. In the 8-cell proembryo, there are two tiers of four cells, the **upper tier** destined to give rise to the SAM, epicotyl and cotyledons, and the **lower tier** destined to give rise to the hypocotyl and radicle (*Fig. 2*). During the late globular/early heart stage, the cells derived from these two tiers begin to behave distinctly, undergoing different and characteristic patterns of division. The lower tier of cells also induces the most apical suspensor cell to form the RAM. The *MONOPTEROS* gene is probably required to specify the lower tier of cells, since in the (basal) *monopteros* mutant, the lower tier behaves like an additional upper tier and the hypocotyl, radicle and RAM are not specified. The apical mutant *gurke* shows the opposite phenotype, in which the cotyledons and SAM are deleted. The defect in this case appears to be in the control of cell division in the apical region of the embryo, during the transition from the globular embryo to the early heart stage. In weaker *gurke* alleles, rudimentary cotyledons may be produced, but no SAM is formed. Another apical mutant, *doppelwurzel*, results in the deletion of the apical region and its replacement with a mirror image basal region (doppelwurzel means 'double root'). The *FÄCKEL* gene appears to specify the central region, as the hypocotyl is deleted in *fäckel* mutants.

Genes that specify the different organ-forming regions are expressed early in development, and should be distinguished from those expressed later, which maintain the organs. For example, the *SHOOT MERISTEMLESS* gene discussed in Topic N4 is required to maintain the population of SAM cells once it has been established. As the name suggests, the phenotype of *shoot meristemless* mutants includes the absence of the SAM. However, this is not because the SAM is never specified, but because all the cells differentiate in the embryo rather than persisting as a proliferative population in the adult plant.

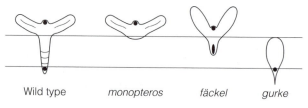

Wild type   *monopteros*   *fäckel*   *gurke*

Fig. 3. Embryo axis patterning mutants in Arabidopsis.

Radial patterning and cell differentiation

In the axial patterning mutants discussed above, particular organ-forming regions are not specified in the embryo. In each case, however, the normal radial organization of the embryo is maintained and all three cell layers are formed. In another mutant called *raspberry*, the embryo fails to initiate organ formation at all, with the result that development arrests at the globular stage. Interestingly, despite its failure to form organs, the *raspberry* mutant also shows normal cell differentiation, producing the three characteristic cell layers seen in normal embryos by the late globular/early heart stage. It therefore appears that patterning the apical–basal axis and the specification of radial cell fates are largely independent processes.

As the embryo is organized into radial bands of cells, it is interesting to ask whether radial patterning, like axial patterning, can be uncoupled from cell differentiation. The isolation of mutants such as *knolle* and *keul* suggest it can. In normal embryos, the outer L1 layer (future epidermis) comprises flattened cells that undergo only anticlinal divisions and therefore remain as a single-layered cell sheet. In *knolle* mutants, cell division in the outer cell layer is irregular so that the sheet becomes disorganized, with some cells even penetrating the underlying L2 and L3 layers. In *keul* mutants, the cells of the outer layer are fatter than normal. In both cases, however, the outer cell layer can be recognized as L1 by its expression of particular marker genes, such as *ARABIDOPSIS THALIANA MERISTEM LAYER 1 (ATML1)* and in certain genetic backgrounds, by its accumulation of anthocyanins.

Somatic embryogenesis

As discussed in Topic N1, plants have the remarkable ability to produce embryos from vegetative tissue, a process called **somatic embryogenesis**. Superficially, somatic embryos are very similar to zygotic embryos. They pass through the same characteristic developmental stages, and give rise to normal seedlings. However, closer examination reveals that the pattern of cell divisions in the somatic embryo is distinct from that in the zygotic embryo. Furthermore, the somatic embryos lack a suspensor. It is therefore likely that there is a maternal influence on plant embryo development, but that this can be mimicked to a large extent by permissive environmental conditions. In support of this, a small number of maternal effect embryo-patterning mutants have been identified in *Arabidopsis*. For example, plants homozygous for the *short integument* mutation produce embryos with axial patterning defects even if they are artificially crossed with plants producing wild-type pollen.

In both animals and plants, the series of events following fertilization is complex, and it therefore seems remarkable that the same course of events can be initiated in cultured somatic plant cells by relatively simple modifications to the environment such as imposing a certain regime of plant hormones. It remains unknown how somatic embryogenesis is initiated at the molecular level, although it is clear that every plant cell is primed with the entire developmental program. An important component of the molecular pathway is the receptor kinase **SERK**, which is not expressed generally in somatic cells but is induced under appropriate culture conditions and marks the onset of competence to undergo somatic embryogenesis. SERK continues to be expressed in somatic embryos until the globular stage. Its precise role is unclear and no other components of this putative signaling pathway have been identified.

N3 DEVELOPMENT OF THE SEEDLING

Key Notes

Seedling development	Following embryogenesis, the mature embryo becomes desiccated and is stored in the seed, where it may remain dormant for some time until favorable environmental conditions stimulate germination. During germination, the embryonic shoot emerges from the seed and grows towards the soil surface. The storage products that accumulated in the cotyledons during embryonic development are used up in this process. The growing seedling responds to many environmental signals, including light and gravity.
Transition to germination	Germination is inhibited by the plant hormone abscisic acid and promoted by gibberellic acid. Therefore, mutations in many of the genes involved in the synthesis of these hormones, or in their signaling pathways, show either viviparous (precocious germination) or delayed germination phenotypes. For genes of the *LEAFY COTYLEDON* class, a viviparous mutant phenotype is combined with the partial transformation of cotyledons into leaves. These genes encode regulatory proteins, acting downstream of the hormones, which inhibit seedling development and promote embryonic development.
Skotomorphogenesis and photomorphogenesis	Seedling development begins in darkness with the extension of a pale, spindly (etiolated) shoot. This is known as skotomorphogenesis. Upon emergence from the surface of the soil, the seedling is exposed to light, and switches to photomorphogenesis. This involves activation of the shoot apical meristem, allowing the development of the first true leaves. Chlorophyll also begins to accumulate and the plant begins to undergo photosynthesis.
Light-perception mutants	*Arabidopsis* light-perception mutants are either light-insensitive (constitutively etiolated) or constitutively photomorphogenic. These mutants identify genes involved in the light-perception pathway. Most light-insensitive mutants represent genes encoding photoreceptors, except *hy5*, which represents a downstream regulatory gene encoding a transcription factor. The constitutively photomorphogenic mutants identify genes whose products interfere with the function of *HY5*. It appears that photomorphogenesis is the default pathway, in which the transcription factor HY5 activates numerous downstream light-dependent genes. In the dark, HY5 is inhibited by the nuclear regulator COP1, which acts in concert with a small group of other proteins. In the light, COP1 is excluded from the nucleus.

Related topics	Developmental mutants (B2)	Plant *vs.* animal development (N1)
	Signal transduction in development (C3)	Development of the plant embryo (N2)

Seedling development

Plant embryogenesis (Topic N2) is followed by **germination**, a remarkable process during which the embryonic shoot emerges from the seed and burrows upwards through the soil. The **seedling** can be thought of as a transition state between embryonic plant development and the post-embryonic development of the mature plant. The later stages of embryogenesis are concerned with the deposition of storage products that are used by the germinating seedling as a source of nutrients before the onset of photosynthesis. Special storage organs, **cotyledons**, are set aside in the embryo for this purpose. Germination therefore marks a switch from an anabolic phase of nutrient accumulation to a catabolic phase of nutrient consumption and growth. These two stages are separated in most plant species by a period of **quiescence**, where the embryo becomes desiccated and is stored in the seed. In some species, germination is activated immediately in response to a favorable environment, whereas in others the seed may remain **dormant** for some time, even if the environment is suitable for germination. Once germination begins, the seedling must respond to a number of environmental influences. Perhaps the most important of these are gravity (which ensures the shoot grows upwards and the root downwards) and light (which enables the seedling to extend through the soil in the dark and then switch to vegetative growth once it has emerged at the surface). In this Topic, we consider the molecular basis of the switch from embryogenesis to germination, and the signaling pathways involved in the response of seedlings to light.

Transition to germination

The transition from embryogenesis to germination is controlled largely by two plant hormones, **abscisic acid (ABA)** and **gibberellic acid (GA)**. Generally, GA promotes germination while ABA inhibits it. Many mutations affecting the transition to germination have identified genes encoding components of the biosynthetic pathways of these hormones, or the signal transduction pathways they activate (*Table 1*). The term **viviparous** means precocious germination, revealed by premature activation of the shoot apical meristem and the differentiation of shoot cells in the embryo. A number of *aba* **mutants** with viviparous phenotypes have been identified in *Arabidopsis* and maize, and these show much lower levels of ABA than wild type embryos, suggesting a defect in ABA synthesis. Other mutants (called *abi* **mutants**) show insensitivity to normal levels of ABA, while the *enhanced response to aba 1* (*era1*) mutant shows the opposite phenotype.

Table 1. Some genes involved in the transition from embryonic to seedling development in Arabidopsis thaliana

Process	Arabidopsis genes
ABA synthesis	*ABA* class (ABA biosynthesis) *ABA1 ABA2 ABA3 ABA4 ABA5*
ABA perception	*ABI* class (ABA insensitivity) *ABI1 ABI2* *ERA* class (enhanced response to ABA) *ERA1*
GA signaling	*GA* class *GA1 GA2 GA3*
Dormancy	*RDO* class (reduced dormancy) *RDO1 RDO2*
Specification of post-germination characters	*LEC* class (leafy cotyledon) *LEC1 LEC2 FUS3*

These genes are likely to encode components of the ABA signal transduction pathway. Mutants representing genes in the GA biosynthesis pathway and the response to GA signaling (*GA1, GA2, GA3*) show delayed germination phenotypes.

A number of other genes have been identified that specifically affect the transition to germination when mutated, but are not involved in hormone synthesis or perception. Instead, these appear to encode regulatory proteins that control downstream embryo-specific and seedling-specific gene expression. Mutations in these genes cause the appearance of trichomes on the cotyledons (trichomes are epidermal hairs usually found only on leaves) and the accumulation of anthocyanins. This represents a partial transformation of cotyledons into leaves, and the genes are therefore grouped as the **LEAFY COTYLEDON (LEC) class**. Two of these genes, *LEAFY COTYLEDON 1* and *FUSCA 3*, show viviparous mutant phenotypes as well as leaf-like cotyledon morphology and reduction in the accumulation of storage proteins.

Insights into the molecular basis of the transition to germination have come from the cloning and characterization of some of the above genes. The *ABA3*, *LEC1* and *FUS3* genes all encode transcription factors with temporally overlapping expression patterns during embryogenesis. They appear to support the expression of each other and to upregulate genes involved in embryonic development, while suppressing genes involved in post-embryonic development. Certain genes that are inactive during embryonic development but expressed during seedling development are de-repressed in *aba3*, *lec1* and *fus3* mutant embryos. Furthermore, the overexpression of *LEC1* in germinating plants leads to the ectopic development of embryo-like structures (although the same phenotype is not caused by the overexpression of either *ABA3* or *FUS3*). The dual function of the ABA3 transcription factor may be mediated by its tandemly arranged activation and repression domains; these presumably act on different downstream genes.

Skotomorpho-genesis and photomorpho-genesis

The first stage of germination is called **dark development** or **skotomorphogenesis**, and is characterized by the rapid extension of the shoot upwards through the soil. The shoot is pale and spindly, a condition described as **etiolated**. When the seedling emerges from the soil, it is exposed to light. This causes a switch to **light development** or **photomorphogenesis**: the seedling turns green as it begins to accumulate chlorophyll, photosynthesis begins, the shoot apical meristem is activated and true leaves begin to form.

Light-perception mutants

The genetic basis of seedling development has been investigated by the characterization of mutants that continue to undergo skotomorphogenesis in the light, and those that undergo photomorphogenesis in the dark (immediately after the onset of germination). *Arabidopsis* has provided an excellent system in which to isolate such mutants because the mutant phenotypes are distinct and easy to identify in a background of wild type seedlings, and many hundreds of germinating seedlings can be screened at the same time.

Light-insensitive, constitutively etiolated *Arabidopsis* seedlings were originally designated as *long hypocotyl* (*hy*) mutants. Several different mutants have been isolated and most of these represent genes encoding photoreceptors that respond to specific wavelengths of light (red, far-red, blue, ultraviolet etc). The analysis of double mutants has shown, however, that the *HY5* gene acts downstream of these photoreceptors, and probably coordinates their signals. Thus, many photoreceptors feed into the same signal transduction pathway, the central component of which is the product of the *HY5* gene, a bZIP transcription factor.

Seedlings with the opposite phenotype (photomorphogenesis in the dark) are described as *de-etiolated* (*det*) or *constitutive photomorphogenic* (*cop*) mutants. It has been found, however, that many of the *DET* and *COP* genes identified from the analysis of these mutants are identical to the *FUSCA* genes involved in the transition to germination discussed above. The *det/cop/fus* mutants have short hypocotyls and open cotyledons in the dark, although they are not green because chlorophyll biosynthesis is also light dependent. Double mutants, where each *det/cop/fus* mutant is combined with *hy5*, have shown that the *DET/COP/FUS* genes function downstream of *HY5*. This indicates that photomorphogenesis is the default pathway, and that it is actively inhibited in the dark by the function of ten or more *DET/COP/FUS* genes (*Fig. 1*). In the light, signals from photoreceptors activate the pivotal HY5 transcription factor, which acts as a selector of light-dependent developmental pathways. The basis of this regulation may involve nuclear transport. It has been shown that the product of the *COP1* (*FUS1*) gene is localized in the cytoplasm of hypocotyl cells in the light, whereas in the dark it is localized in the nucleus, where it interacts with HY5. It appears that COP1 acts as a repressor of HY5 in the nucleus by sequestering it into an inactive heterodimer. In the light, COP1 is transported to the cytoplasm, releasing HY5 and allowing this transcription factor to activate downstream light-activated genes. Interestingly, plant signaling molecules called **brassinosteroids** are also involved in skotomorphogenesis as the application of exogenous brassinosteroids such as brassinolide to certain *det/cop/fus* mutants can restore the normal etiolated phenotype when the mutants are grown in the dark.

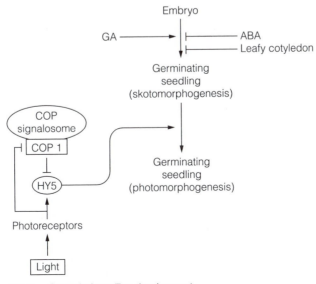

Fig. 1.   Control of seedling development.

N4 SHOOT AND ROOT MERISTEMS

Key Notes

What are meristems?	Meristems are small, self-renewing populations of proliferating cells that produce all the adult organs of a flowering plant. Two meristem populations are established in the embryo, the shoot apical meristem or SAM (which gives rise to the stem, as well as associated organs such as leaves and flowers) and the root apical meristem or RAM (which gives rise to the roots).
Summary of shoot and root development	The shoot develops from the SAM as a series of modules called phytomers, defined by the periodic initiation of lateral organ primordia that give rise to leaves and axillary buds. The root develops continuously and has no lateral organs. Secondary roots may branch from the primary root at a later stage, but this does not involve the RAM.
Organization of the SAM	The SAM is divided into two radial components, a thin outer tunica (comprising layers L1 and L2) which undergoes predominantly anticlinal divisions, and an inner corpus (comprising layer L3) where cell divisions occur in all planes. The SAM is also organized into zones. The central zone contains slowly proliferating stem cells (central mother cells) while the peripheral zone contains rapidly proliferating initials. Differentiation occurs at the flanks of the meristem, giving rise to the lateral organs, and below the meristem, giving rise to the cell files of the stem.
Specification of cell fates in the SAM	Fate maps of the SAM can be generated by clonal analysis in mericlinal chimeras, and can be flattened out, so that the SAM is represented by a circle. The periphery of the fate map contains cells that are determined to contribute to the early (basal) phytomers of the plant. The center of the fate map contains cells that are loosely committed, and clones of marked cells from this region cover several phytomers towards the apex of the plant. However, the SAM is capable of considerable regulation and its organization is thus controlled predominantly by cell–cell interactions.
Molecular basis of SAM development	Genes that specifically affect cell proliferation in the SAM have been identified by the isolation of mutants that do not produce normal shoots, but do produce normal roots. Two genes, *SHOOT MERISTEMLESS* and *CLAVATA1*, appear to have major and antagonistic roles in the control of meristem proliferation and therefore determine and maintain the size of the SAM during shoot development.
Organization of the RAM	The RAM is a complex structure that gives rise to the radially-arranged cell layers of the root as well as the root cap. The base of the RAM is called the promeristem. This has three tiers of cells that generate the pattern of cell types in the root and produce the root cap. The remainder of the RAM is made up of proliferating initial cells whose fate has already been specified.

Specification of cell fates in the RAM	Unlike the SAM, cell divisions in the RAM are ordered and stereotypical, suggesting that lineage may play a significant role in cell specification. However, laser ablation studies and the analysis of mutants that produce extra cell layers have shown that cell fates in the RAM are still largely dependent on position rather than lineage.
Molecular basis of RAM development	Genes that specifically affect cell proliferation in the RAM have abnormal roots but normal shoots. A number of RAM proliferation mutants have been identified, revealing genes such as *ROOT MERISTEMLESS 1* and *2*, whose role in development appears analogous to that of *SHOOT MERISTEMLESS* in the SAM. A number of further RAM mutants have been identified that fail to form particular cell layers. In each case, the deficiency can be traced back to particular cell divisions in the promeristem, and may involve a failure to generate the required number of cell layers, or a failure to specify particular cell types.

Related topics	Cell fate and commitment (A2) Early development by single-cell
	Mechanisms of developmental specification (G1)
	   commitment (A3) Plant *vs.* animal development (N1)
	Pattern formation and
	   compartments (A6)

What are meristems?

The organ systems of the plant embryo (axis and cotyledons) are concerned mainly with the completion of germination and are established in the embryo. In contrast, the adult organs of the flowering plant are formed post-embryonically and are derived entirely from two small populations of rapidly proliferating cells laid down in the embryo, called the shoot and root **meristems**. The **shoot apical meristem** (**SAM**) gives rise to all the aerial parts of the plant, while the **root apical meristem** (**RAM**) gives rise to the root network. As the shoot grows, further meristematic populations are generated, and under the appropriate conditions these can give rise to lateral branches.

Meristems contain populations of slowly dividing central **stem cells** and rapidly dividing peripheral **initial cells**. Infrequent divisions of the stem cells are thought to replace initial cells that are lost from the meristem. Almost all of the cell divisions that take place during post-embryonic plant development occur in and around the meristems. However, the size of the proliferating cell population remains constant, so that peripheral cells continually leave the meristem area and differentiate to form the mature structures of the growing shoot and root. In this way, plant meristems are rather like the progress zone in the vertebrate limb bud (Topic M1). In both systems, proliferation and differentiation must be carefully balanced.

Summary of shoot and root development

During vegetative growth, the **SAM** gives rise to a series of primordia that form **lateral organs** (leaves and axillary buds). The spatial arrangement of these primordia governs the **phyllotaxy** of the stem, i.e. the manner in which leaves are arranged (see Topic N5). There is a delay between the initiation of successive primordia, and this is known as a **plastochron**. Just as counting the number of somites in a vertebrate embryo normalizes development for the effects of external conditions such as temperature, expressing the progress of plant development in terms of plastochrons allows normalization for the many outside conditions that affect plant growth. The intervening plastochron allows the shoot to extend and gives each primordium time to grow before the next one is initiated. Shoots are therefore composed of a series of modules called **phytomers**, each comprising a **node** (with an associated leaf and axillary bud) and an intervening **internode** (*Fig. 1*). Furthermore, phytomers show a developmental gradient along the apical–basal axis of the shoot, with the youngest organs at the apex and the oldest at the base. The growth of each phytomer occurs through a combination of cell proliferation and cell expansion. Each axillary bud contains additional meristem tissue that can give rise to a lateral branch of the shoot.[1]

The root is derived from the **RAM** by continuous growth. There is no periodic initiation of lateral organ primordia in the root and hence no division into repetitive phytomers as seen in the shoot. Lateral roots may grow from the primary root, but these are derived from differentiated tissue further up the root, not from the RAM. Another difference between the SAM and RAM concerns the position of the meristem population relative to the apical extremity of the plant. The SAM is found at the tip of the shoot, and differentiated cells are laid down laterally and behind the SAM as the shoot grows. Conversely, the RAM not only lays down differentiated cells in its wake, but also produces a **root cap**, a structure that completely covers the root tip and protects it from mechanical abrasion as it pushes through the soil. The cells of the root cap are sloughed off as the root extends, and are continually replaced by cells from the RAM.

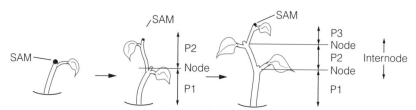

Fig. 1.   Arial growth of plants in phytomers. Each phytomer comprises a node and associated internode.

Organization of the SAM

The shoot and its associated organs contain cells from all three of the fundamental cell layers established during embryonic development (Topic N2). These three cell layers, L1, L2 and L3, must therefore be present in the SAM. Morphologically, the SAM is divided into two bands of cells. The outer band is called the **tunica**, which comprises layers L1 and L2. Both these layers are one cell thick, and the cells undergo predominantly anticlinal divisions therefore expanding the tunica but maintaining its thickness. The inner band is called the **corpus**, which comprises layer L3. Cells in this region undergo both anticlinal and periclinal divisions.

---

[1]Branching is initiated by the proliferation of axial meristems, and this is typically suppressed by hormonal signals from the apical meristem. Only when the growth of the shoot is sufficient to distance the apical meristem from the axial buds can lateral shoot growth occur. This mechanism of suppression, known as **apical dominance**, also explains why lateral branches grow when the apex of the plant is removed. The gardener's trick of making a plant more bushy by pruning is based on the principle of apical dominance.

The layered organization of the SAM is superimposed upon its division into central and peripheral zones. The **central zone** contains **central mother cells**, which are large, vacuolated, slowly-dividing stem cells. The **peripheral zone** contains rapidly proliferating **apical initial cells**, which are small and cytoplasmically dense. The meristem is surrounded by the **morphogenetic zone**, where differentiation occurs. At the flanks of the meristem, the morphogenetic zone comprises the lateral organ primordia, while beneath the apical dome it comprises the **rib meristem**, in which files of cells are laid down to extend the stem as the shoot grows. The relationship between these cell layers and zones is shown in *Fig. 2*.

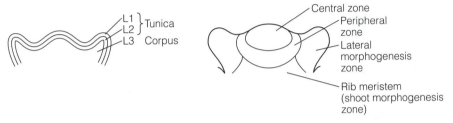

Fig. 2. Organization of the SAM into layers and zones.

Specification of cell fates in the SAM

Cells leaving the SAM give rise to the extending shoot and the repetitive series of lateral organ primordia, each made up of the three fundamental cell layers, L1, L2 and L3. It is interesting to ask whether the organization of the cell layers and the arrangement of the organ primordia are predetermined in the SAM, as would be expected in mosaic development, or whether the SAM can reorganize itself to compensate for missing or repositioned parts, as would be expected in regulative development. These questions have been addressed by clonal analysis, the generation of fate maps and experiments involving the surgical manipulation of meristem tissue. As the population of meristem cells is small and inaccessible, traditional cell labeling techniques such as the injection of dyes or the application of vital stains cannot be used. Instead, cells can be identified with genetic markers, such as mutations that affect pigmentation. Such experiments have shown that cell fates in the meristem depend mostly on position rather than lineage, as would be expected for regulative development. Indeed, the SAM is capable of quite considerable regulation. In many plant species, the SAM can be cut into several pieces, each of which will regulate to form a complete functional meristem.

Specification of the cell layers
The lineage of the three cell layers in the SAM has been established using **periclinal chimeras**, i.e. plants in which cells in different layers have different genotypes allowing them and their descendants to be easily recognized. Periclinal chimeras can be generated by inducing mutations that abolish pigmentation in one layer, or using mitotic inhibitors to induce polyploidy in one layer (polyploid cells can be recognized by their larger nuclei, which contain more nucleoli than usual). More recently, it has also been possible to use endogenous layer-

specific marker genes, such as *ATML1* in *Arabidopsis*, which is expressed exclusively in L1.

Experiments such as these have shown that the lineages of the three cell layers are generally separate from the time they arise in the embryo until their final differentiation in the shoot. However, if a cell is displaced from one layer into one of the others, its fate can be respecified. It is not possible to perform single cell grafts in plants as it is in animals, because of the way plant cells are cemented together. However, the cells in the outer L1 and L2 layers occasionally undergo unusual periclinal divisions, and in such cases the displaced cells appear to be absorbed by the adjacent layers and respecified through interactions with their neighbors. Therefore, while lineage may play a role in the specification of the cell layers, the fates of each cell are not determined and can be changed by interactions with other cells.

Specification of the organ primordia

Fate maps of the SAM can be produced using **mericlinal chimeras**, i.e. plants in which single meristem cells are genetically marked by inducing a mutation that abolishes pigmentation, so that a discrete sector of unpigmented tissue is generated in the mature plant. Cell divisions in the SAM are not stereotypical, as they are in the RAM (see below), so fate maps are probabilistic rather than accurate.

Meristem fate maps have been produced in maize and *Arabidopsis*. The growth of maize plants is **determinate**, i.e. there is a predetermined number of phytomers (18 plus a terminal spike and tassel, representing the male reproductive organs). Conversely, the growth of *Arabidopsis* is **indeterminate**, i.e. the number of phytomers is not fixed. In both cases, however, the SAM fate map appears to be organized in a similar way. If the SAM is flattened out, the fate map comprises a series of concentric circles, with the outer rings representing the oldest (lowest) phytomers and the inner rings representing the youngest (highest) phytomers of the mature plant (*Fig. 3*). At each developmental stage, most cells in the meristem are committed to form older phytomers because the outer rings contain more cells and the fate map is more detailed compared to the inner rings. At any particular stage of development, marked cells from the center of the fate map give rise to wide clones that may span several apical phytomers, while marked cells from the periphery of the fate map give rise to narrow clones that are constrained within a particular phytomer lower down the stem. Loosely committed cells in the center of the fate map therefore become progressively more restricted as they move towards the edges, and regions of the fate map corresponding to individual organ primordia become segregated into developmental compartments representing individual phytomers. The SAM fate map also indicates the phyllotaxy of the plant. The maize fate map is a series of concentric circles commensurate with alternating phyllotaxy, while the *Arabidopsis* fate map is more like a helix, commensurate with spiral phyllotaxy. Phyllotaxy is discussed in more detail in Topic N5.

While cells in the meristem are specified to give rise to certain organ primordia, these fates are not determined until just before the primordia begin to develop. Experiments in which the newest leaf primordium is removed result in the repositioning of subsequent primordia to compensate for the missing leaf (this is discussed in more detail in Topic N5). Therefore, the allocation of cells to particular organ primordia is controlled by signals produced by existing primordia. In determinate plants such as maize, the meristem produces a finite number of primordia. However, the signal to stop producing new primordia

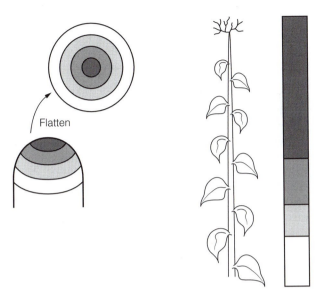

Fig. 3.   Fate map of the maize SAM.

appears to come from the lower phytomers and not from the meristem itself. This can be shown by removing the SAM from a maize plant at some stage during its growth. The cultured SAM is 'deprogrammed' and gives rise to another entire plant, not just the remaining portions of the plant from which it was removed. The positional identity of the cells in the meristem is therefore labile, and maintained by signals from earlier phytomers. In this way, the meristem is different from the progress zone of the vertebrate limb bud, in which the positional identity of the cells is fixed (Topic M2).

Molecular basis of SAM development

A number of mutants have been identified showing unusual patterns of cell proliferation or differentiation in the SAM, and the expression patterns of the corresponding genes have been instrumental in elucidating the molecular basis of SAM function. It is important to differentiate between genes specifically affecting SAM function and those encoding general regulators of the cell cycle or DNA replication, which influence the proliferation of all cells (e.g. the *Arabidopsis* gene *PROLIFERA*). Several mutants specifically fail to maintain cell proliferation in the SAM, while the RAM is unaffected. In *Arabidopsis*, the *shoot meristemless* mutant does not produce a recognizable SAM in the embryo and no leaves are produced in the seedling, although the hypocotyls and cotyledons are normal. Weaker alleles produce single leaves, and very weak alleles produce a number of leaves. These results suggest that the function of the **SHOOT MERISTEMLESS** gene is to maintain the population of proliferating cells in the SAM. In strong mutant alleles where the gene function is abolished, the very early SAM cells do not proliferate at all and are probably recruited to the cotyledons, so no leaves are formed. In weaker alleles, a small population of meristem cells is generated but these are not maintained in a proliferative state and hence differentiate to form one or a few leaves. Concordantly, the product of *SHOOT MERISTEMLESS* is expressed in the SAM and other aerial meristematic tissue, but is strongly downregulated in emerging organ primordia. Further evidence for the role of *SHOOT MERISTEMLESS* has emerged from analysis of the related maize gene, *knotted-1*. As in *Arabidopsis*, this gene is expressed in meristem tissue but downregulated in emerging organ primordia. If maize *knotted-1* is overexpressed in tobacco, small ectopic shoots appear on the leaf surface resulting from increased cell division.

In *clavata* mutants ('clavata' meaning club-shaped), the balance is tipped in the opposite direction and proliferating cells accumulate resulting in larger than usual SAM. However, organ formation outside the expanded SAM is normal, indicating that the process of organ formation itself is not affected, and suggesting the normal role of the *CLAVATA* genes is to restrict cell proliferation. It appears that the size of the SAM is controlled by the competitive activities of the *CLAVATA* genes and *SHOOT MERISTEMLESS*. The expression domains of *CLAVATA1* and *SHOOT MERISTEMLESS* overlap, and it is conceivable that the graded activity of these proteins could define the proliferative (meristem) and morphogenetic (differentiating) zones of the apical dome. A possible target for these genes is *WUSCHEL*, whose activity is required for cell proliferation in the SAM (in *wuschel* mutants, the proliferating cells are committed to differentiation too quickly so that the meristem dwindles and plant growth stops after the production of a few leaves). It appears that *CLAVATA1* can inhibit *WUSCHEL* whereas its activity is stimulated by *SHOOT MERISTEMLESS*. *SHOOT MERI-STEMLESS* encodes a homeodomain transcription factor while *CLAVATA1* encodes a receptor kinase. Both genes are expressed only in the inner cell layers, not the outer layer L1, of the meristem, but in both cases mutations affect the L1 layer. This indicates that cells expressing *CLAVATA1* and *SHOOT MERISTEM-LESS* can communicate with cells that do not express these genes. It has been shown that the maize transcription factor encoded by *knotted-1* is physically transported to L1 cells through plasmodesmata, so it is possible that the product of *SHOOT MERISTEMLESS* acts in the same manner.

Organization of the RAM

The tip of the growing *Arabidopsis* root is organized into a series of concentric cell layers, comprising an outer epidermis, and inner cortical, endodermal and peri-cycle layers surrounding a vascular core (*Fig. 4*). The number and arrangement of the cells in these layers is invariant. The root is also divided into several morpho-logical zones along the longitudinal axis. The most distal part of the root tip is the **root cap**, which protects the root as is pushes through the soil. Behind the root cap are zones of **cell proliferation** (the **RAM**), **elongation** and **differentiation**. Cells arising in the RAM begin to elongate once they are a certain distance from the growing tip. Differentiation occurs after elongation, and can be seen with respect to the epidermal cell layer by the growth of root hairs (*Fig. 4*).

The RAM contains several important populations of cells and its fine struc-ture is therefore very important. The majority of the RAM is made up of prolif-erating cells whose radial organization has already been specified, and which therefore extend the root by adding to the files of pre-existing cells in the epidermal, cortical, endodermal and pericycle layers. At the base of the RAM is the so-called **promeristem**. This comprises three tiers of initial cells that generate the radial pattern of the root and produce the root cap (*Fig. 4*). The lower tier comprises a central group of 12 cells that give rise to the **root cap columella** (the tip of the root cap). The columella initials are surrounded by a ring of 16 cells that give rise to the **lateral root cap** and the outer cell layer of the root, the **epidermis**. The middle tier of cells comprises a central group of four large cells known as the **quiescent center**, which is thought to contain the RAM's stem cells and therefore to be equivalent to the central zone of the SAM.

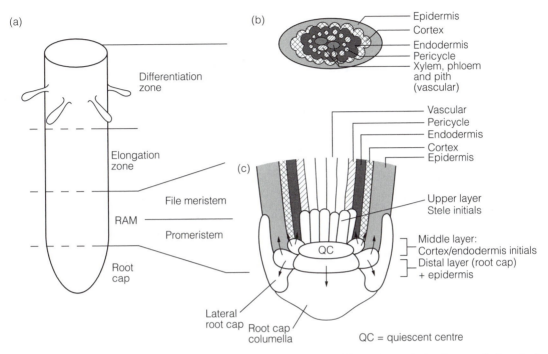

Fig. 4. (a) Gross structure of the Arabidopsis root. (b) Transverse section of the mature root, showing organization of cell layers. (c) Detailed structure of cell organization in the RAM, longitudinal section.

This is surrounded by a ring of eight **endodermal/cortical initials** that give rise to the cortex and endodermis. The cells in the upper tier give rise to the **pericycle** and **vascular bundles**.

Specification of cell fates in the RAM

Cell lineage analysis during *Arabidopsis* root development has shown that the organization of the root is generated by a stereotypical series of cell divisions in the promeristem. Similarly, a series of invariant cell divisions occurs during the development of the nematode *C. elegans*, where the entire developmental lineage has been determined from the egg to the adult (Topic G2). However, although patterns of cell division may be invariant, this does not mean that the fate of each cell is determined from the outset of development, as both *C. elegans* embryos and *Arabidopsis* roots provide evidence that cell–cell interactions are also important. In both systems, laser ablation experiments have been extensively used to demonstrate the dependence of certain developmental processes on the proximity of adjacent cells. For example, endodermal/cortical initial cells in the second tier of the promeristem normally give rise to daughter cells that divide to produce cells of the cortex and endodermis (*Fig. 5*). If one initial cell is ablated with a laser beam, the adjacent pericycle cell undergoes an unexpected division and replaces it. If one daughter cell is ablated, the underlying endodermal/cortical initial cell undergoes an additional division to generate a new one. Oddly, if the three cortical daughter cells are ablated, the endodermal/cortical initial cell underlying the middle daughter cell is no longer able to produce any more daughter cells that divide. This experiment shows that signals from the cortical daughter cells control the developmental potency of the underlying initial cells.

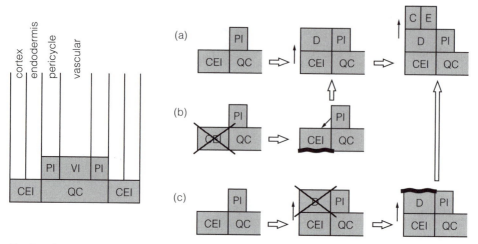

Fig. 5.   Cell ablation in root development. Left panel shows schematic representation of the RAM (CEI = cortical/ endodermal initial; QC = quiescent center; PI = pericycle initial; VI = vascular initials). Corresponding cell layers in the file meristem are shown above. (a) In normal development, each CEI gives rise to a daughter cell (D), which divides to form cortical and endodermis cells (C, E) adding to existing files. (b) If the CEI cell is ablated, an unexpected division of the PI cell replaces it. (c) If the daughter cell is ablated, the CEI undergoes an extra division to replace it. Wavy lines indicate squashed remains of ablated cells.

Molecular basis of RAM development

Several *Arabidopsis* mutants have also been identified that interfere with the developmental potency of the endodermal/cortical initial cells. In the *short root* mutant, the endodermis is missing because the cortical daughter cells fail to divide, and produce a single file of cells that differentiates as the cortex. The endodermal/cortical initials also fail to divide in *scarecrow* mutants – in this case they produce a single file of cells which is intermediate in character between the cortex and the endodermis. In *pinocchio* mutants, the initials again fail to divide, but in this case the endodermis is present while the cortex is missing.

Most laser ablation experiments result in normal development of the root because the ablated cells are crushed and replaced by the division of an adjacent cell. This shows that, despite the stereotyped divisions of the promeristem, cell–cell signaling has a very important role to play in root development. This is confirmed by the *fass* mutant, which has fat roots with many more cells than usual and extra cell layers ('fass' means barrel). However, the overall organization of the root in *fass* mutants is not disrupted. Indeed the *fass* mutation can partially compensate for the *scarecrow* mutant phenotype when both mutants are combined. The deficiency in *scarecrow* appears to be in the generation of additional cell layers, since the double *fass + scarecrow* mutant has both a cortex and endodermis. Conversely, despite the extra cell layers that form, double *fass + short root* mutants still lack an endodermis. The defect in the case of *short root* must therefore be in the molecular pathway that determines endodermal cell identity rather than a simple lack of cell layers.

Another class of root mutants show phenotypes that are analogous to *shoot meristemless*, but specifically affect cell proliferation in the RAM. The corresponding genes have been appropriately named *ROOT MERISTEMLESS 1* and *2*. These show no defects in SAM proliferation, confirming that the defect is not a general problem with the cell cycle. Interestingly, cell proliferation is also abolished in the *short root* mutant, indicating that the continued proliferation of the RAM is in some way linked to the ability of the endodermal/cortical initials to divide and produce endodermis.

N5 LEAF DEVELOPMENT

Key Notes

Leaf structure and development

Leaves are diverse in structure but generally show bilateral symmetry, with pronounced dorsoventral and apical–basal polarity. Dicot leaves comprise a broad blade (lamella) joined to the stem by a petiole and leaf base, while monocot leaves have a long thin blade joined to the stem by a sheath. Leaves develop from lateral organ primordia periodically initiated on the shoot apical meristem.

Leaf arrangement – phyllotaxy

The spatial arrangement of leaves around the stem is called phyllotaxy, and corresponds to the spatiotemporal pattern in which the lateral organ primordia arise in the shoot apical meristem. The ideal, even spacing shown by leaves suggests lateral inhibition is involved in the specification of the primordia, and this is supported by experiments in which newly arising primordia are removed, leading to the repositioning of future primordia. The maize *terminal ear 1* gene provides insight into the molecular basis of this process.

Leaf determination

Newly arising lateral organ primordia are initially uncommitted, and can give rise to lateral shoots as well as leaves. Leaf determination requires signals from the SAM, and results in the polarization of the dorsoventral axis. Mutants that abolish dorsoventral polarity produce ventralized needle-like leaves that resemble stems.

Leaf growth and morphogenesis

Leaf growth begins with changes in the rate and plane of cell divisions in the emerging primordium, extending the petiole and spreading the leaf base. Clonal analysis indicates that the pattern of cell divisions changes during leaf development, and certain mutations indicate that such changes may be instrumental in shaping the leaf blade. However, the *tangled-1* mutant in maize, which causes disorganized cell divisions, does not alter leaf morphology. Cell expansion is also important in leaf development, and *Arabidopsis* mutants with defects in this process have been isolated.

Pattern formation in the leaf – trichome spacing

The dorsal surface of the leaf contains specialized epidermal structures called trichomes. These develop individually in wild type plants, but mutants such as *glabrous 1* and *transparent testa glabrous* can have clusters of trichomes. This suggests that the *GLABROUS1* and *TRANSPARENT TESTA GLABROUS* genes mediate lateral inhibition during trichome development, and the phenotypes of transgenic plants overexpressing these genes support this hypothesis. Both genes encode bHLH transcription factors, which have been shown to be involved in the spacing of neuroblasts in *Drosophila* and vertebrates.

Related topics

Pattern formation (A6)
Morphogenesis (A7)
Neurogenesis (J5)

Plant *vs.* animal development (N1)
Shoot and root meristems (N4)
Flower development (N6)

Leaf structure and development

Leaves are determinate structures that arise from the lateral organ primordia of the shoot apical meristem. The leaves of dicot and monocot plants are distinct (*Fig. 1*). Dicot leaves comprise a **leaf blade** or **lamina** with a single central vein (**midrib**) and subsidiary veins running off at oblique angles. The lamina is joined to the stem via a stalk (**petiole**) and **leaf base**. Dicots show great diversity in leaf shape and size. In many plants the lamina is a simple ovoid shape, while in others it may be lobed or serrated. Many dicot plants have **simple leaves**, where a single lamina spreads out from the petiole, but others have **compound leaves**, in which a number of leaflets can be attached to the end of the petiole, or to a rachis. Monocot leaves are long and thin, with multiple parallel veins running along the long axis of the lamina. The lamina is attached to a **sheath** via a **ligule** and a pair of **auricles**, and the sheath is attached to the stem.

Leaves are bilaterally symmetrical structures that are polarized along two axes (*Fig. 2*). The **apical–basal axis** runs proximal to distal, i.e. from stem to leaf tip. The **dorsoventral axis** runs from the upper surface of the leaf (which is usually relatively smooth and bears specialized epidermal hairs called **trichomes**), to the lower surface (which is characterized by protruding veins and regularly-distributed gaseous exchange organs called **stomata**). The dorsoventral axis is also called the **adaxial–abaxial axis** because when the leaf forms, the dorsal surface is innermost, wrapped around the stem, while the ventral surface is furthest away from the stem. Leaf development begins with the emergence of a **leaf primordium** from the shoot apical meristem. The primordium expands laterally so that the leaf base partially encircles the stem, and extends longitudinally to produce the petiole and midrib vein. Growth of the lamina occurs later, initially by the proliferation of meristem cells flanking the midrib and later by the expansion of cells in the young blade.

In this Topic, we discuss some of the molecular mechanisms underlying leaf morphology. First, however, we look at how the global arrangement of leaves on the stem is controlled.

Leaf arrangement – phyllotaxy

Phyllotaxy is the spatial arrangement of organs such as leaves around the shoot. Many plants, including *Arabidopsis*, show **spiral phyllotaxy**, where the leaves

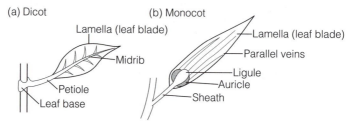

Fig. 1.   *Comparison of dicot and monocot leaves.*

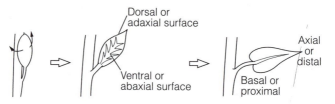

Fig. 2.   *Leaf axes. Since dicot leaves originate adjacent to the stem, the upper (dorsal) surface is also termed adaxial, and the lower (ventral) surface abaxial.*

are arranged in either a left- or right-handed helical pattern when viewed from the apex of the plant. Other species, such as maize, show **alternating phyllotaxy**, where there is one leaf per node, but alternate leaves project in opposite directions. Another variation is **decussate phyllotaxy**, where there are two leaves per node (one on each side) and alternating pairs are perpendicular (*Fig. 3*). Remarkably, it has been shown that the phyllotactic patterns of many plants have ideal **divergence angles** (angles that provide maximum separation between successive primordia) and that the positions of new primordia can be predicted precisely using mathematical algorithms.

The precise and ideal spacing of successive leaf primordia suggests that existing leaf primordia might inhibit the development of new ones. This possibility has been addressed by grafting experiments in which the positions of developing leaf primordia are artificially changed. For example, the removal of the newest leaf primordium from the apical dome of the lupin does not affect the position of the next primordium, but it does affect the one after that, which is shifted towards the site of the excision. This indicates that signals from existing primordia can influence the development of later ones. In the example above, it appears that these inhibitory signals are produced when the new leaf primordium has already emerged. This is too late to affect the position of the next primordium, as by this stage its position is already determined. However, the third primordium is sensitive to these signals, and if they are abolished, the fate of cells nearer to the site of excision is respecified (*Fig. 4*).

Evidence of the molecular basis of phyllotaxy comes from the analysis of a maize mutant called *terminal ear 1*, in which the normal alternating phyllotaxy is interspersed with decussate, spiral and other variations. In wild type plants, the *terminal ear 1* gene is expressed in a horseshoe pattern in each phytomer, with the open end of the horseshoe corresponding to each leaf primordium. The

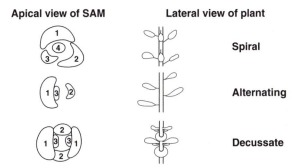

Fig. 3.   Spiral, alternating and decussate phyllotaxy. The left panel shows the positions and order of emergence of the leaf primordia. The right panel shows the corresponding mature plants.

Fig. 4.   Ablation of leaf primordia in the lupin. If a newly emerging leaf primordium (1) is ablated, the position of the next primordium (2) is not affected, but that of the third (3) is shifted towards the site of excision. This shows that signals from primordium 1 affect the position of primordium 3. Primordium 2 is not affected because its position is already determined by the time primordium 1 has emerged.

orientation of the horseshoe alternates along the stem concomitant with the phyllotaxy of the leaves. The product of the *terminal ear 1* gene may therefore repress leaf development, thus representing a component of the inhibitory signal that facilitates optimal leaf spacing during plant development.

Leaf determination

The primordium that gives rise to a leaf can also give rise to a branch of the stem, an inflorescence, a flower, or in some plants, a variant organ such as a thorn. Special processes are required to produce floral organs (Topic N6) and in this sense, the leaf appears to be the default pathway for lateral organ development. However, the fate of the organ primordium is not specified until after it has emerged. This can be shown by interposing an impermeable barrier between the leaf primordium and the shoot apical meristem, which results in the growth of fewer (determinate) leaves but more (indeterminate) lateral branches. Therefore, there appears to be a phase just after the primordium emerges during which its fate is undecided. In the presence of signals from the SAM, the primordium becomes determined as a leaf, while in the absence of those signals, it may develop as a shoot. The timing of this commitment can be investigated by explanting individual primordia at different times after emergence and seeing whether they are able to generate leaf tissue in culture (remarkably, once a leaf primordium is determined, it becomes autonomous and can develop into an entire and morphologically normal leaf when placed on nutritive medium). The results of such experiments show that plants vary in the timing of leaf determination, but that it occurs very soon after emergence of the primordium.

The first process that distinguishes leaf development from lateral branch development is the appearance of dorsoventral polarity. The *phantastica* mutant, identified in both *Antirrhinum* and *Arabidopsis*, shows dorsoventral polarity defects in emerging leaves. It appears that the upper leaves on mutant plants are ventralized, and form needle-like structures reminiscent of stems in their organization. The *PHANTASTICA* gene is therefore an important component of a dorsalizing signal early in leaf development.

Leaf growth and morphogenesis

The outgrowth of leaves and lateral branches is initiated by changes in the rate and orientation of cell divisions in the emerging primordium. Morphological observations show that leaf growth begins with the extension of the petiole, and the lamella grows later. Clonal analysis using pigment-deficient mericlinal chimeras has supported this view (*Fig. 5*). In many cases, mutant cells generated in very early leaf primordia give rise to clones that cover the petiole, midrib and part of the lamina, indicating that all cells in the leaf are derived from the same apical initial cells in the SAM. Other clones cover just the petiole and midrib, indicating that these structures are established early in leaf development before the lamella begins to expand. Chimeras generated in more advanced leaf primordia give rise to unpigmented sectors of tissue within the leaf blade, not at its edges, and this shows that the leaf blade grows by intercalary cell division rather than growth at the expanding edge. The largest mutant sectors are obliquely angled and elongated, indicating that early cell divisions expand the lamella parallel to the subsidiary leaf veins. Smaller clones are more regular in shape, indicating that later development consists of cell divisions in alternating perpendicular directions.

These clonal analysis experiments suggest that the rate and orientation of cell divisions are critical for morphogenesis in the leaf blade. This is supported by

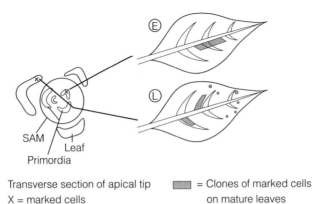

Transverse section of apical tip ▨ = Clones of marked cells
X = marked cells on mature leaves

Ⓔ Early – few large regular clones
Ⓛ Late – many small irregular clones

Fig. 5. Clonal analysis of tobacco leaf development. Marked cells in emerging leaf primordia give rise to clones of marked cells in mature leaves. The shape and size of the clones reveal how patterns of cell division change during leaf development.

the existence of mutants with abnormal leaf morphology, which show defects in cell division. For example, the tobacco mutant *lam-1* fails to form the proliferative meristem populations on either side of the central vein, resulting in the absence of a lamella and the growth of rod-shaped unexpanded leaves. In another tobacco mutant called *fat*, there is abnormal cell division in the plane of the dorsoventral leaf axis resulting in the growth of thickened leaves with extra layers of cells. In both *lam-1* and *fat* mutants, the outgrowth of the petiole is normal, suggesting that the apical–basal organization of the leaf is not dependent on the correct growth of the lamella. While these mutants suggest that regulated cell divisions are essential for leaf blade morphogenesis, the *tangled-1* mutant of maize provides contrary evidence. In normal maize plants, cell divisions in the leaf blade are orchestrated so that the cells form regularly-arranged parallel files. In *tangled-1* mutants, these divisions are randomly orientated resulting in a chaotic arrangement of cells. Despite this, the overall morphology of the leaf is not affected. One possible explanation for these seemingly contradictory results is that there is a global mechanism to generate the correct leaf morphology that controls the balance between cell divisions in different planes. In mutants such as *fat*, extra cell divisions occur in one plane, resulting in altered leaf morphology. However, in mutants such as *tangled-1* cell division may be affected equally in all planes so the overall leaf morphology is maintained.

Mutations have also been identified that affect cell expansion in the leaf rather than cell division. In *Arabidopsis*, two mutants have been isolated that affect leaf expansion in perpendicular directions. The *augustifolia* mutant has narrow leaves, reflecting a defect in lateral cell expansion. Conversely, the *rotundifolia* mutant has short round leaves, reflecting a defect in apical cell expansion. Double mutants are both narrow and short, indicating that expansion of the leaf in each direction is controlled independently.

The development of characteristic patterns at the leaf margin (such as lobes and saw-tooth serrations) is a very late event in leaf morphogenesis, as the early leaf primordia of all dicot plants have simple ovoid shapes. Insights into the basis of morphogenesis at the leaf margin have come from the analysis of transgenic *Arabidopsis* plants in which the normally simple leaves are transformed into lobed ones. For example, this can be achieved by the overexpression of the *Arabidopsis* gene *KNAT 1*. Progressively higher expression levels generate leaves with more profound lobular structures, and at the highest levels, inflorescences can form at the base of each lobe. Variation in the expression levels of such genes may have played a role in the evolution of different leaf morphologies.

Pattern formation in the leaf – trichome spacing

The leaf contains specialized epidermal hairs called trichomes on its dorsal surface. In *Arabidopsis*, the trichomes are three-branched structures derived from normal epidermal cells by a complex process involving cell enlargement, genome duplication, branching and lignification. In normal leaves, the trichomes are not regularly spaced but they are found as individuals rather than clusters. The existence of mutants in which this spacing pattern is disrupted suggests that the positions of the trichomes may be defined, in part, by lateral inhibition.

Mutants that abolish the formation of trichomes include *glabrous 1* and *transparent testa glabrous*. These genes act early in the trichome development pathway because the mutants are epistatic to other trichome mutants that have been identified. Weaker mutant alleles of these genes do not abolish trichome development, but result in a disrupted pattern where clusters of trichomes can form. This suggests that the genes are required both to initiate trichome development and to control the spacing pattern. Interestingly, both genes encode bHLH transcription factors. Such transcription factors are also involved in the regulation of spacing patterns in the *Drosophila* and vertebrate nervous systems (Topic J5). Although exact parallels to the proneural and neurogenic genes in animals have not been identified, the appearance of trichome clusters when the activity of these genes is reduced suggests that lateral inhibition plays a predominant role in generating the spacing pattern. If *GLABROUS 1* is overexpressed in otherwise normal *Arabidopsis* leaves, the spacing pattern of the trichomes is not altered. However, if this gene is overexpressed in plants heterozygous for the *transparent testa glabrous* null mutation, clusters of trichomes develop. This suggests that *GLABROUS 1* and *TRANSPARENT TESTA GLABROUS* encode inhibitors of trichome development that may antagonize each other. If either gene is deleted, the inhibitory effect of the other is unrestrained and no trichomes form at all. If *GLABROUS 1* is overexpressed in plants with two copies of the *TRANSPARENT TESTA GLABROUS* gene, its activity is still inhibited resulting in a normal trichome pattern. If one copy of *TRANSPARENT TESTA GLABROUS* is missing, the inhibitory effect is reduced and clusters of trichomes can form. Similarly, in weak *transparent testa glabrous* mutants, there is a reduced inhibitory effect on normal levels of *GLABROUS 1*, resulting in the formation of clusters of trichomes. It is not known whether these two genes negatively regulate each other's expression directly, or act though an indirect pathway.

N6 FLOWER DEVELOPMENT

Key Notes

Overview of flower development

Vegetative growth in plants produces shoots and leaves, while reproductive growth produces flowers. The floral initiation program is activated in many plants by signals from the environment, especially day length, which convert the leaf-producing SAM into a flower-producing inflorescence meristem. Floral initiation involves the respecification of an indeterminate inflorescence meristem as a determinate floral meristem, and induces the expression of floral patterning genes that generate the four specialized organs of the flower.

Floral organs

Typical hermaphrodite flowers comprise four types of floral organ (sepals, petals, stamens, carpels) arranged in four concentric whorls. Some plants have flowers of separate sexes, in which case the unwanted floral organs either do not form, or form but do not function.

Specification of floral organ identities

The identities of the floral organs are specified by the expression of homeotic selector genes, which give each whorl a positional value in the flower. Mutations in these genes, which cause particular whorls to be given the wrong positional values, result in homeotic transformations. There are three classes of homeotic gene, each class expressed in a particular pair of adjacent whorls. Positional values are generated by particular combinations of the gene products, which are MADS box transcription factors. The floral homeotic genes function in much the same way as the HOM-C/*Hox* genes of animals.

Control of floral organ formation

A number of *Arabidopsis* mutants show abnormal numbers, as well as types, of floral organs in the flower. These mutants represent genes with various functions in flower development, including control of cell proliferation in the floral meristem, the establishment of compartment boundaries, and possibly lateral inhibition between different organ primordia.

Transition to flower development (floral initiation)

Floral organ formation and patterning occur in the determinate floral meristem, which arises from the indeterminate inflorescence meristem during the floral initiation program (FLIP). This program begins with the expression of three genes – *LEAFY, APETALA1* and *CAULIFLOWER* – which specify the floral meristem by inhibiting indeterminate growth, and activate the downstream floral patterning genes. Many of the floral homeotic genes that pattern the emerging flower have earlier roles in floral initiation.

Transition to inflorescence development in *Arabidopsis*

In many plants, activation of the floral initiation program is dependent on the environment, although in facultative plants such as *Arabidopsis*, flowering also occurs eventually in the absence of external influences. *Arabidopsis* mutants showing delayed or late flowering under all conditions, or specifically in response to long days, have identified genes involved in this activation process.

Related topics	Mechanisms of developmental commitment (A3)	Homeotic selector genes and regional specification (H5)
	Pattern formation and compartments (A6)	Plant *vs.* animal development (N1)
	Gene expression and regulation (C1)	Shoot and root meristems (N4)
		Leaf development (N5)
	Signal transduction in development (C3)	

Overview of flower development

Vegetative growth in flowering plants produces organs such as shoots and leaves, but not flowers. Flowers represent a fundamental change from vegetative to reproductive development, involving the specification of a determinate **floral meristem** that gives rise to a series of specialized organs found only in the flower (sepals, petals, stamens and carpels). Flowers develop on **inflorescences** (flower stalks), which are morphologically distinct from the vegetative shoot that gives rise to leaves. The growth of the inflorescence requires the **shoot apical meristem** to be respecified as an **inflorescence meristem**, a process that is often dependent upon signals from the environment. In species with a single flower, the transition to flowering marks the end of vegetative growth – the inflorescence meristem becomes the floral meristem generating a stalk with a terminal flower. In other species, including *Arabidopsis*, the inflorescence meristem gives rise to a succession of indeterminate branches or **paraclades**. In many of these branches, the inflorescence meristem is converted into a floral meristem to generate a flower, while some inflorescence meristems continue to grow indeterminately.

In this Topic we consider three major stages of flower development. First, we look at the molecular basis of pattern formation in the flower, which involves the generation of specific patterns of organ primordia in the floral meristem followed by regional specification to identify the different floral organs. Second, we discuss the molecular basis of floral initiation, during which indeterminate inflorescence meristems are converted into determinate floral meristems. Finally, we discuss how floral initiation is activated in the first place. In most plants this first stage is either influenced by or dependent upon environmental signals, especially day length.

Floral organs

Flowers show great diversity in shape and structure, but the simplest hermaphrodite flowers (such as those of *Arabidopsis thaliana*) are approximately radially symmetrical and comprise four sets of **floral organs – sepals**, **petals**, **stamens** (male reproductive organs) and **carpels** (female reproductive organs). These are arranged as four concentric circles, or **whorls**, with the sepals on the outside and the carpels on the inside (*Fig. 1*). In other species (such as *Antirrhinum majus*), dorsoventral polarity is imposed on this radial pattern, but mutants exist in which dorsoventral patterning is abolished, generating radial flowers organized similarly to those of *Arabidopsis*.

Specification of floral organ identities

Flowers comprise four concentric whorls of floral organs, and an important phase of flower development concerns the specification of organ identities. This is revealed by the existence of **homeotic mutants**, where particular floral organs are transformed into the likeness of another. In *Arabidopsis*, at least five

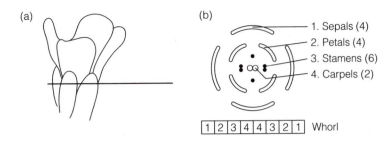

(a)

(b)

— 1. Sepals (4)
— 2. Petals (4)
— 3. Stamens (6)
— 4. Carpels (2)

| 1 | 2 | 3 | 4 | 4 | 3 | 2 | 1 | Whorl

Fig. 1.   Structure of the Arabidopsis *flower. (a) Lateral view. (b) Transverse section, showing concentric whorls of floral organs. These are labeled 1–4, with whorl 1 outermost.*

homeotic selector genes (floral organ identity genes) have been identified through the isolation of different types of homeotic mutant (*Table 1*). For example, in the mutant *apetala2*, the identities of whorls 1 and 2 are changed, so that carpels develop instead of sepals in whorl 1 and stamens develop in place of petals in whorl 2. Indeed, the *Arabidopsis* homeotic mutants fall into three classes according to the whorls that are misidentified, and similar mutants have been identified in *Antirrhinum* (*Table 1*). In each class, two adjacent whorls are affected. In **class A mutants**, the identities of whorls 1 and 2 are changed. In **class B mutants**, the identities of whorls 2 and 3 are changed. Finally, in **class C mutants**, the identities of whorls 3 and 4 are changed. The mutant phenotypes are summarized in *Table 2*.

The mutant phenotypes can be explained by the simple **ABC model**, in which floral organ identity is specified by a particular combination of homeotic selector genes. In this model, the class A genes are expressed in whorls 1 and 2 while the class C genes are expressed in whorls 3 and 4. The domains of the class A and class C genes do not overlap because these genes inhibit each other; this type of interaction has been termed **cadastral** as it limits the expression of each gene (a *cadastre* is a register of property showing the extent of ownership, for taxation purposes). The class B genes are expressed in whorls 2 and 3, in a domain that overlaps both the class A and class C domains. This would divide the flower into four organ-forming zones defined by different combinations of homeotic genes: A = sepals, A+B = petals, B+C = stamens, and C = carpels

Table 1.   Floral organ identity genes (homeotic selector genes) grouped by class in Arabidopsis thaliana *and* Antirrhinum majus

Class (and expression)	*Arabidopsis* genes	*Antirrhinum* genes
Class A (whorls 1 and 2)	*APETALA1 (AP1)* *APETALA2 (AP2)*	*SQUAMOSA (SQA)*
Class B (whorls 2 and 3)	*APETALA3 (AP3)* *PISTILLATA (PI)*	*DEFICIENS (DEF)* *GLOBOSA (GLO)*
Class C (whorls 3 and 4)	*AGAMOUS (AG)*	*PLENA (PLE)*

(*Fig. 2*). The mutant phenotypes displayed by each of the homeotic genes can be explained in terms of this model (*Table 2; Fig. 2*). For example, in the *apetala2* mutant, class A gene function is lost in whorls 1 and 2. This allows class C genes to become active in those whorls, generating the following domain pattern: C = carpels, B+C = stamens, B+C = stamens, C = carpels.

Table 2.   Explanation of Arabidopsis *floral homeotic mutant phenotypes in terms of the ABC model of floral organ specification*

Genotype	Phenotype				Expression domain pattern			
	whorl 1	**whorl 2**	**whorl 3**	**whorl 4**				
Wild type	sepal	petal	stamen	carpel	A	A+B	B+C	C
apetala1 or *apetala2*	carpel	stamen	stamen	carpel	C	B+C	B+C	C
apetala3 or *pistillata*	sepal	sepal	carpel	carpel	A	A	C	C
agamous	sepal	petal	petal	sepal	A	A+B	A+B	A
apetala2+apetala3+agamous	leaf	leaf	leaf	leaf	0	0	0	0
APETALA3, PISTILATA	petal	stamen	stamen	petal	A+B	A+B	B+C	B+C

Note that genotypes in lower case represent loss of function mutations. In the final row, genotypes in upper case represent transgenic plants in which the functional genes are overexpressed in all four whorls.

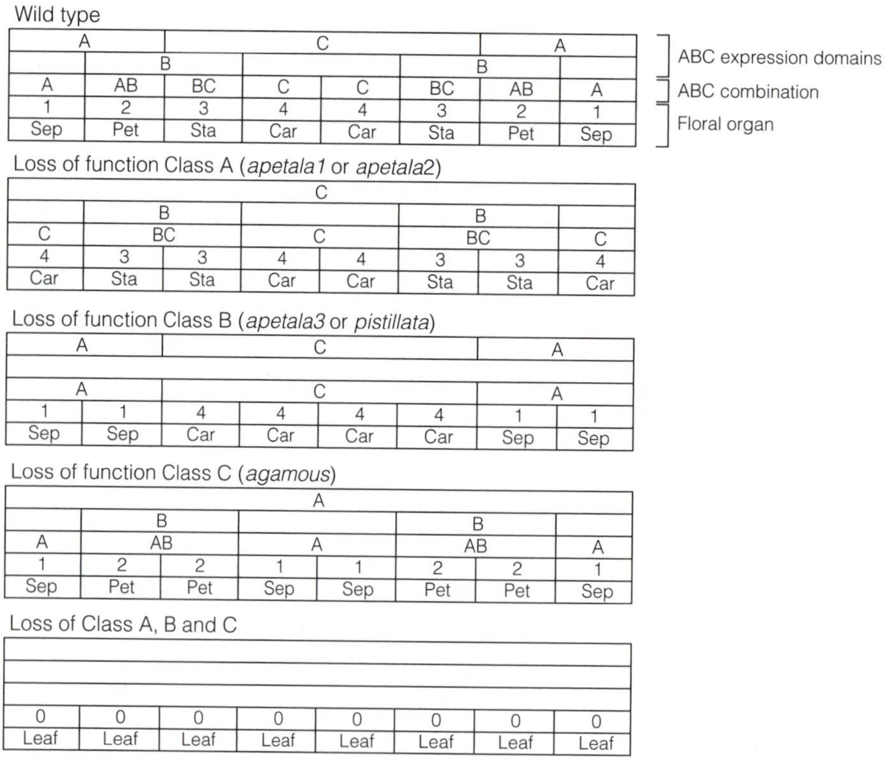

Fig. 2.   The ABC model of floral organ specification.

The expression patterns of the homeotic selector genes have been revealed by *in situ* hybridization, and fit well with the ABC model. In wild type flowers, all the genes are expressed in those whorls that are transformed in loss-of-function mutants. Furthermore, while the class C gene *AGAMOUS* is normally expressed only in whorls 3 and 4, it is expressed in all four whorls in *apetala2* mutants as would be expected based on its proposed cadastral interaction with the *APETALA2* gene. Similarly, the expression of *APETALA2*, which is normally restricted to whorls 1 and 2, expands to all four whorls in *agamous* mutants (*Fig. 2*).

The activity of the floral homeotic genes is therefore very similar to that of the HOM-C/*Hox* complex genes that specify positional values in animal embryos (Topic H5, Topic J2, Topic K2). The HOM-C/*Hox* genes all encode transcription factors containing a conserved DNA-binding domain, the homeodomain. The floral homeotic genes also encode related transcription factors, although these contain a different DNA-binding module called a **MADS box**. MADS box transcription factors are also present in animals, but they play no role in regional specification (for example, the MEF2 transcription factors that regulate muscle development contain MADS boxes). Similarly, homeodomain transcription factors are found in plants but are not involved in patterning the flower – the SHOOT MERISTEMLESS protein is a homeodomain-containing transcription factor. In animals and plants, positional values reflect a combinatorial code of homeotic gene products, and in both cases, there are examples of cooperative as well as negative (cadastral) interactions. For example, there are two class B genes in *Arabidopsis*, but the loss of either one gene generates a mutant phenotype indicating that both genes cooperate to establish class B functions. This has been explained by looking for potential interactions among the gene products. It appears that the APETALA3 and PISTILLATA proteins must form heterodimers to bind to target sequences in DNA, while neither homodimer can do so. Cadastral interactions in *Arabidopsis*, characterized by the mutually exclusive patterns of class A and class C genes, also have parallels in animals. The HOM-C/*Hox* genes show **posterior dominance**, i.e. genes expressed more posteriorly inhibit the expression of those with more anterior boundaries of expression (Topic H5; Topic J2; Topic K2). The analysis of compound HOM-C mutants in *Drosophila* also reveals a default or ground state identity for each segment. For example, when the entire Bithorax complex is deleted, all abdominal parasegments are respecified as parasegment 4 (Topic H5). Similarly, if all three classes of homeotic genes are inactivated in *Arabidopsis*, the four concentric whorls of floral organ primordia all develop as leaves. In molecular terms, the flower therefore appears to be a modified series of leaves.

In *Drosophila*, the HOM-C genes are activated by upstream patterning genes expressed transiently in the embryo, but their expression is maintained by the modification of chromatin structure, a function carried out by genes of the *Polycomb* and *trithorax* families. Similarly in plants, the expression of floral homeotic genes is induced by regulators of the floral initiation program (see below) but maintained by chromatin-remodeling proteins, such as the product of the *Arabidopsis CURLY LEAF* gene. Remarkably, *CURLY LEAF* is related to the *Drosophila* gene *Polycomb*.

Control of floral organ formation

The floral organ identity genes discussed above play a major role in determining the particular organs that form in each whorl of the flower, and this reflects their expression in the corresponding primordia that emerge from the floral meristem. However, just as the correct regional specification of segments in *Drosophila* requires the prior formation of a particular number of segments, the correct regional specification of floral organ primordia in *Arabidopsis* requires the prior formation of the correct number of primordia. The identification of mutants in which the number of organs in each whorl has changed from the normal number of four sepals, four petals, six stamens and two carpels, provides some insight into the molecular basis of this process.

One factor that plays an important role in the formation of flowers with normal proportions is the size of the floral meristem. As discussed in Topic N4, the products of the *SHOOT MERISTEMLESS* and *CLAVATA1* genes compete with each other to maintain the size of the meristem. In *clavata1* mutants, the meristem is much larger than normal, and *clavata1* floral meristems produce flowers with supernumerary whorls and additional floral organs. This is reminiscent of the extra digits that form in the developing vertebrate limb bud when the progress zone is wider than usual. In each case, extra organs form rather than the disorganized tissue that might be expected. This indicates that floral meristems, like the limb bud, may possess an underlying spontaneous pattern-forming capacity, which may reflect some form of reaction diffusion mechanism (Topic A6).

Since lateral inhibition plays an important role in the positioning of vegetative primordia in the shoot apical meristem (Topic N4) it is possible that the numbers of organ primordia in each whorl of the flower, and the relative positions of different primordia in different whorls, are also specified by some form of lateral inhibition. In wild type *Arabidopsis* flowers, the petals arise in the whorl adjacent to the sepals, but each petal forms at the boundary between two sepals. Similarly, the stamens form in the next whorl but at the boundaries between petals (*Fig. 1*). This alternating pattern is consistent with a mechanism in which the primordia of each whorl prevent the formation of further primordia in the same and adjacent whorls using relatively short-range inhibitory signals. The *Arabidopsis* mutant *perianthia* has five sepals, five petals and five stamens, but the alternating pattern of the first three whorls is maintained. The formation of one extra primordium in each whorl may reflect a reduction in the strength or range of such an inhibitory signal.

Several mutations appear to affect the central floral organs, producing greater numbers of stamens and carpels than wild type flowers. For example, in the mutant *floral organ number 1 (fon1)*, whorls 1, 2 and 3 are normal, but the fourth whorl is expanded, and gives rise to more stamens and supernumerary carpels. The function of the *FON1* gene may therefore be to limit the proliferation of floral meristem cells in the presumptive central region of the flower. In another mutant, called *superman*, all the carpels are replaced by stamens. This is not a classic homeotic transformation and *SUPERMAN* is not another homeotic gene. *SUPERMAN* is initially expressed at the boundary of whorls 3 and 4. Its role may be to limit cell divisions at this boundary, thus generating two developmental compartments. In *superman* mutants, the cells in whorl 3 continue to divide and generate a larger compartment, which may inhibit the formation of the compartment corresponding to the central fourth whorl.

Transition to flower development (floral initiation)

Floral patterning takes place in the context of a determinate floral meristem, so the transition from indeterminate inflorescence meristem to determinate floral meristem is a critical phase in flower development. Many of the homeotic selector genes discussed above have an earlier role in this **floral initiation** program (FLIP), which involves the specification of a floral primordium and the regulated activation of downstream floral patterning genes (*Fig. 3*).

In *Arabidopsis*, there are three primary FLIP genes called **LEAFY, APETALA1** and **CAULIFLOWER**, whose expression in the inflorescence meristem is required at the earliest stage of floral initiation. All three genes encode transcription factors. In *leafy* and *apetala1* loss-of-function mutants, there is transformation of flowers to inflorescence-like structures, showing that these genes are required for floral development. Conversely, the overexpression of either gene using a strong constitutive promoter results in the conversion of inflorescences into flowers, showing that each gene can independently activate floral development. The phenotypes of the loss of function mutants are not identical, suggesting the two genes have distinct although overlapping roles. In loss of function *leafy* mutants, the early flowers develop as inflorescences and the later flowers develop with certain inflorescence-like characters. In *apetala1* mutants, the flower is incorrectly patterned (this reflects the additional role of *APETALA1* as a floral homeotic selector gene; see above) and multiple floral meristems appear at the base of the flower. In both mutants there appears to be a *gradient of inflorescence to floral transition towards the apex of the plant*; the significance of this will be discussed below. In *leafy + apetala1* double mutants, flower development is blocked completely. The *cauliflower* mutant has a subtle phenotype, but the importance of this gene can be seen when the *cauliflower* and *apetala1* mutants are combined in the same plant. In these double mutants, all floral meristems behave as (indeterminate) inflorescence meristems, which proceed to give rise to multiple, higher-order lateral branches. The resulting densely packed clusters of branched meristems resemble the head of a garden cauliflower[1]. The transcription factor encoded by the *CAULIFLOWER* gene is closely related to APETALA1 and it is thought that the two products cooperate in the floral initiation program.

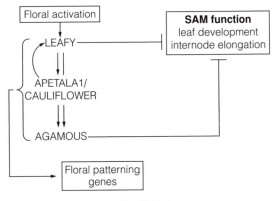

Fig. 3.   Summary of the floral initiation program.

---

[1]This has prompted investigators to look for the equivalent *CAULIFLOWER* gene in the cauliflower itself. The garden-variety cauliflower was compared to a related flowering species. Remarkably, the gene in the garden cauliflower was found to be nonfunctional, indicating that the characteristic structure of the cauliflower head may result from a defect in inflorescence to flower transition.

The data from mutant phenotypes, expression patterns and the effects of transgene overexpression suggest a molecular model for floral initiation in which the FLIP genes maintain each other's expression and cooperate to activate downstream floral patterning genes. First, *LEAFY* is expressed in the inflorescence meristem and predicts the appearance of the floral primordium. The initial molecular function of the LEAFY transcription factor is to suppress genes involved in vegetative growth and upregulate the expression of *APETALA1*, *CAULIFLOWER* and *AGAMOUS*, which are expressed weakly in the inflorescence meristem. The role of *AGAMOUS* in floral initiation is discussed later. The initial role of the APETALA1 and CAULIFLOWER transcription factors is to maintain the expression of *LEAFY*. This self-sustaining loop is revealed by the analysis of gene expression levels in mutant flowers and direct evidence of molecular interactions. APETALA1 expression is delayed in *leafy* mutants and there is evidence that LEAFY binds to the *APETALA1* promoter. Moreover, *LEAFY* mRNA levels are reduced in *apetala1 + cauliflower* double mutants.

LEAFY and APETALA1/CAULIFLOWER then act on different downstream genes involved in floral initiation, including *APETALA2*, *AGAMOUS* and *UNUSUAL FLORAL ORGANS* (*UFO*). This level of regulation helps to establish the expression patterns of the homeotic selector genes discussed above. Note that all the class A and class C homeotic genes have earlier roles in floral initiation, as shown by their ability to promote floral meristem identity when expressed in vegetative shoots. AGAMOUS is also thought to *maintain* the identity of the floral meristem, preventing it from reverting to a vegetative state if flower-inducing conditions are removed. LEAFY promotes the expression of class B and class C homeotic genes, requiring the cooperation of UFO to activate the class B genes, thus directing petal and stamen development. APETALA1/CAULIFLOWER promotes the expression of *APETELA2*, and these homeotic genes cooperate to activate sepal and petal development. Finally, both LEAFY and APETALA1/CAULIFLOWER promote the expression of *AGAMOUS*, the class C homeotic gene, to direct stamen and carpel development.

Transition to inflorescence development in *Arabidopsis*

Arabidopsis is a **facultative long day plant**, which means that flowering is actively promoted by increased day length but will also occur under short day conditions. The first stage of flower development in *Arabidopsis*, the specification of the inflorescence meristem, is therefore influenced by environmental signals but also has an intrinsic developmental program. The components of this intrinsic program have been identified by isolating mutants showing delayed or early flowering phenotypes under both inductive (long day) and non-inductive (short day) conditions. Conversely, components of the environmental signaling pathway have been identified by isolating mutants that show delayed or early flowering phenotypes specifically under long day conditions (*Fig. 4*).

The floral repressor concept

In wild type *Arabidopsis* plants, the main stem shows a sudden transition from leafy shoots to inflorescences bearing flowers. Under long day conditions, this transition occurs earlier in development so that there are fewer leaves at the base of the stem (*Fig. 4*). In *leafy* and *apetala1* mutants, however, this sudden transition is converted into a gradient, with successive phytomers showing increasingly floral properties. The *leafy* and *apetala1* mutant phenotypes suggest each successive phytomer has a greater competence to produce flowers than the last. The wild type floral initiation program (i.e. when both *LEAFY* and

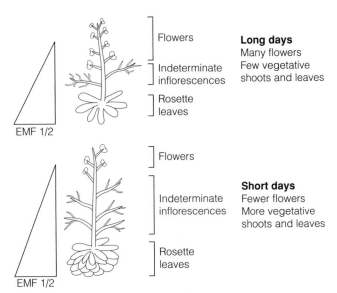

Fig. 4. *Activation of floral initiation in* Arabidposis *under long and short days. A gradient of floral repressor, probably involving the EMF1 and EMF2 gene products, is sharper under long days resulting in the production of more flowers and less vegetative shoots. Under short days, the gradient is longer, resulting in more vegetative shoots and indeterminante inflorescences.*

APETALA1 genes are functional) ensures that the transition to flowering occurs suddenly and completely once the appropriate level of competence has been achieved. The underlying gradient of competence could reflect the gradual increase in levels of a **floral activator** or the gradual decrease in levels of a **floral repressor** in the shoot apical meristem as development proceeds. The existence

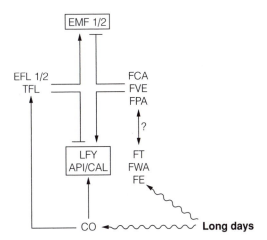

Fig. 5. *Simplified model for activation of the floral initiation program. Genes involved in the constitutive (short day) response (e.g. FCA, FVE, FPA) and the induced (long day) response (e.g. CONSTANS) are thought to work by inhibiting the floral repressor (EMBRYONIC FLOWER 1 and 2) and activating FLIP genes (LEAFY, APETALA1 and CAULIFLOWER). Their activity is balanced by floral inhibitors such as EARLY FLOWERING 1 and 2, and TERMINAL FLOWER.*

of mutants such as *embryonic flower 1* and *2*, where there is no vegetative growth at all and even the rosette leaves at the base of the stem are converted into floral primordia, suggests that competence depends on a diminishing gradient of a floral repressor. Genes that affect floral initiation by advancing or retarding flowering may work by either inhibiting or stimulating this repressor, as the *embryonic flower* mutations are epistatic to all other mutations affecting flower development. However, a number of such genes have been shown to regulate the expression of downstream FLIP genes such as *LEAFY* and *APETALA1*. Therefore, activation of the floral initiation program may involve both the induction of FLIP genes and the regulation of a floral repressor that controls the competence of the shoot apical meristem to respond to floral initiation.

Genes that regulate flowering time under both long- and short-day conditions

A number of *Arabidopsis* mutants have been identified that show either early or late flowering phenotypes under all environmental conditions.

Mutants with constitutive early flowering phenotypes identify genes that may promote the activity of the floral repressor, and therefore help to maintain vegetative growth. Such genes include *EARLY FLOWERING 1*, *EARLY FLOWERING 2* and *TERMINAL FLOWER*. In the *terminal flower* mutant, all shoot meristems are converted into floral meristems. The *TERMINAL FLOWER* gene is expressed at a low level in the shoot apex, but is strongly expressed in inflorescence meristems that grow indeterminately. It is thought that the TERMINAL FLOWER protein may be involved in the transduction of signals from the floral repressor and may inhibit floral initiation genes such as *LEAFY* and *APETALA1*. There is no evidence for any direct effect of TERMINAL FLOWER on the expression of these genes, but it may interfere with the function of their products.

Mutants with constitutive late flowering phenotypes identify genes whose normal function may be to inhibit the floral repressor. Such genes include *FCA*, *FLC, FPA, FVE, FY* and *LUMINDEPENDENS*. The effect of many of these mutations can be suppressed by vernalization (exposure to prolonged low temperature) or GA. This suggests that GA is required for flowering under short day conditions. Consistent with this, mutations in the GA signal transduction pathway, such as *gai* (gibberellic acid insensitive) are unable to flower under short day conditions, while constitutive GA signaling mutants such as *spindly* are early flowering under all conditions. Both *FCA* and *FVE* are thought to activate *LEAFY*, as the constitutive expression of a *LEAFY* transgene in *fca* and *fve* mutant plants can restore the normal flowering time.

Genes that regulate flowering time in response to long days

A number of genes show late flowering mutant phenotypes only under long day conditions. Genes in this pathway include *CONSTANS, GIGANTEA, FD, FE* and *FHA*. The *CONSTANS* gene is expressed only under long day conditions and has been studied in detail. CONSTANS is a zinc finger transcription factor and is thought to regulate at least one FLIP gene, *LEAFY*, as well as the *TERMINAL FLOWER* gene that inhibits floral initiation. However, plants constitutively expressing *CONSTANS* flower much earlier than those constitutively expressing *LEAFY*, indicating that *CONSTANS* has a number of roles in addition to its activation of *LEAFY*. One of these roles could be the inhibition of the floral repressor, therefore increasing the competence of the meristem to respond to *LEAFY*.

FURTHER READING

Student reading lists often contain a formidable mix of obscure book chapters long since out of print, dated papers, and long detailed reviews that can be difficult for the newcomer to get into. To avoid this, I have restricted much of the reading list below to recent short reviews, such as those featured in the *Trends* and *Current Opinion* journals. These are generally written in an accessible and entertaining style, and for those wishing to delve deeper, usually contain references to more detailed reviews and the primary literature. The first three sections of this book are introductory, and cover a lot of general material. For these sections, I have therefore listed some other books that provide good comprehensive sources of further information, as well as recent reviews relevant to development. There are also some excellent websites that are frequently updated with current research, and these are cited below where appropriate.

General reading   The following books are excellent introductory texts and have served students well.

Slack, J.M.W. (1991) *From Egg to Embryo* (second edition), Cambridge University Press, Cambridge UK.

Wolpert, L. (1998) *Principles of Development*, Oxford University Press, Oxford, UK.

Gilbert, S. (1997) *Developmental Biology* (fifth edition), Sineaur Associates Inc., Mass. This book is very detailed and contains an extensive reading list at the end of each chapter. The sixth edition has just been published, and judging by its predecessors, should be a very useful text.

More advanced reading

Section A   Christian, J.L (2000) BMP, Wnt and Hedgehog signals: how far can they go? *Curr. Opin. Cell. Biol.* **12(2)**, 244–249.

Chuang, P.-T. and Kornberg, T.B. (2000) On the range of Hedgehog signaling. *Curr. Opin. Genet. Dev.* **10(5)**, 515–522.

Dahmann, C. and Basler, K. (1999) Compartment boundaries: at the edge of development. *Trends Genet.* **15(8)**, 320–326.

Hogan, B.L.M. (1999) Morphogenesis. *Cell* **96(2)**, 225–233.

Maves, L. and Schubiger, G. (1999) Cell determination and transdetermination in *Drosophila* imaginal discs. *Curr. Top. Dev. Biol.* **43**, 115–151.

Perrimon, N. and McMahon, A.P. (1999) Negative feedback mechanisms and their roles during pattern formation. *Cell* **97(1)**, 13–16.

Podos, S.D. and Ferguson, E.L. (1999) Morphogen gradients: new insights from DPP. *Trends Genet.* **15(10)**, 396–402.

Selleck, S.B. (2000) Proteoglycans and pattern formation: sugar biochemistry meets developmental genetics. *Trends Genet.* **16(5)**, 206–212.

Steinberg, M.S. and McNutt, P.M. (1999) Cadherins and their connections: adhesion junctions have broader functions. *Curr. Opin. Cell. Biol.* **11(5)**, 554–560.

Streuli, C. (1999) Extracellular matrix remodelling and cellular differentiation. *Curr. Opin. Cell. Biol.* **11(5)**, 634–640.

Wolpert, L. (1996) One hundred years of positional information. *Trends Genet.* **12(9)**, 359–364.

Section B   Anderson, K.V. (2000) Finding the genes that direct mammalian development.
                        ENU mutagenesis in the mouse. *Trends Genet.* **16(3)**, 99–102.
                        Bard, J.B.L. (ed.) (1994) *Embryos: A Color Atlas of Development*. Wolfe, London.
                        Hansen, G. and Wright, M.S. (1999) Recent advances in the transformation of
                        plants. *Trends Plant Sci.* **4(6)**, 226–231.
                        Ihle, J.N. (2000) The challenges of translating knockout phenotypes into gene
                        function. *Cell* **102(2)**, 131–134.
                        Müller, U. (1999) Ten years of gene targeting: targeted mouse mutants, from
                        vector design to phenotype analysis. *Mech. Dev.* **82(1–2)**, 3–21.
                        Old, R.W. and Primrose, S.B. (1994) *Principles of Gene Manipulation* (5th edition),
                        Blackwell Science, Oxford, UK.
                        Porter, A. (1998) Controlling your losses: conditional gene silencing in
                        mammals. *Trends Genet.* **14(2)**, 73–79.
                        Rossant, J. and Spence, A. (1998) Chimeras and mosaics in mouse mutant
                        analysis. *Trends Genet.* **14(9)**, 358–363.
                        Stern, C.D. and Holland, P.W.H. (1993) *Essential Developmental Biology: A
                        Practical Approach*. IRL Press, Oxford, UK.

Section C   These books and chapters are a good source of general information on
                        molecular and cell biology relevant to development:

                        Alberts, B., Bray, D., Lewis, J., Raff, M., Roberts, K. and Watson, J.D. (1994).
                        Molecular Biology of the Cell (3rd edition), Scientific American Books, WH
                        Freeman, NY.
                        Lewin, B. (1997) *Genes VI*, Oxford University Press, Oxford, UK.
                        Lodish, H., Baltimore, D., Berk, A., Zipursky, S.L., Matsudaira, P. and Darnell, J.
                        (1995) *Molecular Cell Biology* (3rd edition), Scientific American Books, WH
                        Freeman, NY.
                        Twyman, R.M. (1998) Signal transduction, in *Advanced Molecular Biology: A
                        Concise Reference*. BIOS Scientific Publishers, Oxford, UK, pp. 425–442.

                        Aza-Blanc, P. and Kornberg, T.B. (1999) Ci: a complex transducer of the
                        Hedgehog signal. *Trends Genet.* **15(11)**, 458–462.
                        Giancotti, F.G. and Ruoslahti, E. (1999) Transduction – Integrin signaling.
                        Science **285(5430)**, 1028–1032.
                        Gould, A. (1997) Functions of mammalian *Polycomb* group and *trithorax* group
                        related genes. *Curr. Opin. Genet. Dev.* **7(4)**, 488–494.
                        McMahon, A.P. (2000) More surprises in the Hedgehog signaling pathway. *Cell*
                        100(2), 185–188.
                        Metzstein, M.M., Stanfield, G.M. and Horvitz, H.R. (1998) Genetics of
                        programmed cell death in *C. elegans*: past, present and future. *Trends Genet.*
                        14(10), 410–416.
                        Ruiz i Altaba, A. (1999) Gli proteins and Hedgehog signaling: development and
                        cancer. *Trends Genet.* **15(10)**, 418–425.
                        Vaux, D.L. and Korsmeyer, S.J. (1999) Cell death in development. *Cell* **96(2)**,
                        245–254.
                        Wodarz, A. and Nusse, R. (1998) Mechanisms of Wnt signaling in development.
                        Ann. Rev. Cell. Dev. Biol. **14**, 59–88.
                        Wrana, J.L. (2000) Regulation of Smad activity. *Cell* **100(2)**, 189–192.
                        Zhang, P. (1999) The cell cycle and development: redundant roles of cell cycle
                        regulators. *Curr. Opin. Cell Biol.* **11(6)**, 655–662.

Section D

Brown, J.M. and Firtel, R.A. (1999) Regulation of cell-fate determination in *Dictyostelium*. *Dev. Biol.* **216(2)**, 426–441.

Hoch, J.A. (1998) Initiation of bacterial development. *Curr. Opin. Microbiol.* **1(2)**, 170–174.

Thomason, P., Traynor, D. and Kay, R. (1999) Taking the plunge: terminal differentiation in *Dictyostelium*. *Trends Genet.* **15(1)**, 15–19.

Section E

Forbes, A. and Lehmann, R. (1999) Cell migration in *Drosophila*. *Curr. Opin. Genet. Dev.* **9(4)**, 473–478.

Lebel-Hardenack, S. and Grant, S. (1997) Genetics of sex-determination in flowering plants. *Trends Plant Sci.* **2(4)**, 130–136.

Lin, H. (1998) The self-renewing mechanism of stem cells in the germline. *Curr. Opin. Cell Biol.* **10(6)**, 687–693.

Meyer, B.J. (2000) Sex in the worm. Counting and compensating X-chromosome dose. *Trends Genet.* **16(6)**, 247–253.

Venables, J.P. and Eperon, I.C. (1999) The roles of RNA-binding proteins in spermatogenesis and male infertility. *Curr. Opin. Genet. Dev.* **9(3)**, 346–354.

Wassarman, P.M. (1999) Mammalian fertilization: Molecular aspects of gamete adhesion, exocytosis, and fusion. *Cell* **96(2)**, 175–183.

Wylie, C. (2000) Germ cells. *Curr. Opin. Genet. Dev.* **10(4)**, 410–413.

Section F

Ettensohn, C.A. (1999) Cell movements in the sea urchin embryo. *Curr. Opin. Genet. Dev.* **9(5)**, 461–465.

Kodjabachian, L., Dawid, I.B. and Toyama, R. (1999) Gastrulation in zebrafish: What mutants teach us. *Dev. Biol.* **213(2)**, 231–245.

Leptin, M. (1999) Gastrulation in *Drosophila*: The logic and the cellular mechanisms. *EMBO J.* **18(12)**, 3187–3192.

Narasimha, M. and Leptin, M. (2000) Cell movements during gastrulation: come in and be induced. *Trends Cell Biol.* **10(5)**, 169–172.

Section G

Di Gregorio, A. and Levine, M. (1998) Ascidian embryogenesis and the origins of the chordate body plan. *Curr. Opin. Genet. Dev.* **8(5)**, 457–463.

Golden, A. (2000) Cytoplasmic flow and the establishment of polarity in *C. elegans* 1-cell embryos. *Curr. Opin. Genet. Dev.* **10(4)**, 414–420.

Kemphues, K. (2000) PARsing embryonic polarity. *Cell* **101(4)**, 345–348.

Satou, Y. and Satoh, N. (1999) Developmental gene activities in ascidian embryos. *Curr. Opin. Genet. Dev.* **9(5)**, 542–547.

Section H

Brody, T.B. (1995) The Interactive Fly. This is an excellent resource for those interested in fly development. URL = http://flybase.bio.indiana.edu/allied-data/lk/interactive-fly/aimain/1aahome.htm

Lawrence, P.A. (1992) *The Making of a Fly*, Blackwell Science, Oxford, UK.

LeMosy, E.K., Hong, C.C. and Hashimoto, C. (1999) Signal transduction by a protease cascade. *Trends Cell Biol.* **9(3)**, 102–106.

Morata, G. and Sanchez-Herrero, E. (1999) Patterning mechanisms in the body trunk and the appendages of *Drosophila*. *Development* **126(13)**, 2823–2828.

Rivera-Pomar, R. and Jackle, H. (1996) From gradients to stripes in *Drosophila* embryogenesis: Filling in the gaps. *Trends Genet.* **12(11)**, 478–483.

Section I

Bachvarova, R.F. (1999) Establishment of anterior–posterior polarity in avian embryos. *Curr. Opin. Genet. Dev.* **9(4)**, 411–416.

Beddington, R.S.P. and Robertson, E.J. (1999) Axis development and early asymmetry in mammals. *Cell* **96(2)**, 195–209.

Capdevila, J., Vogan, K.J., Tabin, C.J. and Belmonte, J.C.I. (2000) Mechanisms of left–right determination in vertebrates. *Cell* **101(1)**, 9–21.

Niehrs, C. (1999) Head in the WNT: the molecular nature of Spemann's head organizer. *Trends Genet.* **15(8)**, 314–319.

Ramsdell, A.F. and Yost, H.J. (1998) Molecular mechanisms of vertebrate left–right development. *Trends Genet.* **14(11)**, 459–465.

Schier, A.F. and Talbot, W.S. (1998) The zebrafish organizer. *Curr. Opin. Genet. Dev.* **8(4)**, 464–471.

Sokol, S.Y. (1999) Wnt signaling and dorsoventral axis specification in vertebrates. *Curr. Opin. Genet. Dev.* **9(4)**, 405–410.

Supp, D.M., Potter, S.S. and Brueckner, M. (2000) Molecular motors: the driving force behind mammalian left–right development. *Trends Cell Biol.* **10(1)**, 41–45.

Section J

Arendt, D. and Nubler-Jung, K. (1999) Comparison of early nerve cord development in insects and vertebrates. *Development* **126(11)**, 2309–2325.

Chitnis, A.B. (1999) Control of neurogenesis – lessons from frogs, fish and flies. *Curr. Opin. Neurobiol.* **9(1)**, 18–25.

Crair, M.C. (1999) Neuronal activity during development: permissive or instructive? *Curr. Opin. Neurobiol.* **9(1)**, 88–93.

Eislen, J.S. (1999) Patterning motoneurons in the vertebrate nervous system. *Trends Neurosci.* **22(7)**, 321–326.

García-Castro, M. and Bronner-Fraser, M. (1999) Induction and differentiation of the neural crest. *Curr. Opin. Cell Biol.* **11(6)**, 695–698.

Ghysen, A. and Dambly-Chaudière, C. (2000) A genetic programme for neuronal connectivity. *Trends Genet.* **16(5)**, 221–226.

Harland, R. (2000) Neural induction. *Curr. Opin. Genet. Dev.* **10(4)**, 357–362.

Hobert, O. and Westphal, H. (2000) Functions of LIM-homeobox genes. *Trends Genet.* **16(2)**, 75–83.

Holt, C.E. and Harris, W.A. (1998) Target selection: invasion, mapping and cell choice. *Curr. Opin. Neurobiol.* **8(1)**, 98–105.

Le Dourain, N.M. and Halpern, M.E. (2000). Discussion point. Origin and specification of the neural tube floor plate: insight from the chick and zebrafish. *Curr. Opin. Neurobiol.* **10(1)**, 23–30.

Lumsden, A. and Krumlauf, R. (1996) Patterning the vertebrate neuraxis. *Science* **274(5290)**, 1109–1115.

Matsuzaki, F. (2000) Asymmetric division of *Drosophila* neural stem cells: a basis for neural diversity. *Curr. Opin. Neurobiol.* **10(1)**, 38–44.

O'Leary, D.D.M. and Wilkinson, D.G. (1999) Eph receptors and ephrins in neural development. *Curr. Opin. Neurobiol.* **9(1)**, 65–73.

Simeone, A. (2000) Positioning the isthmic organizer where *Otx2* and *Gbx2* meet. *Trends Genet.* **16(6)**, 237–239.

Streit, A. and Stern, C.D. (1999) Neural induction – a bird's eye view. *Trends Genet.* **15(1)**, 20–24.

Tanabe, Y. and Jessell, T.M. (1996) Diversity and pattern in the developing spinal cord. *Science* **274(5290)**, 1115–1123.

Tear, G. (1999) Neuronal guidance: a genetic perspective. *Trends Genet.* **15(3)**, 113–118.

Tessier-Lavigne, M. and Goodman, C.S. (1996) The molecular biology of axon guidance. *Science* **274(5290)**, 1123–1133.

Section K

Bodmer, R. and Venkatesh, T.V. (1998) Heart development in *Drosophila* and vertebrates: conservation of molecular mechanisms. *Dev. Genet.* **22(3)**, 181–186.

Burke, A.C. (2000) *Hox* genes and the global patterning of the somitic mesoderm. *Curr. Top. Dev. Biol.* **47**, 155–181.

Davies, J.A. and Brandli, A.W. (1997) The Kidney Development Database. An excellent resource of up to date kidney research. URL = http://www.ana.ed.ac.uk/anatomy/database/kidbase/kidhome.html

Edlund, H. (1999) Pancreas: how to get there from the gut? *Curr. Opin. Cell Biol.* **11(6)**, 663–668.

Grapin-Botton, A. and Melton, D.A. (2000). Endoderm development – from patterning to organogenesis. *Trends Genet.* **16(3)**, 124–130.

Kimelman, D. and Griffin, K.J.P. (2000) Vertebrate mesendoderm induction and patterning. *Curr. Opin. Genet. Dev.* **10(4)**, 350–356.

Kuure, S., Vuolteenaho, R. and Vainio, S. (2000) Kidney morphogenesis: cellular and molecular regulation. *Mech. Dev.* **92(1)**, 31–45.

Lechner, M.S. and Dressler, G.R. (1997) The molecular basis of embryonic kidney development. *Mech. Dev.* **62(2)**, 105–120.

McGrew, M.J. and Pourquie, O. (1998) Somitogenesis: segmenting a vertebrate. *Curr. Opin. Genet. Dev.* **8(4)**, 487–493.

Naya, F.J. and Olson, E. (1999) MEF2: a transcriptional target for signaling pathways controlling skeletal muscle growth and differentiation. *Curr. Opin. Cell Biol.* **11(6)**, 683–688.

Richardson, M.K., Allen, S.P., Wright, G.M., et al (1998) Somite number and vertebrate evolution. *Development* **125(2)**, 151–160.

Smith, J. (1999) T-box genes: what they do and how they do it. *Trends Genet.* **15(5)**, 154–158.

Zaret, K.S. (2000) Liver specification and early morphogenesis. *Mech. Dev.* **92(1)**, 83–88.

Section L

Dahmann, C. and Basler, K. (1999) Compartment boundaries: at the edge of development. *Trends Genet.* **15(8)**, 320–326.

Kornfeld, K. (1997) Vulval development in *Caenorhabditis elegans*. *Trends Genet.* **13(2)**, 55–61.

Thomas, B.J. and Wassarman, D.A. (1999) A fly's eye view of biology. *Trends Genet.* **15(5)**, 184–190.

Section M

Dudely, A.T. and Tabin, C.J. (2000) Constructive antagonism in limb development. *Curr. Opin. Genet. Dev.* **10(4)**, 387–392.

Manouvrier-Hanu, S., Holder-Espinasse, M. and Lyonnet, S. (1999) Genetics of limb anomalies in humans. *Trends Genet.* **15(10)**, 409–417.

Ng, J.K., Tamura, K., Buscher, D., et al. (1999) Molecular and cellular basis of pattern formation during vertebrate limb development. *Curr. Top. Dev. Biol.* **41**, 37–66.

Schwabe, J.W.R., Rodriguez-Esteban, C. and Belmonte, J.C.I. (1998) Limbs are moving: Where are they going? *Trends Genet.* **14(6)**, 229–235.

Tsonis, P.A. (2000) Regeneration in vertebrates. *Dev. Biol.* **221(2)**, 273–284.

Section N

Bowman, J.L. (2000) Axial patterning in leaves and other lateral organs. *Curr. Opin. Genet. Dev.* **10(4)**, 399–404.

Bowman, J.L. and Eshed, Y. (2000) Formation and maintenance of the shoot apical meristem. *Trends Plant Sci.* **5(3)**, 110–115.

Colasanti, J. and Sundaresan, V. (2000) 'Florigen' enters the molecular age: long-distance signals that cause plants to flower. *Trends Biochem. Sci.* **25(5)**, 236–240.

Costa, S. and Dolan, L. (2000) Development of the root pole and cell patterning in *Arabidopsis* roots. *Curr. Opin. Genet. Dev.* **10(4)**, 405–409.

Fletcher, J.C. and Meyerowitz, E.M. (2000) Cell signaling within the shoot meristem. *Curr. Opin. Plant Biol.* **3(1)**, 23–30.

Holdsworth, M., Kurup, S. and McKibbin, R. (1999) Molecular and genetic mechanisms regulating the transition from embryo development to germination. *Trends Plant Sci.* **4(7)**, 275–280.

Irish, V.F. (1999) Patterning the flower. *Dev. Biol.* **209(2)**, 211–220.

Langdale, J.A. (1998) Cellular differentiation in the leaf. *Curr. Opin. Cell Biol.* **10(6)**, 734–738.

McSteen, P., Laudencia-Chingcuanco, D. and Colasanti, J. (2000) A floret by any other name: control of meristem identity in maize. *Trends Plant Sci.* **5(2)**, 61–66.

Ng, M. and Yanofsky, M.F. (2000) Three ways to learn the ABCs. *Curr. Opin. Plant. Biol.* **3(1)**, 47–52.

Pidkowich, M.S., Klenz, J.E. and Haughn, G.W. (1999) The making of a flower: control of floral meristem identity in *Arabidopsis*. *Trends Plant Sci.* **4(2)**, 64–70.

Reeves, P.H. and Coupland, G. (2000) Response of plant development to environment: control of flowering by daylength and temperature. *Curr. Opin. Plant Biol.* **3(1)**, 37–42.

Scanlon, M.J. (2000) Developmental complexities of simple leaves. *Curr. Opin. Plant Biol.* **3(1)**, 31–36.

Szymanski, D.B., Lloyd, A.M. and Marks, M.D. (2000) Progress in the molecular genetic analysis of trichome initiation and morphogenesis in *Arabidopsis*. *Trends Plant Sci.* **5(5)**, 214–219.

GLOSSARY OF ACRONYMS

As in other biological sciences, the terminology of developmental biology is rich in acronyms. To a newcomer, these can be confusing, so this glossary provides a quick explanation of common acronyms used in this book.

AEC. Apical ectodermal cap. The covering of ectoderm that forms over the stump of a regenerating limb in urodele amphibians. Analogous to the apical ectodermal ridge of the developing limb. Topic M3.

AER. Apical ectodermal ridge. The ridge of ectoderm that forms over the developing limb bud in vertebrates. A source of fibroblast growth factors, this structure is responsible for maintaining the proliferating cells of the progress zone and the posterior organizing center, the zone of polarizing activity. Topic M2.

ALC. Anterior-like cells. These are scattered among the prestalk cells in the migrating *Dictyostelium* slug, but have the potential to replace prespore cells that are lost during migration. Topic D3.

ANT-C. *Antennapedia* complex. Part of the homeotic complex of the fruit fly *Drosophila melanogaster*, the complex of genes that controls positional information along the anteroposterior axis. Topic H5.

AVE. Anterior visceral endoderm. The embryonic structure responsible for head induction in the mammalian embryo. Topic I1.

BMP. Bone morphogenetic protein. Any of a family of proteins, most of which are members of the transforming growth factor b superfamily of signaling molecules, with various patterning roles in development. Related to the *Drosophila* protein Decapentaplegic. Topics C3, H6, I2, K1.

BX-C. Bithorax complex. Part of the homeotic complex of the fruit fly *Drosophila melanogaster*, the complex of genes that controls positional information along the anteroposterior axis. Topic H5.

EGF. Epidermal growth factor. A signaling molecule involved in epidermal differentiation in mammals. Homologs in other species control varied processes including eye development in *Drosophila* (Topic L3) and vulval development in *Caenorhabditis* (Topic L1).

ES cells. Embryonic stem cells. Pluripotent cells derived from the inner cell mass of the mouse blastocyst, which are particularly amenable to genetic manipulation by homologous recombination, and are therefore used to generate targeted mouse mutants. Topic B3.

FGF. Fibroblast growth factor. Any of a family of signaling molecules with multiple and diverse roles in development.

FLIP. Floral initiation program. The developmental program that initiates flower development. Topic N6.

HOM-C. Homeotic complex. The complex of genes that control regional specification along the anteroposterior axis in *Drosophila*, equivalent to the *Hox* gene complexes of other animals. Topic H5.

ICM. Inner cell mass. The inner part of the preimplantation mammalian blastocyst, which gives rise to the embryo. The outer part forms the trophectoderm. Topic F3.

LRD. Left-right dynein. The product of the mammalian *iv* gene, which is involved in left-right axis specification. Topic I3.

MBT. Midblastula transition. The landmark in animal development, usually occurring during cleavage, in which cell divisions become asynchronous and major zygotic gene expression begins, coinciding with the depletion or inactivation of key maternal gene products. Topic F1.

NIC. Node inducing center. A region of the vertebrate embryo responsible for maintaining but not initiating the organizer. Topic I2.

PMC. Primary mesenchyme cells. The first cells to ingress during gastrulation in the sea urchin embryo, originally the vegetal micromeres. Topic F2.

PMZ. Posterior marginal zone. Region of the chicken blastodisc responsible for inducing the organizer. The position of the PMZ is specified by gravity. Topics I1, I2.

RAM. Root apical meristem. The part of the plant root just behind the tip. A proliferative cell population responsible for producing the cells of the mature root and the root tip. Divided into the promeristem and file meristem. Topic N4.

REMI. Restriction enzyme-mediated integration. A technique in which restriction enzymes are used to introduce foreign DNA into isolated nuclei. The only available procedure for generating transgenic frogs. Topic B3.

SAM. Shoot apical meristem. The apical tip of the plant shoot. A dividing population of cells responsible for producing the cells of the mature shoot as well as lateral organs such as leaves. Topics N4, N5.

SMC. Secondary mesenchyme cell. Cells that differentiate from the archenteron roof during sea urchin gastrulation. Topic F2.

TGF. Transforming growth factor. A superfamily of signaling molecules including the TGF-a, TGF-b, activin/inhibin, BMP and nodal-related protein families, with multiple roles in development.

VPC. Vulval precursor cell. One of six cells in the *C. elegans* embryo with the potential to form part of the vulva. Topic L1.

ZPA. Zone of polarizing activity. Part of the mesenchyme in the posterior of the vertebrate limb bud, which is required to maintain the apical ectodermal ridge and plays an important role in anteroposterior patterning. Secretes the signaling molecule Sonic hedgehog. Topic M1.

INDEX

Somatic embryogenesis 409
Somites
    axial patterning 332
    differentiation 337–40
    Hox genes 332
    origins 331
sonic hedgehog
    dorsoventral neuraxis 288
    gut development 355
    left-right axis 266
    limb development 378, 384
    somite differentiation 338
Sox genes 73, 153–4, 352–3
spätzle 247
Specification 11
    autonomous 21
    conditional 22
    single cell 187–9
Spemann organizer 258
Spermatogenesis 137–8
Spermiogenesis 138
Splotch 305
Sporulation 107–111
SRY 153
Steel 305
Stem cells 10
string 163
SUPERMAN 434
Suspensor 406
Symmetry-breaking 252–6
Synapse 319
Syncytium 206–209

T gene (see *Brachyury*)
tailless 217
T.ap lineage 195
Teleoplasm 203
Telson 212
Terminal differentiation 10
Terminal regions (*Drosophila*) 212

Testosterone 153
TGF-β family 90, 237
Toll 297
torpedo 248
Torso 217
torso-like 217
Totipotent 26, 401
Transcription 72
Transcription factors 75, 403
Transdetermination 25
Transdifferentiation 25
transformer genes
    C. elegans 157
    Drosophila 154
Transgenic animals 57–63
    Drosophila 61
    mammals 59–61
    Xenopus 63
Transgenic plants 62
Translation 75, 217
Tribolium 210
Trophectoderm 171, 183
trithorax 243
twist 176, 248

Ultrabithorax 238, 241
UV treatment (*Xenopus*) 65, 253
Ureteric buds 344
Urodeles 393

Veg-T 326
Vegetal pole 165
Vertebrates (see mammal, mouse, chick,
    Xenopus, zebrafish)
Vg-1 326
Visual cortex 326
Vulval development 358–60

WT-1 346
Wing development

chick (see Limb development)
    Drosophila 365
wingless 91, 291
Wnt signaling 91, 278
Wnt1 283
Wnt4a 154
Wnt7a 386
Wnt11 153

X chromosome
    C. elegans sex determination 156
    dosage compensation 83
    Drosophila sex determination 154
    inactivation 83
Xenopus
    axis specification 253
    cleavage 168
    gastrulation 178
    model organism 47
Xist 83
XO lethal (*Xol-1*) 157
Xwnt8 280, 326, 328
Xwnt11 281

Y chromosome 153
Yeast 112
Yellow crescent 198
Yolk, effect on cleavage 165
Yolk sac 181, 185

Zebrafish
    cleavage 168
    gastrulation 180
    model organism 48
    organizer 181
Zona-binding proteins 148
Zona pellucida 140
Zone of polarizing activity (ZPA) 378
Zygote 150
Zygotic genes 23